Plant Evolution in the Mediterranean

Plant Evolution in the Mediterranean

John D. Thompson
UMR 5175 Centre d'Ecologie Fonctionnelle et Evolutive, CNRS, Montpellier

This book has been printed digitally and produced in a standard specification in order to ensure its continuing availability

OXFORD
UNIVERSITY PRESS

Great Clarendon Street, Oxford OX2 6DP

Oxford University Press is a department of the University of Oxford.
It furthers the University's objective of excellence in research, scholarship,
and education by publishing worldwide in

Oxford New York

Auckland Cape Town Dar es Salaam Hong Kong Karachi
Kuala Lumpur Madrid Melbourne Mexico City Nairobi
New Delhi Shanghai Taipei Toronto

With offices in

Argentina Austria Brazil Chile Czech Republic France Greece
Guatemala Hungary Italy Japan South Korea Poland Portugal
Singapore Switzerland Thailand Turkey Ukraine Vietnam

Oxford is a registered trade mark of Oxford University Press
in the UK and in certain other countries

Published in the United States
by Oxford University Press Inc., New York

Reprinted 2010

ISBN 978-0-19-851534-0

Printed and bound in Great Britain by CPI Antony Rowe,
Chippenham and Eastbourne

Preface

My fascination for the Mediterranean, and its plants, began when I arrived in Montpellier for a one-year post-doctoral position in 1989. Since then I have become more and more interested in the ecological, genetic, and historical causes of diversity in the flora. My research has been focused on the ecology and evolution of just a few groups of plants in the flora, which have allowed me to pursue my main interests concerning how plant species respond to, and cope with, spatial variation in their environment. This work naturally made me curious about the history of the region, its climate and its flora, including domesticated and invasive species. In this book I have thus treated several subjects which are outside of my primary research themes. I trust that I have discussed them to a level which satisfies others more competent and knowledgeable than I in these fields.

I did not fully realize what I needed to write this book until it was almost finished. First, I of course needed a good story with a solid scientific basis. I trust I have supplied a text which is convincing. Second, you need endless motivation. My interest for the subject held me strong here. The third and final ingredient is the encouragement, help, and support of my colleagues. That I managed to write this book attests to the excellence of the help and encouragement I have received over the four years since I began this project.

It is to the people in Montpellier to whom I intend a special thanks. Their help has come in a variety of ways. Bertrand Dommée, Isabelle Olivieri, and Denis Couvet for their warm welcome in the late 1980s and their stimulating company as I settled into a new chapter of my scientific life. Other colleagues at the Centre d'Ecologie Fonctionnelle et Evolutive (CNRS) laboratory in Montpellier (particularly José Escarré, Rosylene Lumaret, and Max Debussche) provided me with informative and critical discussion of different aspects of functional and evolutionary ecology in a Mediterranean context. Here, the encouragement and advice of Jacques Blondel was very important as my project to write a book unfolded. Next, the PhD and post-doctoral students with whom I have interacted, hosted, or supervised in one way or another (François Bretagnolle, Christophe Thébaud, Laurence Affre, Michèle Tarayre, Christophe Petit, Thierry Pailler, Laurence Humeau, Anne Charpentier, Perrine Gauthier, Angélique Quilichini, Sebastien Lavergne, Bodil Ehlers, Adeline Césaro, Justin Amiot, Isabelle Litrico, and Emilie Andrieu). They have kept me on my toes and allowed my research to progress. Then, all the people who have helped with the plants themselves, particularly Christian Collin, Marie Maistre, Annabelle Dos Santos, and Alain Renaux, the administrative support staff at CEFE and Geneviève Debussche who kept my computer working. Finally a particular thanks to Max Debussche, who first took me to see a wild population of *Cyclamen*. He has shared with me his interest in, and knowledge of, the ecology and natural history of Mediterranean plants and has never been too shy to provide an alternative explanation for something I had written. His eye for detail and extensive knowledge have greatly contributed to improve my own understanding of plant evolution in the Mediterranean.

During the writing phase I received much positive criticism and guidance. In this respect, I thank Gideon Rosenbaum who put me straight on the geological history of the Mediterranean, Richard Abbott for his knowledge of the phylogeography of Mediterranean taxa, Yan Linhart for his advice on differentiation patterns, Spencer Barrett for his enthusiastic and constructive criticism of my discussion of plant reproduction, and once again Max Debussche for his attentive remarks. In addition, many thanks to Geneviève Debussche who drew several of the figures. I am also greatly indebted to all the people who have provided me with figures, photos, and unpublished and published manuscripts, and who have discussed with me and advised me about the plants they study. My apologies to those whose work is not cited, I had to make a choice in several places, and could not include everything. Finally, my thanks to Ian Sherman at Oxford University Press for his encouragement, advice, and his endless patience.

When I came to Montpellier I received a warm welcome and entered into a stimulating atmosphere in which to work. Since then I have continually benefited from the experience and advice of several close colleagues who have shared their wide-ranging knowledge of the Mediterranean environment and its plants. There was just one thing missing: a book which introduced me to the evolutionary ecology of plants in the region. The absence of such a text was the primary reason motivating me to write this book on plant evolution in the Mediterranean. I hope that newcomers to the region and its flora will find in this book a broad-based introduction to the evolution of plants in the Mediterranean Basin. I also hope that those researchers who know the region well will find something new and interesting. For those who have never been, take this book as an invitation to come.

I have written this book with many people in mind. Therein lies my major problem: satisfying the curiosity of different groups with very different backgrounds. I hope that there is something new for molecular phylogeographers who study the evolution of distribution patterns in relation to climate change and for plant population geneticists and ecologists interested in adaptation, plant reproduction, and the processes and consequences of landscape change. I also hope to develop themes which interest those whose research bears on the conservation of plant diversity in the region and botanists who study and classify Mediterranean plants. Finally, I trust that this book will find an audience in the large community of naturalists and botanical society members who, always keen to see a rarity or something new, have greatly improved our knowledge of Mediterranean plants and their distributions. I hope that the ecological and evolutionary themes I develop will thus stimulate people with a wide range of backgrounds as they pursue their diverse interests.

I have cited a small number of general references throughout the book to set the context for my exploration of plant evolution in the Mediterranean region and to lead the reader to recent key papers from which they can base a literature survey. I have also broadened my discussion where possible to include examples from other Mediterranean-climate regions. To avoid littering the book with too many names, I have kept family names and species authorities to the species list at the end of the book. I have often included reference to my own unpublished data and observations. My aim has been to cover where possible the extensive literature on Mediterranean plants and to extract and discuss in some detail a smaller number of case studies that I feel are particularly pertinent to the themes of the book.

When sat in front of my computer screen, more than once my thoughts drifted back in time to my high school days when I first became interested in plants through my mother's passion for the plants in her garden and my Aunty Lynne who brought me the latest in biology textbooks. Thanks to their encouragement, the fascination of a schoolboy for the natural world became a passion and a career. Since starting to write this book it has been almost every day that Marie-Andrée has given me that so-much-needed encouragement to keep on and finish it.

John D. Thompson
Montpellier, June 2004

Contents

Introduction: Themes, structure, and objectives 1

1 The historical context of differentiation and diversity 10
- 1.1 Geology, climate, and human activities: the mould and sculptors of plant diversity 10
- 1.2 A meeting of continents: a complex geological history 12
- 1.3 Two seasons: the history of the climate and the vegetation 18
- 1.4 Diversity and unity in the Mediterranean flora 30
- 1.5 Centres of diversity: concordance with history 32
- 1.6 Conclusions 36

2 The biogeography and ecology of endemism 38
- 2.1 Narrow endemism: the cornerstone of Mediterranean plant diversity 38
- 2.2 Endemism in the Mediterranean: patterns and classification 40
- 2.3 Endemism in the Mediterranean: community composition and biogeography 43
- 2.4 The biology and ecology of endemic plants 56
- 2.5 Conclusions 64

3 The evolution of endemism: from population differentiation to species divergence 67
- 3.1 Endemism and evolution: the processes and scale of differentiation 67
- 3.2 Population variation in endemic plants 68
- 3.3 Climatic rhythms and differentiation 71
- 3.4 Divergence in peripheral and marginal populations: isolation, inbreeding, and ecology 77
- 3.5 Hybridization and chromosome evolution 91
- 3.6 Conclusions 106

4 Trait variation, adaptation, and dispersal in the Mediterranean mosaic 109
- 4.1 Ecological constraints and adaptation in the Mediterranean 109
- 4.2 Summer drought and nutrient stress: functional traits and their variability 111
- 4.3 The phenology of flowering and fruiting 120
- 4.4 Dispersal and establishment: the template of local differentiation 130
- 4.5 Variation and adaptation in aromatic plants 144
- 4.6 Conclusions 165

5 Variation and evolution of reproductive traits in the Mediterranean mosaic 167
- 5.1 Reproductive trait variation: the meeting of ecology and genetics 167
- 5.2 Specialization and generalization in a mosaic pollination environment 168

5.3 Attracting pollinators . . . but avoiding herbivores 177
5.4 Mating system and gender variation 180
5.5 Pollination ecology and the evolution of style-length polymorphisms 194
5.6 Conclusions 204

6 Ecology and evolution of domesticated and invasive species **207**
6.1 Migration with man 207
6.2 The evolutionary history of domesticated plants 208
6.3 Invasive species in a Mediterranean environment 223
6.4 Conclusions 238

Conclusions: Endemism, adaptation, and conservation **240**

Species list **246**

Bibliography **253**

Index **291**

Introduction: Themes, structure, and objectives

... areas of mediterranean climate afford not only repositories for relict plant families but great natural laboratories for students of evolution ...

P.H. Raven (1971: 132)

The primary goal of this book is to blend information from diverse domains into a synthetic account of evolutionary ecology in which the central theme is differentiation, both among and within plant species. To do so, I provide illustrations of principal evolutionary processes and emphasize the relative roles of spatial isolation and ecological variation. The central theme is developed by highlighting how population-level processes not only provide the template for differentiation but also the stimulus for species evolution and by firmly setting trait variation and evolution in the context of spatial habitat variation.

The subject material of this book is the contemporary flora of the Mediterranean Basin. This flora inhabits a region with a complex history and a highly heterogeneous landscape. The evolution of plant diversity in this flora has been greatly influenced by its geological history, the oscillations of the climate, and the impact of human activities.

On a map of the world, one can see the Mediterranean, not just as an inland sea, but more as a region where continents meet. The complex geological history of this meeting has decorated the Mediterranean Sea with islands, which vary from tiny fragments of previous land-bridge connections which barely keep their heads above water, to the big islands with their massive mountains and violent volcanoes. Plunging to vast depths in the centre of its diverse basins, in many places the Mediterranean Sea is reduced to shallow sills which further belie the history of land connections around the region. Almost all the way around its shores are the mountains. This remarkable geology has been instrumental in shaping patterns of plant species distribution.

The fundamental element of the Mediterranean region is its highly seasonal climate. The essential and defining characteristic of this seasonality is that the warmest season is associated with an effective drought which limits plant growth. Although the length and intensity of summer drought vary spatially, and its onset is fairly recent, the occurrence of this climatic regime has had fundamental implications for the ecology and evolution of plants in the region. Since the initial onset of the Mediterranean climate, many parts of the region have acted as a refuge during periods of Quaternary glaciation. Climatic oscillations caused plant species ranges to contract and then to expand again as the climate warmed. These oscillations opened the way for hybridization and evolution in new environments and have been fundamental for patterns of differentiation and diversification in many groups of plants.

The Mediterranean is also the home of many human civilizations. Human activities have been modifying natural habitats and the spatial distribution of species for thousands of years and have thus played a key role in shaping recent and contemporary evolutionary pressures in natural populations. The impact of human activities stems from their effects on both the ecological conditions within habitats, which shape natural selection pressures and adaptive variation, and the spatial configuration of habitats in the landscape, which determines gene flow and seed dispersal. By modifying the action of selection and gene flow, human activities have become a key element of the process of population differentiation.

Mayr (1982) recognized that the interests of evolutionary biologists range from those whose primary interest lies in the study of diversity (speciation in fact) and those for whom 'adaptation ... holds first place in their interest' (p. 358). These two themes, diversity and adaptation, provide the framework for my discussion of plant evolution in the Mediterranean. My purpose is to firmly place the evolutionary processes which shape plant evolution into the context of the three main historical influences on vegetation in the Mediterranean region, that is, geology, climate, and human activities. I thus attempt to write a story of plant evolution in the context of regional history.

To write about the evolution of plants that inhabit the lands around the Mediterranean Sea requires the conception of a certain unity which holds together the immense diversity present in the flora. This unity has both a spatial and temporal context. To delimit a biogeographic region, and thus its flora, one has to have reliable boundary lines. The iso-climatic area proposed by Daget (1977*a*, *b*) is fairly well accepted but actually extends away from the Mediterranean Basin to the Canary islands, south into sub-Saharan Africa, south-east into Arabia, and north-east into other parts of western Asia. At the other end of the extreme, classifications based on the distribution of particular species, such as olives, or the distribution of sclerophyllous vegetation, all fall short of a true estimation of the spatial extent of the Mediterranean region. In accordance with previous studies (Quézel and Barbero 1982; Médail and Quézel 1997; Quézel and Médail 2003), I use a delimitation of the Mediterranean region which falls between these different extremes and which essentially covers the region where an effective drought occurs in the warmest part of the year (Fig. I.1).

The critical defining characteristic of the Mediterranean region is thus that summer is the driest season, and that this dry season involves a period of drought, that is, is biologically dry (Emberger 1930*c*; Quézel 1985). The high mountains that fringe the shores of the Mediterranean and dominate many of its islands as well as some of the steppe formations that spread across the Anatolian peninsula and large

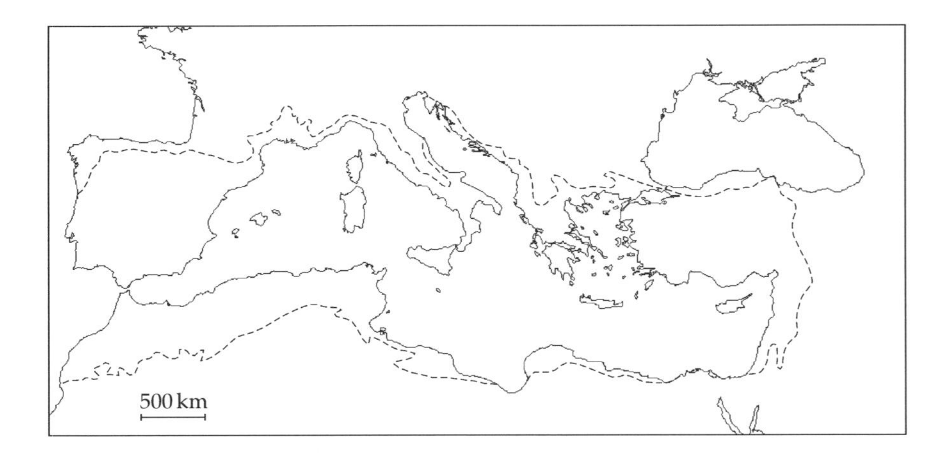

Figure I.1 The delimitation of the Mediterranean region (redrawn from Quézel and Médail 2003).

parts of north-west Africa clearly have their place here (di Castri 1973; Quézel and Médail 2003). On the summits of the Moroccan Atlas or the Taurus mountains of Greece, or in the Sierra Nevada of southern Spain, summer temperatures are not particularly high, however, there is a prolonged dry period at this time of the year. Hence, the flora of such areas can be considered to be 'unequivocally of a Mediterranean type' (Quézel and Médial 2003: 25, my translation). Some of the examples which I use occur on the margins of this geographic delimitation, but in ecological settings which closely resemble those within the strict confines of the Mediterranean-climate region. It is my belief that species which inhabit such peripheral Mediterranean areas, in particular those whose distribution crosses the border into more temperate, continental, or desert climates provide ideal situations for the study of plant evolution in the Mediterranean.

A melting pot of geological activity, climatic evolution, and human civilizations, the Mediterranean Basin is a hot spot of plant biodiversity. The flora of the Mediterranean Basin contains ~24,000 plant species in a surface area of about 2.3 million km^2 (Greuter 1991), that is, 10% of known plant species in really what is only a small part of the world. In contrast, non-Mediterranean Europe covers about 9 million km^2 but only has around 6,000 plant species. In 17 countries with a Mediterranean component to their territory, a large fraction of all their plant species occur in the Mediterranean part of their territory. Even countries with a small percentage of their territory in the Mediterranean region have a high percentage of their total species complement which is present in the Mediterranean region (Fig. I.2). For example, although only ~10% of the territory of continental France occurs in the Mediterranean region, 66% of all species that occur in France occur in the Mediterranean zone. One single administrative '*département*' of southern France (l'Hérault) contains 2,400 species of vascular plants, that is, 55% of the flora of continental France in just 1.1% of the total surface of the country. Even in Mediterranean forests, where endemism is low, woody species richness is twice that of temperate European forests (Quézel and Médail 2003). In addition, ~60% of the native species in the Mediterranean flora are endemic to the region (Quézel 1985; Greuter 1991) making it one of the world's 'hot spots' of species diversity (Myers *et al.* 2000).

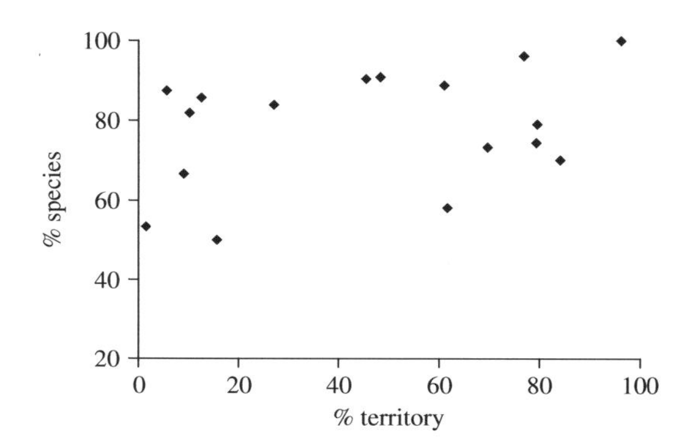

Figure I.2 Percentage of land area and species richness in the Mediterranean part of the territory of 17 countries on the shores of the Mediterranean Sea (drawn from data in Médail and Quézel 1997).

In Chapter 1, I build a framework for our present understanding of the biogeographic origins and diversity of the Mediterranean flora. To do so, I reconstruct the geological history of connections and isolation among different land masses and islands, the development of the Mediterranean-type climate that currently reigns, and the history of human activities in the last few millennia. These three regional features will be used to structure my discussion of plant evolution throughout the book.

The Mediterranean flora shows extremely high rates of narrow endemism in many regions, particularly in the mountains and on islands (Greuter 1991; Médail and Quézel 1997). This narrow endemism is a key ingredient of plant biodiversity in the Mediterranean flora, and also the other Mediterranean-climate regions where ecological specialization and geographic isolation have been primary determining factors (Cowling *et al.* 1996). A primary question motivating Chapters 2 and 3 concerns the role of regional history and spatial environmental variation in the evolution of endemism. In the Mediterranean, narrow endemism often involves disjunct distributions among closely related taxa, creating an ideal setting to link the study of population differentiation with that of species divergence (Thompson 1999). In Chapter 2, I describe and explore the biogeography of endemism in the Mediterranean flora and assess whether ecological characteristics and biological traits are associated with narrow endemism. In Chapter 3, I illustrate the diversity of evolutionary processes acting on variation at the population and species level in relation to the history of the region described in Chapter 1. To explore and assess the history, ecology, and evolution of species divergence and endemism I insist on the need to link population differentiation to species divergence.

In Chapters 4 and 5, I switch to the theme of trait variation and adaptation in a spatially heterogeneous environment. The focus here is on the ecological and historical factors which determine trait variation and evolution and the ecological basis of adaptation in the highly heterogeneous Mediterranean mosaic environment.

The Mediterranean region is where landscapes vary dramatically, often over short distances, with perhaps the most original and fascinating aspect of this spatial variation being the mosaic-like aspect of the vegetation in the landscape. Anybody who has hiked across the Peloponnese peninsula in southern Greece, the Sierras of southern Spain, or one of the big islands such as Crete or Corsica will appreciate this point. Such landscapes and islands contain a diversity of ecosystems which harbour rich floras with striking local variation in community structure and the presence of individual species: mountain pine forests, pungent arid garrigues and phrygana, humid canyons, deep and dense oak forests, savannahs where nothing moves out of the shade of isolated trees in mid-summer, ... the list goes on. Sharp cliffs, deep gorges, vast sedimentary basins, and meandering rivers all multiply the effects of substrate diversity and climatic stress. Human activities have further added to this spatial complexity, reinforcing environmental variation in a landscape already structured by spatial heterogeneity of geology, soils, and climate in many areas. In some zones human activities have varied dramatically in their effects, due to constraints on their action in association with environmental variation. For example, zones with deeper soils would have been the first to be cultivated as domesticated plants were dispersed across the Mediterranean region. As a result, species more prevalent in open rocky habitats, and in and around cliffs, may have been more persistent as human activities developed and spread.

I will thus insist repeatedly on the importance of spatial heterogeneity. In a quantitative classification of the Mediterranean mosaic, Dufour-Dror (2002) recognized 55 different vegetation types in Israel of which 30 were present in the Mt Carmel region. In Turkey, steppe vegetation occurs on an immense diversity of substrate types (limestone, gneiss, ultrabasic rocks, and schists) creating a myriad of selection pressures on plant populations within this type of vegetation. As Quézel and Médail (2003) point out, in such situations, the nature of the substrate plays a primary role in the composition and dynamics of local plant communities. Substrate may also contribute to patterns of endemism and the immense diversity of the flora. Even on single islands (Fig. I.3(a) and (b)) the combination of geological variation and altitude, along with strong climatic variation among different slopes can create marked heterogeneity in the ecological forces acting on the evolution of plant diversity. Such variation can also occur on a highly localized spatial scale in continental regions, such as the landscape depicted in Box I.1. Here the geology varies dramatically over a few kilometres, creating a mosaic of substrate types. Local variation in soils (which are deeper and more humid in the sedimentary basin), climate (the sedimentary basin is a frost hollow in winter), and human activities further accentuate spatial variation in ecological conditions across this landscape, where genetic differentiation among populations of common species has been reported (Chapter 4). This is but one example of localized mosaic vegetation, which is common to many Mediterranean landscapes (Fig. I.4). The point here is that tectonic activity and edaphic variation have created a template for plant evolution, which has been further modulated by climate and then more recently by human activities.

(a)

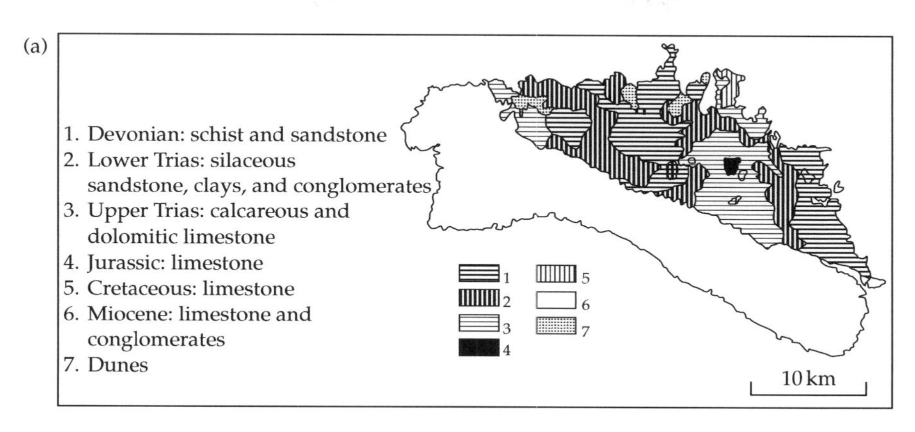

(b)

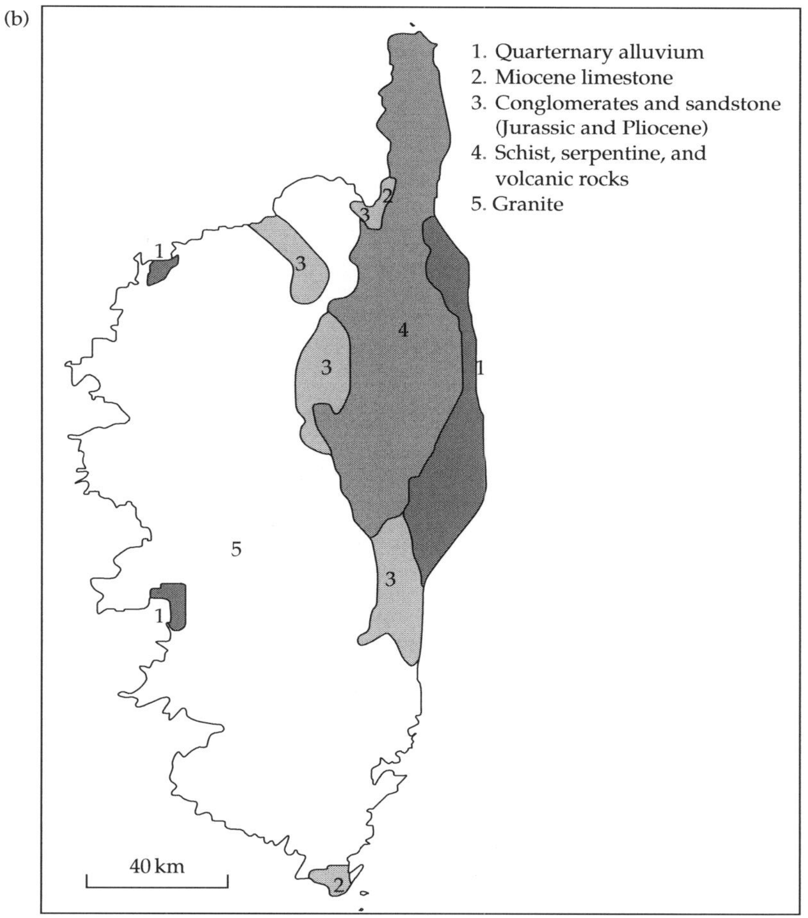

Figure I.3 The mosaic of substrate types on individual islands as illustrated by the simplified geology of the islands of (a) Minorca and (b) Corsica (redrawn from de Bolòs and Molinier 1970 and Gamisans 1999, respectively).

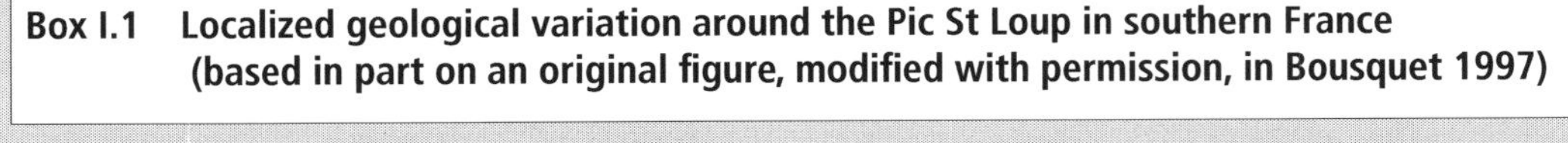

Box I.1 Localized geological variation around the Pic St Loup in southern France (based in part on an original figure, modified with permission, in Bousquet 1997)

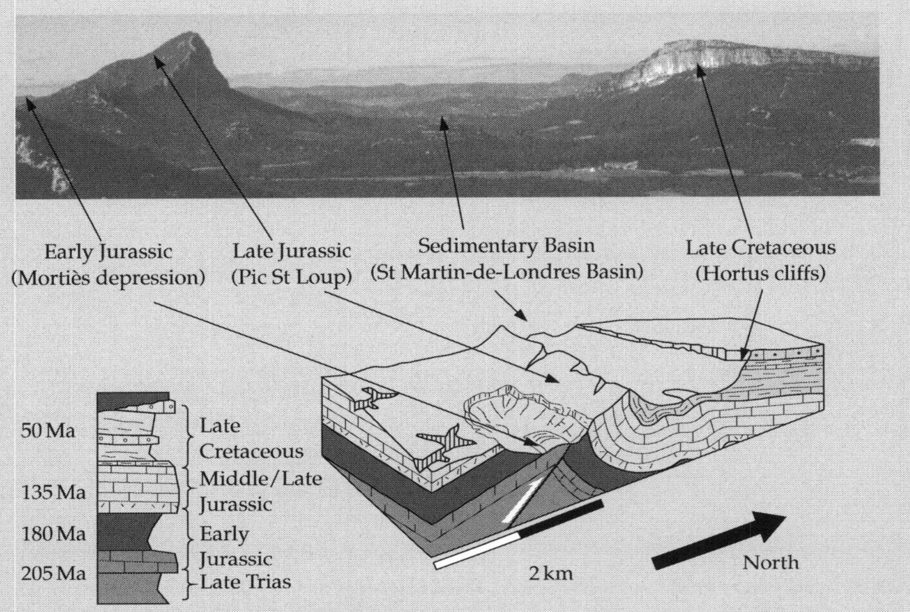

The history and variety of geological formations and orogenic movements in this area of southern France have been such that the Late Jurassic limestone of the Pic St Loup (left) stands 'face to face' with the Late Cretaceous cliffs of the Hortus (right). Subsidence and erosion has revealed the black marls of the Early Jurassic to the south of the Pic St Loup, in a basin with a small area of Early Jurassic limestone. To the north and east a sedimentary basin (the St Martin-de-Londres Basin discussed in Chapter 4) rich in marine fossils covers large expanses. This sedimentary basin is in a frost hollow where the winter climate is much colder than the surrounding higher elevation lands. The deeper and better water retention capacity of soils in the basin have promoted agricultural exploitation of many areas, modifying the spatial configuration and size of semi-natural habitats in the landscape.

The mosaic landscape is thus shaped by the interplay of geology, climate, and human activities. In fact, as Lepart and Debussche (1992) illustrate, abiotic spatial heterogeneity has been a major determinant of the nature and impact of human activities on natural habitats in the Mediterranean region.

The evolution of plants in such a spatially heterogeneous landscape requires an understanding of regional history and the ecological differences which occur at a variety of spatial scales. The habitat mosaic is associated withsharp and local variation in selection pressures and regulates gene flow by modulating pollen and seed dispersal among patches of favourable habitat. Spatial heterogeneity is thus at the heart of my discussion of plant evolution in Chapters 4 and 5 of the book. This discussion is strongly motivated by my conviction, that to understand plant evolution requires an

Figure I.4 Mosaic habitat variation in the Mediterranean: (a) shrub–woodland interface in the Sierra de Cazorla in southern Spain (photo kindly supplied by C. Herrera), (b) *Pinus brutia* forest pocket in the mountains of Crete, (c) oak woodland, limestone cliffs, scree slopes, and open garrigue vegetation adjacent to cultivated fields and abandoned cultivated areas in southern France.

understanding of how spatial variation in ecological processes regulates dispersal, thus creating a template for differentiation, and how trait variation influences establishment and reproduction, and thus affects long-term population dynamics and evolution.

The central theme of Chapters 4 and 5 concerns variation and adaptation within and among local populations in relation to the environmental constraints and selection pressures that natural populations encounter in the Mediterranean mosaic landscape. The emphasis is thus on intraspecific variation. In Mediterranean-climate regions, plant form and function has traditionally been interpreted as a response to climatic and edaphic constraints. I thus explore and evaluate evidence of adaptive trait evolution in this context. A key issue I develop is that local populations occur in habitats that are part of a highly heterogeneous mosaic of environmental variation in the landscape. Population differentiation is a balance between the local ecological and population processes acting on the genetic variation in a habitat patch and the regional processes which determine gene flow and migration among patches. I address this issue in Chapter 4 where I discuss functional trait variation, adaptation, and dispersal patterns, and in Chapter 5 where I discuss the ecology, spatial dynamics, and evolution of reproductive strategies.

Then, in Chapter 6, I discuss the ecology and evolution of species whose distributions have been modified as a result of human-induced dispersal. The focus of this final chapter is evolution under domestication and cultivation and the population ecology of invasive species. By moving plants around the Mediterranean Basin, and into and out of the region, humans have not only created exasperatingly complex problems for the conservation of differentiation diversity but have also set up experimental populations ready for the study of plant evolution in a new environment.

The Mediterranean Basin, along with parts of south-western Australia, the south-western Cape of South Africa, western California, and central Chile is one of five Mediterranean-climate regions of the world. These five regions of the world only occupy ~5% of the land surface but harbour 20% of known vascular plant species (Cowling *et al.* 1996). They also contain a large number of endemic species and show strong patterns of localized or regional differentiation. The similarities and differences of vegetation in these Mediterranean-climate regions (Dallman 1998) have given rise to much interest in the possible convergent evolution of vegetation and traits in these different regions. Rather than write a single chapter on the comparative ecology and evolution of floras in the different Mediterranean regions of the world, I have repeatedly broadened my discussion to compare patterns with those in the other Mediterranean-climate regions. In Chapter 2, I extend my exploration of the biology and ecology of endemic plants in the Mediterranean to encompass patterns observed in South Africa and Australia. In Chapter 3, the importance of climate change in different Mediterranean regions for the evolution of endemism is discussed. In Chapter 4, I discuss variation in the occurrence of traits associated with sclerophylly and resprouting ability in different Mediterranean-climate regions. In Chapter 6, I compare patterns and processes of invasion in different Mediterranean regions. My purpose has not been to provide a comprehensive comparative examination of convergence in different Mediterranean regions but to illustrate those features of plant evolution in the Mediterranean flora which are common to other Mediterranean-climate regions and those which are more closely tied to specific aspects of the regional history of the Mediterranean Basin. I have thus selected specific examples from other Mediterranean-climate regions to illustrate general patterns. In addition, I have included sections which gives each chapter a broad-based conceptual framework.

Finally, the richness of endemic plants in the Mediterranean has lured botanists into the region for centuries. In more recent years, the population ecology and genetics of Mediterranean plants have received growing attention, with much interest directed towards understanding the ecology and evolution of natural plant populations. There now exists a large body of information concerning various aspects of the biology of Mediterranean plants. In this book my aim is to draw together such information in a general synthesis of evolutionary ecology in which evolutionary processes are discussed in relation to regional history. To conclude, I recapitulate some of the main themes and issues of plant evolution in the Mediterranean within the context of the conservation of endemic plants in the region. I argue that more emphasis should be placed on conservation strategies which explicitly integrate ecological processes and evolutionary potential.

CHAPTER 1

The historical context of differentiation and diversity

The mountains and basins of the Mediterranean have been called the Enigma Variations of tectonic geology. Certainly it is a symphony of the earth that is not easy to understand.

J.M. Houston (1964: 51)

1.1 Geology, climate, and human activities: the mould and sculptors of plant diversity

To provide a framework for my discussion of plant evolution, I have outlined what I consider to be the three dominant factors which have been instrumental in shaping the evolutionary forces acting on plant variation and diversity in the Mediterranean region in a historical triptych (Box 1.1). Geological and climatic histories have greatly impacted on species distributions, isolating individual populations or localized groups of populations and bringing into reproductive contact previously isolated but closely related taxa. Human activities have modified selection pressures and the potential for pollen and seed dispersal across the landscape. Although these three factors are presented in separate panels, in reality there are no sharp boundaries. Indeed, I will emphasize throughout this book that plant evolution has been greatly influenced by the interaction among the three different elements of this triptych.

First, the Mediterranean region has a complex geological history. The Mediterranean is the largest inland sea in the world. From Gibraltar in the west, the Mediterranean Sea stretches eastwards for just over 3,500 km. Its width is highly variable: whereas ~750 km separate the south of France from Algeria, in some places, such as across the Straits of Gibraltar or where Italy sleeks down via Sicily towards Tunisia, only a few kilometres separate the northern and southern shores. Trapped in a collision zone between the African and Eurasian plates, the Mediterranean Sea has only one narrow natural outlet, via the Straits of Gibraltar, which provides an exchange with the oceans outside. This setting represents one of the most geologically complex areas of the world and a unique example of a sea surrounded by different continents. It is in this context that plants have diversified. The geological complexity has untold ramifications for our understanding of the origins of the flora in the Mediterranean Basin and provides a fascinating setting for the study of plant evolution. As Oleg Polunin (1980: 1) pointed out in the opening sentence of his book on the flora of the Balkans, 'The geological history of the Balkans is perhaps the most important single factor contributing to the diversity of the present-day flora'. I will illustrate how similar statements could be made for other regions around the Mediterranean Sea.

Second, the Mediterranean region has a characteristic climate with two main seasons. The essential characteristic of this climate is the occurrence of hot and dry summers which impose an effective drought on the plants. There is also a cool or cold moist season in which unpredictable and often intense rainfall events occur from autumn through spring. Close to sea, this season is mild compared with inland, where freezing temperatures commonly occur in winter. As I discuss later in this chapter, the relative length of the summer drought and the amount and timing of rainfall in the moist

Box 1.1 Historical triptych

What are considered in this book to be the three key factors which have modulated the action of the main processes impacting on plant evolution in the Mediterranean are depicted here in the panels of a historical triptych. Each panel illustrates the timing of some important events.

Human activities	Geological history	Climate
Last 150 years Abandonment of traditional rural land-use and reforestation on northern shores	*Pliocene (~1.5–2 Ma)* Alpine orogeny (uplift and folding)	*Last 100 years* Gradual warming
Holocene Major forest clearance	*Miocene (5.3–6 Ma)* The Messinian crisis	*Holocene* Increased aridity
Prior to 8,000 bp Early harvesting and cultivation of the wild relatives of domesticated cereals, legumes, and fruit trees in the Near East	*<25 Ma* Migration of Cyrno-Sardinian microplate	*12,000* BP Late glacial warming
	Oligocene (25–30 Ma) Alpine orogeny	*Quaternary* The major Pleistocene glaciations
	Jurassic/Cretaceous Apulian plate contacts Europe. Rotation and north-east migration of African plate	*Pliocene (~2–3 Ma)* Onset of the Mediterranean climate
		Middle Miocene (~15 Ma) Mild seasonal climatic contrasts begin to develop
		Early Miocene and beyond... Subtropical conditions

season show much spatial variation around the Mediterranean Basin. It is the alternation of these two seasons which unifies the region, its landscape and its flora. The climate we now experience is a relatively recent phenomenon and has oscillated repeatedly. Its evolution and oscillations have, within the constraints of land connections and dispersal limitation, caused species' range sizes to contract and expand at repeated intervals. As some species disappeared from the landscape, others expanded their range. Plant diversity in the Mediterranean tells this tale of climate change.

Third, nowhere else in Europe has there been such a long history of human presence and activity. Harvesting, cultivation, and domestication began early, particularly in the eastern Mediterranean. Since the Neolithic, the impact of human activities on the landscape has been dramatic in terms of the spatial configuration and size of natural habitats. Such impacts have created new opportunities for colonization in some species and caused others to retract into isolated patches. As a result, human activity should be viewed as an integral ecological feature of the Mediterranean scene, modifying not only the spatial configuration of habitats in the landscape, with consequent effects on gene flow and the potential for differentiation, but also the local selection pressures and constraints that determine plant establishment, persistence, and evolution.

This chapter traces the history of the Mediterranean flora in relation to the three factors presented in Box 1.1. My objective is to lay the foundation for my subsequent exploration and evaluation of patterns of differentiation and divergence in Chapters 2 and 3.

1.2 A meeting of continents: a complex geological history

The Mediterranean has been fashioned by the meeting of Eurasia and Africa. The precise geological history of the Mediterranean is far from being completely understood, and the account I give in this chapter attempts to synthesize (largely from the geological literature) the extent of current knowledge, some of which remains hypothetical since some interpretations require further confirmation. My impression is that our understanding of the geological history of the Mediterranean is at a stage similar to that of an evolutionary biologist staring at a molecular phylogeny of a large genus based on one or a small number of gene(s). Although the framework of the tree is no doubt close to its true form, several species may not be in their correct clades. This analogy should be kept in mind as one reads through my interpretation of geological history.

1.2.1 From ancient Tethys to a series of basins

The Mediterranean Sea has an ancestor named Tethys, whose history is complex. Most evidence points to the existence of an equatorial ocean, or Paleotethys, between the northern and southern continents of Pangea during the Triassic (Maldonado 1985). This ocean was wedge-shaped, open to the east and closed to the west, where a Hercynian continent linked what is now Africa to north-western Europe (Fig. 1.1). Paleotethys closed in the early Mesozoic due to the overall northward motion of continental blocks that rifted away from Gondwana and collided with Eurasia (Sengör 1979). This produced Neotethys, a Permian to Jurassic ocean that is widely known from remnants of oceanic crust (ophiolitic structures) now found in the Alpine Mediterranean belt. The whole of the eastern wedge of Tethys began to disappear as a result of subduction during the early Mesozoic as the 'Eurasian' continent spread.

The configuration of the ancestral Mediterranean Sea, a series of closed basins in the Oligocene and Miocene, was thus closely related to the structural relationships between the major tectonic belts of Africa and Eurasia. With the opening of the

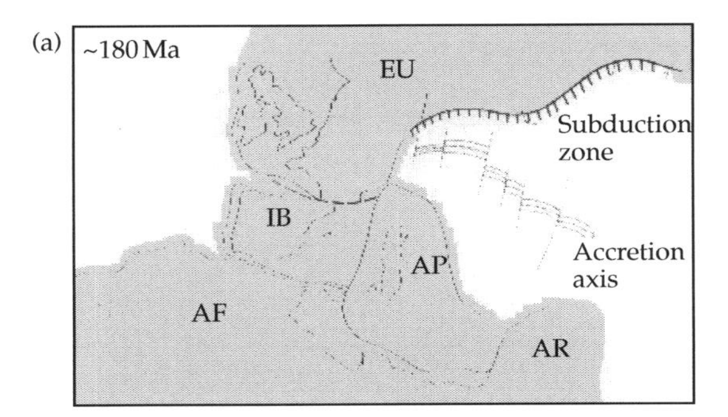

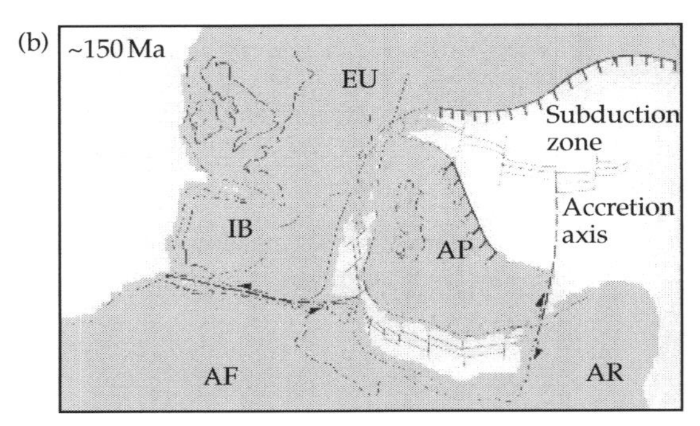

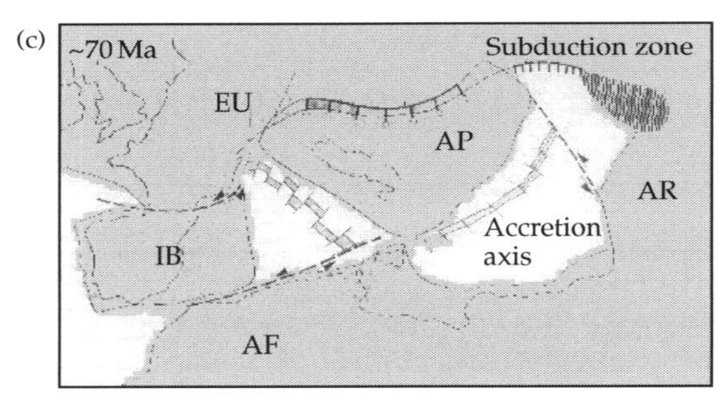

Figure 1.1 The ancient Tethys and the historical positions and movements of microplates during the development of the Mediterranean. AF: African plate, AP: Apulian microplate, EU: European plate, AR: Arabian plate, IB: Iberian microplate. Arrows represent plate movements (redrawn from figures in Biju-Duval *et al.* 1976).

Atlantic Ocean in the Early and Middle Jurassic, that is, at 165 Ma (used to signify 'Mega Annum' in the geological literature, this abbreviation gives us a timescale in millions of years), Eurasia and Africa began convergence motion which was to shape the early formation of the Alps and the Mediterranean Basin (for details of what follows see: Biju-Duval *et al.* 1976; Dewey *et al.* 1989; Rosenbaum *et al.* 2002*b*). During the Late Jurassic and Early

Cretaceous (170–120 Ma), the two plates showed left-lateral strike-slip motion and ~200 km of displacement (Fig. 1.1(b)). Then, in the Cretaceous (120–80 Ma), plate convergence brought Africa and Europe closer together (Fig. 1.1(c)) and Alpine orogenesis began. Collision may have commenced in the Early Tertiary (~65 Ma) although this remains unsure (Rosenbaum *et al.* 2002*b*). After a period of relative quiescence, more convergence occurred during the Eocene and Early Oligocene (55–45 Ma) as Africa rotated by more than 50° relative to Europe, swinging from a north-east/south-west tilt to its present position, face to face with Europe. The African plate is still moving.

The approach of the European and African plates produced two characteristic features of the Mediterranean landscape.

First, the Mediterranean Sea contains a series of deep basins bordered by relatively shallow sills (Fig. 1.1; Box 1.2). The different basins are small in terms of their surface area, and the continental margins (i.e. the transition zone between continental and oceanic crust) cover more than 100 km in many areas and less than 10% of the surface of the different basins lies more than 100 km from the continental plateau. The different continental margins thus touch each other, adding to the unity which makes up the contemporary configuration of the Mediterranean region. The formation of the different basins has been closely associated with the configuration of adjoining land masses and, along with its almost lack of any tidal regime (except in some restricted areas) and its high salinity, is one of the principal characteristics of the Mediterranean sea.

Second, the Mediterranean region has many mountains. For example, the Atlas Mountains of North Africa (geologically speaking: the Rif and Maghrebides), the Sierra Nevada (or Betic Cordillera), the Pyrenees, Appenines, Dinarides, Taurus, and Anatolian chains and Mt Liban all form an imposing backdrop. In some areas these mountains drop directly into the sea, while in other parts of the Mediterranean Basin the transition is more gradual through low hills and a coastal plain. Centres of diversification, many of the mountainous areas represent hot spots of endemism (see later in this chapter). Their formation was closely associated with Alpine orogeny which occurred in two main periods. The first occurred in the Cretaceous and Early Tertiary when compression and mountain building produced the initial socle of mountains in many areas. The second followed later in the Pliocene and Pleistocene and involved vertical uplift and fracturing. Quaternary processes were thus important elements in the fashioning of the current day landscape both in the western (Houston 1964) and eastern (Zohary 1973) Mediterranean.

1.2.2 Micro-plate configuration: dispersal and contact

Since at least the Tertiary, and perhaps during the earlier stages of Alpine orogenesis (G. Rosenbaum, University of Mainz, personal communication), microplate individualization and dispersal have played a major role in the tectonic evolution of the Mediterranean Basin (Alvarez *et al.* 1974; Biju-Duval *et al.* 1976; Rosenbaum *et al.* 2002*a*, 2004). The three most well studied are the Iberian microplate, Adria (or Apulian microplate comprising Italy, the Balkans, and Greece), and the Cyrno-Sardinian microplate.

The Iberian microplate occupied a key position in the geological evolution of the Mediterranean Basin due to its position at the western extremity of the contact zone between the African and European plates (Fig. 1.1). The geological evolution of this region exhibits a complicated interplay of orogenic processes and plate movements (Rosenbaum *et al.* 2002*b*). The Iberian microplate was initially attached to Europe, albeit further west than at present. The movement of the African plate pushed it north-eastwards from the Late Jurassic to the Late Cretaceous (~70 Ma), causing the uplift of various mountain ranges, notably the Pyrenees.

The Apulian plate or Adria represents the continental crust bridging the continental masses of Africa and Eurasia across the central Mediterranean where it separated the eastern and western basins (Rosenbaum *et al.* 2004; Fig. 1.1). This plate was centred on what is now the Adriatic Sea. Connected to the African plate, perhaps as a promontory rather than a detached fragment, this microplate came into contact with the southern part of the European plate.

Box 1.2 Geological history of the basins under the Mediterranean Sea

The Mediterranean Sea is subdivided into individual basins (black) separated by shallow sills, some of which represent ancient arcs of mountains now under the sea (dark grey). Several islands are poised on these sills. In the east, the Gibraltar sill maintains a degree of isolation from the Atlantic, with implications for sea currents and the climate of the Mediterranean region.

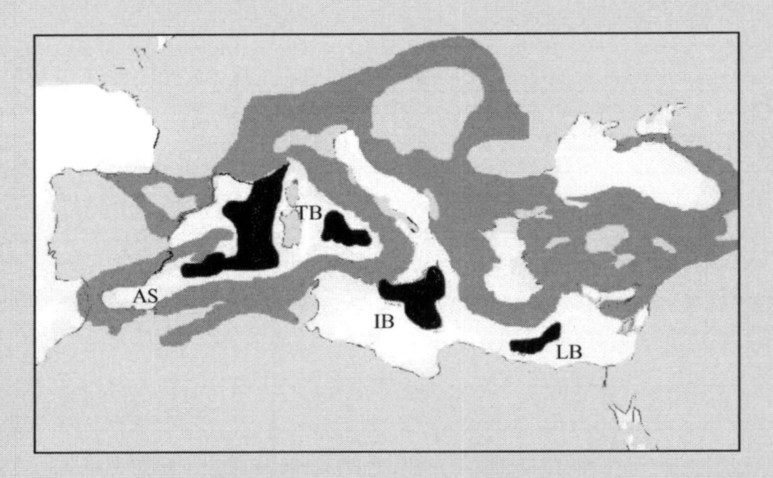

The different basins contain the different localized seas within the Mediterranean, that is, the Tyrrhenian, Alboran, Ionian, Adriatic, and Aegean Seas.

The Alboran Sea (AS) probably opened during the north-westward movement of the Apulian microplate (Maldonado 1985). Collision of Africa and Europe led to the formation of a south-western Mediterranean microplate (Araña and Vegas 1974). At this time the Alboran Sea was, with the Betic Cordillera, above sea level. Continued westward movement of the microplate caused the formation of the Gibraltar arc as the plate was over-thrusted on collision with the Atlantic continental margin. The Calabrian and Hellenic arc formations probably formed in this way (Biju-Duval *et al.* 1976; Maldonado 1985).

The Tyrrhenian 'back arc basin' (TB) was more recently created (in the Late Miocene) by extensional rifting and subduction (Cherchi and Montadert 1982; Mascle and Rehault 1991; Robertson and Grasso 1995).

The Aegean Sea developed in the Pliocene and Quaternary (Maldonado 1985; Robertson and Grasso 1995), primarily as a result of northward subduction, continued back arc extension and volcanism. This development, like that of the western Mediterranean, involved a reconfiguration of ancient rocks.

The Ionian Basin (IB), which plunges to 5000 m, and the Levant Basin (LB) are the only areas in the Mediterranean region where remnants of ancient Neotethys oceanic crust underlie sea floor sediments (Rosenbaum *et al.* 2002*a*). The shallow submarine sill linking Calabria, Sicily, and Tunisia (at depths of <600–700 m) which subsided at the end of the Tertiary, represents an important north–south historical connection.

Analysis of paleomagnetic, geophysical, and geological data point to a relatively coherent motion of Adria and the African plate since the Jurassic, albeit with some independent rotation. The detachment of Adria and Africa involved the opening of the Ionian Sea perhaps as early as the Permian. Movement of the African plate and the collision of its Arabian margin with Europe caused the Mediterranean Sea to become closed in the east, and the rotation of the Iberian microplate allowed for only a small outlet in the west.

The history of the *Cyrno-Sardinian microplate* (see Rosenbaum *et al.* 2002*a*) is critical to our understanding of endemism in the western Mediterranean (as we shall see in Chapter 2). In the Late Oligocene (35–30 Ma), a Hercynian massif connected the Pyrenees to the outer crystalline massifs of the Maures—Esterel, and ultimately the Alps via what are now the cliffs on the north-east tip of Minorca, Corsica, and Sardinia (Fig. 1.2(a)). The latter two islands were then part of a continental environment, with Corsica 30° and Sardinia a little over 60° north-west of their present position and orientation (Hsu 1971; Westphal *et al.* 1976; Cohen 1980; Cherchi and Montadert 1982). Corsica and the Esterel (now part of continental France) were contiguous (Cohen 1980). Based on their geological similarity, Alvarez (1976) postulated that north-eastern Corsica, Calabria, the Kabylies (in North Africa), and the Betic Cordillera were also linked to one another in an Alpine Belt that extended around the southern edge of the Hercynian massif (Fig. 1.2(a)).

In the Late Oligocene (Fig. 1.2(b)), from an initial closed position against southern France and north-west Italy, this microplate began to rotate south-eastwards (Alvarez 1974; Rosenbaum *et al.* 2002*a*). The dispersal and fragmentation of the Tyrrhenian islands and Calabria on a single microplate probably started due to the rifting-off of the European continental margin to produce the Cyrno-Sardinian microplate (Cherchi and Montadert 1982; Robertson and Grasso 1995). The Balearic Basin opened behind the rotating microplate. Once Corsica collided with the crust of the northern Appenines (~20 Ma) it could no longer rotate (Fig. 1.2(c)). As a result, the depression where the Straits of Bonifacio now occur opened as Corsica became separated from Sardinia and Calabria, which continued to rotate towards the south-east. According to Alvarez (1974) the rotating plate collided with the Tunisian margin of North Africa at ~14 Ma, a collision which stopped the rotation of Sardinia. In the Middle Miocene, Sardinia became separated from Calabria and north-east Sicily. The opening of the Tyrrhenian Sea occurred in two stages. From 9 to 5 Ma (Fig. 1.2(c),(d)) an opening to the north appeared and then after the Messinian (Fig. 1.2(d),(e)) an opening to the south developed. Corsica, Sardinia, and the other fragments of the initial microplate have thus been isolated from the Balearic islands and southern France since the Miocene. The Balearic islands have however had repeated connections among each other (Minorca and Majorca had their latest connection in the Pleistocene) and with the Iberian peninsula.

As Corsica and Sardinia rotated south-eastwards during the Miocene, the Kabylies broke away from the Balearic islands in a southerly direction (Fig. 1.2(b),(c)) and collided with the African margin (Fig. 1.2(c)). During this period, the Betic Cordillera became separated from the eastward migrating fragments and accreted (~10 Ma) to the south-eastern tip of Spain (Fig. 1.2(b)–(d)). Calabria and north-east Sicily continued their rotation till the Pliocene when they arrived in their present position (Fig. 1.2(d),(e)).

1.2.3 When the Mediterranean salted up

In 1961, a newly developed type of echo-sounding used by oceanographers produced what was then a startling finding: the Mediterranean Sea floor is underlain by an array of pillar-like structures. Some of these exceed several kilometres in diameter, reach 1,500 m in height, and protrude as knolls on the sea floor. Today, some are exposed on land, the best examples being on Sicily (Robertson and Grasso 1995). The resemblance of these structures to salt-domes put geologists onto the idea that vast salt deposits are currently hidden beneath the floor of the Mediterranean. This was the first hint that in the Messinian stage of the Late Miocene, the Mediterranean dried up (Hsü 1972; Hsü *et al.* 1973, 1977; Cita 1982; Duggen *et al.* 2003). This 'Messinian salinity crisis' is now known to have begun at 5.96 Ma (Krijsman *et al.* 1999).

In the Late Miocene (~8 Ma), marine passages in southern Spain and northern Morocco linked the Mediterranean Sea to the Atlantic Ocean. Although a global drop in sea level occurred at about this time, this marine gateway from the Mediterranean to the Atlantic was probably closed as a result of uplifting

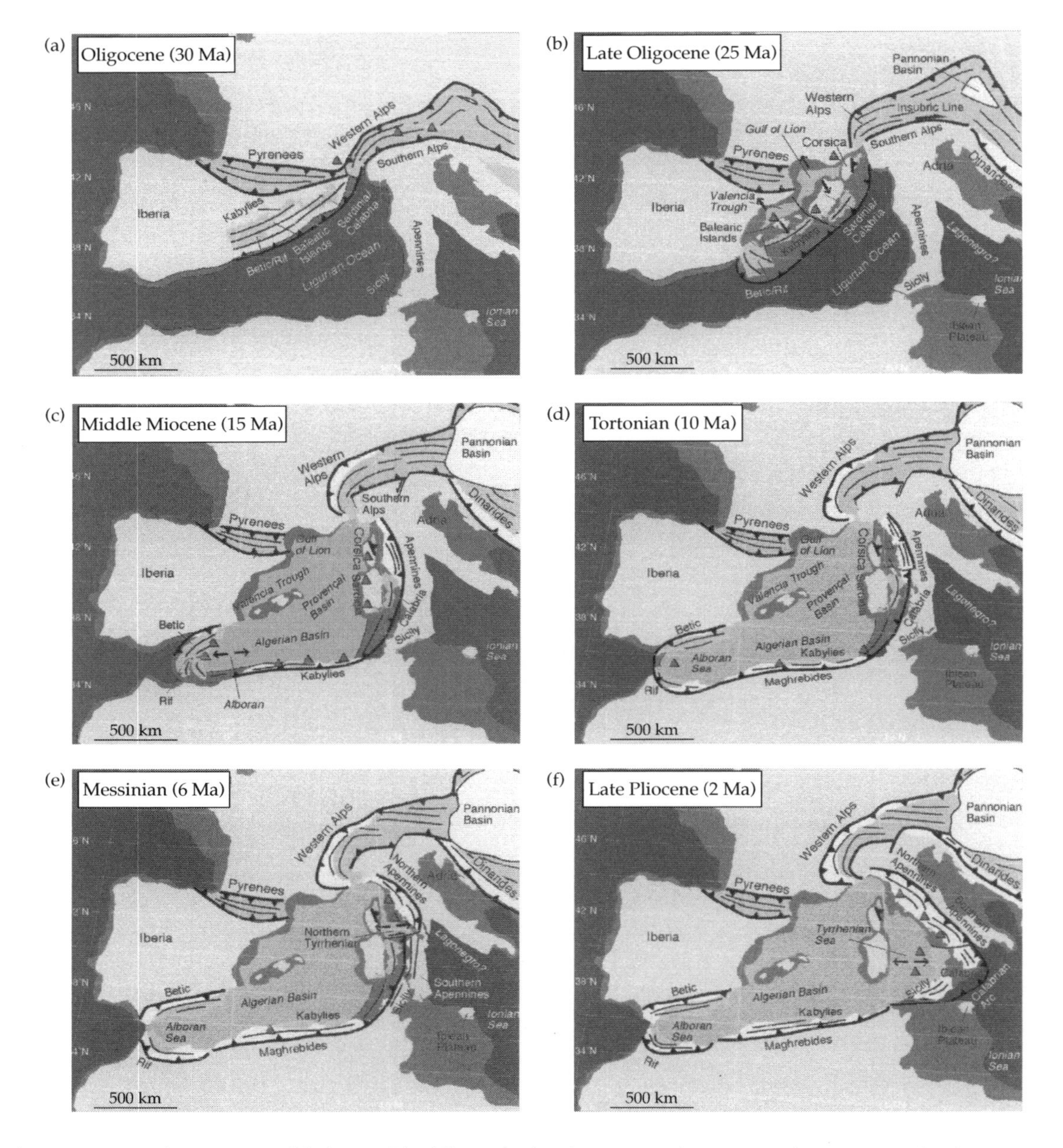

Figure 1.2 Historical reconstruction of the history of the different islands and continents in the western Mediterranean since the Oligocene. (reproduced with permission from Rosenbaum *et al.* 2002*a*).

along the African and Iberian plate margins in association with mantle processes (Duggen *et al.* 2003). Marked regional aridity led to high levels of evaporation from the closed Mediterranean Sea, which led to a basin-wide lowering of sea level as the other sea levels lowered. The Mediterranean Sea became a disjunct mosaic of large lakes in which thick and extensive evaporites precipitated, particularly in the deepest parts of the basins. The presence of fossilized remains of light-demanding cyanobacteria, which usually develop in shallow water, indicate that although the precipitation occurred within the confines of deep basins, it occurred in relatively shallow water (Hsü *et al.* 1973). As these authors discuss, the depth of the evaporite in some areas could not have occurred from a single desiccation since there is simply not enough salt in sea water to produce the immense salt deposits currently under the Mediterranean. In fact, cycles of desiccation–inundation probably repeated themselves ~8–10 times during the Messinian.

The Messinian salt crisis was, to quote Duggen *et al.* (2003: 602) 'one of the most dramatic events on Earth during the Cenozoic era'. During this period, the Mediterranean became a desert (Hsü 1973). There were land-bridge connections between, for example, Corsica, Sardinia and the north of Italy; connections linking Sicily with southern Italy and perhaps North Africa, and connections from continental Greece (a) to the south-east across the small Aegean islands (perhaps to Rhodes) and (b) to Crete via the Peloponnese. Several authors (Bocquet *et al.* 1978; Cardona and Contandriopoulos 1979) have thus discussed whether such land-bridge connections permitted plant migration, indeed there is good evidence for the migration of animals during this period (Alcover *et al.* 1999). I do not doubt the occurrence of land connections during this period. What remains questionable, however, is how suitable such connections may have been for plant life, and thus the migration of sedentary organisms. A problem here is that the climate is not thought to have been markedly different during the Messinian compared to the preceding part of the Late Miocene and the subsequent Early Pliocene (see below). In the absence of a change in climate it is difficult to envisage how vegetation could have dropped in altitude in order to allow for migration among newly connected areas (Suc 1989; Quézel 1995).

The end of the Messinian occurred suddenly at 5.33 Ma (Krijgsman *et al.* 1999). The distinct separation between Messinian evaporites and Pliocene marine sediments confirms this abrupt end, which coincided with the opening of the Gibraltar Straits and the establishment of a permanent connection between the Atlantic and the Mediterranean. Mantle-related causes may have created a new marine connection to the Atlantic (Duggen *et al.* 2003). Since then the precise location of the Mediterranean coastline has developed its present configuration, with an extension of coastal areas as sea levels declined during the different Quaternary glaciations. For example, sea level was ~150 m lower than the present sea level, during the last glacial maximum (Kaiser 1969), allowing for many land-bridge connections in various parts of the Mediterranean (Corsica with Sardinia, Majorca with Minorca, and among different Aegean islands).

1.2.4 Recent geological history

Geological activity has remained a major feature of the Mediterranean region in recent history and of course continues. The volcanoes under and around the Mediterranean Sea have different magmatic origins (Rosenbaum and Lister 2004), being related to either crust extension (those in the centre of the basin), convergence and subduction (the chain of volcanoes in the Aeolian islands), or intra-plate magmatism. Subduction continues where plates meet in many areas such as under the island of Cyprus (Robertson and Grasso 1995) and in the Calabrian Arc. Volcanism has remodelled the Greek island of Santorini and repeatedly caused extinction and re-colonization of plant communities on Vesuvius and Etna (where more than 100 eruptions have been signalled in the last 2,500 years). Volcanic activity has never ceased, and continues to shake southern Italy, north Africa, and Turkey.

Another important recent event concerned the contemporary Mediterranean coastline. Four main types of erosion have been important here: (a) mechanical action in the high mountains, (b) linear erosion by rivers, (c) lateral erosion in semi-arid areas, and (d) aeolian erosion in more arid regions. Marine terraces are an important feature of this coastline (Houston 1964; Biju-Duval *et al.* 1976), and are particularly prominent along the Ionian coast of Italy, Corsica, Tunisia, and parts of southern France and Spain.

I have detailed this complex history of land masses in the Mediterranean region to illustrate how geological history has no doubt been closely associated with the limitation of species distributions and the creation of phylogeographic divisions within the Mediterranean flora. Since the end of the Miocene, two other historical factors have become decisive elements in the shaping of plant species distributions and endemism around the Mediterranean: the onset of a summer drought and, more recently, the development of human activities.

1.3 Two seasons: the history of the climate and the vegetation

1.3.1 The contemporary climate

The Mediterranean region has a climatic regime which is characterized by two main seasons. The first, a hot dry summer, is the essence of the Mediterranean climate region, whose definition relies on the regular occurrence of an effective drought at the hottest time of the year (Quézel 1985). During the Mediterranean summer, the main weather centre of influence is the Atlantic anticyclone of the Azores which imposes fairly uniform sunshine and a quasi-absence or scarcity of rainfall. Although the onset of summer can be fairly gradual, its end is usually abrupt and accompanied by intense rainfall events. Thus begins the second or 'wet' season whose moist and cool climate assures plant development. This cooler period lasts from 5–10 months, depending on the region. During this period, intense cold may occur in some regions and limit plant development. Such winter stress may limit species distribution, as noted for some sclerophyllous species (Mitrakos 1982) and thus represents a second constraint on the phenology and growth of Mediterranean plants (Chapter 4). To sum up, the Mediterranean climate imposes a double constraint on plant growth, lack of moisture in summer and cold temperatures in winter.

During the cool moist period the Mediterranean region lies between a cold anticyclone present in Asia and the Atlantic anticyclone, that is, in a zone of low pressure with minimum values over the main sea basins, which are the centres of frequent cyclogenesis. The advection of air streams from these sources dominate weather conditions and create important events and periods of rainfall. There are four important characteristics of rainfall in the Mediterranean (Houston 1964). First, annual rainfall is concentrated into a small number of events. Second, in a given region, there is enormous interannual variation in the timing and intensity of rainfall events. Third, Mediterranean depressions rarely have clear-cut fronts and bare little resemblance to the Atlantic depressions that regularly swing across north-west Europe. Fourth, the majority of depressions originate within the Mediterranean Basin itself.

An important feature of rainfall in the Mediterranean is the marked regional variation in annual levels and timing of peak rainfall events. Rainfall may be concentrated in the autumn, winter, or spring, depending on the region. In the western Mediterranean, the peak rainfall occurs in the autumn. A smaller peak occurs in spring around the coastal areas of the northern rim of the Mediterranean, in winter from the southern tip of Spain (Andalousia) across north Africa and Sicily to southern Italy (Calabria), and in spring in the centre of the Iberian peninsula, the Moroccan Atlas and the high plateau of Algeria. Overall, rainfall declines from west to east. More important, seasonal aridity becomes longer and more severe in the south than in the north, and in the east compared with the west. These trends occur on three spatial scales: basin-wide, on individual pieces of continents such as the Balkans and Greece (where annual rainfall more than doubles as one moves from the eastern fringes of north-west Greece to the western Balkans), and on individual islands such as Crete (where once again annual rainfall in the west is twice that of the eastern tip of the island). Detailed graphs of sunshine and precipitation are well documented elsewhere (e.g. Houston 1964; Zohary 1973; Blondel and Aronson 1999; Grove and Rackham 2001; Quézel and Médail 2003), where they illustrate clearly the spatial heterogeneity in climatic regime in terms of the length of the summer drought and the timing and amounts of rainfall.

The contemporary Mediterranean climate is a recent phenomenon and three main chapters of climatic history provide the backdrop to the constitution of the contemporary vegetation.

1. Prior to the Middle Pliocene (i.e. >3 Ma).
2. Recent alternation of summer drought and cold temperatures associated with glaciation.
3. Climate change in the presence of human activities, that is, since the last glaciation.

Box 1.3 Pollen analyses and vegetation history

The study of pollen composition in cores of sediment is a major tool in the study of vegetation history. Nonetheless, important caveats in the use and interpretation of pollen data for assessing community composition require appreciation (see Bazille-Robert *et al.* 1980; Pons and Suc 1980; Reille *et al.* 1996; Grove and Rackham 2001):

1. A bias due to the marked variation among taxa in their preservation and their likely input to sediments due to differences in pollen production and release. It is highly unlikely that insect pollinated species will input as much pollen to a sediment as wind pollinated species, even if the two were at equal abundance in the landscape.
2. It is difficult to distinguish congeners—which may have very different ecological requirements. In mixed oak forests, deciduous oak pollen may be more abundant than evergreen oak pollen in soil samples because they grow on deeper soils.
3. Only particular conditions (such as in peat bogs) allow preservation. Mediterranean ecosystems away from the coast are, for the most part, far from being peat bogs. Such conditions are uncommon and only occur in isolated spots. Mediterranean taxa may thus be underestimated in historical reconstruction.
4. Some areas of the Mediterranean have been intensively studied (Spain, southern Italy, and the South of France) while others have received almost no attention.

The historical study of Mediterranean vegetation has a long tradition (de Saporta 1863). Much of what is known is based on analysis of pollen frequencies in cores, which, for diverse regions, should be interpreted with some caution (Box 1.3) with additional information coming from fossilized plant parts and charcoal remnants.

1.3.2 The onset of the Mediterranean climate

The Early Tertiary
During the Early Tertiary the accuracy of vegetation analysis based on pollen remains is poor since only few tree species are identifiable. It is, however, generally thought that south of Tethys, the vegetation was essentially tropical (forest and savanna) and different to that to the north (Quézel 1995), where sclerophyllous vegetation, akin to that in western North America, was present: the so called Madro-tertiary Geoflora (Axelrod 1958) or 'Madre-Tethyan' sclerophyll vegetation (Axelrod 1975).

The Late Tertiary
In the Late Tertiary (Oligocene and Miocene, 33–5 Ma) evergreen rainforest and laurel forests were the two most important vegetation types on the European plate (Mai 1989). In North Africa, temperate rainforest occurred in the Saharan zone and subtropical woodland savanna (with a species composition that suggests the occurrence of a dry season) existed in the area where present day Mediterranean vegetation occurs, that is, the three countries of the Maghreb and coastal areas of Libya and Egypt (Quézel 1978). The decline of this flora occurred during the cyclic periods of cooling from the Late Miocene onwards. Relicts of the Tertiary flora include the following: laurel forest vegetation (on the Macaronesian islands) in a few ancient forests in relatively humid areas (e.g. *Laurus, Prunus, Persea, Daphne,* and *Ocotea*), *Rhododendron* in oceanic parts of the Iberian peninsula, the presence of *Liquidambar, Parrotia,* and *Pterocarya* in eastern Turkey, *Zelkova abelicea* and *Phoenix theophrastii* on Crete, and *Nerium* in stream beds on Corsica. The findings of fossil leaves of several species have been particularly instructive for our knowledge of the overall vegetation types of this and subsequent periods of climate history (Vernet 1997).

Otherwise, only small amounts of fossil material are available to help elucidate the origins of the Mediterranean sclerophyll forests. Some species (*Nerium, Olea, Cupressus, Punica, Pyracantha,*

Jasminum) could have had their origins in the laurophyll floras of the Eocene (>33 Ma), while others (*Ceratonia, Cercis, Phillyrea, Pistacia*, and the evergreen oaks) may have belonged to the mixed mesophytic forest flora of the Oligocene (33–23 Ma). But these taxa were, towards the end of the Oligocene (~25 Ma), only sporadic elements of a very different vegetation to that in which they now occur (Medus and Pons 1980). The Mediterranean meso-xerophytic sclerophyll forest in its contemporary composition and structure is a relatively recent phenomenon, and could only have become established after the disappearance of the laurophyll vegetation.

The analysis of fossil macrofloras (Palamarev 1989) attests to the appearance of a ligneous sclerophyll vegetation in the Late Eocene. These sclerophyllous species were secondary elements of the widespread subtropical woody vegetation. Their abundance increased during the Oligocene when they formed more complete xerothermic communities. The best-represented species were in the genera *Quercus, Arbutus, Pistacia, Ceratonia, Acer, Periploca, Smilax*, and *Pinus*.

The Early Miocene

In the Early Miocene (~23 Ma) pollen spectra indicate that the flora of the northern sector of the western Mediterranean was rich in subtropical species and families (Bessedik *et al.* 1984; Bessedik 1985). In short, the climate was tropical, with little seasonal change in temperature and fairly high levels of summer rainfall. Major elements of this subtropical vegetation included representatives of the Taxodiaceae, Bombacaceae, Hamamelidaceae, Juglandaceae, Melicaceae, Melastomataceae, Menispermaceae, Oleaceae, Restionaceae, Sapindaceae, Sapotaceae, and Simaroubaceae. In addition, in many coastal areas of the western Mediterranean, mangrove swamps dominated by the genus *Avicennia* were present.

The overall vegetation was highly heterogeneous and some parts of the western Mediterranean, in particular the low plains (<500 m elevation), hosted a semi-arid open vegetation (Bessedik 1985). Some of the elements of the current day Mediterranean vegetation, for example, *Olea, Pistacia, Nerium*, and *Rhamnus*, and species of *Prosopis, Vitis*, and *Cistus*, were present along the northern shores of the Mediterranean (Pons and Suc 1980; Bessedik *et al.* 1984; Bessedik 1985). Elements of contemporary Mediterranean-type vegetation were present, but only in complex vegetation associations that no longer exist. Semi-arid elements were present as a secondary component of an ancient Mediterranean landscape dominated by tropical and warm temperate, evergreen and deciduous elements, with a mangrove coastal vegetation.

The Middle Miocene

In the Middle Miocene (16–14 Ma) seasonal contrasts in the temperature regime developed, perhaps as a consequence of glaciation in northerly latitudes and the loss of water connections to the Indian Ocean. Tropical elements began to disappear from pollen diagrams during this period (Pons *et al.* 1995) and floristic richness declined due to the loss of whole taxonomic groups from many regions. As a result, the flora began to resemble contemporary vegetation. Pollen analyses from cores in southern France show that this was the case for *Bombax* (Bombacaceae), *Alchornea* (Euphorbiaceae), two genera of Icacinaceae, Simaroubaceae, *Avicennia* (Avicenniaceae), *Rhodoleia* and *Eustigma* (Hamamelidaceae), and *Gunnera* (Gunneraceae). For example, *Avicennia* disappeared from the Languedoc in southern France (14 Ma), from Sicily at 5 Ma, and is now extant on the Red Sea coastline (Suc *et al.* 1992). Extinctions were thus primarily in taxonomic groups with high temperature and humidity requirements. Several of the groups that became extinct from the western Mediterranean, currently occur in tropical and subtropical regions of south-east Asia, Africa, central America and the Neotropics (Bessedik 1985).

The Late Miocene

By the Late Miocene (10–6 Ma), prior to the Messinian salinity crisis, important concentrations of Palaeo-Mediterranean species began to develop as more tropical elements were lost. Bocquet *et al.* (1978) proposed that the drop in sea level during the Messinian was associated with a drier climate during the Messinian salinity crisis, and a greater possibility for plant migration as a result of land connections. However, it is probable that the climate was not particularly dry during this period relative to the preceding period (Hsü 1973), and vegetation does not show major changes during this period (Suc and Bessais 1990; Suc *et al.* 1992, 1995; Bertini 1994; Fauquette *et al.* 1998). Mangroves (*Avicennia*) continued to disappear, although increased salinity, rather than aridity, may have been the cause. Communities of a Mediterranean sclerophyllous-type, during this period, contained evergreen oaks and pines, along with *Arbutus*, *Ceratonia*, *Olea*, *Phillyrea*, *Pistacia*, and *Pyracantha* (Palamarev 1989). In the mountains of North Africa, sclerophyllous evergreen forests occurred (Quézel 1978). Various pollen evidence suggests the presence of *Quercus* pollen in the Mediterranean region in this period (Barbero *et al.* 1992).

The Early and Middle Pliocene

In the Early Pliocene (~4 Ma), the climate was probably a few degrees warmer and slightly more humid than at the present time (Fauquette *et al.* 1998). Pollen diagrams show that tropical elements had disappeared from the flora of the northern shores of the Mediterranean but were still present to the south. In the north-western Mediterranean, pollen analysis suggest that coastal vegetation was dominated by Taxodiaceae (with *Myrica*, *Symplocos*, and *Nyssa*) and Lauraceae, while drier inland areas had many Junglandaceae and Hamamelidaceae (e.g. *Liquidambar*). A type of evergreen broad-leaved forest prevailed at low altitudes, as did rainy summers. Pignatti (1978) hypothesized that at the end of the Pliocene altitudinal transitions occurred on Mediterranean mountain slopes, from Laurophyllous forests at low altitude, through *Ilex-Taxus* forests (with *Buxus*, *Ruscus*, and *Daphne*) and mountain conifers (*Picea*, *Abies*, *Cedrus*) to spiny shrubs (*Astragalus*, *Genista*) at high altitude. The latter two belts can be observed on many southern massifs of the Mediterranean, while the Laurophyllous forests have disappeared and the *Ilex-Taxus* belt now only occurs in relictual formations.

During this period vegetation in the northern sector of the western Mediterranean had three main components (Pons 1984):

- a temperate vegetation similar in composition to that present in non-Mediterranean Europe;
- groups of taxa now present either in North America or the far east;
- an ancestral Mediterranean element with *Abies*, *Cedrus*, *Nerium*, *Parrotia*, and *Quercus*.

During this period, vegetation was spatially heterogeneous. For example, open xeric assemblages were probably more common in Catalonia (north-east Spain) compared to the Languedoc of southern France (Suc *et al.* 1992). South of Barcelona, xeric herbs were increasingly important with *Ceratonia* and Palmae, while Taxodiaceae, *Quercus*, and other more mesophilous taxa show decreased abundance. So moving south, the mixed forest was replaced, by open, xeric, Mediterranean-like communities (Bessais and Cravatte 1988).

The decline and disappearance of the Taxodiaceae was not synchronous across western Europe. In the Mediterranean region, the decline set in during the Middle Pliocene whereas in the rest of Europe this decline did not occur till the Late Pliocene and Early Pleistocene (Michaux *et al.* 1979). Indeed by the Middle Pliocene, taxa in the Hamamelidaceae and Juglandaceae diminished and the Taxodiaceae were completely lost from the coastal areas. The loss of the Taxodiaceae and depletion of other groups was only part of a more general increase in rates of extinction during the Middle Pliocene (Bessedik *et al.* 1984). In addition to the loss of Taxodiaceae, this period witnessed the disappearance from the western Mediterranean of several genera in the Sapotaceae, Restoniaceae, Agavaceae, *Leea* (Leeaceae), *Embolanthera* and *Hamamelis* (Hamamelidaceae), *Rhioptelea* (Rhiopteleaceae), *Symplopus* (Symplocaceae), *Microtropis* (Celastraceae), and *Nyssa* (Nyssaceae). These extinctions were not simultaneous in different regions; Taxodiaceae, *Symplocus*, *Nyssa*, and a few others remained for longer periods in Catalonia than they did in the Languedoc. The progressive appearance of the Mediterranean climatic regime, and in particular the precipitation regime was the key element in these extinctions. While setting the scene for the evolution of one of the world's most diverse contemporary floras, the evolution of the Mediterranean climate thus also caused high levels of extinction in the pre-existing flora.

It was thus in the Pliocene (~3 Ma) that a gradual but profound climatic change in the Mediterranean region began (Suc 1984). This change did not occur in a parallel fashion elsewhere in temperate Eurasia or in the tropics of Africa. At this time, the temperature regimes began to drop significantly, introducing a marked seasonality, not just in temperature, but also in the establishment of a marked and prolonged dry season, which became more and more severe as its association with the newly established warm season developed.

As coastal forests thinned out and disappeared and more xeric vegetation developed, a complex mosaic of vegetation types established in the landscape of the Mediterranean (Suc 1984; Combourieu-Nebout 1993; Pons *et al.* 1995). The summer drought stabilized at ~2.8 Ma and by ~2.3 Ma some of the oldest signs of extensive Mediterranean vegetation can be detected in pollen cores (Suc 1984). The origin of the Mediterranean climatic regime is thus very recent and its onset had, as its main element, a fluctuation in the rhythm of rainfall, rather than a strong contrast in temperature (Suc 1984).

The onset of the highly seasonal climate showed marked spatial variation. For example, the first signs of a summer drought appear from the Late Miocene (10–6 Ma) in North Africa (Bachiri Taoufiq 2000), the Early Pliocene (~4 Ma) in Calabria and Sicily (Bertoldi *et al.* 1989), and ~3.5 Ma in the north-west of the Mediterranean Basin (Suc 1984).

The Late Pliocene

In the Late Pliocene (2.5–2.1 Ma) pollen spectra indicate strong fluctuations between two main vegetation types: (a) forest communities rich in deciduous species no longer present in the region (e.g. *Carya*, *Pterocarya*, and *Parrotia*) and (b) steppe communities dominated by *Artemisia* accompanied by the genus *Ephedra* and a range of Amaranthaceae and Chenopodiaceae. The presence of steppe vegetation attests to drier (and slightly cooler) climatic conditions than in the Middle Pliocene (Suc and Cravatte 1982). The presence of taxa such as *Cistus and Phlomis* in some sites indicates that the temperature had not cooled to a great degree, but had become much drier. There was also much variation among sites in the composition of pollen spectra: well-developed Mediterranean communities appear to have already occurred in southern Italy and Languedoc at this time whereas a more steppe-like vegetation occurred in north-west Spain (Bazile-Robert *et al.* 1980; Pons *et al.* 1995). Geographic variation in the development of Mediterranean vegetation thus probably occurred. The fluctuations of these two types of pollen spectra suggest rapid climatic cycling of short duration but intense amplitude.

Altitudinal zonation of the vegetation was present in the Late Pliocene (Suc 1984). For example, pollen spectra from Calabria contain pollen of different vegetation associations, some of which no longer coexist in the Mediterranean and others which are now absent from the region (Combourieu-Nebout 1993). These spectra suggest a succession of deciduous forest (with abundant *Quercus* associated with *Acer*, *Carpinus*, *Celtis*, and others), subtropical humid forest (Taxodiaceae and *Cathaya* which now occur in western Asia), high-altitude coniferous forest (*Tsuga*, *Cedrus*, *Abies*, and *Picea*) and open steppe vegetation (composition as above). This sequence most likely represents the vegetation response to climatic change from warm and fairly humid interglacial periods to colder, drier glacial periods (Combourieu-Nebout 1993). Fairly rapid climatic oscillations, mostly due to changes in rainfall, and less the result of temperature variations, probably caused the shifts from subtropical forest to herbaceous open vegetation. The presence of strong altitudinal gradients around the shores of the Mediterranean may have allowed different associations to locally persist and thus rapidly track climate change. In the Late Pliocene, floristic differences between the western and eastern Mediterranean regions were already apparent (Palmarev 1989).

As the climate changed in the Pliocene a summer drought climatic zone was formed between 37°N and 45°N where prominent sclerophyllous woody ecosystems developed. The important presence in the pollen spectra of this period of Cupressaceae, *Pinus*, *Quercus*, *Olea*, *Phillyrea*, *Cistus*, *Helianthemum*,

Rhus, and *Rhamnus* suggests that these initial Mediterranean plant communities developed on low hillsides with a dry calcareous soil. Their direct descendants constitute much of woodland vegetation of the contemporary Mediterranean flora.

1.3.3 Climatic oscillations associated with glaciation

Late Pliocene to Early Pleistocene

During this period, the gradual cooling and drying of the climate occasioned the extinction of many species from various regions. In some cases, whole genera, sometimes the sole representatives of particular families, became extinct in particular regions of the Mediterranean. In southern France these losses included genera such as *Carya*, *Pterocarya*, and *Juglans* (Juglandaceae), *Eucommia* (Eucommiaceae), *Elaeagenus* (Elaeagenaceae), *Zelkova* (Ulmaceae), and genera such as *Parrotia*, *Parrotiopsis*, and *Liquidambar* in the Hamamelidaceae (Bazile-Robert *et al.* 1980; Bessedik *et al.* 1984). The loss of such groups was again variable between regions, occurring gradually later towards the east (Bessedik *et al.* 1984). Aridity, associated with reduced temperature, was thus a key factor causing what was to be a third wave of enhanced extinction rates.

The Pleistocene

In the Pleistocene (1.8 Ma–15,000 BP) forest and steppe continued to alternate, at a rhythm of ~100,000 years (Pons *et al.* 1995). Steppe vegetation covered large expanses of the landscape as glaciers moved south across the northern parts of Europe. In contrast to some of the previous colder periods, steppe vegetation comprised species indicative of cooler temperatures. In the western Mediterranean, the most common pollen types were *Artemisia* and various Chenopodiaceae (Bazile-Robert *et al.* 1980). Trees were not absent from the landscape, pines were abundant, albeit with much spatio-temporal variation, *Juniperus* are recorded in the early parts of this period, as are, *Betula* and *Hippophae* during the coldest periods. Steppe vegetation occurred in southern Spain (Pons and Reille 1988) and a savannah-like vegetation may have persisted in sheltered areas.

During such cool periods, strong ecological differences developed between the eastern and western Mediterranean regions. In parts of the eastern Mediterranean such as Israel (Horovitz 1979), pollen analyses suggest the occurrence of oaks and olives in a steppe vegetation lacking *Artemisia* and more reminiscent of western Mediterranean garrigues and maquis (*Pistacia*, *Cupressus*, *Rosaceae*, *Poaceae*). The coniferous forests (with fir and scots pine and some *Corylus*, *Acer*, *Carpinus*, *Buxus*, *Tilia*, and *Fagus*) present in southern Europe during the Quaternary also showed an east–west variation in composition, with *Betula* and *Hippophae* in the west and deciduous oaks in the east.

The warmest periods saw the localized development of Mediterranean forest vegetation, containing *Quercus*, various *Oleaceae*, *Pistacia*, *Cistus*, *Ostrya*, *Vitis*, *Juglans*, and *Pinus*. In the Late Quaternary, the climate dried and the vegetation took on a more Mediterranean aspect with evergreen oaks and *Cistus* becoming more abundant (Bazile-Robert *et al.* 1980; Pons and Suc 1980; Brenac 1984).

The last glacial maximum

The last glacial maximum in southern Europe occurred around 20,000 BP. This was a dramatic moment for Mediterranean vegetation which declined to what Pons (1984) termed '*état zéro*'. Temperature depression in southern Europe is thought to have been something of the order of 5–7°C; markedly less than the 15–16°C depression at the southern limits of permafrost further north (Kaiser 1969). The seasonality of rainfall was also more marked than at the present time (Prentice *et al.* 1992). Mediterranean plants persisted through this period in isolated glacial refugia which were to be the sources for future re-colonization (Box 1.4).

Start of the late glacial

The start of the late glacial occurred from ~15,000 BP in southern Europe. The pattern of late glacial vegetation development in southern Europe showed much spatial heterogeneity in relation to local climatic conditions and soil type and moisture (Turner and Hannon 1988; Reille *et al.* 1996; Carrión 2001). The spread of vegetation in association with this

Box 1.4 Survival during '*état zero*': types of glacial refugia and where they occurred

In Mediterranean Europe, major regions of refuge occurred in the southern Iberian peninsula, Greece, and the Balkans. Other refugia no doubt occurred in the Middle East and North Africa. In addition to the major zones of refuge other smaller refugia may have occurred across the southern extremes of Europe. Refugia would have been located in a landscape whose vegetation was otherwise dominated by grasses and *Artemisia*, then widespread over southern Europe. This was essentially a steppe vegetation associated with an arid climate where lack of rainfall, perhaps as much as low temperature, put a severe restriction on tree growth (Kaiser 1969; Van Campo 1984).

Exactly where Mediterranean vegetation persisted during the glacial periods and whether such persistence occurred in more than just a few isolated and small pockets in protected rocky areas around the coast is not completely known, although in some precise locations long-term persistence of tree cover has been clearly demonstrated (Tzedakis 1993). In the western Mediterranean, glaciation may have had more severe effects on plant distribution than further east, and evergreen oak forests probably only persisted in the southern tips of Spain and Italy, being otherwise displaced into large areas of north Africa and the south-east margin of Europe. Mixed deciduous forests would have been fairly extensive across the southern half of the Iberian peninsula, down the east and west coasts of Italy, and perhaps in a small coastal band around parts of southern France. Otherwise many species probably persisted in warmer and more humid localized pockets of the landscape (Pons and Suc 1980; Pons 1984; Hermenger *et al.* 1996).

Pons (1984) provides several elements of response to the question of precisely what types of habitat acted as refugia in a bleak and barren Mediterranean landscape.

1. In the western Mediterranean on south-facing slopes above the arid plains at 400–800 m elevation. During this period the treeline was situated at around 800–1000 m on the northern shores of the Mediterranean (but reached 1,500 m in north Africa). The existence of such pockets of vegetation on south-facing slopes above the plains helps explain why the reforestation of Mediterranean mountains occurred so quickly in the first few thousand years (particularly between 13,000–11,000 BP) after the glaciers began to beat a retreat (Peñalba 1994; Reille *et al.* 1996, 1997).
2. In the western Mediterranean isolated trees and shrubs persisted in the lower parts of ravines and river gorges—many of these sites have probably since been submerged.
3. Pockets of species-poor deciduous oak forest on the southern shores of the Mediterranean, and probably in southern tips of the continent on the northern shores.
4. Open deciduous oak forest with pines and a few other Mediterranean taxa near the sea in the eastern Mediterranean.
5. Highly isolated refugia probably dotted the landscape in sheltered valleys and near the coast.

Cliffs are a conspicuous feature of the Mediterranean landscape and probably played an important role as a refuge, particularly maritime cliffs and those with a southerly exposure. In cliffs, open vegetation typical of contemporary garrigues, phrygana, and maquis vegetation may have persisted alongside strict chasmophytes during the glacial maxima (Davis 1951; Snogerup 1971). The fact that not all the species that occur on limestone cliffs are chasmophytes supports the idea that at least a few elements of the Mediterranean vegetation found a refuge in cliffs during the Quaternary glaciations. For example, in the Aegean, some species (e.g. *Anthyllis hermanniae*) occur as a chasmophyte in parts of their range and in phrygana vegetation elsewhere and several forest species (e.g. *Quercus ilex, Pistacia terebinthus*, and *Rhamnus alaternus*) can be observed in cliffs (Snogerup 1971). Likewise in limestone cliffs of the Iberian peninsula, many common species are generalist chamaephytes (e.g. *Saxifraga monocayensis* and *Silene saxifraga*) which represent pioneer colonists, or nanophanerophytes (e.g. *Rhamnus alpinus* and *Lonicera pyrenaica*).

warming varied in relation to latitude and the existence of local refugia which would have acted as additional sources for expansion. Whereas oaks began to spread form their probably quite vast and patchy network of refugia in southern Spain from 13,200 to 12,000 BP, their expansion is only recorded from 10,000 BP in northern Spain (Peñalba 1994), where refugia would have been few and far between. In north-west Syria, vegetation of a Mediterranean type is thought to have been an important component of the landscape by about 11,000 BP with assemblages of *Quercus*, *Pistacia*, and *Olea* present in the plains and *Cedrus*, *Carpinus*, *Ostrya*, and *Quercus* on the mountain slopes (Niklewski and van Zeist 1970).

The initial warming of the climate was disrupted by short but intense cold periods, such as in the recent Dryas (11,000–10,000 BP). During these cold snaps, forests retreated again and steppe vegetation spread (Pons and Reille 1988), *Betula* pollen showed a marked decline in some sites (Turner and Hannon 1988). In the drier regions, such as in southern Spain, *Quercus ilex* showed a marked decline during this period (Reille *et al.* 1996). On Corsica, a marked increase in *Artemisia* pollen and the absence of *Pinus nigra* subsp. *laricio* pollen suggests that this characteristic tree in upland forests was not then present on the island (Reille *et al.* 1997). However, even during these cold periods, when *Artemisia* pollen dominated with various Chenopodiacae, Poaceae, Apiaceae, Asteraceae, and *Ephedra*, a xerophytic vegetation similar to that which is now present, persisted in sheltered lowland sites on Corsica (Reille *et al.* 1997). In the last period of climatic oscillations, that is, 18,000 to 10,000 BP, *Prunus* pollen became more abundant in some regions, as did *Juniperus* and *Cistus*. Bushes and small trees thus appeared to predominate in the landscape, suggesting a fairly cool climate.

From 10,000 BP onwards

From 10,000 BP onwards a more definitive warming began. At this time, deciduous oak forests covered large areas on the slopes of many of the Mediterranean mountains (Pons *et al.* 1995; Grove and Rackham 2001). Present in these forests were *Corylus*, *Alnus*, *Fraxinus*, *Betula*, *Ulmus*, and *Tilia*, all of which now occur more commonly at higher latitudes and in cooler and wetter parts of the Mediterranean landscape where they represent relicts of a once more widespread distribution. High frequencies of *Pistacia* in southern areas and *Corylus* elsewhere point to the existence of a fairly open forest vegetation in some areas. This forest diversified and in some areas became more dense and closed by about 8,000 BP (Pons *et al.* 1995).

There was marked geographic variation in the dominant species present in these forests: evergreen oaks in drier areas of southern Spain, deciduous oaks in southern France and in Italy, firs in the northern Appenines, pines in the Maritime Alps and the eastern Pyrenees, and *Pinus nigra* subsp. *laricio* forests on Corsica (Reille *et al.* 1996). Pollen cores in southern Spain attest the presence of a Mediterranean-type vegetation from around 10,000 BP in some sites, for example, *Pistacia* pollen has been recorded from 9,500 BP and cork oak (*Quercus suber*) from 6,800 BP in the Sierra Nevada (Peñalba 1994; Grove and Rackham 2001). However there is a great deal of heterogeneity among pollen sequences from different sites in Spain, where even in the semi-arid south-east, pollen records show much variation in the timing of different vegetation stages, perhaps as a result of local variation in topography and microclimate or time lags in vegetation development linked to the vegetation present at a site prior to any climate modification (Carrión 2001). Since ~10,000 BP trees were present in the large majority of the landscape, forest was less abundant in the eastern and southern parts of Mediterranean Europe than in the north-west part of the basin. In southern Spain, the return of forest vegetation was rapid, probably because of the numerous localized refugia that may have dotted the landscape during the '*état zero*' of glacial maxima (Box 1.4).

1.3.4 Ever since glaciation: climate change and human activities

In the Holocene (<10,000 BP), the natural ecology of all circum-Mediterranean regions came under the influence of a new ecological factor, namely human activities (Triat-Laval 1979; Pons 1984; Barbero *et al.* 1990; Pons *et al.* 1995; Quézel and Médail 2003).

Humans have been present in the Mediterranean for longer than anywhere else in Europe, as the skeleton dated at ~400,000 years discovered in a cave near the village of Tautavel in the Roussillon of southern France illustrates all too well. As the cave paintings of the western Mediterranean and its periphery also illustrate, Cro-magnon humans were present in the southern parts of the Europe for long periods during the Quaternary.

Since the last glacial maximum, the composition and spatial relationships of Mediterranean ecosystems have been greatly modified by more extensive human activities. However, in the same time spell, the climate has also warmed and become drier. There has thus been some debate as to whether major modifications to Mediterranean plant communities in the last 6–7,000 years are more the result of human activities than they are of climate change. Since the intensity and timing of human activities were different in different places there is also much spatial heterogeneity in the anthropogenic element of pollen sequences (Carrión 2001). Pons and Quézel (1985) and Quézel and Médail (2003) argue that although climate change probably drove vegetation change up till ~10,000 BP, human activities have since then become the determining factor influencing Mediterranean forest cover, beginning in the east and moving west.

Around 10,000 BP, human-induced forest clearance was dotted around the landscape in the form of small, temporary clearings in an otherwise forested landscape (Pons and Thinon 1987). Since then, the impact of human activities has increased dramatically (Pons 1984). One can identify the following phases: (a) the development and diffusion of agriculture in the Neolithic with cereal cultivation on plains and low-altitude plateaux, (b) more generalized forest clearance, a little after 3,200 BP in north-west Greece, 2,800 BP in Provence, 2,700 BP on the Dalmatian coast, and 2,500 BP on Corsica (i.e. during Greek Antiquity and the Roman Empire), (c) political events and socio-economic changes (e.g. medieval changes in land ownership, wars, and population movements and other demographic changes) up to the end of the nineteenth century, and (d) twentieth-century rural depopulation, coastal development and the evolution of environmental perception and conservation. Evidence for human-induced decline in forest vegetation up till the last 200 years or so has been presented for a diverse array of situations, most of which involve changes in the distribution and composition of forest communities, and their reversion to shrubland over large areas, the spread of evergreen oaks at the expense of deciduous oak and the spread of pines (Box 1.5). Human activities have clearly had a major impact on the Mediterranenan landscape (Lepart and Debussche 1992).

One way in which human activities have had a dramatic influence on vegetation in the Mediterranean region has been via the use of fire. Fires are thought to have occurred naturally since at least the Miocene (Dubar *et al.* 1995). With the onset of a highly seasonal association of high summer temperatures, drought and often strong winds, fires probably became more frequent and may have been an important feature of the ecology of natural vegetation in the primeval Mediterranean forests. This natural selection pressure has been greatly modified as human activities learnt its use and started to create pastures and enrich soils for cultivation. In the current-day Mediterranean landscape, forest fires are almost exclusively linked to human activities, hence their consideration as an anthropogenic disturbance in the landscape (Moreno and Oechel 1994).

In historical times, burning was a constant feature of land clearance and settlement in the Mediterranean, where the use of fire began much earlier than elsewhere in Europe. Some of the oldest evidence for a human presence in the Mediterranean, such as the cave settlements near Tautavel (southern France) which date to ~400,000 BP, attest to the, albeit probably limited, use of fire. Some time in the Neolithic (~8,000 BP), fires became more frequently used over a period of about 2,000 years, and since ~4,500 BP they have become general practice and widely used, as the Early Holocene abundance of cork oak pollen suggests (Pons and Thinon 1987; Pons and Reille 1988; Grove and Rackham 2001). According to Pons and Thinon (1987: 10) the importance of fires associated with human activities became such that

Box 1.5 The diverse types of human impact on forest vegetation in the Mediterranean

Several lines of evidence for human impact on the Mediterranean landscape, and in particular the destruction of forest and changes in tree/shrub species composition have been proposed.

1. A reduction in cover and thinning out of deciduous oak forests and their replacement by evergreen oaks, for example, in the Moroccan Rif where *Quercus canariensis* and *Q. toza* have been replaced by *Q. ilex* and *Q. suber* (Reille *et al.* 1996). In Provence, *Q. ilex* increased naturally in abundance, from a scattered tree in open communities dominated by *Juniperus*, to a more open woodland as climate warmed between 15–10,000 BP (Triat-Laval 1979). This expansion occurred as *Juniperus* declined, not at the expense of deciduous oaks, and prior to significant human impacts in the region. It was only after the cutting of deciduous oak forests from 7,000 bp onwards that evergreen oak spread into what was previously deciduous forest in this region. From then on, the two types of oak show negative correlations in abundance, suggesting the replacement of deciduous oaks by evergreen oaks as human activities expanded. In southern France this may have been due to the occurrence of deciduous oaks on deeper soils which were the most valuable for cultivation (Lepart and Debussche 1992).
2. The regression of mountain forests, for example, *Cedrus atlantica*, *C. libani*, and *Juniperus thurifera* in the Rif and Atlas (Pons 1984; Pons *et al.* 1995).
3. Invasion of other forest types by *Q. ilex* and pines: On Corsica, Reille (1992) showed that *Q. ilex*, although present prior to human presence, has only become a dominant component of forests as human activities (e.g. colonization by the Romans and the Republic of Genova) developed in the last few thousand years.
4. Expansion of woody shrub and herbaceous vegetation and in particular the replacement of evergreen oak woodland by maquis (Pons *et al.* 1995).
5. Contemporary extension of pine forest, for example, *Pinus pinaster* in the High Atlas and *P. halepensis* in the Languedoc of southern France (Quézel and Médail 2003).
6. The extension of Kermes oak garrigues in association with human fires and intense grazing (Triat-Laval 1979).
7. The stopping of *Fagus* (and perhaps also *Carpinus*) expansion in a westerly direction in northern Spain (Peñalba 1994).
8. Species introductions, such as on Corsica where two species which are now important vegetation elements stem from human introduction: *P. halepensis* introduced as late as the nineteenth century (Reille 1992) and *Castanea sativa*, totally absent from the pollen spectra prior to the sub-Boreal (Reille *et al.* 1997).
9. Pollen analyses on Corsica (Pons and Reille 1988; Reille and Pons 1992) show that the marked rise in abundance of *Q. ilex* and *Q. suber* only happened as development of human activities proceeded, that is, after 6,000 bp. Prior to this, deciduous oaks were the dominant tree species. Although *Q. ilex* has been naturally present in many areas, it was probably not the dominant tree species, except in areas with a semi-arid climatic regime and/or on shallow soils.

'the changes induced by . . . [fire were] . . . so great that an analysis of the present relationships between ecosystems and environmental factors can be significant only if it takes into prior consideration all the anthropogenic past of the ecosystems'. This is nevertheless a hefty statement, given the strong influence of drought and nutrient stress and the impact of land-use changes. For example, the spread of *Pinus halepensis* in southern France is more due to the abandonment of cultivation and pastoralism than the direct result of fires (Barbero *et al.* 1987*b*).

Although recent fire statistics are often difficult to interpret (see Grove and Rackham 2001), the fire regime in the Mediterranean region has changed. In a general survey of Mediterranean forest fires over a 30 year period, Le Houerou (1987), documented the gradual increase in the spatial extent of burnt areas from an average 200,000 ha/year (1960–71) to 470,000 ha/year (1975–80), and 660,000 ha/year (1981–85). The number of fires has increased in parallel fashion. Since the 1980s, data for Mediterranean France indicate that the surface burnt by fire has stabilized (albeit with much annual variation) while the number of fires continues to increase (Quézel and Médail 2003). Analysis of fire regimes in Catalonia, where detailed inventories exist for the medieval period (1370–1462) and the late twentieth century (1966–96) indicates that although there has not been an increase in fire frequency between the two periods, the surface burnt by individual fires has greatly increased and the annual number of summer fires has increased (Lloret and Marí 2001). As these authors illustrate, the occurrence of a small number of very large fires has become, in the last 50 years, an integral part of the fire regime in Mediterranean forests. In the light of the recent massive and numerous summer fires of 2003, the conclusion made by Le Houerou (1987: 22) that 'the ever increasing build up of fuel in Mediterranean forests and shrublands as a result of rural depopulation and the abandonment of marginal lands . . . will sooner or later make new legislations necessary as well as the adoption of new methods of prevention', remains pertinent.

Diminished human activity in forests, for example, glass-making, tanning, and charcoal burning have become (almost) obsolete, is an essential element of the fuel build-up and the occurrence of a small number of very large fires, or conflagrations. In what were once actively exploited forests there has been an increase in tree height and density and thus a dramatic increase in woody biomass (e.g. Debussche *et al.* 1999). In the absence of fire, the understorey vegetation of pine forests in the Mediterranean changes gradually as woody shrubs such as *Phillyrea*, *Viburnum*, and *Rhamnus* increase in abundance before the eventual appearance of oaks (Barbero *et al.* 1987*a*). This increase in woody biomass has no doubt contributed to the increased risk of fire in many Mediterranean forests (Le Houerou 1987; Moreno *et al.* 1998), a trend which climate change may exacerbate in the future (Piñol *et al.* 1998).

The prevailing paradigm of landscape destruction as a result of human activities has been challenged by a number of authors, who argue that human activities are not the major cause of landscape degradation in the Mediterranean region. In their recent book on the ecological history of the Mediterranean landscape, Grove and Rackham (2001) argue at length against the idea that the contemporary Mediterranean landscape is a 'degraded' landscape. Indeed, in many areas open landscapes are not the result of human activities. Here are some examples.

1. Many endemic species do not occur in forest habitats (Chapter 2), suggesting that the latter has not been a ubiquitous landscape feature in the Mediterranean.
2. The well-developed semi-arid floras of south-east Spain and south-east Crete attest to the historical existence (throughout the Holocene and preceding interglacial periods) of areas too dry for forest, in particular deciduous oak forest. On the Cyclades, tree species are few and well-developed garrigues or maquis are rare due to the poor soils and arid climate. In these areas, climate may have played a more important role in the lack of forest cover than human-induced impacts.
3. On Corsica, maquis vegetation now omnipresent in many areas on the island (e.g. in the Agriates and the Cap Corse) was a well-established feature of the landscape prior to the onset of human activities, as pollen spectra for *Erica arborea* attest (Reille 1992). The abundance of *E. arborea* was probably due to the chance absence of potential dominant tree species on this island and the nature of the soils formed on very compact acid rocks. This landscape is thus not a degraded landscape linked to the removal of the forest, although the spread of *Q. ilex* is probably due to human activities and opening of the deciduous oak forest (Reille 1992).

Grove and Rackham (2001) enumerate numerous other examples to back up their doubt about 'any theory that the normal state of wildwood was trees upon trees upon trees . . .' (p 153). For these authors the critical changes in vegetation change during recent history involved two main processes: the advance of agriculture and pasturage in the

Bronze Age followed by a gradual drying of the climate, whose effects on the vegetation were complete 2,000–3,000 BP. Evidence for Holocene vegetation changes synchronous with climatic changes support this idea (Beug 1967, 1975). There is no doubt an element of truth in their argument. It is difficult to precisely identify the structure of a pristine landscape prior to human impacts because of the spatial and temporal variability of the abiotic environment already present prior to the arrival of humans. Landscape patterns were already natural mosaics prior to human activities, and, as Lepart and Debussche (1992: 79) point out 'in the Mediterranean region, it is a myth to think that homogeneous forest was the essential element of the landscape before the arrival of humans'. But since then it is clear that in many areas humans have had a critical effect on the flora and patterns of vegetation (Box 1.5).

Perevolotsky and Seligman (1998) argue that although traditional grazing practices may well have caused forest destruction in many Mediterranean areas, such activities may be an efficient and ecologically sound form of ecosystem management, one that in no way implies degradation. For these authors (pp. 1007–1008), 'even heavy grazing by domestic ruminants on Mediterranean rangelands is a relatively benign factor in ecosystem function and seldom in itself irreversibly destructive to the soil or the vegetation'. The long history of grazing in the Mediterranean region (not the case in other Mediterranean climates where the recent introduction of mammalian grazers has had dramatic impacts on the ecology of natural systems) means that such areas may not be fragile to such grazing. In fact, where grazing and agricultural practices have been abandoned, Mediterranean woodlands are rapidly spreading, with important consequences for the maintenance of a traditional Mediterranean mosaic landscape of open vegetation and woodland (Debussche *et al.* 1999; Lepart *et al.* 2001). Their capacity to recover, even after hundreds of years of heavy grazing, is high, although understorey herbs with limited dispersal and current distributions restricted to small isolated pockets of forest, may take a long time to follow the spread of woodlands.

During the twentieth century, no one can deny that changes in human activities and land-use patterns closely tied to socioeconomic developments and political changes have had major impacts on the vegetation and ecology of many regions around the Mediterranean. Since the late nineteenth century, human activities have had variable consequences depending on whether one observes vegetation on the southern or northern shores of the Mediterranean and depending on whether one observes littoral of hinterland vegetation. On the northern shores of the Mediterranean, forests are spreading in the back country (e.g. Barbero and Quézel 1990; Debussche *et al.* 1999; Arianoutsou 2001; Chapter 4) whereas littoral vegetation goes under concrete due to peri-urban and holiday resort development and sprawl. To the south of the Mediterranean Sea, the search for arable land is ongoing, reducing natural forests to isolated trees and contributing to continued degradation of natural ecological systems.

The low-lying hills and upland plateaux around the northern shores of the Mediterranean have incurred much rural depopulation, in some places local populations have declined to less than one-fourth of their numbers at the end of the nineteenth century. Punctuated collapses of the market for some important crops due to silk worm diseases and *Phylloxera* on vines at the end of nineteenth century, for example, and other more general socio-economic causes have led to the abandonment of agricultural practices (abandon of intensive terrace cultivation and extensive sheep and goat grazing and the reduction of wood cutting) around the northern shores of the Mediterranean. In association with these changes there has, over the last 100 years, been a marked change in perception of forest cover and open vegetation (Lepart *et al.* 2000). All these changes are recent, occurring in the twentieth century and most dramatically since the end of the Second World War.

Clearly human activities have greatly impacted on Mediterranean vegetation, and continue to do so. I will leave others to argue about the generality of this phenomenon and its relative importance compared to climatic variation. What is of concern in this book is where and how changes in land use, species introductions, and other activities have modified both ecological processes acting within natural plant populations and their spatial distribution in the landscape.

1.4 Diversity and unity in the Mediterranean flora

The contemporary Mediterranean climate, in which annual precipitation varies from 100 mm in the most arid parts of the region, to ~3,000 mm on some of the mountains (which nevertheless have a dry summer), and in which temperature varies greatly within and among regions, has stimulated various classifications of climatic diversity in the Mediterranean. Two main classifications can be identified. The first involves a series of bioclimatic types in relation to a rainfall–temperature coefficient (Q_2) developed by Emberger (1930a, b, c). This bioclimatic coefficient relates mean annual precipitation (P) to the mean minimum temperature in the coldest month (m) and the mean maximum temperature in the hottest month (M) by the following equation: $Q_2 = 2{,}000P/(M^2 - m^2)$. This coefficient allows for the delimitation of six bioclimatic types:

(1) per-arid: $P < 100$ mm, 11–12 dry months;
(2) arid: $P = 100$–400 mm, 7–10 dry months;
(3) semi-arid: $P = 400$–600 mm, 5–7 dry months;
(4) subhumid: $P = 600$–800 mm, 3–5 dry months;
(5) humid: $P = 800$–$1{,}200$ mm, 1–3 dry months;
(6) per-humid: $P > 1{,}200$ mm, <1 dry month (and thus barely 'Mediterranean').

Most recent work on the Mediterranean excludes areas which fall into the first of these types, which have a climate and a vegetation which is more typical of a desert (Médail and Quézel 1997; Joffre and Rambal 2002; Quézel and Médail 2003). These six types can be further subdivided in relation to winter temperatures.

A second classification, developed by phytosociologists, describes Mediterranean vegetation as a series of *'étages'* in relation to thermal differences along altitudinal gradients (Table 1.1).

So wherein lies the unity? A first step in defining the unity of the Mediterranean flora can be taken with reference to the work of Pierre Quézel, the French botanist and ecologist who has

Table 1.1 Vegetation classification in relation to altitude in the Mediterranean region

Etage	m (°C)	T (°C)	Principal vegetation and locations
Infra-Mediterranean	>7		Arid communities: *Argania spinosa* and *Acacia*. Only in western Morocco.
Thermo-Mediterranean	>3	>17	Sclerophyllous communities: with *Olea europaea, Ceratonia siliqua, Pistacia lentiscus, Pinus halepensis, Pinus brutia* and *Tetraclinis articulata*. Circum-Mediterranean. Often as a narrow band near the sea and in valleys with *Nerium*, but can reach 800 m in North Africa.
Meso-Mediterranean	0–3	13–17	Sclerophyllous forests of *Quercus ilex* (western and central) or *Quercus calliprinos* (eastern) and pines. Littoral to 400 m to the north of the Mediterranean Sea, ~400 to ~1,000 m to the south.
Supra-Mediterranean	−3–0	8–13	Deciduous oak forests dominant in the humid bioclimate (with *Ostrya* and *Carpinus*), sclerophyllous oaks in zones with low rainfall. 400–900 m to the north of the Mediterranean Sea, up to 1,500 m on the south side.
Mountain-Mediterranean	−7 to −3	4–8	Upland coniferous forests with *Pinus nigra* and Mediterranean firs and cedars. 900–1,400 m to the north of the Mediterranean Sea, 1,400–2,000 m to the south.
Oro-Mediterranean	<−7	<4	Open vegetation with xerophytic shrubs, *Juniperus* and sometimes open pine forest. This belt is not always composed of Mediterranean taxa. Mostly above 2,000 m (Atlas, Taurus).
Alti-Mediterranean			Dwarf chamaephytes—Atlas and Taurus mountains above 2,200 m.

Notes: m: mean minimum temperature of the coldest month; T: mean monthly annual temperature.
Source: Ozenda (1975), Quézel (1985), Médail and Quézel (2003).

probably done more than anyone else to shape ideas on the biogeographic origins and distribution limits of Mediterranean vegetation. In a review of this subject, Quézel (1985: 18) proposed that the Mediterranean flora be viewed as 'a heterogeneous entity associated with a region that is largely defined by climatic criteria'. Many elements of contemporary Mediterranean vegetation existed prior to development of the current-day climate, others became important components of the flora since the onset of the Mediterranean climate in the Pliocene. These different climatic origins introduce much diversity into the flora (Quézel 1985; Quézel and Médail 2003), which has several biogeographic elements.

Taxa with subtropical affinities make up a sizeable proportion of the Mediterranean flora (Raven 1973; Quézel *et al.* 1980; Quézel 1985). Some of these taxa have probably evolved from ancestral stocks present prior to the opening of the North Atlantic and separation of the southern continental plates, for example, *Borderea* and *Dioscorea* (allied to taxa in South Africa or South America), *Tetraclinis* and *Aphyllanthes* (whose affinities lie with Australian taxa), and *Cneorum* which has related species in eastern North America. Alternatively, others had ancestors present in the circum-Mediterranean flora of the Oligocene and Miocene (e.g. *Ceratonia, Chamaerops, Jasminum, Olea, Phillyrea,* and *Nerium*). Many of these taxa represent palaeo-tropical relicts that have persisted and evolved *in situ* as the climate became Mediterranean. There are also clear links with the more semi-arid and arid flora of East Africa and the Cape Province of South Africa. The existence of generic pairs between the Mediterranean and Cape floras (e.g. *Thymelaea* with *Passerina, Echium* with *Echiostachys,* and *Iris* with *Moraea*) support the idea of an ancient origin for many disjunct distributions (Goldblatt 1978; Quézel 1985). Likewise, several genera (e.g. *Pistacia, Anemone, Ceratonia, Coris, Cyclamen,* and *Globularia*) whose centres of diversity occur around the Mediterranean also have isolated species in more arid parts of Africa (Quézel 1995), indicative of historically more widespread distributions and/or migration between the ancestral Mediterranean region and parts of Africa, including the east and as far south as the Cape Province.

Taxa with an autochtonous origin are, not surprisingly, the main constituent of the Mediterranean flora. The strictly Mediterranean elements of the flora developed and differentiated during the Tertiary in association with the existence of isolated microplates and climatic change during this period (Zohary 1973). Some genera are limited to contemporary areas associated with ancient plates and some have greatly diversified within the limits of the zone (Section 1.5). For example, the Iberian peninsula has 16 palaeo-endemic genera and is also centre of diversification for many genera (e.g. *Genista, Narcissus, Linaria, Thymus, Teucrium,* and several Cistaceae). In the Balkans (set on the Apulian plate), diversification of genera such as *Silene* and *Stachys,* to cite but two examples, has been rampant, while in this region other genera, for example, *Jankaea, Petromarula,* and *Haberlea,* contain palaeo-endemic species.

Two other entities which evolved on the eastern and southern borders of the Mediterranean can be added. From the east came the *Irano-Turanian group* which developed during the dry and cold glacial periods of the Pliocene and Pleistocene—the most significant taxa being *Artemisia, Ephedra,* and *Salsola,* and trees such as the Judas tree (*Cercis siliquastrum*), the storax tree (*Styrax officinalis*), and some oaks. Most of the species in the Irano-Turanian element have centres of diversity in the semi-arid steppes of central Asia, that is, a continental climate. The different species in this element probably penetrated the Mediterranean region during episodes of climate change and geological activity since the Tertiary (Zohary 1973). To the south, the *Saharo-Arab element* differentiated from a xerophytic and heterogeneous ancestral stock, and as a result several North African endemics have affinities with Saharan and Arabian taxa.

The final group concerns *Holarctic or Eurasiatic elements* of the flora. One part of this group concerns taxa from the Laurasian flora present prior to the Miocene, which are now localized in parts of the eastern Mediterranean (e.g. *Aesculus hippocastanum, Forsythia europaea, Liquidambar orientalis*) or on islands (e.g. *Zelkova abelicea* on Crete and *Z. sicula* on Sicily) that were little affected by periods of glaciation. A second component includes

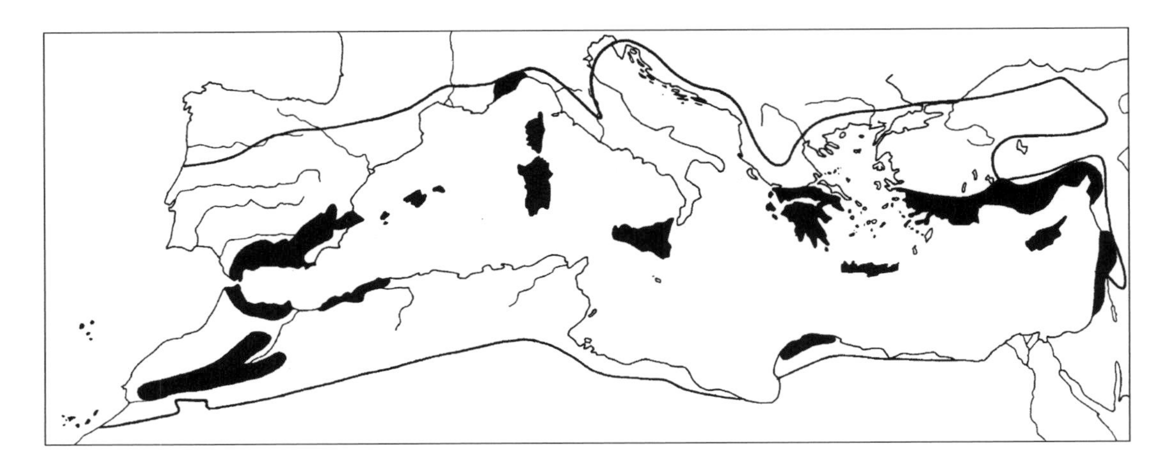

Figure 1.3 The areas of the Mediterranean Basin (shaded areas) recognized as 'hot spots of biodiversity' by Médail and Quézel (1997), where rates of endemism exceed 10% of the local flora (reproduced with permission).

the many temperate Eurasiatic taxa that inhabit the Mediterranean mountains, where they found refuge during periods of glaciation. Another part of this group, the Pontic entity has taxa which evolved under a continental climate with a dry period, for example, *Stipa*, which no doubt migrated into the Mediterranean during periods of glaciation. Finally, a small number of arctic–alpine taxa, rarely with any endemic species in the Mediterranean, can be found on a few high mountain tops (e.g. in the Atlas and Taurus and on Corsica).

To recapitulate, with regard to the diverse origin of the contemporary elements of the Mediterranean flora, I am tempted to cite that old adage of Shakespearian literature (used by other authors in their evaluation of plants and their diverse strategies) . . . while some species were born Mediterranean, others became Mediterranean as they colonized the region, while others had Mediterranean life thrust upon them! Many, but not all, of the latter have now gone. Together these diverse elements make up the contemporary Mediterranean flora.

1.5 Centres of diversity: concordance with history

In a rousing testament to the immense wealth of the Mediterranean flora, Médail and Quézel (1997) proposed the delimitation of ten 'hot spots' of biodiversity within the Mediterranean region (Fig. 1.3). These hot spots are areas where species diversity and endemism are high. The proportion of endemic taxa in these hot spots varies from 7 to 10% on some of the islands to ~20% in Andalucia and the Rif of North Africa, up to 27% in the mountains of Greece, and up to 31% in Turkey. These centres of diversity occur where geological and climatic histories have played crucial roles in shaping plant diversity. Based on data from Médail and Quézel (1997) and as a foretaste of things to come (Chapters 2 and 3), I now describe the main features of plant diversity in relation to history and setting in three major poles of diversity: (a) the Iberian Peninsula, (b) the Balkans and the Aegean, and (c) Anatolia and Cyprus.

1.5.1 The Iberian peninsula

A complex geology, a beautiful and diverse scenery, and a spatially heterogeneous climate characterize the Iberian peninsula. Other than the northern fringe and the north-west corner of the peninsula, the climate is Mediterranean. Spain and Portugal sit on a hard crystalline core of very ancient and now partly metamorphosed rocks. Similar formations also occur in Galicia and the Massif Central of south-central France. Around this ancient core there have been successive periods of uplift, subsidence, folding, and faulting. These processes have moulded the current day geology of the Iberian peninsula and the surface processes which have fashioned the current day landscape (see Polunin and Smythies 1973). In the Late Miocene, the Iberian plate was roughly in its present position and the main mountain ranges were in place. The Balearic islands, a continuation of the Betic Cordillera, were then connected to the mainland extending the peninsula to the north-east.

The contemporary Iberian Peninsula thus has a number of recognizable features.

1. The crystalline massifs in the north-west of the peninsula (particularly in Galicia), most of which do not have a Mediterranean climate and have relatively low rates of endemism.
2. The central plains (mesetas) where more recent Tertiary sediments overlie the ancient core, forming vast plateaux comprising limestone, sandstone, clays, and marls. This area is home to the Spanish salt-steppes and their unique vegetation, whose species often have their closest relatives as far away as the salt-steppes in Turkey, the Caspian Sea, or in North Africa.
3. The old-fold mountains (Iberian mountains, Central Sierras, and Sierra Morena) formed directly from the ancient granite core. They are home to many mountain plants.
4. The new-fold mountains. Once part of the ancient Hercynian massif, these represent the highest summits of the peninsula (almost 3,500 m in the Betic Cordillera). These are young rocks, mostly sedimentary in origin, formed by up-thrusting 50–100 Ma around the margins of the ancient core. Such places often contain alpine plants in refugia where they have either persisted since separation from their closest relatives in the Massif Central or the Alps or diverged more recently.
5. The depressions where subsidence has occurred against the hardcore of the ancient rocks during Alpine orogeny. They contain the main river valleys (Guadalquivir, Tagus, Ebro, Guadiana, and Duero) whose watersheds drain the peninsula. Flooded with Tertiary deposits these now form low-lying plains.
6. The coastal plains, adjacent to and sometimes dissected by coastal mountains. The Iberian peninsula has more than 2,500 km of coastline, however this natural habitat has been decimated by tourist developments.
7. The physical diversity of the peninsula is further enhanced by the highly localized occurrence of rare substrate types such as serpentines and gypsum.

The north-west tip and a narrow band of the Iberian peninsula along the north coast do not have a Mediterranean-type climate. In the extreme south-west, facing the Atlantic, the Mediterranean climate prevails with more than two months of effective drought. In the centre of the peninsula, extreme heat in summer and freezing in winter are locally enhanced where mountains and high elevation allow for extreme cold and create rain-shadows (<300 mm rain may fall in a given year and drought may extend for up to seven months in some parts of the central plateau and the south-east rim of the peninsula).

The diverse geology, substrates, and climate of the Iberian peninsula are paralleled by its botanical richness: 1,258 species or subspecies are endemic to this region (Gómez-Campo and Malato-Beliz 1985). Half of these belong to only 27 genera, while the other half are distributed across no less than 286 genera. This illustrates the phylogenetic overrepresentation of certain genera (in this case *Hieracium, Centaurea, Linaria, Saxifraga, Armeria, Sideritis*, and *Thymus* all have >20 endemic species in the region). To these should be added the ~500 Iberian endemics whose distribution also extends into the Rif in northern Morocco, where the flora has close similarities to that of Andalousia.

1.5.2 Mountain configuration and island isolation in the Balkans and the Aegean

The foundation of the Balkan and Aegean area is an ancient crystalline massif probably uplifted in the Carboniferous period. This massif has two zones (Polunin 1980). First the high mountains of the Rhodope massif (culminating above 2,000 m) which stretch across southern Bulgaria and Macedonia with two southerly extensions, one down through eastern Greece (including Mt Olympus which reaches 2,911 m) into the Euboea peninsula and the other through Thrace to the islands of Thasos and Samothrace in the north-east corner of the Aegean. The second area, formed of a crystalline block, lies in the south-central part of the Aegean, where the remains of the summits and crests of the ancient mountains are now the islands of the Cyclades and some of the continental peninsula.

During Alpine orogeny, new fold mountains composed of Secondary and Tertiary sedimentary limestone, sands and conglomerates were thrown up to create several more recent mountain ranges. These stretch in a north-east to south-west direction from the Dinaric Alps down the east coast of the Adriatic, through the Pindhos of central Greece and the Taiyetos and Parnon ranges in the Peloponnese peninsula, forming an arc across the island of Crete and up across the islands of Karpathos and Rhodes to the Taurus Mountains in Turkey. During the same period the Rhodope massif to the north uplifted to form the Balkan Mountains of Bulgaria and Serbia. Two main groups of rocks occur in these different chains:

1. Non-calcareous shales, sandstones, and marls which originated under shallow water and have in some areas formed slates, quartzites, and conglomerates and which contain many rivers and streams.
2. Calcareous sediments of deeper clearer water that have given rise to the limestone and marble that characterize many areas in this region, notably the old Mesozoic limestone that forms the distinctive Karst scenery in Dalmatia and the many cliff habitats in the islands of the Aegean.

In the Late Tertiary and Early Quaternary, the Tertiary fold mountains underwent much faulting and shattering and further uplift contributed to the formation of many of the escarpments and cliff formations that can be seen in the region. Although some of the Aegean islands are comprised of mainly volcanic rock (e.g. Milos and adjacent islands), most are crystalline schist of ancient origin. Hard crystalline limestone occurs on many islands where they create the cliffs and gorges that are famous in the region. Softer marly limestone is rare and only occurs on the small islands south of Naxos. Granite outcrops occur sporadically, for example, on Naxos, Mikonos, and Tinos.

In the Early Pliocene, the Aegean area resembled a large continent extending from the Greek mainland across to Turkey, via the Cyclades and Rhodes, and an arc of land connected the Peloponnese peninsula, Crete, Karpathos, and Rhodes. The Cyclades and Crete were separated by a sea, and have probably been so ever since, and a large lake covered most parts of the present North Aegean Sea. During the Pliocene, Greece and the Cyclades became isolated from Turkey, Rhodes, and other eastern islands by a sea extending northwards. What are now islands off the east coast of Turkey (such as the large islands of Kos, Patmos, Ikaria and Chios, and the many smaller islands in between) were probably connected to one another and to the mainland. At various times during this period Crete was a series of 3–5 isolated mountain ranges rather than a single island.

The origin of the islands of Crete and Rhodes, on the Tertiary arc of fold mountains that link the Peloponnese peninsula to the mountains of south-west Turkey, around the Aegean margins, is more recent than that of the islands of Corsica and Sardinia (Biju-Duval *et al.* 1976; Mascle and Rehault 1991). The multitudinous islands in the south and east Aegean have had even more recent connections among each other. During the first cold snap of the early Pleistocene (~2–1 Ma) (and perhaps later during successive glacial maxima) further regression of the sea (a drop in sea-level 150–200 m below present levels occurred during the Riss glacial period) made land-bridge connections possible from Greece across the Cyclades, although the big islands probably remained isolated. Again there would have been connections between many of the eastern islands off the west coast of Turkey with each other and the Turkish mainland. During the successive glacial maxima of the Pleistocene, connections between islands where the sea is <100 m deep, that is, many of the Cyclades, no doubt occurred.

So when one looks at a map of the Balkans and the Aegean in relation to Turkey, it is not complete fantasy to envisage three main land-bridge connections across the area: one across the Bosphorus to the north of Turkey, one through the Cyclades in the centre (perhaps with a link as far as Rhodes) and one around the arc of Tertiary Mountains from the Peloponnese to Crete, Karpathos, Rhodes, and south-west Turkey. Deep basins north and south of the Cyclades separated these connections. As for the Iberian peninsula, the Balkans, including Greece and its islands, have an amazing climatic diversity. Annual rainfall varies from <500 mm in the south-east to over 2,000 mm at high elevation in the north-west mountain ranges.

The richness of the flora of this region is one of the most diverse, for a given area, of the whole Mediterranean region, containing 6,500–7,000 native species, of which ~1,500 are endemic to the region (Polunin 1980). The arc of Tertiary fold mountains that delimit the southern part of this region contain particularly high rates of endemism (Strid and Papanicolaou 1985). On Crete, which runs 245 km in an east–west direction along this arc, 70% of species are of Mediterranean origin. This cortège of Mediterranean taxa includes many woodland and maquis species from the western Mediterranean, for example, *Quercus coccifera* (one can see real trees of this species on Crete!!), *Q. ilex*, *Pistacia terebinthus*, *Arbutus unedo*, *Erica arborea*, and *Euphorbia dendroides*. Another 20% have their origin in the Irano-Turanian floristic province. Compared with Cyprus (see below), the importance of the Irano-Turanian element in the island flora is of minor importance. Crete has 150–170 endemic taxa (Zohary 1973; Médail and Quézel 1997), which represent around 12% of the flora. A roughly equal number of these endemic taxa have affinities, that is, purported sister taxa, with either the Greek mainland or Turkey (~25 species in each case). About 40 of these endemic taxa are related to wide-ranging species across the eastern Mediterranean including Greece. Rates of endemism on Crete are almost identical to those on the Peloponnese peninsula, where 12.5% of the flora is endemic.

1.5.3 The Anatolian peninsula and Cyprus

A major centre of diversity and endemism in the eastern Mediterranean concerns the mountains and coasts of southern Turkey. In this region, volcanic activity is still in progress, subduction continues to produce earthquakes and intense faulting and uplift have created some of the highest mountains of the Mediterranean region. These mountains are comprised of long narrow chains mostly oriented either east–west (inland from the Aegean coast) or north–south (to the south-east) and mostly composed of sedimentary strata deposited in the ancient geo-synclinal Tethys Sea. The mountain chains are separated by deep often broad valleys or high plateaux (1,500–2,000 m elevation across vast expanses). Central Anatolia is a rolling plateau which stretches across vast expanses at 900–1,200 m elevation and is made up of Mesozoic hard limestone and Tertiary marls and soft chalks. The complex coastline of the eastern Aegean Sea, comprising many islands and estuaries was formed by the foundering of the central part of the Aegean and complex faulting in the Turkish mountains.

The elevation gradient that exists in the region means that annual precipitation varies more than threefold over small distances (300–1,000 mm in southern Turkey). Within the region there is a distinct decline in rainfall from west to east and from north to south and the Mediterranean vegetation at low altitude is more xerophytic than that of the western Mediterranean. Although most authors include most of Turkey in the Mediterranean region, once one gets over the coastal mountains and moves inland, the climate rapidly resembles a more continental regime. There is thus high spatial variation in the climatic regime in this region, further details of which can be gleaned from Zohary (1973).

The main centres of endemism in Turkey are the mountains, particularly the Taurus, Armenian, and Kurdo-Zargrosian ranges and also the vast expanses of steppe vegetation on the central plateaux areas (Zohary 1973). The vegetation contains many Mediterranean elements and even the transition zone with the Irano-Turanian floristic province includes many species that one can observe in the strictly Mediterranean part of the territory. As a result, in Anatolia the distinction between the Mediterranean territory and the eastern steppes that spread across Asia is perhaps less marked than the distinction between the Mediterranean and Euro-Siberian and Atlantic provinces in the western Mediterranean. The precise delimitation of a Mediterranean region is particularly difficult in some areas of western Turkey where the mountain ranges run east–west and where broad valleys are open to the migration and persistence of steppe elements from the Irano-Turanian floristic province.

The Mediterranean flora of Turkey contains around 5,000 species of which ~30% are endemic to the region, including 12 monotypic, endemic genera (Zohary 1973; Demiraz and Baytop 1985). There are two important features of this diversity. First, the presence of a large number of genera which are very rich in species (sometimes several tens of species), such as *Astragalus*, *Trifolium*, *Centaurea*, *Salvia*, *Stachys*, *Alyssum*, *Silene*, and *Dianthus* to name but a few. In the genus *Ebenus*, all 14 species are endemic to Turkey. As a result of this radiation of a relatively small number of genera, endemism in the Anatolian peninsula reaches one of the highest rates of endemism of any region on earth. Somewhere around half of the endemic taxa have evolved from Irano-Turanian elements that have colonized the eastern Mediterranean. Second, the occurrence of a relatively large number of tree species (more than 300).

The region has one main island, Cyprus. The geological foundation of this island is an extension of the Taurus/Amanus folded system. Cyprus is floristically related to Syria, with affinities to southern Turkey. On the island, 1,620 species occur of which roughly 10% are endemic. Unlike Crete, the Irano-Turanian element is the dominant element of this flora.

1.6 Conclusions

The Mediterranean flora has had a complex and eventful history. The evidence is overwhelming: Mediterranean plant communities have continuously undergone profound compositional changes, some gradual, some more abrupt, some episodic, and some repeated. The vegetation history of the Mediterranean has been strongly marked by successive extinctions of whole taxonomic groups during the Middle Miocene, the Middle/Late Pliocene, and the Early/Middle Pleistocene. A point to note here is that extinctions will have involved not just plants but also their pollination and dispersal agents. In spite of repeated extinctions and glacial impoverishment, the current-day flora is immensely rich, largely because of its heterogeneous origins, notably from the ancient floras of the tropical and temperate regions between which it has formed a sort of 'ecotone' since at least the Cretaceous, and the differentiation of taxa *in situ*. These diverse floristic origins have come together to create a diverse and spatially heterogeneous flora, unique on earth.

The geological history illustrates all too clearly that the probabilities of dispersal and gene flow among populations may, at certain times, have been radically different to what they are at the present time. Major geographical and geological barriers have limited dispersal and created phylogeographic divisions in many plant groups at various moments during the history of the region. In particular, the dispersal and fragmentation of microplates during the Oligocene and Miocene initially provided a template for the evolution of many endemic species in the Mediterranean.

The development of contemporary Mediterranean-type vegetation occurred in the Late Pliocene as the summer-drought regime set in. Prior to this time, Mediterranean sclerophyllous vegetation existed as isolated taxa adjacent to other vegetation types, but not as the characteristic vegetation. Since the Pliocene, the onset and development of the highly seasonal Mediterranean climate has been more closely related to the precipitation regime than to an alternation of temperatures. Even during glacial maxima, aridity probably had a major effect on species distributions. Future climate change may thus have significant effects on Mediterranean vegetation principally as a result of modified precipitation regimes, for example, if the summer drought regime extends for longer periods during the year via the modification of spring rainfall. Strong regional contrasts in precipitation have occurred over a long period of time, even before the onset of a Mediterranean climate, adding to geological spatial heterogeneity. My synthesis of historical events points to the occurrence of a sort of palaeo-mosaic in Mediterranean habitats. A spatial environmental mosaic has indeed probably been present in many landscapes throughout the history of the Mediterranean Basin due to local geological and edaphic variation and spatial variation in climate (Lepart and Debussche 1992). Such historical spatial heterogeneity may thus have had lasting effects on population differentiation and species divergence in the Mediterranean Basin. More recently, human activities have created even more variation in a landscape already structured by spatial heterogeneity of environmental factors. In some areas, the spatial mosaic of environmental factors has determined the extent and impact of human activities, in others, human activities have reinforced the natural variation in the landscape.

The three panels of my historical triptych (Box 1.1) form the essential framework of this chapter. In the rest of this book I will explore how these three components of regional history have moulded and sculpted evolutionary processes acting on distribution patterns, differentiation, and divergence in the Mediterranean flora. Other authors have outlined the roles of geology, climate, and human activities in the shaping of species diversity and distributions in the Mediterranean region (Zohary 1973; Heywood 1995; Quézel and Médail 2003). In this book, I will use these themes as a framework to discuss the process of plant evolution, both within and among species. A complex geological history, which introduced diverse biogeographic origins to the flora and set physical limits on species distribution, combined with the evolution and oscillations of the climate, which forced plant migration, caused successive contractions and expansions of species' distributions and imposed strong selection pressures on plant traits, are the natural foundations of plant variation and evolution in the Mediterranean region. On to these can be added the powerful and longstanding influence of human activities. Ever since Neanderthal spread westwards from Asia Minor, pre-empting the subsequent direction of spread of domesticated plants, the Mediterranean landscape has been deeply altered and re-modelled. Human activities have created massive contemporary changes in the spatial configuration of habitats and the ecological conditions within them, but have also been constrained in their impact by the presence of spatial environmental variation imposed by geology, soils, and climate. The long history of human impacts should not be ignored as we try to understand plant evolution in theMediterranean.

Despite the dividing lines I have placed between the three panels of Box 1.1, the boundaries between them are fluid. Examination of the nature of the processes involved in each of the three panels, and the interactions among them, allows us to understand the development of narrow endemism, which as I will explore in Chapter 2, is one of the most characteristic features of Mediterranean plant diversity.

CHAPTER 2

The biogeography and ecology of endemism

> Correlations between the palaeogeography of the Mediterranean and the flora of the island of Corsica . . . [illustrate] . . . not only the diversity of biogeographic origins but also the remarkable specificity of this flora.
>
> J. Contandriopoulos (1990: 414–415, my translation)

2.1 Narrow endemism: the cornerstone of Mediterranean plant diversity

The closely interrelated questions of what factors limit range size, why species differ in relative abundance within and among sites and which factors regulate species diversity within local communities are central to the discipline of ecology. Over the years, research on these issues has developed a solid understanding and a rich literature. In this chapter, I will primarily be concerned with the first of these questions: what causes some species to have restricted range size while others, even very closely related species, have more widespread distributions? This question has stimulated much interest (Stebbins 1942, 1980; Kruckeberg and Rabinowitz 1985; Gentry 1986; Major 1988), and a number of questions which relate to the causes of range size variation and endemism can be identified. Are endemic distribution patterns simply a result of chance events during the evolutionary history of a species or group of related species? Have endemic taxa evolved via ecological specialization at the range limits of widespread species or are widespread species successful derivatives of their endemic congeners? Do endemic taxa have reduced genetic diversity, perhaps associated with differentiation in an isolated situation? Are endemic species the relicts of previously widespread species or new taxa that have not yet had time to spread?

The first step towards understanding these issues is to recognize that endemism has multiple causes and that a diversity of factors can influence variation in range size. These factors include geological barriers to dispersal, modification of distributions in association with climate change, genetic factors such as hybridization, polyploidy, and mating system evolution which can greatly affect the capacity for adaptation to new ecological conditions, interactions with and/or absence of pollinators and dispersal agents, and species traits which restrict dispersal. In this chapter, I recast some of the above questions in a Mediterranean setting in order to explore and evaluate the historical, biogeographic, and ecological features of endemic plant distribution patterns in the flora. I will take up the issue of how evolutionary processes have acted to create patterns of differentiation that are the template for endemism in Chapter 3.

The Mediterranean region is an ideal place to study plant endemism. What is perhaps the major characteristic of the Mediterranean flora is the fact that its well-known high overall floristic richness (24–25,000 plant species) is to a large part due to the high incidence of species turnover (β-diversity) and regional endemism. Rates of endemism in particular regions within the Mediterranean are frequently above 10% and can exceed 20% of a local flora (Greuter 1991; Médail and Quézel 1997). It is in the different mountain chains, for example, the Betic-Rifan complex on either side of the Straits of Gibraltar, the Maritime Alps of south-east France, the mountains of southern Greece and Turkey, and on several of the islands which straddle previously connected mountain chains (notably Sicily, Crete and Rhodes, Cyprus, Corsica and Sardinia, and the Balearic islands) that rates of endemism (Médail and Quézel 1997) and species diversity (Lobo *et al.* 2001) are at their highest. These zones of high endemism correspond to zones of high tectonic activity and/or microplate fragmentation and isolation and are areas which may have been less affected by more contemporary human activities.

Somewhere close to 60% of all native taxa in the Mediterranean region only occur in the Mediterranean, that is, are endemic to the region as a whole (Greuter 1991). An important trend within those species endemic to the Mediterranean is that close to 60% are narrow endemic species, that is, species whose distribution is restricted to a single well-defined area within a small part of the Mediterranean region (Fig. 2.1). In contrast, only 28% of non-endemic species (i.e. species which also occur outside the Mediterranean region) occur in just a single region within the Mediterranean. So more than one-third of the native flora (~37%) have restricted distribution patterns, they are narrow endemic species. In addition, only ~8% of species endemic to the Mediterranean have a distribution pattern that encompasses more than five of the Med-checklist regions (for details concerning the delimitation of these regions see Greuter 1991). In contrast, 37% of non-endemic species have distributions that extend to more than five regions. Narrow endemism is thus the integral component of endemism in the Mediterranean.

In this chapter, I will focus on narrow endemic plants whose distribution is restricted to a small region, most often one or a few adjacent islands or a small and often distinct region of a continental area. My objectives are as follows:

- First, I discuss the diversity of endemic distribution patterns in the Mediterranean flora.
- Second, I explore the historical and biogeographic associations of endemic plants and set endemism in the context of the ecology and dynamics of the plant communities in which they occur.

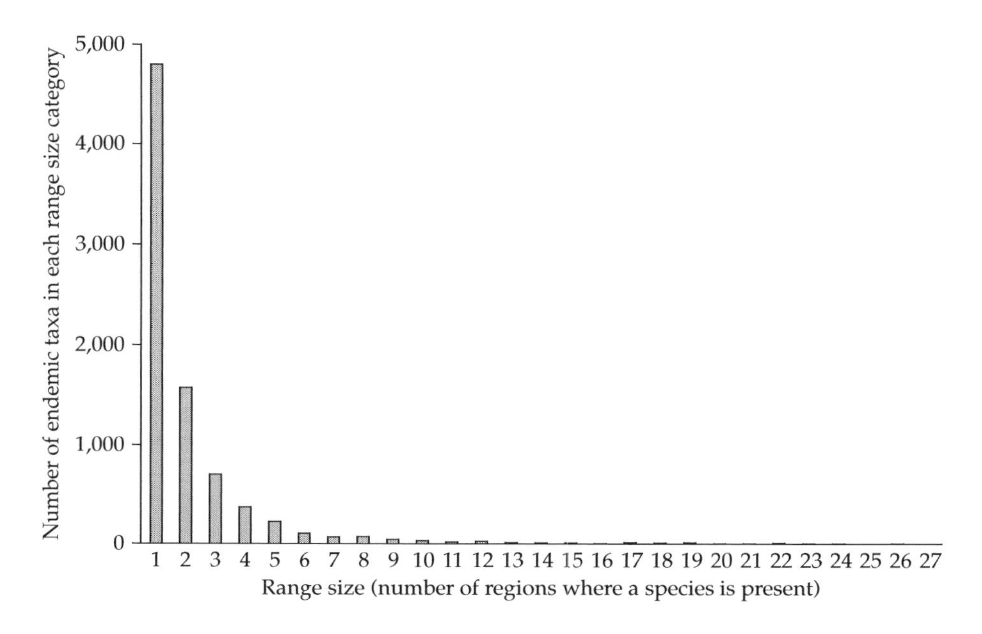

Figure 2.1 The prevalence of narrow endemism in the Mediterranean region (drawn from data in Greuter 1991).

• Third, I evaluate the ecological and biological correlates of endemism. In particular, I examine whether endemic species differ from widespread species in traits linked to ecological function and habitat occupation.

2.2 Endemism in the Mediterranean: patterns and classification

A convenient place to start the treatment of a subject is to define it. Endemism however is one of those enigmatic issues where definitions are difficult to apply. The problem here relates to the relative and subjective nature of the phenomenon—plants have distribution areas of all sizes and degrees of disjunction imaginable (Raven 1972), hence the degree of endemism may increase as the size of an area increases (Major 1988). In consequence, defining endemism becomes highly problematic since 'the scale of the investigation creates the phenomenon' (Favarger and Contandriopoulos 1961: 384, my translation). Endemism is also a relative phenomenon, since taxa may be endemic to a single area (such as an island) or several now disjunct areas that have a common history (e.g. an archipelago of islands which once had land-bridge connections). But let us not stop there, the relative nature of endemism, far from being a major obstacle, actually renders its study all the more interesting because of the novel insights it can provide for our understanding of biogeography and speciation.

Endemism has received several definitions. Several authors have used the term endemism to describe taxa whose distribution is markedly more restricted than the average distribution of taxa of the same taxonomic rank. For Kruckeberg and Rabinowitz (1985: 448) 'narrow endemic taxa are those that occur in one or a few small populations'. However, further on in their review, Kruckeberg and Rabinowitz (1985: 451) recognize that 'the term endemism, in its classical biogeographical usage, does not necessarily imply rarity or even small range'. Major (1988: 117) proposed that 'a taxon is endemic if confined to a particular area through historical, ecological or physiological reasons'. In this chapter, I use this definition in order to examine the biogeographical, ecological, and biological features of narrow endemic plants within the Mediterranean region.

Endemic plants can be relicts or newly formed. These two categories of endemic taxa are commonly referred to as palaeo-endemic and neo-endemic taxa, respectively (Favarger and Contandriopoulos 1961; Stebbins and Major 1965), or as Zohary (1973) preferred, primary (active) and secondary (relict) endemism. I will keep to the former terminology which is more commonly used. Palaeo-endemic taxa are ancient or relict elements of a given taxonomic group, often systematically isolated from other taxa. Neo-endemic taxa are more recently evolved and have extant sister taxa. This duality of endemism, which Stebbins and Major (1965) clearly highlighted in their analysis of endemic plants in California, has been combined with karyotype variation to produce a fourfold classification of endemic taxa (Favarger and Contandriopoulos 1961).

Palaeo-endemics. These are systematically isolated taxa (isolated species in large genera or monotypic genera), which are clearly ancient and usually show little variability. They are probably relict taxa that have persisted through long periods of time. Although they do not necessarily occupy the region in which they originated, they provide illustrations of ancient lineages present in the Mediterranean region, some since the Late Tertiary. Examples include the monospecific genera of *Naufraga balearica* (endemic to Corsica and Minorca), *Soleirolia soleirolii* (endemic to Corsica, Sardinia, Tavolara, and Majorca), and *Nananthea perpusilla* (Corsica and Sardinia), and species with isolated positions within their genus, for example, *Mercurialis corsica* and *Ruta corsica*, both endemic to Corsica and Sardinia. Palaeo-endemics may be of any ploidy level.

Patro-endemics. These are diploid endemic taxa which represent the progenitors of now more widespread polyploid entities. Examples of patro-endemics on Corsica include *Pinguicula corsica*, *Chrysanthemum tomentosum*, and *Arrhenatherum elatius* subsp. *sardoum*. In several patro-endemic species which have given rise to autotetraploids, the diploids and tetraploids are not treated as distinct species, primarily because of their morphological similarity. For example, in both *Dactylis glomerata* on the Balearic islands and *A. elatius* on Corsica unambiguous distinction of diploids from tetraploids requires cytological investigation. Strong reproductive isolation between the two ploidy levels can nevertheless occur due to the formation of sterile triploids when crosses are made between cytotypes. Given that the diploids are ancestral, one should be careful not to treat them as simple varieties or subspecies of an otherwise polyploid taxon (Favarger and Contandriopoulos 1961; Stebbins and Major 1965).

Apo-endemics. These are endemic polyploids whose distribution is a small portion of the range or a disjunct isolate of a more widespread ancestral diploid. Apo-endemics are thus the reverse case of patro-endemics. Sticking to Corsica, one can cite a number of examples, including *Cerastium soleirolii, Viola corsica, Genista corsica*, and *Cymbalaria hepaticaefolia*. In some cases the diploid ancestor is unknown. In addition, polyploid evolution and endemism may be recurrent in particular diploids in different areas. For example, multiple origins of endemic polyploids from a single widespread diploid have been documented in *Plantago subulata* which has endemic polyploid variants in the Atlas Mountains and the mountains of Corsica, and in *D. glomerata* where different endemic diploid subspecies have repeatedly given rise to endemic polyploids in the western Mediterranean (Lumaret 1988).

Schizo-endemics. This class of endemics have undergone differentiation due to the fragmentation of the range of a widespread ancestral taxon producing endemic taxa in different parts of the original distribution. The resulting pattern is one of disjunct distributions of closely related species with the same chromosome number. Schizo-endemic species may be of any age and degree of divergence from the parental stock, which is often identifiable on the basis of morphological and molecular techniques. As Stebbins and Major (1965: 5) quip.... 'The older the divergence, the "better" are the species'. There are multiple examples of this class of endemism in the Mediterranean flora (Fig. 2.2) which illustrate the diverse patterns of isolation and connections among different regions of the Mediterranean.

This classification illustrates a crucial point. Endemic species are not a homogeneous group, other than the fact that they can all be classified as endemics because their distribution is limited to a particular area. The different types of endemic taxa often co-occur in local floras, hence several authors have drawn attention to the importance of the Mediterranean region for both (a) the conservation of ancient taxa whose distribution is relictual and (b) recent and ongoing diversification. In the Mediterranean flora, somewhere close to 75% of endemic taxa are neo-endemics, illustrating the dynamic nature of speciation in the region, a feature which conservation biologists should bear in mind. The classification is however far from being perfect and a number of points should be recognized when analysing patterns of endemism and the evolutionary processes which promote endemism.

First, a given taxon may be classified into more than one of the above classes. The example of polyploid complexes which contain different diploid subspecies with a common ancestor that has differentiated in the different parts of its range could be classified as schizo-endemic taxa since they have the same ploidy level. Most of them are also the progenitors of new polyploids in each of the different regions and thus could also be classified as patro-endemic taxa. Indeed, the dichotomy between old (relict) and recent (neo-) endemism is often presented without conclusive evidence concerning whether a taxon has a relict distribution or is of recent origin (Stebbins and Major 1965). For example, *Cyclamen balearicum*, endemic to the Balearic islands and southern France was considered as a palaeo-endemic species by Contandriopoulos and Cardona (1984). However, its sister species are well known (Fig. 2.2; Chapter 3) and have the same ploidy level, indicative that this is a schizo-endemic species, albeit perhaps quite old.

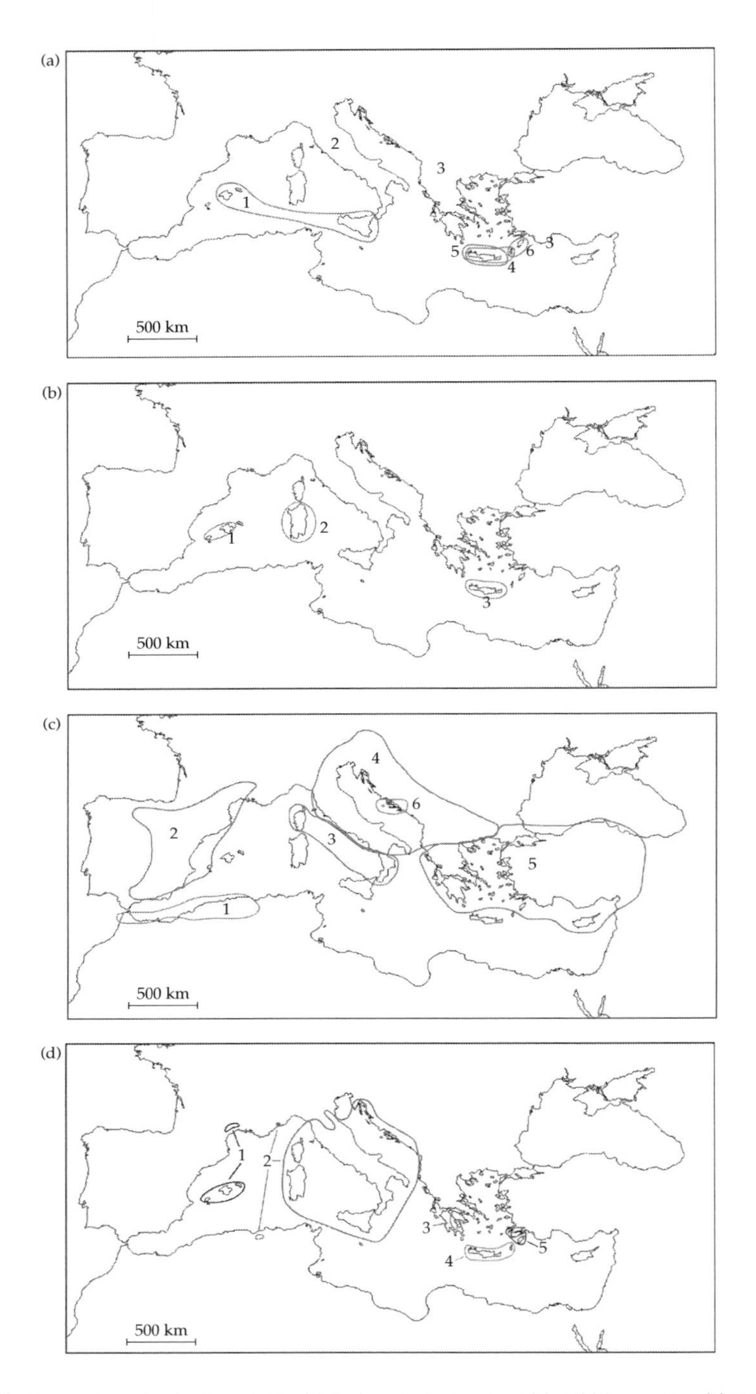

Figure 2.2 Examples of schizo-endemic distribution patterns. (a) Mediterranean *Scabiosa*: (1) and (2) *S. cretica*, (3) *S. hymetia*, (4) *S. minoana*, (5) *S. albocincta*, (6) *S. variifolia*. (b) *Cephalaria squamiflora* subspecies: (1) subsp. *balearica*, (2) as yet an unrecognized variant which the authors call subsp. *mediterranea* (3) subsp. *squamiflora*. (c) *Pinus nigra* subspecies: (1) subsp. *mauretanica*, (2) subsp. *salzmanii*, (3) subsp. *laricio*, (4) subsp. *nigra*, (5) subsp. *pallasiana*, (6) subsp. *dalmatica*. (d) *Cyclamen* subgenus *Psilanthum*: (1) *C. balearicum*, (2)–(5) the different subspecies of *C. repandum*, with (2) subsp. *repandum*, (3) subsp. *peloponnesiacum* (which has two almost allopatric varieties in the Peloponnese peninsula), (4) subsp. *creticum*, (5) subsp. *rhodense*. (a) & (b) Drawn from Contandriopoulos and Cardona (1984), (c) Redrawn from Quézel and Médail (2003), and (d) Redrawn from Debussche and Thompson (2002).

The problem here is that the above classification says little about the processes underlying species differentiation and the evolution of endemic distributions. Some palaeo-endemic species will have originated as a result of similar processes as more recent schizo-endemic or apo-endemic species have done. Some apo-endemic species whose diploid ancestors are extinct could also be considered as palaeo-endemic. Favarger and Contandriopoulos (1961) suggest that the rates of differentiation may vary among classes of endemism, being gradual in schizo-endemics or abrupt in patro- and apo-endemics, a suggestion that is unlikely to be so simple and straightforward. The conventional distinction between palaeo-endemics which are relict taxa and neo-endemics which are recently evolved is of little importance. What should be emphasized is the distinction between neo-endemics, where the genetic and phylogenetic relationships among sister taxa and their contemporary distributions can be established, and palaeo-endemics which are systematically isolated, probably due to the extinction of ancestral and sister taxa, and which also have a contemporary distribution that may not correspond to their site of divergence. Neo-endemic taxa thus provide model systems for the study of differentiation in relation to recent history and contemporary selection whereas palaeo-endemic species provide the chance to study ancient divergence and persistence in relation to geological history.

In addition, endemic distributions may arise in a variety of ways (Stebbins 1950). These include long distance dispersal, localized dispersal following range restriction in almost contiguous areas, migration over historical land-connections, migration of an ancestral stock into two different areas (with possible extinction in different parts of the range), and of course straightforward isolation of two formerly connected areas. The above classification in some ways oversimplifies this diversity of causes. It should not be forgotten that endemism is a feature of distribution which says little about local population characteristics in terms of abundance across the regional landscape and numbers of individuals in local populations. An endemic species is not necessarily rare in terms of local abundance, indeed some endemic species have large populations and/or many populations in the region in which they occur.

Another problem is that in many schizo-endemic groups of species, although ploidy level may be constant, different endemic species may vary in chromosome number, structure, or types of rearrangements (Favarger and Siljak-Yakovlev 1986). This variation in karyotype may be quite common in Mediterranean endemic species (see Chapter 3). Hence, the category of schizo-endemics may encompass a rather diverse and heterogeneous group of endemics.

A final point worth noting here is that in a given region the diversity of endemic species may not only be high, but also may contain a range of endemic species with very different histories, some originating in the Late Tertiary, others evolving as the climate oscillated during the Quaternary. The description of endemism in the Maritime Alps (on the border of southern France and Italy) by Barbero (1967) illustrates all too well the localized co-occurrence of ancient and more recently derived endemic taxa and the potential role of geology and climate during the different episodes of Mediterranean history. Similar analyses elsewhere in the Mediterranean would no doubt produce comparable results. In a given place, different endemic species (even within a single genus) reflect different episodes of Mediterranean history.

2.3 Endemism in the Mediterranean: community composition and biogeography

2.3.1 Endemism and community composition: islands and mountains

Understanding why some species have widespread distributions and others are endemic to localized regions requires appraisal of why species differ in relative abundance within local communities. The abundance of species within samples often show regular patterns of distribution, and communities regularly contain many species at low abundance. This feature of community diversity has long attracted the attention of ecologists and continues to stimulate the development of testable theories (Hubbell 2001). Species which are rare within sites, simply by virtue of their low abundance, may be less common at the regional and geographic scale, and thus have endemic distributions. Within patches of habitat, common species are more common and rare species more rare than one would expect because the latter are both more extinction prone and less likely to colonize new sites or areas where they have gone extinct than common species (Hubbell 2001). Such patterns may occur because of random events associated with dispersal, colonization, and extinction, because rare species have particular ecological requirements and thus only occur in certain types of habitat which are themselves rare in the landscape, or because rarity is associated with traits that reduce the probability of movement across the landscape and thus the potential to expand range size. These issues are the subject matter for the rest of this chapter.

As outlined in Chapter 1, endemism in the Mediterranean is highest in mountain ranges and on islands. A feature of geological activity in the Mediterranean is the discontinuity of land masses and the heterogeneity of the landscape that have resulted from isolation by water, successive phases of mountain formation, and a variety of chemical and physical substrate compositions. These processes have created strong geographic barriers to dispersal.

Islands and archipelago systems have long fascinated biologists, primarily because of their unique and somewhat unusual faunas and floras (Darwin 1859; Wallace 1880; MacArthur and Wilson 1967; Carlquist 1974). Although island floras may actually have low levels of biodiversity in terms of species number, the proportion of endemic species that occur on islands is unrivalled by continental areas, illustrating the importance of isolation for the presence of endemic taxa. In the Mediterranean, islands are for the most part fragments of land that have become isolated due to their separation from continental areas (exceptions include the Aeolian islands off the coast of Sicily and southern Italy). Most Mediterranean islands are thus very different from oceanic islands such as Hawaii, the Canary islands, or the Mascarene islands which have never been connected to continental land masses. The flora of the Mediterranean islands and their adjacent continents were thus very similar prior to isolation, hence, the distance of islands from each other and from continental areas, which may have a strong effect on species number on oceanic islands, is poorly correlated with species richness for Mediterranean islands (e.g. Médail and Vidal 1998).

The rate at which species increase in number with increasing area is one of the fundamental issues in community ecology and biogeography (McArthur and Wilson 1967; Hubbell 2001). Species diversity in any locality is a balance between local abiotic factors and biotic interactions, which act to reduce diversity, and immigration from other habitats and communities, which tends to increase local diversity and maintain diversity at equivalent levels across the landscape (Schluter and Ricklefs 1993). In a given region of relatively uniform climate, there commonly exists a close relationship between the size of the area and the number of taxa which are present (Preston 1962*a*,*b*; McArthur and Wilson 1967). A common pattern is that species number (S) and area (A) are linked by a strong and recurrent relationship with two fitted constants (c and z): $S = cA^z$. A log–log plot of species number against area produces a linear relationship of slope 'z'. This species–area curve is not just an inevitable relationship, it provides a first step to understanding how diversity and endemism are area-dependent and closely related to the regional processes that act on community composition and species distribution.

The flora of local regions in the Mediterranean provides some clear illustrations of the species–area relationship across different spatial scales and in different biogeographic contexts. A significant relationship between species number and area can be seen in the species composition of 48 small islands in the Provence archipelago off the south-east coast of France (Fig. 2.3(a)), for 7 large islands, namely Corsica, Sardinia, Crete, Cyprus, Malta, Sicily, and the Balearic islands (Fig. 2.3(b)), surface area in Andalusia in southern Spain (Fig. 2.3(c)), and in the Mediterranean regions of 17 continental areas in different countries (Fig. 2.3(d)). The study of the Provence archipelago by Médail and Vidal (1998) (Fig. 2.3(a)) showed that the cornerstone of this relationship is habitat diversity, which increases on larger islands, which have more altitudinal variation. As a result, taxonomic diversity is closely correlated not only with area but also with maximum elevation. Of key importance is thus the presence of mountains and in particular the mid-elevation ecotone zones on the slopes of these mountains, as previously suggested (Stebbins and Major 1965).

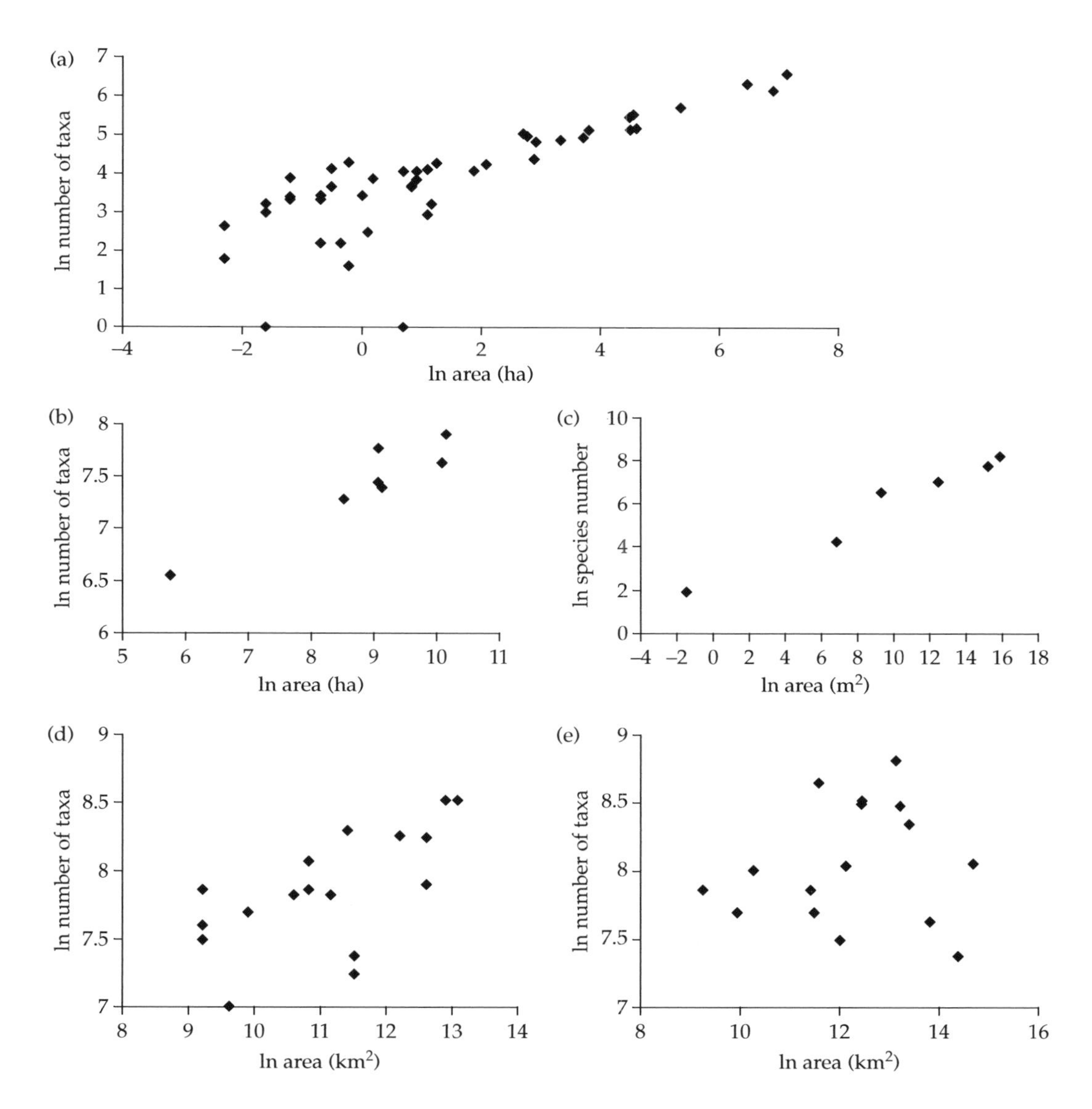

Figure 2.3 Species–area relationships for islands and continental Mediterranean regions: (a) 48 small islands off the coast of Provence, (b) 7 large islands, (c) southern Spain, (d) the Mediterranean sector of different countries, (e) whole countries with part of their land in the Mediterranean. Data are natural logarithms (drawn from data in (a) Médail and Vidal (1998); (b), (d), and (e) Médail and Quézel (1997); (c) Ojeda *et al.* (2000*a*)).

There are several examples which illustrate that the form of the species–area relationship varies across different spatial scales (MacArthur and Wilson 1967; Shmida and Wilson 1985). The slope of the curve is high on local spatial scales, relatively low on intermediate (landscape or regional) spatial scales, and is again high on very large (inter-region or intercontinental) spatial scales. Hubbell (2001) explores how at very local scales the species accumulation curve is particularly sensitive to the abundance of species in local communities, as first common and then rare species are encountered. On this local scale, the sampling size is smaller than most range sizes and dispersal limitation has a negative effect on the rate of species accumulation. As a result, the relationship has a high slope. On regional scales, the rate of encounter of a species depends less on relative abundance and more on rates of dispersal and extinction and thus on species range size, which becomes smaller than the range of area sampled for more and more species. Dispersal limitation thus has a positive effect as newly encountered (more dispersal-limited species) are encountered. The overall result is that the rate of increase in the slope of the species–area curve will decline. On intercontinental scales, species accumulate as one crosses major barriers to dispersal and enters distinct biogeographical zones, with separate evolutionary histories, causing a sharp up-turn in the slope of the *S–A* relation. The Mediterranean examples of the species–area curve also illustrate that the form of the relationship depends on spatial scale and biogeographic context.

First, the relationship observed for the Mediterranean regions of 17 countries (Fig. 2.3(d)) does not occur when diversity of whole countries is used (Fig. 2.3(e)). In the large North African countries (Algeria, Libya, and Egypt are the three points to the right in Fig. 2.3(e)) species richness of the southern arid zones which occupy a large proportion of their territory is low due to climatic constraints. So although these countries have a large area, species number does not increase consistently with area. Hence the importance of having a uniform climate across the range of areas studied. A lower slope in continental regions may thus arise because new habitats with low species diversity are encountered as area increases. If an area is greater than the size of a local hotspot of diversity the slope of the relationship will decline.

Second, dispersal limitation may control variation in the slope of the relationship, particularly for island systems. On Mediterranean islands the rate of increase in taxonomic diversity with area is scale dependent, being greater for the 48 small islands in the Provence archipelago (slope = 0.51) than for the 17 larger Mediterranean regions on the continent (slope = 0.21), in accordance with the pattern described above. Greuter (1991) also reported that within the Mediterranean the slope of the species–area curve is greater than that for a similar range of areas in temperate Europe. In the Mediterranean the high incidence of narrow endemism means that small distribution ranges will allow for a greater slope in the species–area curve for a given range of areas.

Insularity may strongly affect such patterns. MacArthur and Wilson (1967) first pointed out that the insularity prevents common species from being present on a large number of islands since dispersal among islands is less probable than dispersal among contiguous patches of mainland. High dispersal rates will thus distribute common species more widely on continental areas, where overall species richness will be reduced due to a higher rate of extinction of rare species across the landscape (Hubbell 2001). On islands, overall species richness may remain high since common species will not displace rare species everywhere. However, as islands get smaller, the probability of extinction increases, particularly for rare species which are likely to be lost from small islands more rapidly than common species by virtue of their low 'global' abundance and reduced potential to re-colonize sites. The fate of common and rare species is thus context dependent. Three observations from the Mediterranean flora are pertinent to these ideas.

First, palaeo-endemic species may be more prevalent in the endemic flora of islands compared to continental areas (Verlaque *et al.* 1997). This observation suggests that geographic isolation may be important not just because it promotes genetic differentiation and the evolution of endemism (Chapter 3) but also because it reduces immigration of common species and thereby favours the persistence of endemic species on islands.

Second, close inspection of the dataset for the 48 Provence islands obtained by Médail and Vidal (1998) shows that variation in taxonomic diversity among islands tends to decrease as islands increase in size. If one compares the 24 largest islands with the 24 smallest islands, one finds that among island variation in the number of taxa present on the small islands is twice that among large islands ($F = 2.1$; $P < 0.05$). This may reflect greater amounts of random variation in presence/absence of taxa on small islands (which have the least variation in ecological conditions).

Third, in a study of species composition in habitat fragments ('islands') in a landscape of cultivated fields in southern France, Lumaret *et al.* (1997) reported that polyploid perennial herbs occur on a larger number of such 'islands' than diploids, the latter being more often restricted to a single 'island'. They suggest that this result is due to a higher competitive ability and lower rate of extinction of polyploids. In addition, it is possible that polyploids have a greater colonization capacity and thus a more widespread distribution.

Another important issue here is that endemic plants may occur in a biological community which itself has a fairly unique and novel composition (Kruckeberg and Rabinowitz 1985). The communities of chasmophyte species on and around the limestone cliff faces in Mediterranean landscape provide a clear illustration of this. To quote Polunin (1980: 44) 'each gorge or cliff will have its own collection of species and many rarities are only to be found in such habitats'. Mediterranean cliffs often have a high diversity of species, for example, limestone cliffs in northern Spain have higher species diversity than equivalent areas on different substrata in the surrounding region (Escudero 1996).

The islands of the Aegean are the place to see spectacular cliff formations (Fig. 2.4), Karpathos and south-west Crete being the most grandiose and well

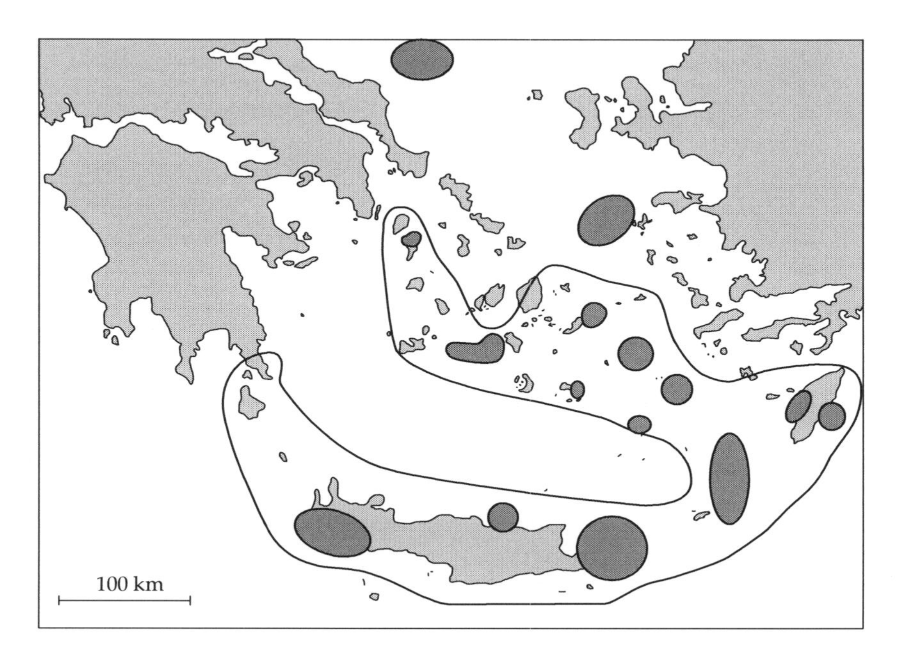

Figure 2.4 The distribution of major cliff formations (shaded areas) in the Aegean archipelago (redrawn from Runemark 1971).

known. Chasmophytes are thus a notable element in the flora of the southern and central Aegean, where the distribution of the habitat and the species shows some intriguing patterns. More than 60 obligate angiosperm chasmophytes are to be seen in these cliffs, ~50 of which occur primarily in hard limestone cliffs. Most species occur in the maritime cliffs not far from the sea, some are endemic to the central Aegean (e.g. *Helichrysum amorginum* and *Campanula calaminthifolia*), most are highly lignified and several represent systematically isolated species (e.g. *Eryngium amorginum* and *Linum arboreum*). Several chasmophyte species have closely related species or occur as a single highly variable species with characteristic races in different parts of the Aegean. On some of the islands, for example, Ikaria, species with affinities to the south, east, and west can be observed together, indicative of a relict flora on an island once connected to different continental areas. The cliff communities on such islands probably served as refugia for chasmophytes and other non-chasmophytes during the Quaternary glaciations.

Runemark (1969) illustrates how across the Aegean Sea, the presence/absence of different chasmophytes on different islands does not fit any particular geographic or ecological pattern. Even islands less than 10–20 km apart show marked differences in chasmophyte species presence. Many adjacent Aegean islands are separated by very short distances (5–20 km in many cases), and the distance between two adjacent central Aegean islands never exceeds 40 km. Despite repeated connections, dispersal limitation thus appears to have played a most effective role in shaping distribution patterns in this areas. Small distances have been effective barriers to dispersal. Very few species reach the Greek mainland, yet many occur across more ancient seas and occur on Crete or less often on the eastern Aegean islands. None of the obligate chasmophytes in the archipelago have an even distribution pattern across the area and some of the most widespread species in the archipelago (e.g. *Dianthus fruticosus* and *Scrophularia heterophylla*) are hardly to be found elsewhere. Runemark (1971) also points out that taxa such as *Inula candida* and *Campanula* subsect. *Quinqueloculares*, which are dominant components of cliff communities in the surrounding geographical regions have only restricted distributions in the Aegean. These distribution patterns, the restriction of many species to limestone cliffs, and the taxonomic isolation of many of these chasmophytes suggest that the cliff flora is all that is left of a flora that once prospered in larger and more connected areas of coastline in the Pliocene.

The fact that several chasmophyte species do not occur on adjacent small islands, but cross phytogeographical divisions which span the region (Box 2.1), for example, with Crete where large areas of suitable habitat occur, suggests that community composition and distribution may have a random component. Runemark (1969) argues that because populations of chasmophyte species are characteristically small, new individuals probably arrive very rarely and the colonization of new sites is likely to occur by only a very small number of propagules. Many species will thus have high local extinction risks because of their small numbers. So, irrespective of the ecological preferences and competitive abilities of the species, the dynamic equilibrium between colonization and local extinction due to climate change and dispersal limitation may create irregular patterns of distribution and abundance.

Runemark (1969) also points out that these seemingly random patterns of distribution in chasmophytes are paralleled by similar patterns in the phrygana and sublittoral vegetation which occur on the Aegean islands. Many of the islands have a normally developed phrygana, while others have little trace of this vegetation type and a sublittoral vegetation with a unique composition. For Runemark (1969) it is not just random extinction following fragmentation that is important, low rates of re-colonization due to 'reproductive drift' and failure of most colonization events to establish a new population also play a crucial role in community composition and distribution patterns. His explanation is appealing and provides an instructive link between the theory outlined above and observations of distribution patterns and community composition on Mediterranean islands.

Box 2.1 Biogeography of the Aegean islands

You do not have to pore over a map for long to realize that the islands and continental margins of the Aegean Sea provide a fascinating situation to study the biogeography and community ecology of endemic plants. The archipelago of multitudinous islands (~200) of different size (~100 have an area <1 km^2) within a sea surrounded on three sides by continents with different histories and biogeographic connections, and across which there has been a history of connections and island isolation (Chapter 1), have created a patchwork system well-adapted to the study of the limitations on species distribution and the biogeography of endemism.

The islands of the Aegean Sea can be grouped into four main sectors which reflect historical connections and migration routes across land-bridge connections and subsequent repeated periods of isolation (redrawn from Runemark 1969).

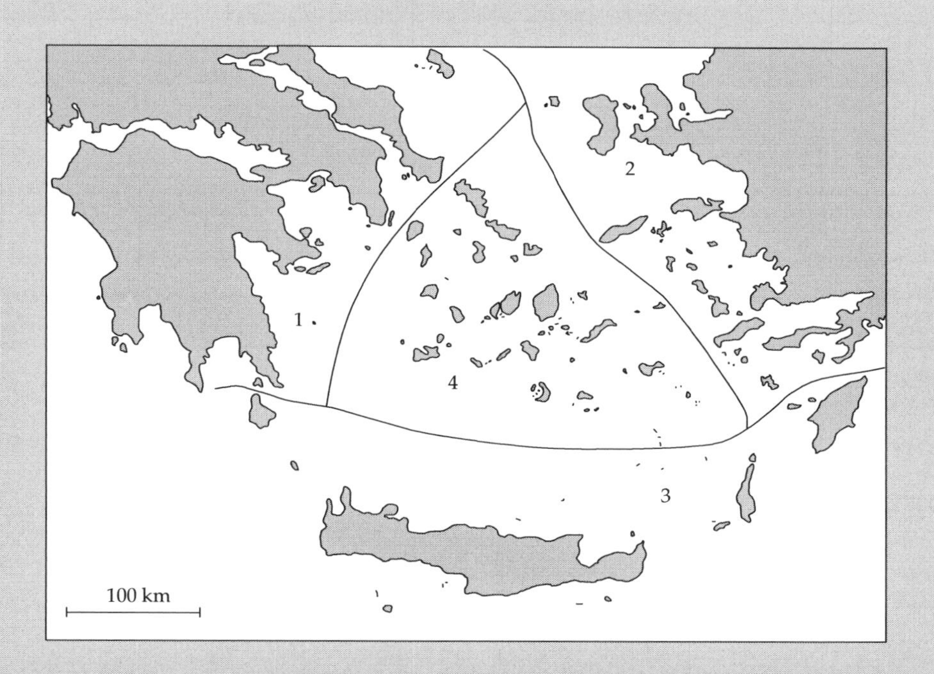

1. The western small islands with very close affinities to the Greek mainland.
2. The eastern islands with affinities to Turkey and Thrace (due to the presence of a deep channel and more permanent sea in a north–south direction in the eastern Aegean.
3. The islands of the southern arc of Tertiary fold mountains (including Crete, Karpathos, and Rhodes) where rates of endemism are high.
4. The Cyclades archipelago which occur on the eroded summits and crests of the ancient mountains that once dominated this region. On these islands elevations reach 1,000 m on Naxos and a little less on Andros (944 m) but elsewhere do not exceed 600 m.

In the Mediterranean it is not islands but regions with high mountains which have the highest rates of endemism (Médail and Quézel 1997; Verlaque *et al.* 1997). For example, endemic taxa on limestone in Mediterranean shrublands in southern Spain are more common above 1,000 m elevation than on similar bedrock at lower altitude in the mountains of southern Spain (Arroyo and Marañón 1990). Where mountains occur as islands in the sea then endemism is even more prevalent, for example, on Corsica, where the percentage of endemism reaches 35% in the mountain flora. The comparison of the affinities of taxa endemic to Corsica and those endemic to both Corsica and Sardinia further illustrates this point (Contandriopoulos 1990). Whereas two-thirds of taxa whose endemism is limited to Corsica are probably of non-Mediterranean origin (most of these taxa have Eurasiatic affinities) and only one-third are of Mediterranean origin, for taxa endemic to both Corsica and Sardinia, the pattern is reversed, the majority have Mediterranean affinities (Fig. 2.5). If Sardinia had mountains similar in elevation to those of Corsica, the affinities between the flora of the two islands would certainly be tighter.

Associated with the increased rates of endemism in the mountain flora of Corsica is a decline in species diversity (Fig. 2.6). On Corsica, endemic taxa have diversified in a relatively impoverished mountain flora. Similar patterns occur on Crete (Greuter 1972) and in the mountains of southern Spain, where most endemic species occur in species poor, open heathlands on mountain ridges (Ojeda *et al.* 1995). It is uncertain if Corsica has had any continental connections (suitable for plant migration) since the Miocene when its rotation with the other parts of the microplate was stopped by collision with the Apulian plate. On Corsica, most of the endemic mountain taxa have affinities to alpine–arctic species, which probably date to periods of climate change in the Late Tertiary (Contandriopoulos 1962). The migration of new taxa to Corsica, particularly the mountains, has thus been very low for a very long period of time. The same may be true of other large islands such as Crete (Greuter 1972). Impoverishment in the mountain floras of such islands is thus most likely due to a long history of extreme isolation during which extinction rates probably exceeded immigration rates, the flora being in a sort of non-equilibrium. Indeed in some places in the mountains of Corsica one can observe taxa that one would not expect to be present in a given community.

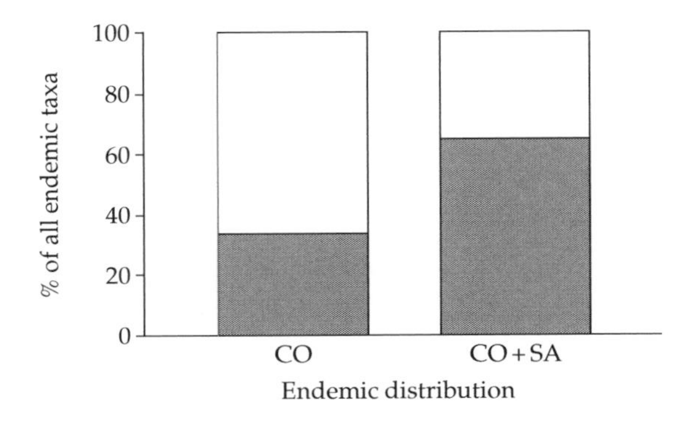

Figure 2.5 Percentage of taxa endemic only to Corsica (CO) or to both Corsica and Sardinia (SA) which have affinities to the Mediterranean (shaded part of bar) or non-Mediterranean (open bars) element of the flora (drawn from data in Gamisans 1999).

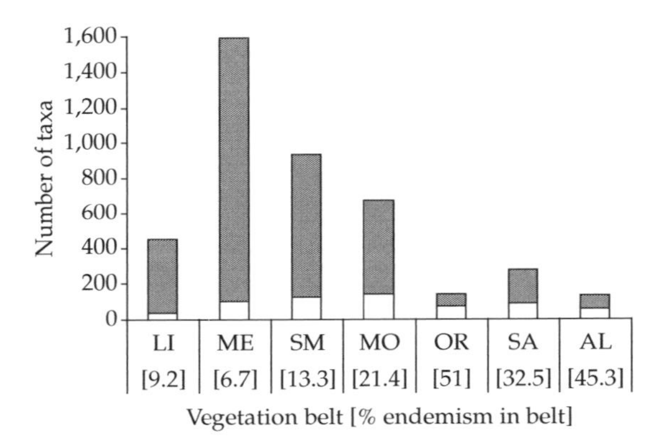

Figure 2.6 Variation in the percentage of endemism in different vegetation belts on the island of Corsica. Each histogram represents the number of endemic (open portion) and non-endemic (closed portion) taxa in each vegetation belt. LI: coastal vegetation, ME: thermo- and meso-Mediterranean, SM: supra-Mediterranean, MO: mountain-Mediterranean, OR: mountain, SA: sub-alpine, AL: alpine (drawn from data in Gamisans 1999).

A certain disharmony may thus reign in the structure of some communities on Mediterranean islands, as proposed for oceanic islands (Carlquist 1974), despite the fundamental differences in how the communities came to be on these two types of islands. A potential consequence of low diversity is reduced competition in some habitats which may allow species to colonize new ecological conditions. Gamisans (1991) has suggested that reduced competition has allowed some endemic species (e.g. *Stachys corsica, Robertia taraxacoides, Cerastium soleirolii,* and *Galium corsicum*) to have a wider ecological amplitude than that of related taxa elsewhere. Such a wider range ecological amplitude on islands has been reported for *C. balearicum* on the Balearic islands compared to continental France (Debussche and Thompson 2003), and is predicted in many texts on the biogeography of island communities.

On a geographic scale, centres of endemism tend to be regions with high species richness (Fig. 2.7; Chapter 1). Hence, there is a complex relationship between diversity and endemism which depends on the balance between (a) local abiotic and biotic interactions which act to reduce diversity and (b) immigration from other habitats and communities which may increase local diversity. Resource diversity and habitat productivity can have a strong effect on species diversity since although theory predicts a positive relationship between resource diversity and the diversity of coexisting species which compete for these resources, decreased availability of nutrients may prevent the most competitive species from excluding others, and thus maintain diversity at higher levels than in highly productive habitats (Tilman 1994). I discuss this issue in more detail later in the chapter.

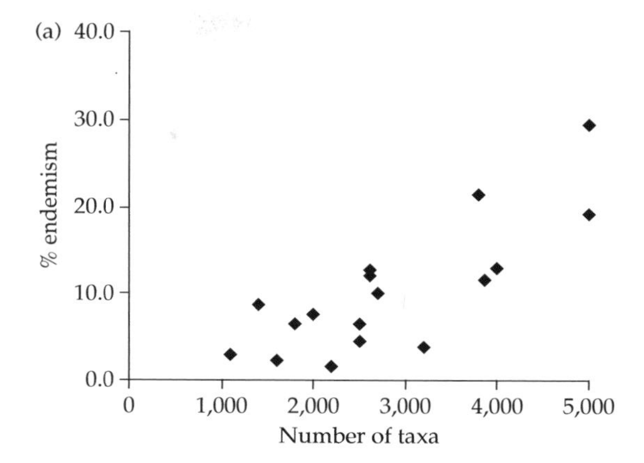

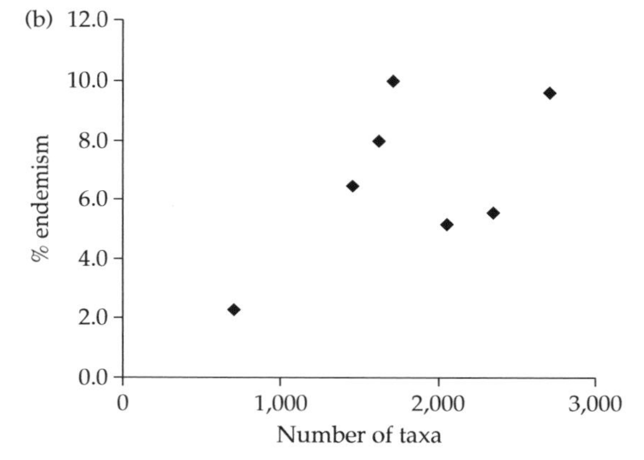

Figure 2.7 Relationship between proportion of endemic species and species diversity in the flora of (a) the Mediterranean area of 17 countries, and (b) 7 large islands in the Mediterranean (drawn from data in Médail and Quézel 1997).

2.3.2 Endemic species with disjunct distributions

The discussion of endemism is often confined to species unique to a single piece of land, be it a mountain range or an island. However, many endemic species have distribution patterns which encompass a small number of disjunct land fragments, a situation which provides insights into the historical component of range size variation.

Endemic plants on the island of Corsica, studied by researchers based primarily in Marseille and Geneva over the second half of the twentieth century illustrate this theme well; details of what follows can be found in Contandriopoulos (1962, 1990), Gamisans (1999), and Gamisans and Jeanmonod (1995). The island of Corsica has a little over 2,100 indigenous vascular plant species plus 400 different subspecies/varieties. Almost 130 of these are endemic to the island, that is, 5% of the total flora. For species on Corsica, endemism extends beyond the boundaries of this island to the other islands and fragments of continent which have historical land connections with Corsica (Chapter 1). In fact, the number of endemic species on Corsica whose distribution does not extend beyond this island is <50% of endemic species on the island (Box 2.2). Two-thirds of the taxa which show endemic distributions that extend beyond Corsica also occur on Sardinia. The majority of these disjunct endemic distribution patterns encompass the different parts of the microplate which fragmented and dispersed in the Oligocene and Miocene (Chapter 1). A smaller number of species also occur on the continental parts of the Hercynian massif, with which they have a longer history of geographic isolation. An important point here is that where disjunct endemism involves the Balearic and Tyrrhenian islands ancient endemism is the prevailing pattern, for example, palaeo-endemic species such as *S. soleirolii*, *N. balearica*, *Cymbalaria aequitriloba*, *Delphinium pictum*). In contrast, no palaeo-endemic species on Corsica have endemic distributions which extend to Sicily and/or Calabria.

A similar pattern of disjunct endemism exists for the endemic plants of the Balearic islands (Contandriopoulos and Cardona 1984; Alomar *et al.* 1997), which are made up of two main groups of islands, the Gymnesian islands of Majorca, Minorca, and Cabrera (plus some very small adjacent islands) and the Pityusic islands of Ibiza and Formentera (plus adjacent small islands). Endemic taxa in this archipelago can be restricted to a single island (e.g. *Daphne rodriguezii* on Minorca, *Globularia cambessedesii* on Majorca, *Viola stolonifera* on Minorca, and *Allium grosii* on Ibiza), a subset of islands, usually to the different Gymnesian islands of Majorca and Minorca and sometimes Cabrera (e.g. *Thymelaea velutina* and *Rhamnus ludovici-salvatoris*), to the Gymnesian islands plus parts of the original microplate (Corsica and Sardinia), or to the Pityusic islands and the eastern Iberian peninsula (e.g. *Silene cambessedesii*). Although there are some exceptions, these distribution patterns illustrate nicely the different biogeographic affinities of taxa on the Gymnesian and Pityusic islands. Whereas endemic taxa on the Gymnesian islands

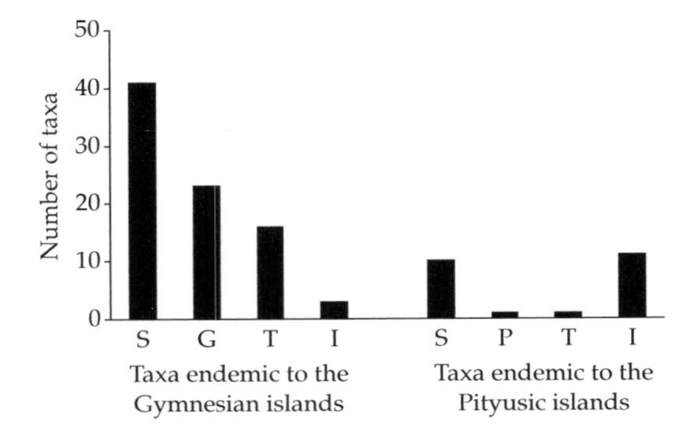

Figure 2.8 The distribution patterns of endemic taxa in the Balearic islands. Not included in this graph are taxa whose endemic distribution encompasses the Gymnesian and Pityusic islands (18 taxa). S: single island endemics, G: taxa endemic to two or more Gymnesian islands, P: taxa endemic to two Pityusic islands, T: taxa whose distribution extends to the Tyrrhenian islands and/or surrounding continents, I: taxa whose distribution extends to the eastern Iberian peninsula (drawn using the distribution patterns of 124 endemic taxa described by Alomar *et al.* 1997).

tend to be endemic to this subset of islands or have distributions which encompass the Tyrrhenian islands and surrounding continents, endemic plants on the Pityusic islands are either endemic to these islands or have distributions which extend to the eastern Iberian peninsula (Fig. 2.8). There is a marked break across the Balearic islands, a sort of Wallace-line type separation between the Gymnesian and Pityusic islands, that is, through the middle of the archipelago (Box 2.2). The stronger historical affinities of the Gymnesian islands to the ancient microplate means that palaeo-endemic taxa occur primarily on the islands of Majorca and Minorca and only extremely rarely on the Pityusic islands.

2.3.3 The east–west divide

East–west vicariance is common in the Mediterranean and involves some characteristic Mediterranean species (Dallman 1998; Quézel and Médail 2003). Despite the separation of North Africa from Europe by the Mediterranean Sea, there is a weaker floristic relationship between the eastern and western Mediterranean Europe than there is between northern and southern shores. This is particularly apparent when one

Box 2.2 Patterns of single-island and disjunct endemism in the western Mediterranean as illustrated by species endemic to Corsica and associated regions

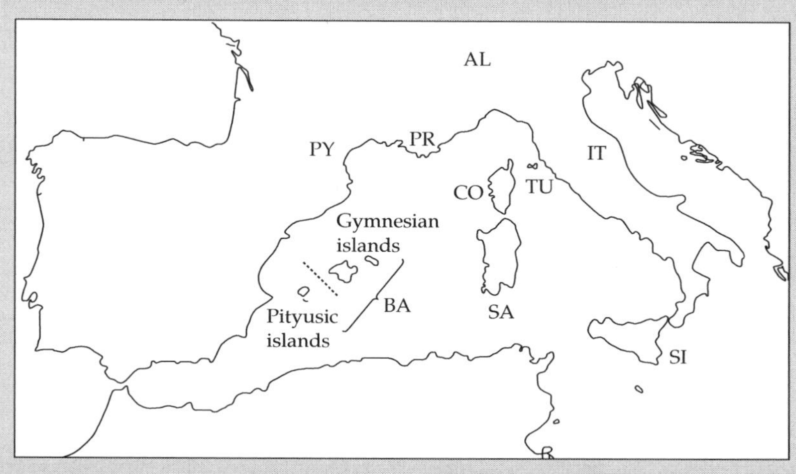

CO: Corsica
SA: Sardinia
SI: Sicily
TU: Tuscan archipelago
IT: Italian peninsula
BA: Balearic islands
PY: Pyrenees
AL: Alps
PR: Provence

The histogram, drawn from data in Gamisans (1999), shows the percentage of all endemic taxa which occur on Corsica as a function of their endemic distribution (e.g. CO + BA is the percentage of all endemic species on Corsica whose distribution extends to the Balearic islands).

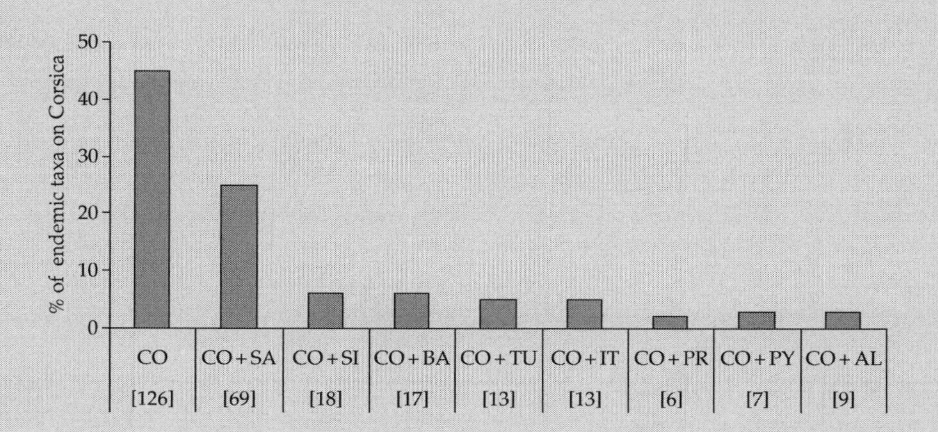

There is a clear concordance between endemic distribution patterns and the geological history of the region since ~30% of endemic taxa on Corsica have distributions which encompass either the different parts of the microplate, which fragmented and migrated in the Miocene, or Continental France and Spain. The most common pattern of 'disjunct endemism' involves taxa which occur on Corsica and Sardinia, and one of the following areas:

- Sicily (e.g. *Berberis aetnensis*, *Carduus cephalanthus*, and *Cardamine chelidonia*);
- the Balearic Islands (e.g. *Naufraga balearica*, *Soleirolia soleirolii*, and *Arenaria balearica*);
- the Tuscan or Hyères islands (e.g. *Teucrium marum* and *Delphinium pictum*);
- Italy (e.g. *Alnus cordata* and *Pinus nigra* subsp. *laricio*).

compares the flora and dominant vegetation of the western (Iberian peninsula–Morocco) and eastern (Turkey–Israel) limits of the Mediterranean. First, in the different altitudinal vegetation belts there is a closer resemblance of dominant tree species between the south-east and north-east and between the south-west and north-west regions, than there is from east to west (Quézel and Médail 2003). Second, the flora of the southern part of the Iberian peninsula and the Moroccan Rif have 75% of plant species in common (Valdés 1991). Messinian land-bridge connections (~5–6 Ma) between the Iberian peninsula and Morocco, a similar climate and ecology on either side of the Mediterranean and the migration of bird dispersers in the autumn have undoubtedly contributed to the homogenization of the flora of the two regions which form the basis of the Ibero-Mauritanian floristic zone, which has more than 500 endemic species (Quézel 1978).

East–west differentiation is particularly apparent when one examines the distribution of closely related species and patterns of differentiation within individual species in the Mediterranean. Patterns of schizo-endemism frequently involve disjunction and divergence in an east–west direction. I have shown some examples in Fig. 2.2, and many more can be seen in other works on the Mediterranean flora, such as those depicting the distribution of tree species in the Mediterranean (Quézel and Médail 2003), for example, the east–west disjunct distributions of *Pinus brutia* and *Pinus halepensis* and *Quercus coccifera* subsp. *calliprinos* and subsp. *coccifera* on either side of the Aegean, and *Quercus ilex* subsp. *rotundifolia* (central and southern Spain) and *Q. ilex* subsp. *ilex* (from Provence to the shores of the Aegean) whose distributions are connected by the presence of an intermediate 'morphotype' present across southern France and north-east Spain.

Three major factors (all discussed in Chapter 1) have no doubt greatly contributed to east–west floristic and species differentiation and thus the patterns of endemism depicted above. First, the climatic history of the northern shores of the Mediterranean has probably been a major element in east–west patterns of divergence due to the retreat of a large number of species into different geographically isolated refugia during the Quaternary glaciations. Second, historical and current day differences in climate, in particular rainfall and seasonal aridity, are more extreme from east to west than they are in a north–south direction. Finally, the eastern Mediterranean flora has been strongly influenced by immigration of Irano-Turanian floristic elements, much more than the western Mediterranean (Zohary 1973). For example, the flora of Crete has a predominantly Mediterranean origin whereas that of Cyprus has a dominant Irano-Turanian element (Chapter 1).

The eastern Mediterranean, particularly in and around the Aegean Sea, has undergone major tectonic changes and configuration of land masses and island formation in more recent history than Corsica and the Balearic islands (Chapter 1). It is thus not surprising to find that the percentage of all endemic species in the four classes of endemism described in Section 2.1 becomes more biased towards neo-endemics as one moves eastwards. In the Peloponnese peninsula and on Crete 83–85% of endemic species are schizo- or apo-endemic taxa and less than 17% are ancient endemics. In contrast, on Corsica, in the Balearic islands, and in south-east France there is a more mixed blend of ancient (28–32%) and neo-endemics (68–72%).

Another feature which distinguishes groups of islands in an east–west direction across the Mediterranean concerns the relative proportion of single island and disjunct endemic species. Whereas the relative proportion of these two types of endemics is roughly equal (48–52%) on Corsica, Sardinia, and the Balearic islands (and Malta), single island endemics are far more common (75–85% of all endemics) on the islands of Sicily, Cyprus, and Crete (Médail and Quézel 1997). The large number of disjunct endemics present in Corsica, Sardinia, and the Balearic islands are related to historical connections (Box 2.2). The higher rates of strict endemism on Sicily, Cyprus, and Crete suggest that either ancestral connections were lost earlier (not so for Crete), that once connected areas are now submerged, or, as is probably the case for Sicily, that parts of these islands were not part of an ancestral connection. The high degree of single island endemism and widespread taxa at their distribution limits on Crete (e.g. *P. brutia*, *Q. coccifera* subsp. *calliprinos*, etc.) highlights the unique position and botanical significance of this island in the southern Aegean.

2.3.4 Endemic relicts from the Tertiary palaeoflora

This discussion of endemism would be incomplete without mention of Mediterranean endemics which represent the last remaining evidence that a few million years ago a luxuriant broad-leaved forest clothed the landscape around the shores of the Mediterranean. Quézel and Médail (2003) identify 26 localities around the Mediterranean where relict vegetation elements occur, notably in the mountains and coastal areas of Morocco, southern Spain, Sicily, Calabria, Crete, the mountains of Greece, south-west and south-east Turkey, and the Amanus and Liban mountains. Scattered trees in these relictual fragments are all that is left of the Tertiary palaeoflora (Table 2.1). Several of these areas illustrate where Tertiary vegetation found refuge as the Mediterranean climate set in and they all may have given refuge to species from a combination of reduced temperature and greater aridity during the glacial maxima of the Quaternary (see Chapter 1). Endemic Tertiary relict species can also be observed in the herbaceous flora of cliffs and crevices, for example, the five species of Gesneraiceae, of which *Ramonda myconi* occurs in the Pyrenees and four others in the Balkans, that is, *Ramonda nathaliae* and *Ramonda serbica*, *Haberlea rhodopensis*, and *Jankaea*

Table 2.1 Examples of relictual populations of 'Tertiary endemics' in the Mediterranean region

Species and family	Contemporary distribution in the Mediterranean	Past distribution (see also Chapter 1)
Liquidambar orientalis	Humid bioclimate (riparian habitat) of the thermo-Mediterranean in south-west Turkey and on Rhodes[a]	An ecologically important genus in the Miocene with many fossil remains[b]. Known from the Late Pleistocene in Italy[c]
Laurus azorica	Atlas Mountains[d]	Neogene ancestors occurred as far north as south-central France[b]
Zelkova sicula	Single locality on Sicily (Monte Lauro)[e]	Widespread genus during the Tertiary, up till the Early Pleistocene (30,000 BP) in the western and central Mediterranean[e]
Zelkova abelicea	Three mountain ranges on Crete[f]	
Phoenix theophrasti	Coastal vegetation in riparian habitat and gorges (<350 m elevation) in the Peloponnese, Crete and south-west Turkey[g]	More widely distributed in Greece and perhaps Turkey and the Aegean islands
Rhododendron ponticum	Disjunct populations of subspecies in the southern Iberian Peninsula and east of the Aegean[h]	Widely distributed throughout southern Europe during the Tertiary[i]. Now an invasive species in the British Isles
Frangula alnus	A relict subspecies has been described in riparian forest vegetation in southern Spain[j] and northern Morocco[k]	Other subspecies still widely distributed in central Europe
Tetraclinis articulata	Semi-arid bioclimate of north-west Africa, southern Spain and Malta[g]	Ancestral taxon present in Europe during the Eocene[l]

[a] Akman *et al.* (1993). [b] Roiron (1992). [c] Combourieu-Nebout (1993). [d] Barbero *et al.* (1981).
[e] Di Pasquale *et al.* (1992). [f] Barbero and Quézel (1980). [g] Quézel and Médail (2003). [h] Mejías *et al.* (2002).
[i] Mai (1989). [j] Hampe and Arroyo (2002). [k] Fennane and Ibn Tatou (1998). [l] Palmarev (1989).

heldreichii which is endemic to Mt Olympus (Polunin and Smythies 1973; Polunin 1980; Vokou *et al.* 1990). These relict populations provide us with a contemporary image of the past flora of the Mediterranean region.

Such relict distributions concern not only individual species but in some cases a whole *'cortège'* of plants. In a study of the flora of south-west Morocco, Médail and Quézel (1999) identified this region as an important refuge for Mediterranean elements of the flora, many of which have dispersed to the Canary islands (e.g. succulent *Euphorbia, Aeonium, Dracaena,* and *Sonchus*), and various elements of the Tertiary palaeoflora. The occurrence of summer fogs, common in other Mediterranean-type ecosystems which flank oceans rather than a closed sea, maintain air humidity at high levels in this region. In addition, the contrasting relief may have permitted species to survive the various periods of major climatic change since the Pliocene. It is noteworthy that several Tertiary endemic taxa occur in riparian and/or coastal areas where the summer drought may be mitigated.

Some relictual populations, for example, *Frangula alnus* subsp. *baetica* (Hampe *et al.* 2003; Chapter 3), contain high levels of genetic diversity which is not encountered elsewhere in the species range, and thus represent an important part of the evolutionary potential of the species. They are not just genetically depauperate relict populations. They are of significance for our understanding of plant evolution and also for the conservation of a representative sample of genetic diversity. However, in the last century or so, many of the species in Table 2.1 have seen their populations further depleted and fragmented as a result of human activities and the conservation of such species has become a pressing problem, whose resolution is problematic. For example, as a result of forest destruction less than 25% of the habitat of *Liquidambar orientalis* known to have existed in the early twentieth century now remains (Quézel and Medail 2003). In *Zelkova sicula*, only ~200 plants now exist in a single population on Sicily, where overgazing, a precarious ecological situation and poor pollen viability in this andromonoecious tree (Nakagawa *et al.* 1998) all threaten the demographic viability of this species (Di Pasquale *et al.* 1992).

2.4 The biology and ecology of endemic plants

2.4.1 The comparison of endemic and widespread species

The study of range size limitation is a central theme in evolutionary ecology. To pinpoint traits associated with endemism is no easy task, and for some authors a fruitless endeavour. Comparisons of rare and common plant species nevertheless indicate that there may exist general differences in the biology of endemic and widespread species (Kunin and Gaston 1993). In plants, endemism may be particularly prevalent in some taxonomic groups, for example, the Ericaceae and Proteaceae of South Africa (Cowling and Holmes 1992). The relative frequency of endemic plants may vary among life-form classes (Major 1988) and nutritionally unusual and rare substrates (e.g. serpentine soils) often harbour high rates of endemic taxa (Kruckeberg and Rabinowitz 1985). Finally, as discussed above, dispersal limitation may be critical in the determination of endemism; after all one of the primary reasons why a species is not present in a given area may be that it has not yet reached the site in question.

A number of important questions concerning the biology of endemic plants can be asked.

1. Is the restricted distribution of narrow endemic taxa in the Mediterranean linked to their ecological characteristics? Are endemic species more ecologically specialized than widespread species? Do endemic species have traits which reflect a lack of competitive ability or a more stress-tolerant ecological strategy?
2. Has human impact been greater on the habitats of endemic taxa?
3. Do endemic taxa have reproductive or dispersal traits which limit dispersal ability?
4. Are endemic plants a taxonomically heterogeneous assemblage or do a small number of particular groups represent the majority of endemic taxa in a region?

In attempts to analyse these questions one must be cautious. First, although the traits of endemic species may indeed be different from widespread species, such differences may be a consequence of restricted distribution, and not the cause. Second, despite observations that geographic range often closely reflects local ecological breadth (Hodgson 1986; Kunin and Shmida 1997) and that there may be the associations between local or regional abundance and geographic range (Hubbell 2001), species with widespread distribution are not always locally abundant, nor do they consistently have a wider ecological breadth. Hence, if a relationship is found between particular traits and estimates of abundance on a local scale, similar relationships may not occur when the traits are examined in relation to range size. The interpretation of differences between widespread and endemic species may thus depend on the scale of investigation (Kunin and Shmida 1997).

Another difficulty concerns the choice of study species. Many comparative studies of widespread and endemic species have involved a single rare species and a single closely related widespread species. This approach provides valuable information for the conservation of particular species but provides only low power for generalization. Comparative evaluation requires large numbers of species, some of which may share traits due to a common ancestral condition. This so-called phylogenetic constraint can be accounted for in two ways.

1. The use of phylogenetically independent contrasts (PICs) in which contrasts of trait values among related species are made at each node of a phylogenetic tree, rather than by using the trait values of the different species (Harvey and Pagel 1991). This method reduces the sample size to account for non-independence associated with phylogenetic history.
2. Multiple contrasts of species pairs in different phylogenetic backgrounds, for example, pairs of widespread and endemic species in different genera.

The importance of such a phylogenetically controlled approach is illustrated by the fact that in different parts of the Mediterranean, endemism is over-represented in certain genera and families, for example, the Iberian peninsula where half of the ~1,250 endemic taxa occur in only 27 genera and the other half are distributed across 286 genera (Chapter 1). A similar concentration of endemic taxa in a small number of genera/families is observed in other Mediterranean-climate regions (Section 2.4.4). Recent studies using phylogenetically controlled approaches indicate how the biology and ecology of Mediterranean endemic taxa may differ from widespread species.

2.4.2 Reproduction and rarity in crucifers

In a study of 52 Mediterranean annual crucifers in Israel, Kunin and Shmida (1997) applied the PICs procedure to test for relationships between reproductive traits and rarity on three scales: (a) local density, (b) regional abundance, and (c) geographic range size in Israel. The reproductive traits they examined involved floral traits (e.g. flower depth, petal length, and floral longevity) and an estimate of whether species are self-compatible. These authors found that species whose populations contain low densities are more likely to be self-compatible (and in most cases capable of spontaneous selfing without any pollinators) than are species whose populations tend to contain dense clumps of plants. This result no doubt reflects the sensitivity of reproductive success to local density in self-incompatible annual plants.

Kunin and Shmida (1997) also found that rare species show more extreme trait values than widespread species. Within self-incompatible taxa, rare species have larger flowers than common species. In direct contrast, in self-compatible taxa, rare species have smaller flowers than common species. Likewise, in self-incompatible taxa, rare species have longer floral longevities than common species, probably due to both their larger flowers and infrequent or rare pollinator visitation. The annual life-history of the study species means that all species require some pollination in the year they flower for them to contribute to subsequent generations. Self-compatible taxa show no requirement to extend floral lifetime since most species were capable of seed set in the absence of pollinators.

2.4.3 A comparison of congeneric endemic and widespread species

In a comparative analysis of endemic and widespread species in 20 genera, Lavergne *et al.* (2004*b*) studied whether endemic taxa in the western Mediterranean differ from widespread congeners in terms of their habitat, community characteristics, traits linked to resource acquisition and allocation, reproduction and dispersal, amounts of herbivory, and maternal fertility (Box 2.3). Surprisingly, levels of herbivory and ecophysiological traits (specific leaf area, leaf dry matter content, leaf nitrogen concentration, and rates of photosynthesis) showed no overall difference between related endemic and widespread taxa. Endemic and widespread species did however differ for a number of ecological characteristics and biological traits, which emphasize three important features of the population ecology of endemic plants: their habitat in terms of abiotic and community characteristics, floral traits and the size and maternal fertility of individual plants. The discussion of these findings is based on Lavergne *et al.* (2004*b*) and illustrated in Box 2.3.

Distinct ecology

Endemic species have a distinct ecology compared to widespread congeners. The principal component of this difference concerns the occurrence of endemic species in rocky and steep slopes in low, open vegetation with low species richness. This indicates that endemic species grow at sites not only for historical reasons which have isolated their distributions and limited their dispersal, but also because of a 'fit' between their ecology and prevailing site conditions. This trend for endemic species to occur in distinct ecological conditions compared to their widespread congeners, in particular the association with open habitats rather than forest vegetation has been commented on elsewhere (Grove and Rackham 2001; Quézel and Médail 2003). Indeed, a survey of different flora illustrates the frequent occurrence of endemic species in rupicolous habitats, often in and around cliffs, for example, in the Balearic islands (Alomar *et al.* 1997), Greece, and the Balkans (Polunin 1980; Strid and Papanicolaou 1985).

A characteristic of open rocky habitats on steep slopes (sometimes on scree slopes, sometimes in and around cliff faces) which may be crucial for the persistence of endemic species is that such habitats are relatively stable, both in relation to vegetation succession and human activities. In such habitats, pioneer species which colonize rocky cliffs may facilitate the establishment of dwarf shrubs. For example, Escudero (1996) suggests that the establishment of nano-phanerophytes in cliffs in the Iberian peninsula (e.g. *Rhamnus alpinus* and *Lonicera pyrenaica*) depends on a process of facilitation whereby colonist chamaephytes (e.g. *Saxifraga moncayensis* and *Silene saxifraga*) cause an accumulation of soil in crevices and hollows. The nano-phanerophytes can grow to much greater size and their root development probably contributes to a shattering of the rock and erosion, allowing for re-colonization by the pioneer chamaephytes. The cliff faces which are home to many endemic species thus appear as a complex mosaic of discrete micro-communities which succeed one another in a colonization, succession, extinction cycle. However, severe environmental constraints on vegetation establishment may halt the successional development of a forest cover. Their inaccessibility and unsuitability for cultivation may have allowed such habitats to serve as a refuge for endemic taxa during periods of intense human-induced landscape modification. Polunin (1980) pointed out that due to the reduced impact of disturbance by humans and grazing animals, the vegetation of Mediterranean gorges and cliffs is often rich in endemic and rare species, which find a refuge there from the competition of aggressive species in the surrounding habitats. Zohary (1973: 315) also commented on how 'the lack of competitive vigour is probably one of the major causes for the stenochory of so many endemics which flourish when artificially grown in an environment free from competition'. It is thus probable that the persistence of endemics may have been favoured by their capacity to grow in rocky habitats where competitive interactions, which may have fundamental effects on diversity and the persistence of endemic plants (Box 2.4), may also be limited.

The question thus arises as to whether the ecological strategies of endemic species, in terms of functional trait variation, differ from those of widespread species. Does trait variation reflect reduced competitive ability, increased tolerance of stress, and/or reduced dispersal ability? Lavergne *et al.* (2004*b*) detected a clear trend for endemic species to be smaller than widespread congeners. However, congeners showed no differences in leaf traits or photosynthetic capacity and thus their capacity to acquire resources. There was thus no evidence for a stress-tolerant ecological strategy in endemic taxa, as suggested by a previous study based on an analysis of life-history strategies in endemic species (Médail and Verlaque 1997). Of course, other unstudied factors may be involved, for example, allocation to root growth, which could contribute to the greater affinity of Mediterranean endemics for rocky, steep sloping habitats in the Mediterranean and cause reduced allocation to above-ground stature.

Floral traits

Endemic species have floral traits which suggest that their populations are more inbred than those of widespread congeners, that is, smaller flowers with a lower pollen : ovule ratio and less herkogamy (i.e. stigmas closer to the anthers) than flowers of their widespread congeners. However, I am inclined to interpret this difference as being a consequence of their restricted range size and small population size rather than a cause of limited distribution. Many invasive and colonizing species are inbred, hence it is hard to see why inbreeding should limit the distribution of endemic species.

Lower maternal fertility

Endemic species have a significantly lower maternal fertility than their widespread congeners. In addition to causes linked to pollen and resource limitation, the male function of hermaphrodite flowers and inbreeding depression, perennial plants may reduce annual reproduction in order to optimize lifetime fitness. It is thus typical to observe low maternal fertility in herbaceous perennial plants in the Mediterranean (J. Herrera 1988, 1991*a*; C.M. Herrera 1993; Thompson and Dommée 1993; Baker *et al.* 2000*a*; Méndez and Traveset 2003). Such low maternal fertility in endemic taxa could contribute to reduced dispersal across the landscape but at the same time may be a strategy that has evolved in association with increased persistence of endemic taxa if there is a negative correlation between annual investment in seeds and vegetative persistence. In a few of the genera studied by Lavergne *et al.* (2004*b*) individuals of the endemic species live longer than those of the widespread species (even though they both show a perennial life history).

Habitat, fertility and population ecology

The differences in ecology and fertility suggest marked differences in the population ecology of endemic and widespread species. Populations of endemic species have traits and recent history that suggest high local persistence and low population turnover. A study of population persistence in the Hérault *département* of southern France from 1886 (based on a detailed local flora published at the end of the nineteenth century) to 2001 (from a database created by the Conservatoire Botanique National Méditerranéen de Porquerolles) by Lavergne *et al.* (2004*c*) illustrates that individual populations of endemic species at a given site have a greater temporal stability than those of widespread species. In contrast, the latter show stability in numbers of populations in the same area but in different sites. Widespread species may thus have populations which are more closely connected to one another by virtue of higher rates of colonization and extinction. Widespread species may thus move around the landscape to a greater extent and function more as typical meta-population systems have been suggested to do. Having said this, the marked spatial and temporal variation in the dynamics of six small populations of *Centaurea corymbosa* (i.e. all known populations of this species which is endemic to an area of $\sim$3 km^2 on a single massif in southern France) over a period of eight years detected by Fréville *et al.* (2004) suggest that such meta-population dynamics may also occur, but on a much more localized scale, in narrow endemic species (B. Colas *et al.*, unpublished manuscript). Differences in the dynamics of other narrow endemic species in relation to microhabitat variation (Albert *et al.* 2001; Chapter 4) further illustrate this mosaic pattern of population dynamics. I am tempted to suggest here that endemic and widespread species may in fact

Box 2.3 The comparison of the biology and ecology of endemic and widespread species in the flora of Languedoc-Roussillon region of southern France (redrawn from Lavergne *et al.* 2004*b*)

Based on one narrow endemic species (upper species in each genus) and one widespread species (lower species) in 20 genera in which the paired species have the same pollination and dispersal modes and growth form, and do not show consistent differences in ploidy level, Lavergne *et al.* (2004) performed multiple comparisons of ecological, ecophysiological, and reproductive traits. Nineteen genera contain herbaceous perennials, one genus involved annuals. The pairs of species were sampled in geographically close sites to minimize climatic variation (38 of the 40 species were sampled in an arc extending from the Rhone Valley to the Pyrenees). Of the 20 endemic species, 11 are protected by law.

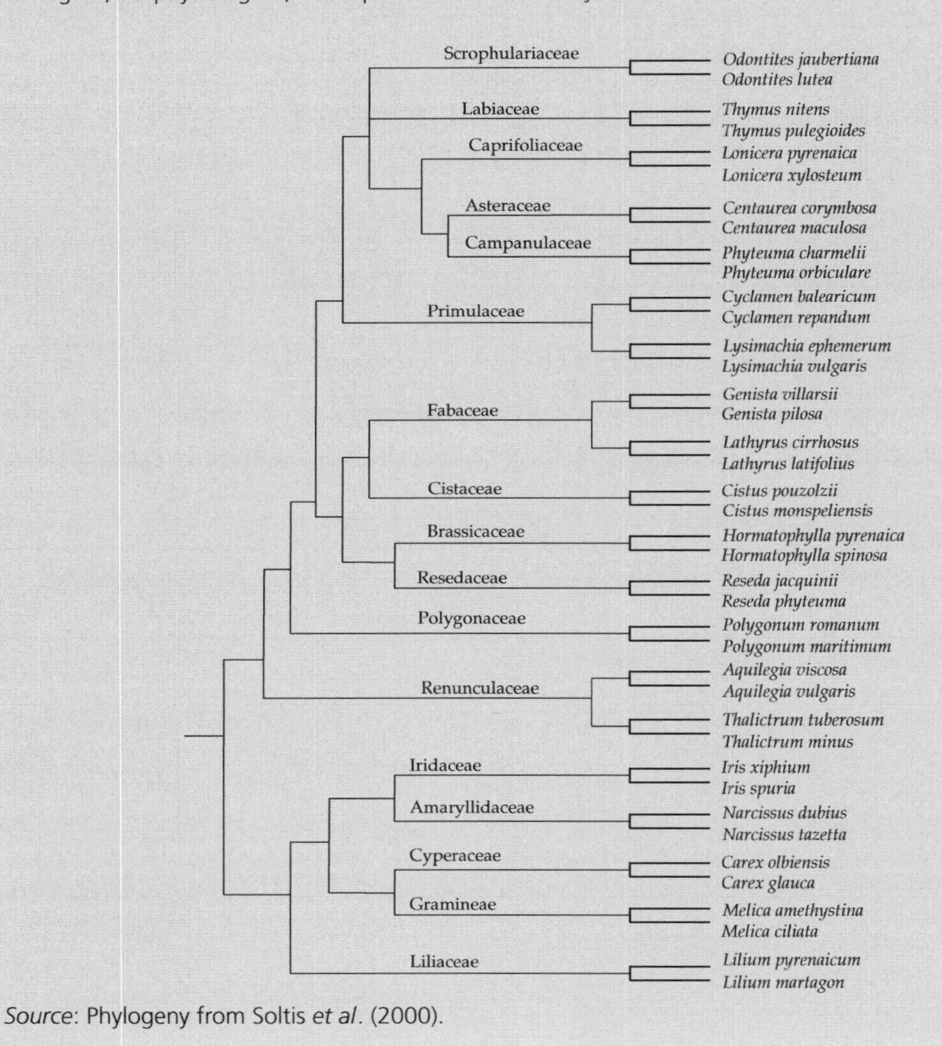

Source: Phylogeny from Soltis *et al*. (2000).

(b) Independent, pair-wise comparisons of ecological variables and biological traits were made for each genera. A point above the bisector means that the value for the widespread species is greater than that of the endemic species in a given genus.

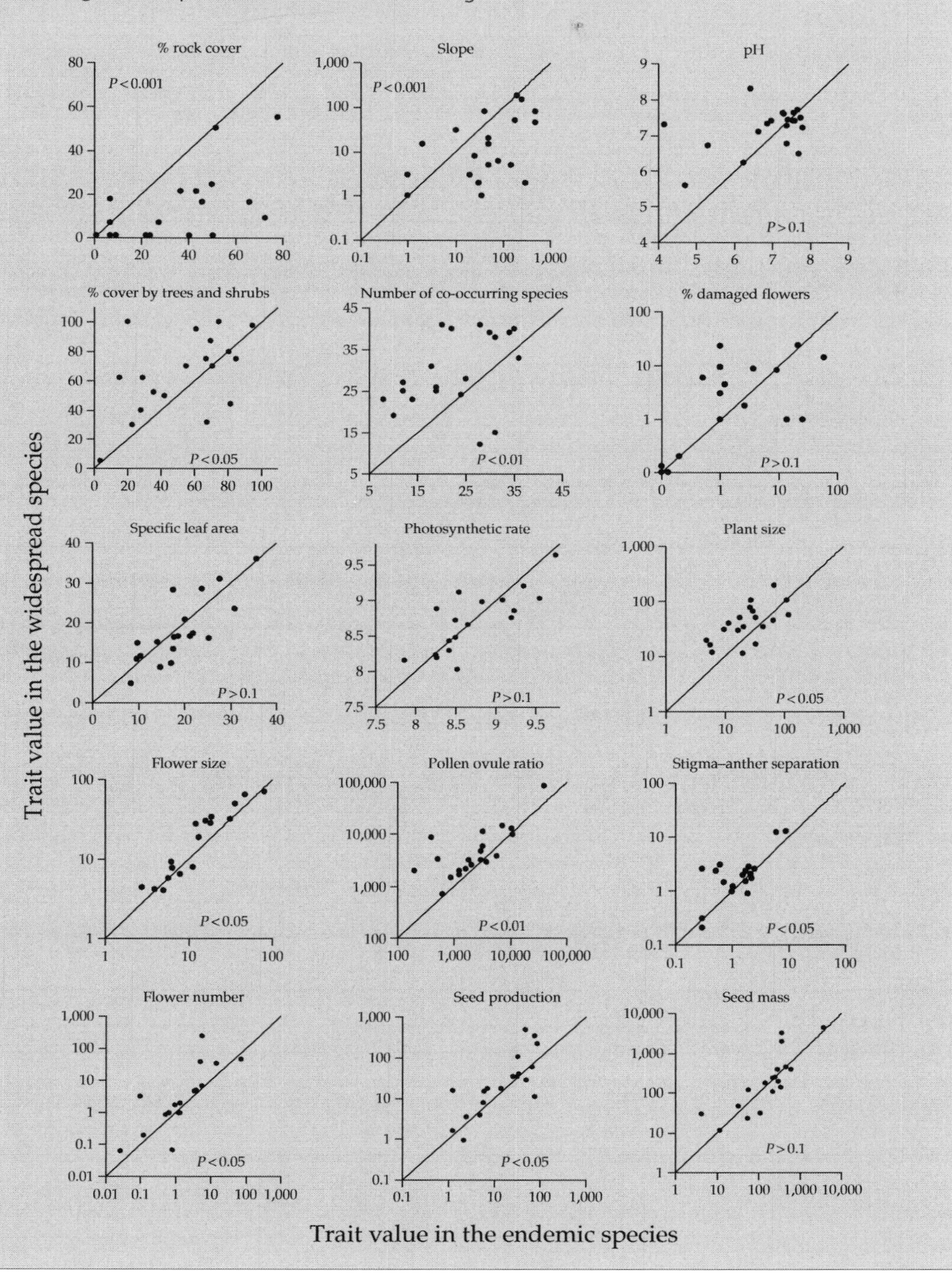

Box 2.4 Species diversity, endemism, and competitive interactions

Tilman and Pacala (1993) discussed how competitive interactions, in relation to the productivity and resource heterogeneity of the local environment, may contribute to patterns of variation in community diversity. These authors suggest that productivity gradients affect species diversity via spatial variation in nutrient supply and light availability along productivity gradients, that is, from habitats with low rates of supply of soil resources and high light penetration to those with high rates of supply of soil resources and low light penetration. In nutrient-poor habitats, all species are limited by soil resource supply, and the best competitor will dominate. At the other end of the gradient the same process, this time in relation to light availability, should cause domination by the best competitor for light. In this manner, low nutrient availability or the large size of organisms lead to marked declines in resource heterogeneity at the two extremes of the productivity gradient. In addition, in the central part of the gradient, different factors may promote coexistence. As a result, species diversity should show a unimodal relationship with productivity (Tilman and Pacala 1993). If the habitats with the necessary conditions for diversity are also the most common habitat type in a landscape or those where disturbance is also at intermediate levels (and thus acting to promote diversity), then species diversity may be further increased.

In Mediterranean grasslands (Puerto *et al.* 1990) and South African Fynbos (Bond 1983) high diversity occurs at intermediate levels of productivity. Similarly, comparison of localities in southern Spain by Ojeda *et al.* (1995) has shown that heathlands on nutrient poor soils and *Quercus canariensis* woodlands in more productive habitats have higher rates of endemism but lower species diversity than cork oak (*Quercus suber*) woodlands, where intermediate levels of fertility occur. Indeed, species richness (measured as the number of woody species) of shrubland and heathland communities in southern Spain is negatively correlated with soil fertility (Arroyo and Marañón 1990). In the latter case, heathland communities on nutrient-poor acid soils have higher diversity than shrublands on nutrient-rich basic soils. There may thus be complex relationships between species richness and rates of endemism.

show little or no difference in their capacity for long-distance colonization (which is a very rare event for both types of species). What may differ however is the spatial scale of meta-population function, which in turn affects geographic range. The probability of natural extinction of any given population may in fact be higher in widespread species than in endemic species.

2.4.4 Consistent patterns in different Mediterranean-climate regions of the world?

The five different regions of the world with a Mediterranean climate (i.e. California, Chile, South Africa, and Australia in addition to the Mediterranean Basin) have attracted a great deal of interest concerning the extent to which the overall similarities one can see in terms of community structure and form, represent functional convergence in similar ecological conditions (Chapter 4). The four other Mediterranean type ecosystems also contain floras which combine immense diversity (Dallman 1998) and high rates of local endemism, for example, almost 30% in the Cape Peninsula (Simmons and Cowling 1996) and the Agulhas Plain (Cowling and Holmes 1992) in South Africa and the Barrens of south-west Australia (Cowling *et al.* 1994). Hence the question: are similar ecological and biological features associated with endemism in the different Mediterranean-climate ecosystems?

In what has become a classic study of endemism and speciation, Stebbins and Major (1965) described how areas with the highest degrees of endemism in the Californian flora are those which have the greatest variety of habitats and ecological conditions and which have escaped glaciation during the Pleistocene. In general, low-lying mountains are centres of endemism. In striking similarity to the Mediterranean examples discussed above, the endemic species in California have very diverse origins. Many endemics have relict distributions while others are recently evolved and, as these authors suggest, may not yet have achieved their maximum possible area of distribution. The former tend to have persisted in regions where the climate has been fairly stable (some of the fairly mesic zones and in the highly arid interior regions) while the latter have probably evolved in regions where spatial heterogeneity has promoted active differentiation.

In the Cape Floristic Region, well known for its diversity of endemic plants in a small surface area, studies of the taxonomic, ecological, and biological characteristics have provided critical and novel insights into the biology and ecology of endemism in this region. In the Agulhas Plain, a rolling coastal plain which occupies 1,600 km^2 at the southern tip of South Africa, rates of endemism are 'extraordinarily high for a lowland continental region' (Cowling and Holmes 1992: 376). The plant communities of this area are the characteristic (and endemic) Fynbos vegetation in which ~30% of the flora is endemic (Cowling and Richardson 1995). This area has no long history of physical isolation with surrounding land and has been submerged more than once in recent geological history, that is, in the Miocene (~15 Ma) and Pliocene (~4 Ma). Diversification typical of physically isolated islands and mountain areas has thus occurred recently and rapidly in a continental setting. This diversity appears to be related to ecological specialization by a small number of taxonomic groups with a similar biological profile.

The geology of the region with a Mediterranean climate in South Africa is complex (Cowling and Holmes 1992), with a diverse number of substrates present in a small area: acidic and rather infertile soils on sandstones and quartzites which make up most of the low mountains of the area, more fertile acidic soils on the shales which separate the main mountains, shallow infertile siliceous soils in pockets across the landscape, well-drained calcareous sands on limestone bedrock in the coastal zone, calcareous aeolian quartz and dunes along the coast and valley, flood plains with alluvial or colluvial topsoils on a clay-base soil. Strong rainfall gradients also exist in this region. Hence, despite a relative lack of physical separation of different land masses, ecological differentiation can occur in what is an intricate mosaic of spatially heterogeneous edaphic environments. Despite the short distances between habitats on these different substrates, Cowling and Holmes (1992), found that more than two-thirds (69%) of regional endemics and 85% of local endemics in the flora in the Agulhas Plain only occur on a single type of substrate. Of the total endemic flora, more than one-third (37%) only occur on limestone. Edaphic specialization is thus rife in endemic plants of this region, suggesting that adaptive differentiation has played an important role in the high rates of endemism in this area.

Cowling and Holmes (1992) also show that endemism is not common to all the plant groups present and is particularly present in species with a certain biological profile. Some families have statistically more regional endemic species than expected by chance, witness the amazing diversity of certain genera in the Ericaceae (45% of speceis in the genus *Erica* are regional endemics), Proteaceae (47% of the genus *Leucadendron*), Polygalaceae (46% of species in the genus *Muraltia*), and Rutaceae (38% of the genus *Agathosma*). Others are under-represented in endemic taxa (e.g. Orchidaceae, Scrophulariaceae, and Poaceae). Second, endemic species fit a biological profile of non-sprouting dwarf shrubs with short-distance (often ant or passive) seed dispersal with (in some families) a symbiotic relationship with soil microorganisms. Cowling and Holmes (1992) argue that lineages of such species may be particularly vulnerable to population reduction and local extinction and at the same time more prone to rapid edaphic specialization. A similar biological profile has been identified in mountain Fynbos (McDonald and Cowling 1995). However, multivariate logistic regression analysis of the relationship between such traits and endemism by McDonald *et al.* (1995) showed that the primary factor determining high rates of endemism is short-distance dispersal, and that the interaction between growth form and regeneration strategy plays only a secondary role.

In a comparison of patterns of endemism in the Cape Floristic Region with matched sites (in terms of historical and contemporary climate, landforms, and soils) in the Barrens in south-west Australia, where species richness and endemism are very high (Crisp *et al.* 2001), Cowling *et al.* (1994) found similar patterns of endemism in the two regions. These patterns are a high degree of edaphic specialization for nutrient-poor acid soils, a broadly similar biological profile of endemic taxa (non-sprouting dwarf shrubs with short-distance dispersal), although serotinous species are also important elements of endemism in south-west Australia (Beard *et al.* 2000), and an over-representation of endemic taxa in certain taxonomic groups (Proteaceae, Epacridaceae, and Myrtaceae). The association between local endemism and the non-spouting growth form has also been detected in genera such as *Arctostaphylos* and *Ceanothus* in the California chaparral (Wells 1969).

The consistency of the association between endemism and nutrient-poor acid soils and shrubby growth form has support from studies in the Mediterranean region. In the heathlands of the Sierras de Algeciras in the Aljibe mountains near the Straits of Gibraltar in southern Spain, endemism is associated with a shrubby growth form and is negatively correlated with substrate fertility (Ojeda *et al.* 2001). This region comprises large areas of acidic soils (pH 4–5) on siliceous sandstone (dating to the Oligocene–Miocene) with low nutrient content and high levels of assimilable aluminium and other heavy metals harmful to normal plant development (Arroyo 1997). The Aljibe mountains probably represented an important glacial refuge for many species during the Pleistocene and currently have a fairly original flora (Arroyo and Marañón 1990; Ojeda *et al.* 2000*a*). In this region, heathland communities are more diverse than those in the non-Mediterranean Europe (Ojeda *et al.* 1998) and show higher rates of endemism than nearby communities in cork oak woodland (Ojeda *et al.* 1995 Box 2.4).

Depending on their substrate preference, endemic taxa in southern Spain show contrasting distribution patterns. Whereas on limestone, endemic species occur only in the Betic cordillera of south-east Spain, the endemic taxa of heathlands on acid soils also occur in North Africa and the western part of the Iberian peninsula where similar habitats on Oligocene siliceous sandstone also exist. In addition, acid soils, on which many narrow endemic species occur in the southern Spanish heathlands, are not common around the Mediterranean. Hence many widespread species which occur on more common and widespread calcareous soils may be absent from the acidic and sandy soils which harbour heathland species. So both heathland communities on acid soils and limestone shrublands tend to occur as edaphic or geographically isolated 'islands' in the landscape, hence the importance of narrow endemism in such communities. Once again specialization on nutrient-poor soils and the spatial occurrence of suitable habitat are critical for the geographic range of endemic species.

2.5 Conclusions

Endemic taxa represent 'something old and something new'. Ancient endemic species provide clues to historical connections while neo-endemic species provide models for the study of more recent and often rapid differentiation in relation to climate and ecology. What is clear from this chapter is that we have gained much ground in our appreciation of the two sets of conditions which favour high rates of endemism: the factors that favour persistence of endemic plants over long periods and the conditions which promote divergence. Endemic species are not all relatively recent in origin and do not all occur at the tips of phylogenetic trees. Nor do they have evolutionary lifetimes shorter than more widespread species. Many Mediterranean endemics are either very old (i.e. palaeo-endemic taxa conserved from the Palaeogene flora of the initial microplates) or, as is the case for patro-endemics, the progenitors of more widespread species. Endemic distribution patterns cover the whole range of spatial scale from taxa endemic to single mountains or islands, groups of disjunct islands or continental areas, to those endemic to whole regions such as the western or eastern Mediterranean. These diverse distribution patterns provide insights into the roles of history and ecology in the determination of species' ranges. Indeed, when associated with the tools of population genetics and systematic and phylogenetic analysis, these patterns can be used to analyse the evolutionary process (Chapter 3).

The occurrence of high rates of endemism in areas with strong geographic barriers to dispersal (islands and mountains) point to the important role of history in shaping species distributions around the Mediterranean. Indeed, for some authors, 'the multitude of discontinuities created by geological processes is perhaps the ultimate cause of local rarity and narrow endemism' (Kruckeberg and Rabinowitz (1985: 465)). In the Mediterranean, to quote Zohary (1973: 320), 'the historical reasons for the outburst of endemism are certainly the most weighty ones'. This author outlined three major historical events which may explain the high rates of local endemism in the eastern Mediterranean: (a) the recession of the Tethys and the expansion of the Irano-Turanian flora across vast areas of new ground; (b) the upheaval of the Tertiary fold mountains which created insular habitats of diverse ecology and produced sharp climatic discontinuities; (c) the onset of summer drought and repeated cold cycles associated with glaciation, which may have created strong selection pressures for adaptation and caused many distribution patterns to contract into isolated refugia.

There are several pieces of evidence which suggest that the organization and structure of plant communities on small islands, and thus potential endemism is linked to random historical processes associated with isolation. Patterns of distribution in small island systems indicate that they are the remnants of a larger flora in which random variation has marked community composition. The process is akin to that of random differentiation in gene frequencies on small island systems which, as I illustrate in the next chapter, may produce striking patterns of differentiation. Indeed, community and population patterns may often show parallel trends, which may have similarity in their underlying causes.

A clear conclusion concerns the association between endemism and ecological conditions. Endemic plants tend to occur on steep rocky slopes and cliffs, and soils which are nutrient poor or have intermediate nutrient levels. Home to many endemic plants, such habitats illustrate both the random elements of variation in species composition due to dispersal limitation and the importance of a suitable habitat for the presence of endemic species. Cliffs and steep rocky areas are stable habitats; their inaccessibility puts a halt to human cultivation and even goat and sheep grazing and, because of constraints on plant development, succession towards a climax forest is prevented. This stability has probably favoured the persistence of endemic plants. The importance of local persistence may have moulded trait evolution in endemic plants, which may have a greater longevity associated with decreased reproductive effort in a given year and whose lower abundance across the landscape reduces the probability of dispersal and colonization of new sites. What remains to be more fully explored is whether Mediterranean endemic taxa have a narrower range of ecological tolerance than their widespread congeners. Although there is some evidence that narrow endemics have a more specialized ecology (Debussche and Thompson 2003), confirmation of the generality of this issue will require detailed and comparative field studies across the range of the distribution of endemic and widespread congeners, accompanied by manipulative transplant experiments.

Endemic species in the Mediterranean flora provide a fascinating material for the study of the factors regulating plant distribution and long-term persistence. I have discussed how endemism and diversity in the Mediterranean landscape depend on the balance between regional processes of immigration and gene flow (and thus the spatial configuration of habitats) and local ecological interactions within habitat patches. Species originate and evolve within the confines of their habitat and, to colonize new habitats, evolutionary change may be required. The evolutionary processes involved in the diversification of endemic species is the subject of the following chapter.

CHAPTER 3

The evolution of endemism: from population differentiation to species divergence

> . . . the processes involved in 'descent with modification', to use Darwin's classic phrase, can be shown clearly to apply to differentiation within species, as well as to the further divergence of species . . . once they have become separated from each other.
>
> G.L. Stebbins (1950: 190)

3.1 Endemism and evolution: the processes and scale of differentiation

To determine how species evolve requires information concerning both the relatedness among different taxa and the microevolutionary forces acting on local populations of individual taxa. The pertinence of such microevolutionary forces for our understanding of plant species divergence was first brought to the fore by one of the great plant evolutionary biologists of the twentieth century, George Ledyard Stebbins. In three books, Stebbins (1950, 1971, 1974) showed all too clearly how the fundamental evolutionary processes of selection and drift, which act to promote genetic differentiation among populations, and migration of pollen and seeds, which counteract differentiation and act to homogenize gene frequencies across the landscape, have shaped not only the contours of differentiation within species but also divergence among species. Onto this foundation he incorporated the natural setting of spatial habitat heterogeneity, the potential for hybridization and reproductive isolation, chromosome evolution, and stochastic factors associated with dispersal requirements. In this chapter, I will make the jump from the ecology of endemism discussed in the previous chapter to explore the evolutionary processes which have caused species divergence and narrow endemism in the Mediterranean flora.

Species are the basic units of plant taxonomy and classification, they allow us to characterize and study the dynamics of biodiversity and to go out in the field and discover new taxa or a species in a new area. It is difficult however to provide a single definition of the term species that would be acceptable to all botanists. I thus do not attempt to provide a species definition in this book. For the range of definitions that can be applied to plant species I refer the reader to Rieseberg and Brouillet (1994) and Levin (2000). My aim in this chapter is to explore how the evolutionary process of divergence has been modulated by the geological and climatic history of the Mediterranean region. I attach particular importance to a discussion of the scale on which species have diverged and endemic patterns of distribution developed, and the role of ecological differentiation in the process of speciation. My premise here is that to understand plant evolution in the Mediterranean requires a knowledge of the role of microevolutionary processes in relation to the historical framework developed in the two previous chapters. Three questions form the basis for this exploration.

- Are endemic species genetically depauperate and do they differ from widespread congeners in the spatial structure of genetic diversity across the landscape?

- What are the evolutionary processes and spatial scales of differentiation and divergence?
- How different are closely related species with disjunct distribution patterns?

3.2 Population variation in endemic plants

Endemic species are rare in the sense that they have a restricted geographic distribution. Their restricted geographic range has long attracted attention from ecologists seeking to determine whether endemism is associated with ecological specialization (Chapter 2). In addition, from a genetic or evolutionary perspective, a long-standing issue concerning endemic plants has been whether they show reduced genetic variation relative to widespread species. Two ideas underlie the issue of reduced genetic variability in endemic taxa. First, if endemic taxa are specialized to local ecological conditions then natural selection will whittle down variation as non-adapted alleles are eliminated. Second, small population sizes associated with founder events and/or genetic bottlenecks may induce rapid rises in the amounts of inbreeding. As a result, genetic variability will decline rapidly within populations which become more and more different from one another. To study the evolution of endemism thus requires an appreciation of levels of population diversity and differentiation in endemic taxa, relative to that in common and widespread taxa.

Stebbins (1942) first addressed the issue of whether rare species show reduced genetic variation compared to more widespread congeners. Indeed he was one of the first plant evolutionary biologists to ask the question: 'why are some plant species widespread and common, while others are rare and local' (p. 241). This question remains all too pertinent today in our search to better understand plant evolution, be it in the Mediterranean or elsewhere. The examples he cites lead to the conclusion that rarity is associated with reduced genetic variation. However, in a somewhat revised and more balanced evaluation of this question some years later, Stebbins (1980) concluded that low levels of genetic variation may be a less important cause of endemism and rarity than ecological preferences. In the last 30 years it has become possible to precisely quantify genetic variation using a number of techniques based on protein analysis with allozyme electrophoresis or more direct quantification of variation in DNA profiles using the tools of molecular biology (Table 3.1). Such methods make it possible to quantify and compare genetic variability based on allele number per locus, the percentage of loci which show a genetic polymorphism (i.e. more than one allele) and heterozygosity (presence of more than one allele at individual loci in individual plants).

Comparisons of 34 pairs of rare and widespread congeners by Gitzendanner and Soltis (2000) have shown that although endemic species have globally significantly lower levels of genetic diversity than their more widespread congeners, in a number of genera the differences are slight, and occasionally reversed. The differences are primarily due to high levels of diversity in a small number of widespread species. In fact in 24–29% of the genera, rare species were at least as variable if not more variable than their widespread congeners and diversity levels in rare species covered a similar range to those of widespread species. Despite the overall trend, it appears that rare species are not a homogenous group with low levels of genetic diversity. In addition, Gitzendanner and Soltis (2000) detected a significant correlation in levels of diversity across genera, a finding which underscores the need for multiple comparisons among closely related rare and widespread species.

In the Mediterranean flora although there are some clear examples of reduced genetic variability in endemic species relative to widespread congeners, there is no clear-cut and consistent trend. In four Mediterranean oaks, genetic diversity in populations of the narrow endemic oak *Quercus alnifolia* (endemic to ultrabasic soils on Cyprus) is less than that in more widespread species such as *Quercus coccifera* (Toumi and Lumaret 2001). Likewise, *Cyclamen balearicum*, which is endemic to the Balearic islands and southern France, shows markedly less within-population variation than its widespread congener *Cyclamen repandum*

Table 3.1 A glossary of terms used in this chapter concerning the analysis of genetic variation and population differentiation

Term	Brief description
F_{ST}, G_{ST}, or θ	Estimation of the proportion of genetic variability in a total sample that is due to genetic differentiation among populations (based on allele frequency variation).
Allozyme electrophoresis	Based on differences in the charge of proteins associated with different enzymes and their relative migration in an electric field it is possible to extract proteins from plant tissues and stain gels with patterns that are interpretable in terms of Mendelian genotypes at particular loci. Allele and heterozygote frequencies can thus be estimated.
RFLP	Restriction fragment-length polymorphism (RFLP) is identified by cutting DNA with enzymes which produce different fragments that can be sorted and visualized on a gel according to their molecular weight. RFLP is a technique in which population differentiation can be assessed based on the length of the fragments produced (which will differ when the DNA is digested with a restriction enzyme). The similarity of the patterns generated can be used to differentiate species and populations.
RAPD	Randomly amplified polymorphic DNA (RAPD) is based on polymerase chain reaction (PCR) which is used to amplify a sequence of DNA. This amplification can produce evidence of polymorphism due to presence of different bands after gel staining. However the target DNA sequence is unknown and RAPDs are dominant in the sense that the presence of an RAPD band does not allow distinction between heterozygous and homozygous states.
AFLP	Amplified fragment-length polymorphism (AFLP) is a PCR-based fingerprinting technology. In its most basic form, AFLP involves the restriction of genomic DNA, followed by ligation of adaptors complimentary to the restriction sites and selective PCR amplification of a subset of the adapted restriction fragments. These fragments are visualized on denaturing polyacrylamide gels. The availability of many different restriction enzymes and corresponding primer combinations provides a great deal of flexibility. A main advantage of this method is the possibility of polymorphism detection at the total-genome level.
cpDNA haplotype	Chloroplast DNA (cpDNA) is transmitted primarily in maternal lineages (gymnosperms are a well-known exception). Genotypes for cpDNA represent non-recombining characters transmitted by female parents through their seeds. They are referred to as haplotypes.

(Affre and Thompson 1997*a*; Affre *et al.* 1997). This trend is not however consistent, *Cyclamen creticum* endemic to the island of Crete does not show less genetic variation than widespread *C. repandum* (Affre and Thompson 1997*b*). In this group it is differences in mating system rather than geographic distribution that occasion the differences in genetic variability (see below). Other examples of high levels of genetic variability in narrow endemic species can be seen in the genus *Antirrhinum* in the Iberian peninsula (Mateu-Andrés 1999). The conclusion is clear, endemic species and their populations do not in general harbour less variation than their widespread congeners.

What about levels of population differentiation? This question is of particular pertinence here because differentiation is the template on which endemic species divergence occurs. Indeed, where widespread species show marked differentiation in geographically isolated and marginal populations, endemism may be a predictable evolutionary outcome (Fréville *et al.* 1998). However, the comparative study by Gitzendanner and Soltis (2000) and examination of population differentiation in Mediterranean plants (Thompson 1999; Table 3.2) show that there is no general difference between rare and widespread species in terms of the spatial organization of genetic variation. This is not surprising since similar processes act on the populations of both endemic and widespread species. All plants are sedentary, and it is only by gene movement that differentiation can be countered. So one expects to observe some level of genetic differentiation in a species, even in rare or endemic species. Rare and endemic species differ (as much as do widespread species) in terms of their life histories, breeding systems, and modes of dispersal, hence they will also show variable amounts of differentiation. Historical

Table 3.2 Examples of genetic differentiation among populations in widespread and endemic Mediterranean plants. Based in part on Thompson (1999).

Species	Location and scale of study	Estimate of differentiation	Reference
Widespread species			
Bromus intermedius	Algeria	$G_{ST} = 0.24$	Ainouche *et al.* (1995)
Bromus squarrosus		$G_{ST} = 0.23$	
Bromus lanceolatus		$G_{ST} = 0.25$	
Bromus hordeaceus		$G_{ST} = 0.06$	
Centaurea maculosa	Southern France	$F_{ST} = 0.26$***	Fréville *et al.* (1998)
Cyclamen hederifolium	Corsica	$F_{ST} = 0.13$***	Affre and Thompson (1997*a*)
Cyclamen repandum	Corsica	$F_{ST} = 0.42$***	Affre and Thompson (1997*a*)
Ecballium elaterium	Across Spain	$F_{ST} = 0.23$ (subsp. *elaterium*)	Costich and Meagher (1992)
		$F_{ST} = 0.96$ (subsp. *dioica*)	
Fagus sylvatica	Balkans	$F_{ST} = 0.01$	Gömöry *et al.* (1999)
Medicago sativa	Spain	$F_{ST} = 0.05$***	Jenczewski *et al.* (1998)
Medicago truncatula	Corsica and southern France,	$F_{ST} = 0.51$***	Bonnin *et al.* (1996)
	Subpopulations ~50 m apart	$F_{ST} = 0.32$***	
	Subpopulations ~10 m apart	$F_{ST} = 0.15$***	
Quercus suber	Across the Mediterranean Basin	$G_{ST} = 0.08$	Toumi and Lumaret (2001)
Quercus ilex		$G_{ST} = 0.14$	
Quercus coccifera		$G_{ST} = 0.19$	
Quercus ilex	Western Mediterranean	$G_{ST} = 0.10$	Michaud *et al.* (1995)
Senecio gallicus	Iberian peninsula and France	$\theta = 0.56$* (cpDNA)	Comes and Abbott (1998)
		$\theta = 0.15$* (allozymes)	
Senecio glaucus	Eastern Mediterranean	$\theta = 0.43$*(cpDNA)	Comes and Abbott (1999*b*)
		$\theta = 0.12$* (allozymes)	
Senecio rupestris	Central Mediterranean	$\theta = 0.37$*	Abbott *et al.* (2002)
Senecio vernalis	Eastern Mediterranean	$\theta = 0.05$ (cpDNA)	Comes and Abbott (1999*b*)
		$\theta = 0.04$* (allozymes)	
Thymus vulgaris	Southern France	$F_{ST} = 0.04$*** (allozymes)	Tarayre and Thompson (1997)
		$F_{ST} = 0.24$*** (cpDNA)	Tarayre *et al.* (1997)
Endemic species			
Antirrhinum mollissimum	Southern Spain	$G_{ST} = 0.11$	Mateu-Andrés (1999)
Antirrhinum microphyllum	Southern Spain	$G_{ST} = 0.04$	Mateu-Andrés (1999)
Antirrhinum valentinum	Southern Spain	$G_{ST} = 0.48$	Mateu-Andrés and Segarra-Moragues (2000)
Argania spinosa	Throughout Morocco	$G_{ST} = 0.60$ (cpDNA)	El Mousadik and Petit (1996*a,b*)
		$G_{ST} = 0.25$ (allozymes)	
Brassica insularis	Corsica	$G_{ST} = 0.11$***	Hurtrez-Boussès (1996)
Centaurea corymbosa	Subpopulations 250 m–2.5 km apart	$F_{ST} = 0.34$***	Colas *et al.* (1997)
Cytisus aeolicus	Eolian Islands	$G_{ST} = 0.01$	Conte *et al.* (1998)
Cyclamen balearicum	Among regions in France	$F_{ST} = 0.42$***	Affre *et al.* (1997)
	Among the Balearic islands	$F_{ST} = 0.11$***	
	Among populations (Cévennes)	$F_{ST} = 0.26$***	
	Among populations (Majorca)	$F_{ST} = 0.16$***	
Cyclamen creticum	Crete	$F_{ST} = 0.17$***	Affre and Thompson (1997*b*)

Table 3.2 *(Continued)*

Species	Location and scale of study	Estimate of differentiation	Reference
Fagus moesiaca	Balkans (allozymes)	$F_{ST} = 0.02$	Gömöry *et al.* (1999)
Silene diclinis	Spain	$G_{ST} = 0.06$	Prentice and Anderson (1997)
Narcissus longispathus	Spain	$\theta = 0.15$*	Barrett *et al.* (2004*a*)
Quercus alnifolia	Cyprus	$G_{ST} = 0.06$	Toumi and Lumaret (1991)
Thymus loscosii	North-east Spain	$G_{ST} = 0.03$	López-Pujol *et al.* (2004)
Triticum dicoccoides	100 m transect	$G_{ST} = 0.26$	Nevo *et al.* (1988*a*)
	subpopulations	$G_{ST} = 0.41$	Golenberg (1987)

* $P < 0.05$, *** $P < 0.001$. For some studies I employed reported CI values to test for significant differentiation at $P < 0.05$.

effects will also influence genetic variation in a similar fashion for both endemic and widespread species. Finally, endemic species are not necessarily rare in terms of their population sizes or even in terms of the number of populations in a given area, they can show considerable variation in the size, number, and spatial organization of their populations. For all these reasons, levels of differentiation will vary among endemic species, which as a group, are unlikely to differ from widespread species.

3.3 Climatic rhythms and differentiation

Taxonomic diversity in the different Mediterranean floras of the world has frequently been interpreted as resulting from major climatic changes. For example, a large proportion of the taxonomic diversity of the Cape Flora in South Africa may have been 'generated during massive bursts of speciation associated with the range extensions of the sclerophylous vegetation that survived the drastic climatic changes at the Tertiary-Pleistocene boundary' (Linder *et al.* 1992: 132). Climatic oscillations associated with glaciations may have further promoted speciation in this region during the Pleistocene (Midgley *et al.* 2001). Molecular analysis of gene sequence variation in the genus *Phylica* (Rhamnaceae) from the Cape Floristic Region illustrates how extensive diversification beginning at ~7–8 Ma coincided with marked aridification of the climate in association with changes in ocean currents during that period (Richardson *et al.* 2001). Rapid diversification in *Oenothera* (Klein 1970) in association with chromosome rearrangement and the colonization of contrasting and often extreme ecological conditions since the start of the Pleistocene in the Mediterranean-climatic region and deserts of California also illustrates this pattern.

Within the constraints of geographical barriers to dispersal, climatic change at the end of the Tertiary and during the Quaternary had important effects on species' distributions in the Mediterranean region (Chapter 1). Climatic variation caused local extinctions and genetic isolation among populations and thus the potential for species divergence. Indeed, the impact of Pleistocene glaciation on population genetic structure has been documented in several European forest tree species (e.g. R.J. Petit *et al.* 1997). In this section, I present evidence for the idea that Quaternary climatic rhythms have contributed to species divergence and population differentiation in different glacial refugia, and thus some of the disjunct distribution patterns that characterize groups of related plants in the Mediterranean.

3.3.1 Species divergence in relation to climate change

The isolation of populations in association with climate change may have contributed to two common distribution patterns in the Mediterranean flora (see Chapter 2): (a) the restricted distribution of (schizo-) endemic species that have diverged following fragmentation and isolation of parts of the distribution of more widespread ancestral species and (b) east–west vicariance.

A genus which has provided key information on differentiation and divergence in relation to climate change in the Mediterranean is the genus *Senecio*. In a combined study of genetic variation for cytoplasmic and nuclear genes in 18 species, Comes and Abbott (1999*a*, 2001) illustrate a number of interesting features associated with the diversification of Mediterranean taxa. They report a distinct lack of differentiation among taxa in terms of their cpDNA haplotype profiles and nuclear gene sequence variation. Several species in the section *Senecio* share the same range of cpDNA haplotypes and show a phylogenetic tree (based on nuclear gene sequence variation) with very short terminal branches and a lack of resolution among species (Fig. 3.1). The lack of distinct differentiation among

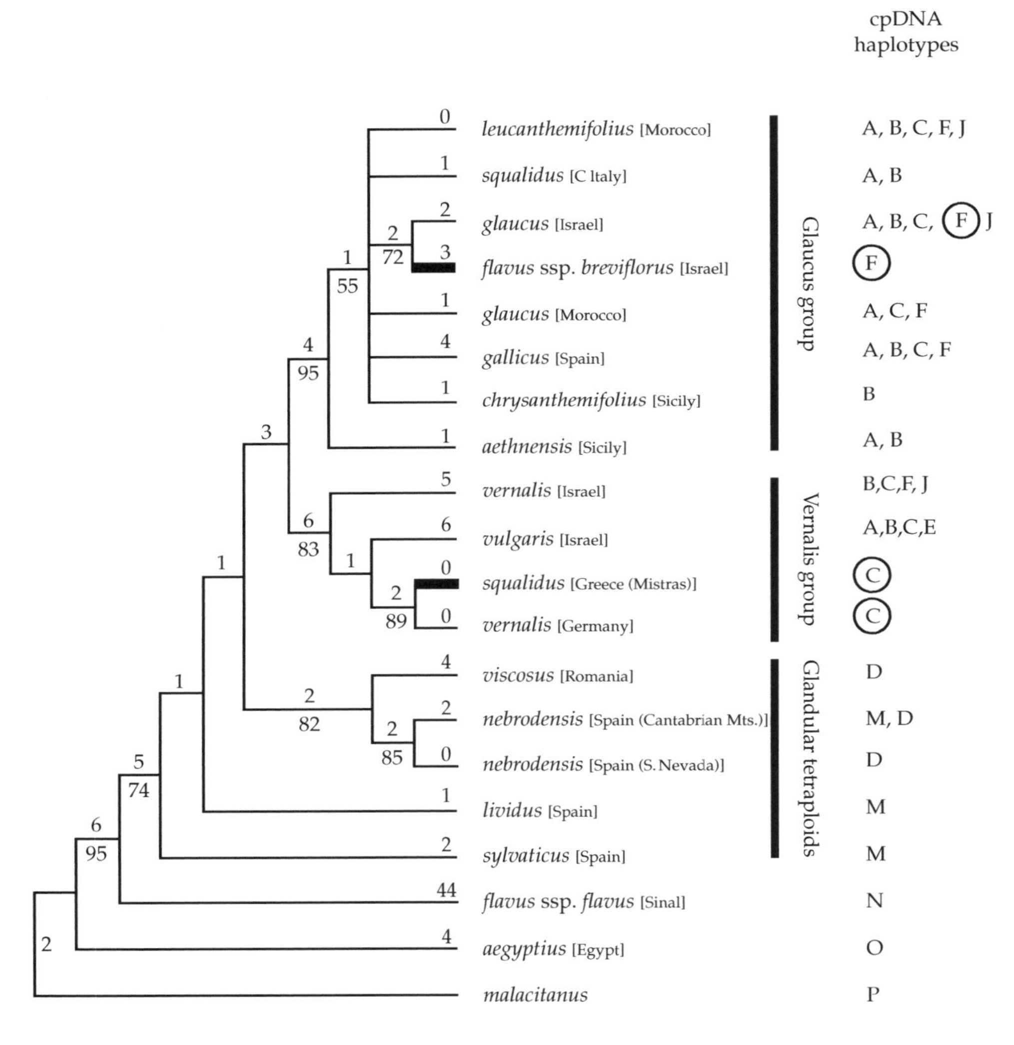

Figure 3.1 Molecular phylogeny of 19 accessions (14 species) of Mediterranean *Senecio* section *Senecio* based on ITS sequence data. CpDNA haplotypes detected in each species/accession are provided for comparison of nuclear and cytoplasmic differentiation (reproduced with permission from Comes and Abbott 1999*a*). Numbers of nucleotide substitutions are shown above the branches, and bootstrap percentages (100 replicates) are shown below.

taxa is particularly apparent for the widespread diploid species which inhabit the western and central part of the Mediterranean region and whose cpDNA diversity includes the range of variation in related endemic diploid taxa (Comes and Abbott 1999*a*). For Comes and Abbott (2001: 1953) this lack of phylogenetic resolution 'reflects a real historical phenomenon of a near simultaneous and relatively recent diversification'. The timescale for differentiation in the Mediterranean *Senecio* species, which probably began at ~1 Ma, is indicative of diversification as a result of rapid and repeated climate change.

Molecular phylogenetic analysis of the silver saxifrages (*Saxifraga* Section *Lingulatae*) by Conti *et al.* (1999) illustrates the evolution of schizo-endemic patterns of distribution in relation to climate change. In this group, narrow endemic species with disjunct distributions, for example, *Saxifraga cochlearis* which is endemic to limestone rocks in a small area of the Maritime Alps and on the Ligurian coast and *Saxifraga crustata* which is endemic to limestone and dolomite in the south-east Alps and the Balkans, appear to have evolved from populations of widespread *Saxifraga paniculata* during periods of isolation in distinct glacial refuge populations. A similar process probably occurred in mountain populations of *Abies* and other tree genera in the Mediterranean region (Quézel and Médail 2003).

Mediterranean oaks provide a well-known example of east–west vicariance (Chapter 2). In the western Mediterranean, two closely related evergreen taxa have been described: *Quercus ilex* with elongated leaves bearing 8–9 nerves and *Q. rotundifolia* with rounder leaves with 6–8 nerves. *Q. ilex* occurs in temperate, subhumid and humid bioclimates whereas *Q. rotundifolia* occurs more often in drier habitats. In southern France intermediate habitats contain both types and a range of variation in morphology. The causal relationship and adaptive significance of this variation remains unknown, and it is probable that the two taxa represent the extremes of morphological variation in a single variable complex (Barbero *et al.* 1992). It is thought that the divergence of *Q. rotundifolia* and *Q. ilex* arose during periods of refuge when the most northern shores of the Mediterranean became too cold under the influence of glaciation for the persistence of the former taxon.

3.3.2 Population differentiation in Mediterranean glacial refugia

The analysis of population differentiation within species provides further evidence for the role of climate change in species divergence. Isolation in glacial refugia will have created the context for genetic differentiation to occur, and in some cases may have pushed populations into new evolutionary trajectories. The degree of such geographic isolation probably varied greatly for different species. Many species probably had highly contracted distributions as they retreated into one or a small number of geographically distinct refugia in southern Europe (Iberian peninsula, Italy, Balkans). Patterns of genetic differentiation in several temperate forest tree species across Mediterranean and temperate regions of Europe (e.g. R.J. Petit *et al.* 1997; Gömöry *et al.* 1999) confirm historical palynological data (Huntley and Birks 1983) which suggest the persistence of tree species in geographically separate Mediterranean refugia located in the southern Balkans, Italy, and the Iberian peninsula. Strong patterns of east–west genetic differentiation for molecular markers detected in two emblematic Mediterranean trees, the cork oak *Quercus suber* (Toumi and Lumaret 1998) and wild olives *Olea europaea* subsp. *europaea* var. *sylvestris* (Besnard *et al.* 2002; see Chapter 6) are consistent with these propositions. The molecular divergence of Turkish and Iberian populations of *Rhododendron ponticum* (Milne and Abbott 2000) is also thought to have occurred during episodes of Pleistocene glaciation since there is good fossil evidence that this species was present in the Alps in the Mindel-Riss interglacial (~0.25 Ma) (Jessen *et al.* 1959). Other species may have persisted in smaller fragments of suitable vegetation in sheltered sites dotted across the landscape on the northern shores of the Mediterranean. Finally, for some species, changes in distribution may have been less dramatic, species distributions becoming fragmented as they persisted in small 'peri-glacial' refugia.

During the Pleistocene, many species will have experienced several periods of expansion and retreat in response to climatic fluctuations. It is thus possible that not all refugia remained independent over the course of the different glacial maxima, since some of the smaller refugia may have been part of colonization routes from major refugia. This scenario has been proposed for *Fagus sylvatica* and related beech tree species whose re-colonization of Europe may have occurred from either non-differentiated or perhaps only one refugia (Demesure *et al.* 1996; Gömöry *et al.* 1999). Beech populations in Calabria are genetically more similar to Balkan beech than to beech distributed across southern and temperate Europe (Gömöry *et al.* 1999). Populations from the central part of the Dinarian region are intermediate between these two southern extremes, indicating historical connections between Calabria and the Balkans, perhaps during the Pliocene. Further evidence for this connection can be seen in the close genetic link among populations of *Senecio rupestris* in central Italy and the southern Balkans (Abbott *et al.* 2002). The latter study also illustrates how populations in refugia are more genetically variable than those in formerly glaciated areas. In maritime pine, *Pinus pinaster*, patterns of population differentiation have also been interpreted in relation to climatic history. Three distinct groups of populations (Atlantic, North African, and Ibero-Tyrrhenian that includes populations from Andalousia to south-east France, Liguria, and the islands of Corsica and Sardinia) have been identified on the basis of molecular genetic and biochemical markers (Petit *et al.* 1995; Salvador *et al.* 2000). Migration out of glacial refugia in Andalousia and the central Spanish mountains is thought to have created these distinct groups in different areas.

Much work has been done with the aim being to identify the source populations, and better understand the process, of post-glacial re-colonization of temperate Europe. Such work has often shown that some refuge populations have not contributed to the genetic diversity of re-colonizing populations. The contemporary populations present in these refuge areas may thus retain a unique store of genetic variation, not present elsewhere in the widespread distribution of the species. An illustrative example of this pattern has been revealed by Hampe *et al.* (2003) in a study of genetic variability in the different parts of the range of alder buckthorn, *Frangula alnus*, a small tree widespread across temperate Europe, whose seeds are dispersed by birds. This species has relict populations in southern Spain, northern Morocco, and in Anatolia. Populations in these disjunct regions are highly differentiated from one another (different subspecies have been described) and the populations in southern Spain have maintained high levels of variation. The latter do not appear to have contributed to post-glacial re-colonization, which has been exclusively from refugia in the Balkans. An important result is thus that relict populations in Spain contain a unique store of genetic variation absent from the rest of the species' range. The conservation value of these refuge populations is thus clear.

The geographic genetic structure of populations of herbaceous plants also provide evidence for differentiation in glacial refugia. Molecular analysis of *Senecio gallicus* across its distribution in the western Mediterranean nicely illustrates how current day patterns of differentiation can provide revealing evidence on the historical process of re-colonization after glaciation (Comes and Abbott 1998, 2000). This widespread outcrossing diploid, which like many *Senecio* species has a marked capacity for relatively long-distance dispersal, occurs in open ruderal habitats, on dunes, river banks, and sandy soils in fields with more natural vegetation, in coastal and inland regions of the western Mediterranean from southern France across the Iberian peninsula. Comparative analysis of cpDNA and nuclear allozyme variability across this distribution has shown that allozyme variation has a relatively random spatial structure with low variation, and cpDNA haplotypes show a significant increase in the frequency of a derived haplotype and a decrease in the number of haplotypes from coastal to inland sites. Different coastal sites showed significant differences among each other in the frequency of cpDNA haplotypes, a pattern not shown by inland populations where the frequency of cpDNA haplotypes showed little variation (Fig. 3.2). Variation in quantitative morphological traits and RAPD markers (Table 3.1) also showed differentiation between coastal and inland populations indicating that the phylogeographic divide for cpDNA also exists for other markers, indicating that the geographical uniformity of allozymes is more a result of low rates of evolution or low genome representation than high contemporary pollen flow. Recent range expansion associated with a more rapid rate of cpDNA lineage sorting may be the cause of the above patterns.

Founder events during colonization and subsequent genetic drift in isolated populations can create sharp patterns of genetic differentiation in plant populations. During glacial maxima *S. gallicus* would have been absent from the north-west and central parts of the Iberian peninsula and its altitudinal range would have been more restricted. The genetic structure of coastal populations suggests that *S. gallicus* persisted in distinct coastal refugia, out of which it has recently spread to cover its present range. Its recent advance may have contributed to the discordance between cpDNA and nuclear allozyme variation. Nevertheless, it is still possible that directional selection has favoured cpDNA haplotype 'G' (Fig. 3.2) in inland habitats, either directly, due to some form of ecological association, or indirectly, via linkage to other genes involved in adaptation to inland habitats.

The population genetic structure of endemic *C. balearicum*, a forest dwelling geophyte on rocky north-facing slopes in the western Mediterranean also provides evidence that glaciation, by virtue of its effects on species distribution and population sizes, may have had a strong impact on levels of population differentiation. This species illustrates the pattern of disjunct endemism discussed in Chapter 2: its distribution encompasses two distinct regions where the scale of isolation is similar, but the history of isolation is different (Fig. 3.3(a)). First, it occurs on several of the Balearic islands where populations on different islands have been isolated from one another since at least the last glaciation period and perhaps longer, particularly those on the island of Ibiza compared to those on the Gymnesian

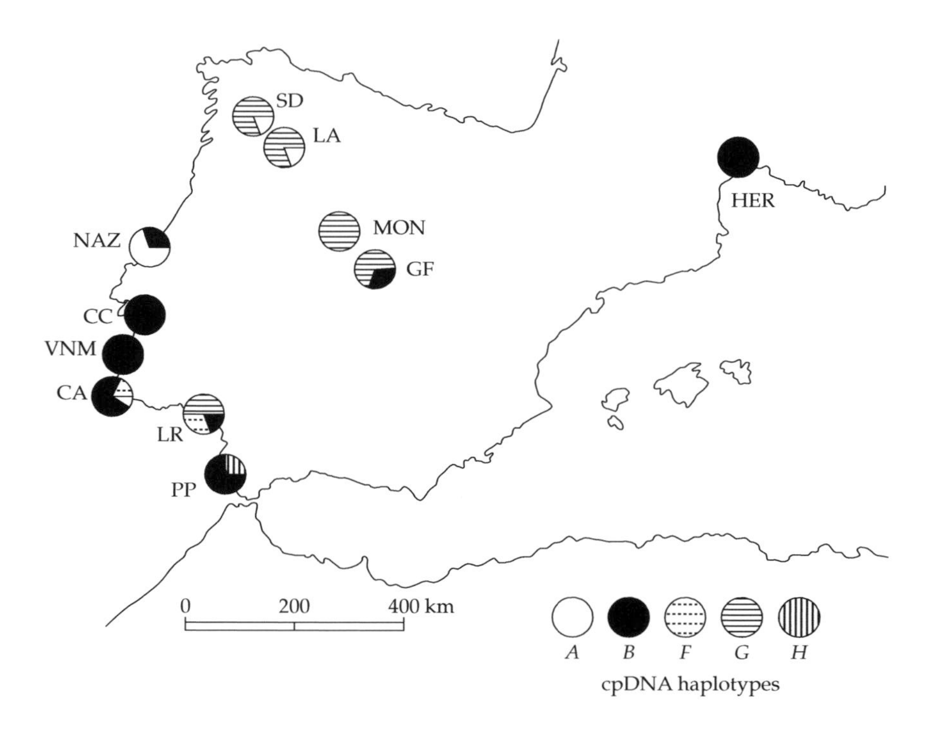

Figure 3.2 Geographic distribution of cpDNA haplotypes in *S. gallicus* in the Iberian peninsula and southern France. Reproduced with permission from Comes and Abbott (1998).

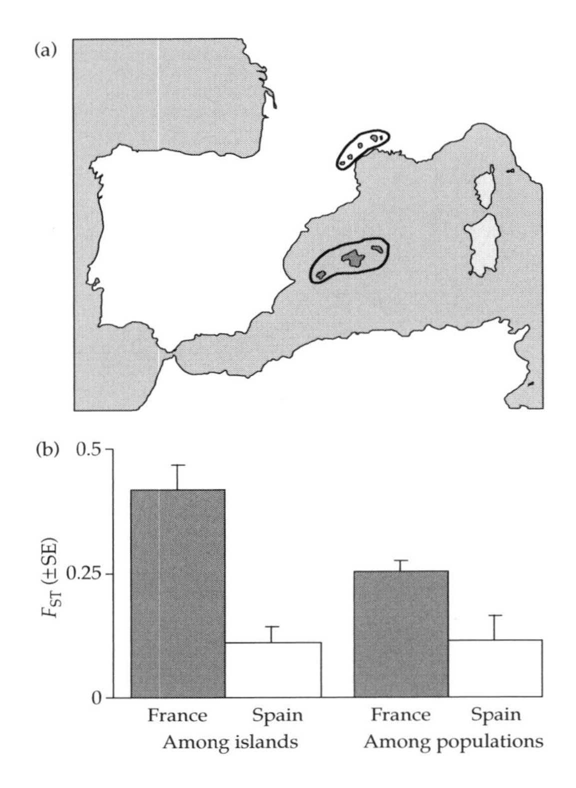

Figure 3.3 The (a) distribution and (b) genetic differentiation (mean F_{ST} values) among populations of *Cyclamen balearicum* on the Balearic islands and populations in habitat fragments in southern France (redrawn from Affre *et al.* 1997).

islands of Majorca, Menorca, and Cabrera. Second, *C. balearicum* occurs naturally in five geographically isolated zones in southern France. Affre *et al.* (1997) reported that genetic differentiation among populations in different habitat islands in southern France was more than twice that among populations on the different Balearic islands (Fig. 3.3(b)). By comparing the amount of differentiation among populations on the largest of the Balearic islands (Majorca) with that among different populations in one region in southern France (the Cévennes), these authors also found that population differentiation was greater among populations within habitat islands compared to on true islands. The most plausible explanation for these patterns is that glaciation hammered population size and number in southern France, reducing the distribution of this species to a small number of isolated pockets in sheltered cliffs. This would have created the template for differentiation to occur. Human activities (forest clearance and grazing) in combination with its ecological habitat requirements (Debussche and Thompson 2003) and reliance on ant dispersal (Affre *et al.* 1995) would have restricted the subsequent spread of this species, maintaining isolation among 'habitat islands' in southern France. Following severe bottlenecks on population size, the highly inbreeding nature of this species, which is rarely if ever visited by pollinators and which can self in the absence of pollinators (Affre *et al.* 1995; Affre and Thompson 1999), will have promoted gene fixation via random drift. In fact, populations in southern France show marked fixation of alleles in each population, whereas populations on the Balearic islands tend to contain more than one allele at each studied locus per population (Affre *et al.* 1997). In contrast, populations on the Balearic islands may have been larger and more connected due to lower sea levels in the Quaternary. As an epilogue, this is a story to be followed, in the wake of forest advance on abandoned terraces in southern France, the species may be currently re-expanding its local distribution in some areas (Chapter 4).

Some Mediterranean mountain species, may have persisted during the Quaternary in peri-glacial refugia, with little change in range size in response to glaciation. An illustration of such distributional stasis has been reported in *Anthyllis montana*, a species which occurs in low mountains around the western Mediterranean (Kropf *et al.* 2002). Phylogenetic analyses across the distribution of *A. montana* revealed a major genetic subdivision between eastern populations (Greece, Balkans, Italy, Maritime Alps) and western populations in France (French Alps and the Cévennes), and Spain (Pyrenees, Cantabrican Mountains, and Sierra Nevada). The western and eastern lineages correspond to two different subspecies which are separated by the Alps where the species is uncommon. The absence of this species from what one would predict to be suitable habitats in the Alps and the distinct pattern of geographic differentiation on either side of the Maritime Alps probably reflect a long-standing separation at the population level and thus stasis in distribution patterns. The extant distribution of some populations in the south-eastern margins of the Alps coincides with an area suggested to represent sites of periglacial refugia for different alpine taxa (e.g. Stehilik 2000). So in some microenvironments a number of Mediterranean mountain taxa may have found refuge during glacial maxima in a fairly wide part of their current distribution.

3.4 Divergence in peripheral and marginal populations: isolation, inbreeding, and ecology

3.4.1 Morphological differentiation in disjunct populations

In Chapter 2, I outlined how closely related widespread and endemic taxa can show significant levels of variation in biological traits and ecological requirements. Precise quantitative investigation of the degree of morphological differentiation among disjunct but closely related taxa or among disjunct populations of a single taxon have however rarely been performed.

Several studies attest to a lack of population differentiation in morphology among highly disjunct populations. Despite a long history of isolation on small fragments of what were once larger microplates or connections of land, many species with disjunct endemic distributions show little or no morphological differentiation on the now disjunct fragments or islands. For instance, in the southern Aegean, 23 species have distributions that encompass western Crete and Andikithira (between Crete and the Peloponnese) or eastern Crete and Karpathos, that is, across what are thought to be important phytogeographical barriers (Greuter 1972). In 21 of these cases, no significant morphological differentiation was observed, that is, no varieties or subspecies on the different islands, and just two groups showed differentiation on Crete and nearby islands: *Campanula saxatilis* on western Crete has subsp. *cytherea* on Andikithira (and Kithira) and *Silene ammophila* on eastern Crete has subsp. *carpathae* on Karpathos. Such morphological stasis is surprising given the length of time that such islands have been isolated. Likewise in the western Mediterranean, several endemic species with geographically disjunct populations (e.g. *Arum pictum*, *Cyclamen balearicum*, *Orchis insularis*, *Soleirolia soleirolii*, and *Arenaria balearica*) show a lack of distinct geographic variation in morphology among the geographically disjunct parts of their range.

Other taxa with disjunct and endemic distributions do however show evidence of distinct morphological differentiation among closely related species (e.g. (a) *Erodium corsicum* (endemic to Corsica and Sardinia) and *Erodium reichardii* (endemic to Majorca and Minorca) and (b) *Pastinaca latifolia* (endemic to Corsica) and *Pastinaca lucida* (endemic to Majorca and Minorca)) and at the subspecies level (e.g. (a) *Helleborus lividus* subsp. *corsicus* (endemic to Corsica and Sardinia) and subsp. *lividus* (endemic to Majorca and Cabrera), (b) *Sesleria insularis* subsp. *cordata* (endemic to Corsica) and subsp. *insularis* (endemic to Sardinia and Majorca), (c) *Alnus viridis* subsp. *viridis* (Alps) and subsp. *suaveolens* (Corsica), and (d) *Herniaria latifolia* subsp. *latifolia* (Pyrenees and Spain) and subsp. *litardierei* (endemic to Corsica and Sardinia)). Several species with disjunct distributions on the Pityusic islands (Ibiza and Formentera) and the Iberian peninsula also show differentiation in morphological traits suggesting recent geographical differentiation (Contandriopoulos and Cardona 1984).

Precise analysis of morphological variation within and among closely related endemic taxa may also blur taxonomic distinctions. In a study of quantitative morphological variation in a group of three closely related *Cyclamen* species (Plate 1), Debussche and Thompson (2002) showed that *C. creticum* (endemic to Crete) should be considered as a geographic subspecies of the widespread *C. repandum*. A conclusion which concords with a phylogeographic study of this group (see below). This raises the question of how different closely related endemic species really are and illustrates a central theme of this chapter: what differs among species often varies considerably within species.

3.4.2 The geographic and local scales of speciation

In this section, I will focus on the evolution of schizo-endemic distribution patterns, that is, where endemic species diverge in the different parts of the range of an ancestral species, to create a contemporary pattern of disjunct distributions among taxa with the same chromosome number. This evolution is assumed to be gradual in the different parts of the range of an ancestral taxon (Favarger and Contadriopoulos 1961). The fact that many closely related species have disjunct distributions around the shores of the Mediterranean is no doubt the source and inspiration for this idea. Favarger and Contadriopoulos recognize that such endemism may evolve as a result of isolation followed by differentiation and divergence in allopatry or as a result of initial differentiation in a variable taxon in the different parts of its range. In the latter case, 'geographic isolation of the different parts of the ancestral distribution follows the initial differentiation, producing schizo-endemic taxa' (Favarger and Contadriopoulos 1961: 398, my translation). There is in both cases an implicit assumption that speciation is geographic, although in the latter case the authors imply that the initial impetus for differentiation was not reproductive isolation in allopatry. They were unable to identify the causes of this process of initial divergence, despite reference to some cases of ecological differentiation among Mediterranean and Alpine taxa.

The reliance on a role for gradual accumulation of genetic differences in allopatry reflects a traditionally accepted view of the importance of spatial isolation in species divergence (Levin 2000). This is no doubt an important process in long-lived species with efficient mechanisms of gene flow and high effective population sizes. However, if one analyses speciation at the scale of the population-level processes that promote differentiation or homogenize variation, several features of plant population ecology and evolution suggest that this mode of speciation may be less important than in animals. First, gene flow is often spatially limited in plants, hence the spread of novel genes over large population systems may require extremely long periods of time. Recent theoretical work attests to the possible evolution of phylogeographic breaks without geographical barriers to gene flow in species with low average individual dispersal distances and small population sizes (Irwin 2002). Indeed, local genetic differentiation is common in plants (Table 3.2; Linhart and Grant 1996) and may be an important step towards speciation (Levin 2000). However, it is well known that advantageous genes may spread rapidly, and thus even small amounts of gene flow may allow for geographic speciation to occur (see Morjan and Rieseberg 2004). Second, in plants there is enormous potential for sympatric or peripatric speciation via rapid genetic change linked to hybridization, polyploidy, inbreeding, and local adaptation on fine spatial scales. Finally, geographic speciation requires fairly uniform selection pressures across the area in which divergence occurs, an assumption which is difficult to uphold for plant populations in the Mediterranean landscape (see Chapter 4).

Local speciation relies on the operation of the two microevolutionary processes which promote population differentiation. First, random genetic drift in gene frequency and the fixation of new gene combinations due to founder events and genetic bottlenecks will favour genetic drift and random differentiation and thus provide a template for local speciation, in the absence of geographic isolation (Levin 1993). Second, novel variants may be favoured by selection if they have an adaptive advantage. This may occur for selfing variants if they provide reproductive assurance in the absence of pollinators, favour seed set in novel ecological conditions or provide reproductive isolation from ancestral species. In other words, new gene combinations (that appear by random fixation) may in fact have an adaptive value for colonization and persistence in novel ecological conditions. However, theoretical work illustrates that the disruption of gene flow in peripheral populations may allow for selection to cause speciation, even in the absence of contributions from random genetic drift (García-Ramos and Kirkpatrick 1997).

This twofold process may be particularly important where species colonize islands or in peripheral populations of a widespread species. Populations at the geographic margins of species' ranges will be particularly prone to the evolution of random and adaptive differentiation due to their small size, potential isolation, reduced gene flow, and faster population turnover (extinction and colonization) than in the central part of the range (Antonovics 1976). Genes with no major impact on fitness in large populations in the central part of the range may confer a selective advantage in the novel conditions experienced in peripheral isolates on the edge of the species range, where the breakdown of developmental canalization due to rapid rise in levels of inbreeding and strong selection will allow previously masked variation to be expressed (Levin 2000). The significance of an abrupt shift to intense inbreeding following dramatic reductions in population size and differentiation in response to novel ecological conditions has in fact long been stressed in the literature (Lewis 1962, 1966; Raven 1964). This form of speciation may thus often be associated with a shift from outcrossing to selfing, that is, one of the major transitions in the reproductive system of plants (Barrett 2002*a*).

In the local speciation model, a new species will begin with a limited distribution. From then on a variety of processes may occur. Some new species may rapidly go extinct, others may establish large populations but only persist in a limited area, and still others may expand well beyond the site of origin (in allopatry or within the range of the progenitor). Speciation in geographically peripheral and ecologically marginal isolates followed by range expansion in allopatry (and the subsequent development of geographic barriers to dispersal) could thus produce a contemporary distribution pattern similar to that which would occur following geographic speciation.

Several plant groups in the Mediterranean illustrate how random loss and fixation of alleles in disjunct parts of a species range can create sharp patterns of differentiation among populations of a single species, with important implications for understanding the process of species divergence.

In *Pinus maritima*, Petit *et al.* (1995) documented significant geographic differentiation ($G_{ST} = 0.17$) for isozyme loci among the fragmented parts of the geographic range of this species in the western Mediterranean (i.e. Corsica, Sardinia, Italy, Spain, Portugal, and south-western France). In contrast, on a more regional scale, from northern Morocco to northern Spain, González-Martínez *et al.* (2002) reported much lower genetic differentiation at isozyme loci among populations ($G_{ST} = 0.05$).

In *C. balearicum*, population differentiation in southern France is thought to have evolved as a result of gene fixation via random drift in a small population system as a result of climatically induced habitat fragmentation (see above). Comparison of the genetic distance among populations of *C. balearicum* and its closely related congeners *C. repandum* on Corsica and *C. creticum* from Crete indicates that random fixation has caused allele frequencies in one population of *C. repandum* to resemble more those of populations of other species than other conspecific populations (Fig. 3.4(a)). The single population of *C. repandum* that is genetically distinct from the other Corsican populations is one with a very high level of homozygosity. In this group of species it is probable that inbreeding causes populations to diverge and show gene frequencies more typical of other closely related species. A similar pattern has been reported for populations of three *Centaurea* taxa in Mediterranean France (Fig. 3.4(b); Fréville *et al.* 1998).

In both *Cyclamen* and *Centaurea*, localized endemism is frequently correlated with the appearance of albino floral variants. Such flower-colour variation may be a common feature of random fixation and species divergence in isolated populations and closely related taxa. For example, in *Satureja cordata* which has two subspecies in the Balearic islands, one is an erect plant with purple flowers (subsp. *rodriguezii*) and one is a prostrate plant with white flowers (subsp. *filiformis*). *Paeonia* and probably many other groups show similar patterns of flower colour variation in association with the isolation of taxa on Mediterranean islands.

Random changes in gene frequency due to genetic drift in small and isolated populations may be the cause of marked differentiation among populations of the chasmophyte *Brassica insularis* on Corsica (Hurtrez-Boussès 1996) and the patchy distribution of rare and infrequent alleles among subpopulations and populations in the rare *Silene diclinis*, endemic to a small area of south-east Spain (Prentice 1984; Prentice and Andersson 1997). The results of these authors (including a strong relation between population size and allele number) are consistent with the idea that the loss of gene diversity will be slower than the loss of alleles in a population that has

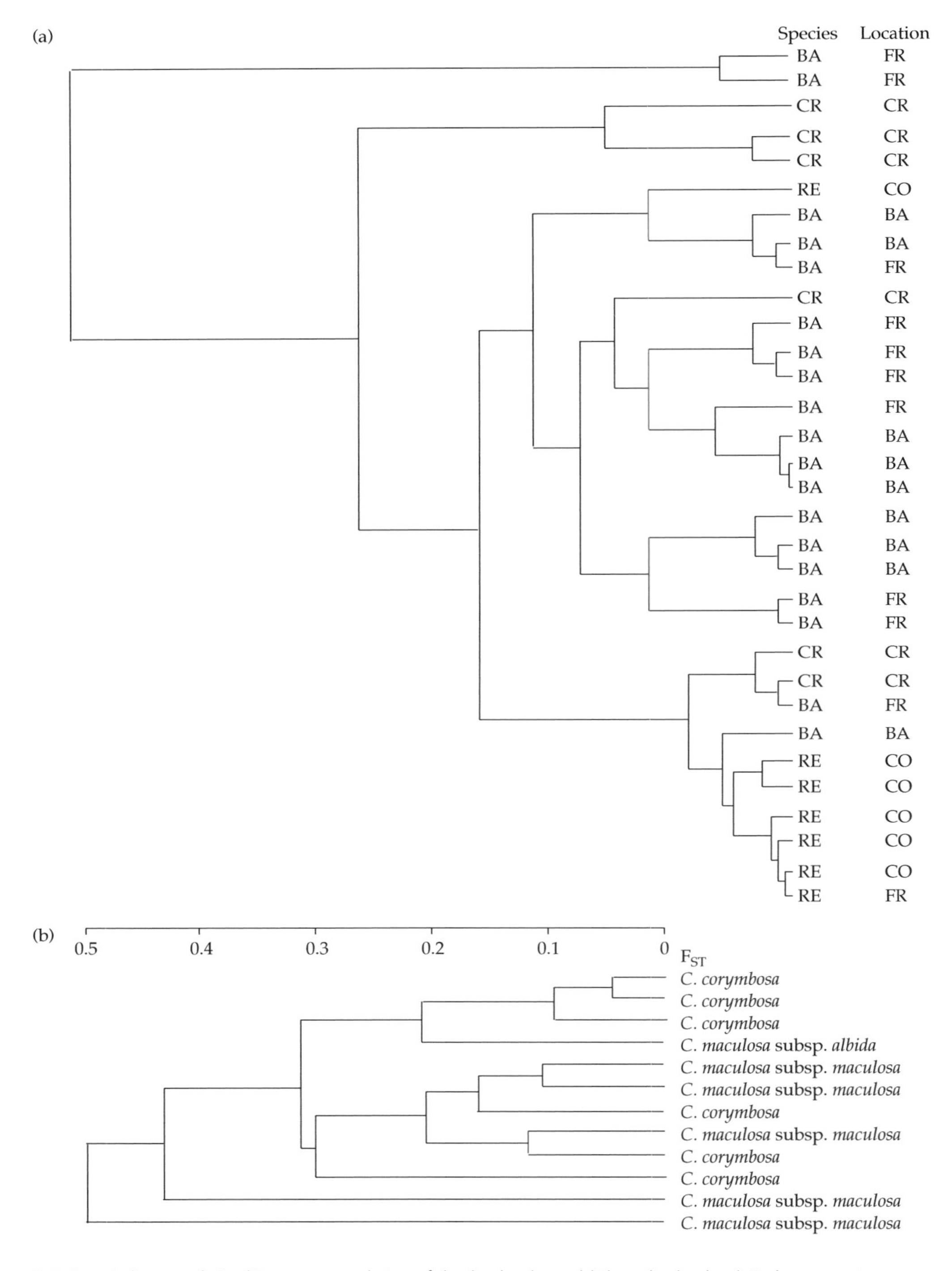

Figure 3.4 Genetic distance relationships among populations of closely related taxa. (a) Three closely related *Cyclamen* species (BA—*C. balearicum*, CR—*C. creticum*, RE—*C. repandum*) from Crete (CR), Corsica (CO), southern France (FR), and the Balearic islands (BA). Based on data in Affre *et al.* (1997) and Affre and Thompson (1997*a*,*b*). (b) Three closely related taxa of *Centaurea* in southern France (redrawn from Fréville *et al.* 1998).

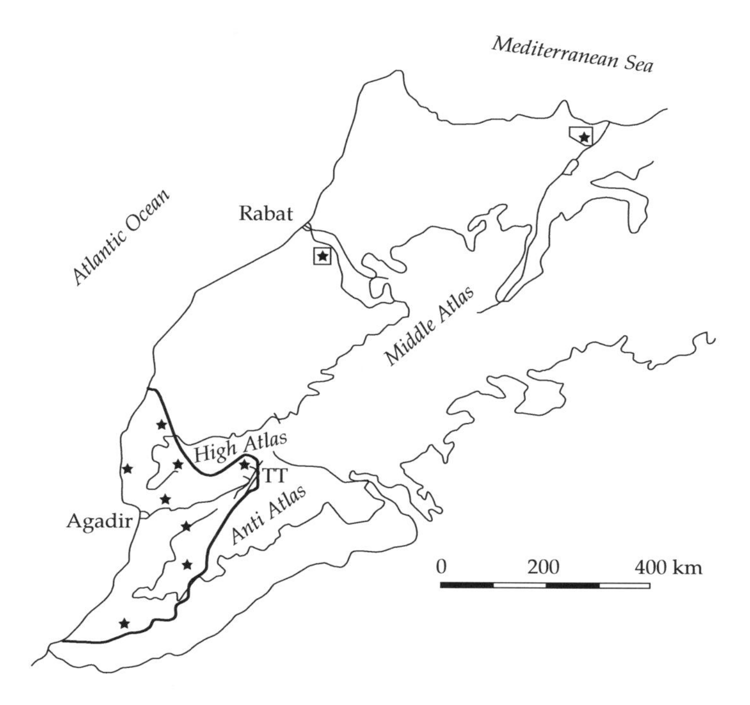

Figure 3.5 The distribution of *Argania spinosa* in North Africa (reproduced with permission from El Mousadik and Petit (1996*b*)).

undergone a rapid reduction in size. Work on isolated populations of the argan tree (*Argania spinosa*) endemic to south-west Morocco has also shown that rare alleles have a more scattered distribution than common alleles (El Mousadik and Petit 1996*a*,*b*). The distribution of these rare alleles and their concentration in the more isolated populations of this species (Fig. 3.5) cause the latter to contribute more to total diversity, despite their reduced allelic richness, than populations in the central portion of the range.

Random population differentiation in isolated populations, and in the peripheral parts of the range of a species can thus be commonly observed and invites the question: is there a link to local speciation in geographically peripheral populations?

3.4.3 Species divergence in western Mediterranean *Cyclamen*

The different taxa of *Cyclamen* subgenus *Psilanthum* (Plate 1) have a distribution pattern which invites study of the processes causing species divergence at range limits (Fig. 3.6(a)). Previous work has generally recognized three distinct species in this subgenus, *C. repandum*, *C. balearicum*, and *C. creticum* (Grey-Wilson 1997). In recent studies, the classification shown in Chapter 2, with *C. creticum* reduced to a subspecies of *C. repandum*, has been suggested (Debussche and Thompson 2002). The different taxa in this subgenus form a closely related and distinct unit in the genus on the basis of molecular and morphological data (Anderberg 1993; Anderberg *et al.* 2000), have the same diploid chromosome number, and all flower in the spring. These charming little plants can produce unmistakable carpets of small single-stemmed flowers with characteristically reflexed petals in many woodlands of the western Mediterranean in the spring. In contrast, their fruits are almost invisible and very difficult to find. Due to a coiling of the pedicel after fertilization, the fruits are brought to ground level where they mature in leaf litter at soil level and open to liberate the seeds, and then are dispersed primarily by

(a)

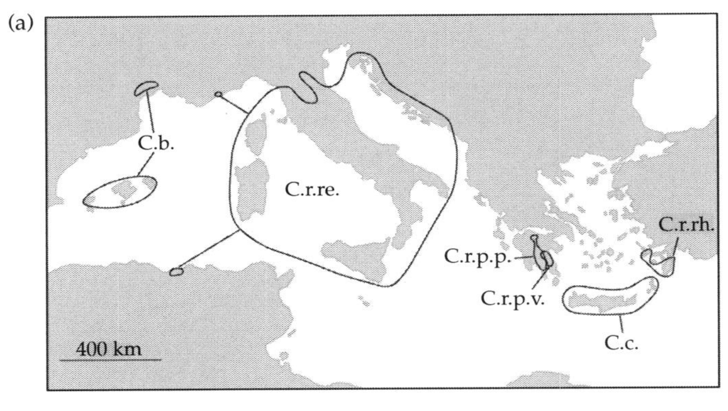

(b)

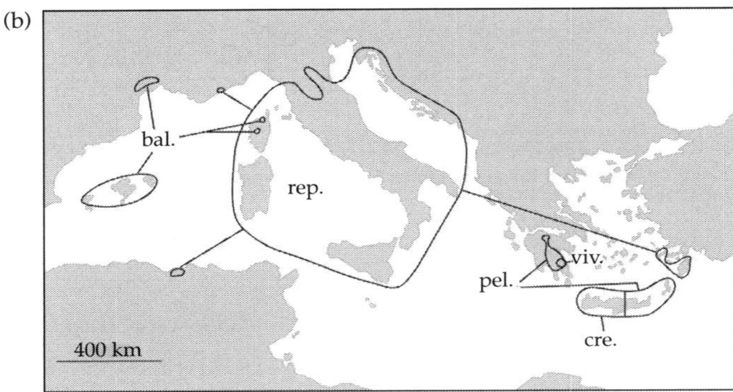

(c)

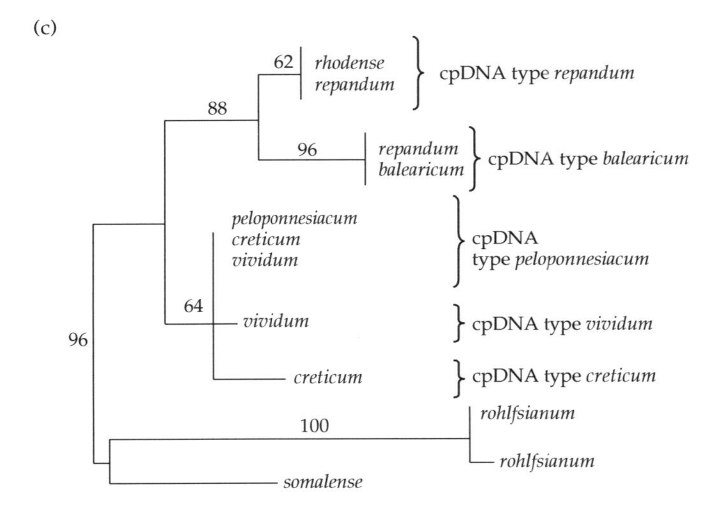

Figure 3.6 (a) Geographic distribution of the various taxa in *Cyclamen* subgenus *Psilanthum*. Abbreviations are as follows: C.b.—*Cyclamen balearicum*, C.r.re.—*Cyclamen repandum* subsp. *repandum*, C.r.rh.—*Cyclamen repandum* subsp. *rhodense*, C.r.p.p.—*Cyclamen repandum* subsp. *peloponnesiacum* var. *peloponnesiacum*, C.r.p.v.—*Cyclamen repandum* subsp. *peloponnesiacum* var. *vividum*, and C.c.—*Cyclamen creticum* (now thought to be a subspecies of *C. repandum*. (b) Tentative geographic distribution of the five cpDNA types in the phylogenetic tree shown in (c). CpDNA types are: bal.—*balearicum*, rep.—*repandum*, pel.—*peloponnesiacum*, viv.—*vividum*, and cre.—*creticum*. (c) The single most parsimonious tree for cpDNA haplotypes of *C. repandum* and its two allopatric congeners, *C. creticum* and *C. balearicum*. Numbers adjacent to branches represent bootstrap values obtained from 1,000 replications. Reproduced with permission from Gielly *et al.* (2001).

ants (Affre *et al.* 1995). Disjunct distribution patterns thus represent the consequences of allopatric fragmentation and vicariance, rather than long-distance colonization, in these species.

Using cpDNA *trn*L (UAA) intron sequence analyses on samples from across the entire distribution of the different taxa in this group, Gielly *et al.* (2001) obtained a phylogenetic tree composed of five haplotypes regrouped in two main clades (Fig. 3.6(b),(c)). The two clades suggest divergence in the disjunct parts of the range of *C. repandum*.

One clade contains samples of *C. repandum* subsp. *repandum* from Croatia, Italy, southern France, Corsica, Sardinia, and Sicily, *C. repandum* subsp. *rhodense* from Rhodes and Kos and *C. balearicum* from the Balearic islands and southern France. In this clade, the divergent position of different samples of *C. repandum* subsp. *repandum* suggests that this subspecies has diverged in the different geographic limits of its range to produce *C. repandum* subsp. *rhodense* and *C. balearicum*.

The second clade contains the samples of *C. repandum* subsp. *peloponnesiacum* (Peloponnese peninsula) and *C. creticum* (Crete). This clade suggests an important phylogeographic separation of taxa in the Peloponnese peninsula and Crete from those elsewhere in the distribution of this subgenus (including Rhodes and Kos).

Despite the fact that in many groups a major floristic division splits the Cyclades from the eastern islands close to Turkey (Chapter 2), *C. repandum* subsp. *rhodense* is more closely related to *C. repandum* subsp. *repandum*. Hence, although *C. creticum* appears to have diverged in allopatry from *C. repandum* subsp. *peloponnesiacum*, *C. repandum* subsp. *rhodense* does not appear to have evolved at the end of a chain from the Peloponnese peninsula across Crete and Karpathos to Rhodes. *C. repandum* subsp. *rhodense* appears to have evolved as a result of geographic isolation following the loss of land-bridge connections across the Cyclades, perhaps in the Pliocene (Chapter 1).

The nested position of *C. balearicum* is typical of a phylogenetic pattern expected by local speciation (Rieseberg and Brouillet 1994). Such nested phylogenetic patterns have been shown to occur for other narrow endemic species in the Mediterranean, such as *Saxifraga* (Conti *et al.* 1999; Vargas *et al.* 1999). Whereas *C. creticum* flowers resemble *C. repandum* in overall size, stigma–anther separation, and pollen–ovule ratio, they more closely resemble *C. balearicum* in their colour (Affre and Thompson 1998; Debussche and Thompson 2002). Pair-wise comparisons of the three species show that the flowers of *C. repandum* and *C. balearicum* are very different from each other (Fig. 3.7). This floral variability is typical of differences regularly observed between outcrossing (*C. repandum*) and selfing (*C. balearicum*) species. Indeed, controlled experiments in an insect-free glasshouse by Affre and Thompson (1999) have shown that *C. balearicum* is capable of autonomous selfing (in the absence of pollinators), while *C. repandum* and *C. creticum* require an external vector to assure high levels of seed production (Fig. 3.7(e)). Field observations of pollinators have revealed that insect visitation to *C. balearicum* is almost non-existent, whereas bumble-bees are frequently observed visiting flowers of *C. repandum* on Corsica. *C. repandum* and *C. balearicum* also show a distinct difference in their habitat conditions and overall ecology (Debussche and Thompson 2003). Whereas *C. repandum* on Corsica is primarily found in either coniferous or deciduous woodlands on a range of bedrocks and with an important litter cover, *C. balearicum* (throughout its entire distribution) occurs almost exclusively on rocky limestone substrates, evergreen shrublands, and open woodlands (Fig. 3.8). To sum up, divergence of *C. balearicum* at the margins of the distribution of *C. repandum* has probably been closely linked to increased inbreeding and ecological specialization.

The link between species disjunction, reproduction, and ecology and the possibility of local speciation in this group has become a distinct possibility in the light of a recent discovery of what appears to be a very old and disjunct population of *C. balearicum*. As Stebbins (1942: 77) commented, 'every field botanist can recall the thrill of excitement that comes with the discovery of . . . a well known species far outside of its normal range of distribution'. Such 'peripheral isolates' are of key importance to the study of plant evolution in a biogeographic setting. It was thus with much excitement that my colleague

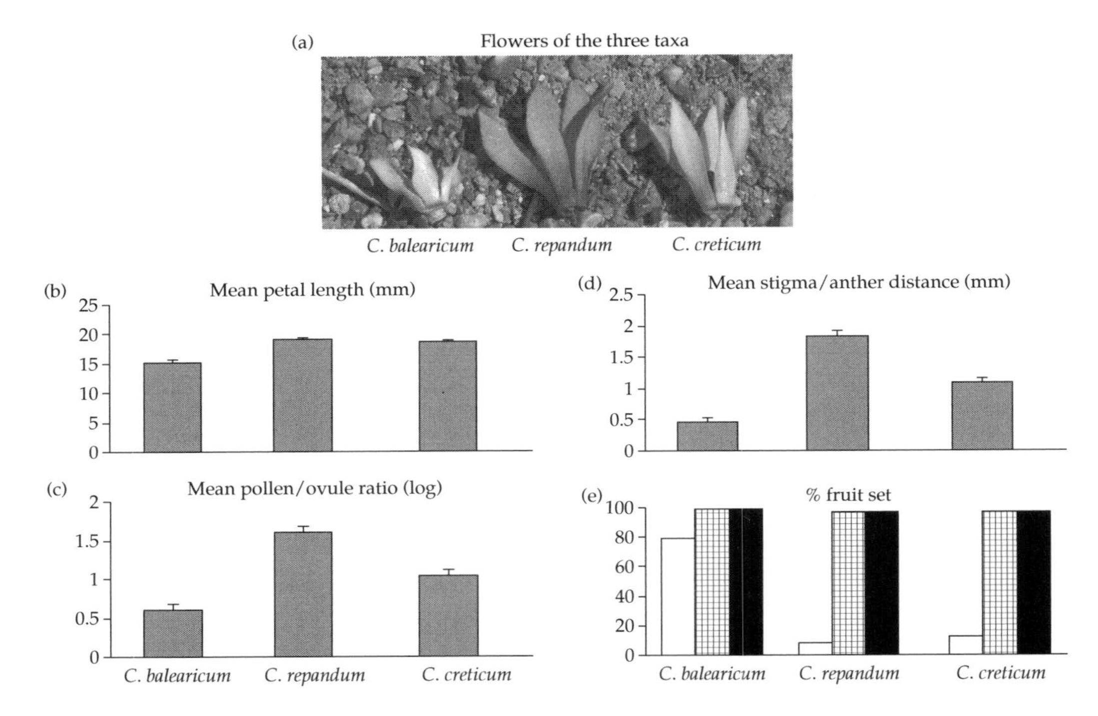

Figure 3.7 (a)–(d) Floral trait variation and (e) capacity for autonomous self-pollination in three taxa of *Cyclamen* subgenus *Psilanthum* (Primulaceae) (drawn from data in Affre and Thompson 1998, 1999). In (e) open bars are autonomous selfing, hatched bars are manual selfing and black bars are outcrossing.

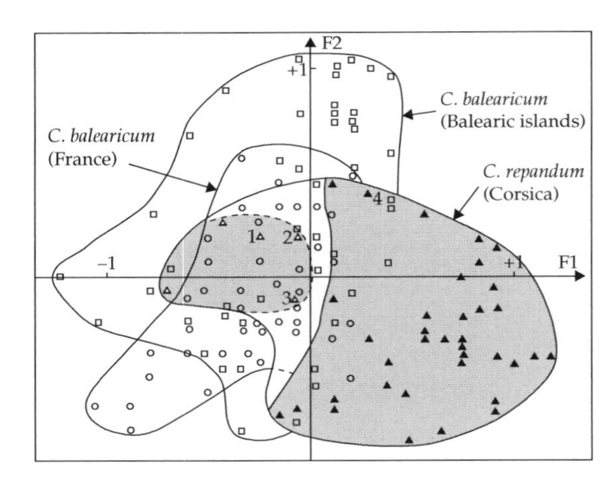

Figure 3.8 Multivariate analysis of ecological habitat differentiation of *C. repandum* and *C. balearicum*. The small area of habitat space occupied by *C. repandum* with a dashed line corresponds to three populations on a limestone massif near St Florent in Corsica where many plants have the morphology of *C. balearicum*. Reproduced with permission from Debussche and Thompson (2003).

Max Debussche and I recently reported that *C. balearicum* may also occur outside of its previously documented range on Corsica, on a limestone massif near the village of St Florent, very close to populations of *C. repandum* (Debussche and Thompson 2000). The cpDNA haplotypes characteristic of *C. balearicum* and *C. repandum* subsp. *repandum* both occur in this site (Fig. 3.6(b)). The size, pollen–ovule ratios, and stigma–anther separation of white-flowered plants in the St Florent populations, where white-flowered plants are at a high frequency, is more similar to *C. balearicum* than (white-flowered) *C. repandum* (Box 3.1). The ecology of this site is typical of *C. balearicum* habitats in the Balearic islands and on the French mainland, and is thus quite distinct from other *C. repandum* populations on Corsica (Fig. 3.8). This site contains an immense diversity of floral forms, in terms of the combination of flower colour, size, and stigma–anther separation. Plants can have flowers which resemble either

Box 3.1 *Cyclamen* on Corsica

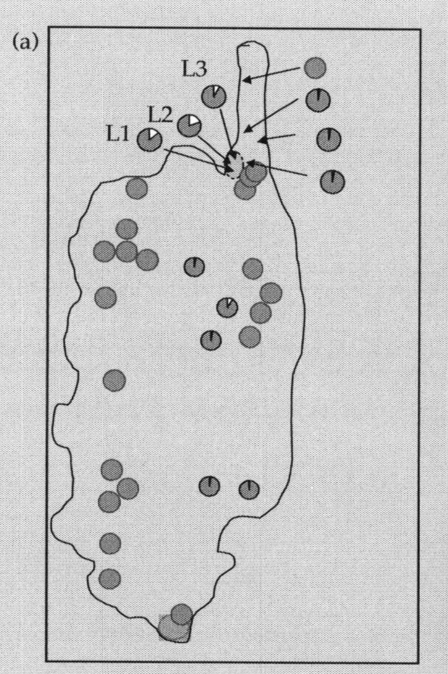

The occurrence of a high percentage of white-flowered plants in three populations (L1–L3) of *Cyclamen repandum* on limestone in northern Corsica (b) compared to other sites on Corsica which occur on granite or schist (c). In (a), each circle is a population and the frequency of white-flowered plants is the open sector of the circle. The high frequency of white-flowered plants in the ecologically marginal sites is accompanied by the presence of plants with the attractive silvery grey leaf markings typical of the dark green leaves of *Cyclamen balearicum*, which contrast to the lighter green and yellow patches on the leaves of *C. repandum*. A population of *C. balearicum* may thus have been historically present on this island. Genetic (d) and morphological (e) data strongly support this idea. (d) The percentage of polymorphic AFLP loci in plants on limestone (BIC, bicoloured flowers; BSI, *C. balearicum*-like flowers; BSE, white flowers with exerted style; RSE, *C. repandum*-like flowers) compared to allopatric populations of *C. balearicum* (B) and *C. repandum* (R) (M. Gaudeul unpublished data). (e) In the three populations on limestone (6–8), white flowers (open bars) are morphologically (size, colour, stigma–anther separation, and pollen–ovule ratio) more similar to *C. balearicum* than to pink-flowered *C. repandum* (filled bars).

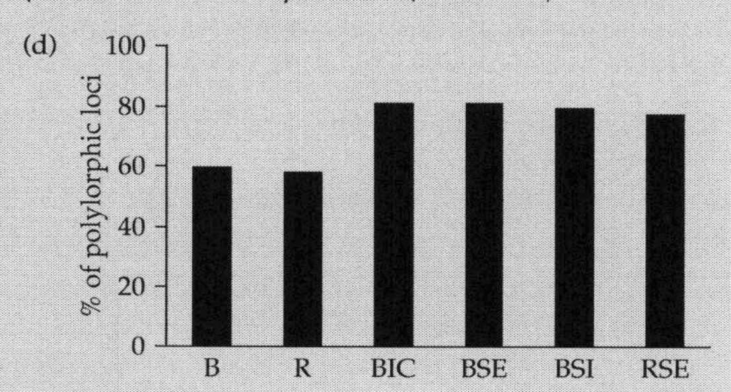

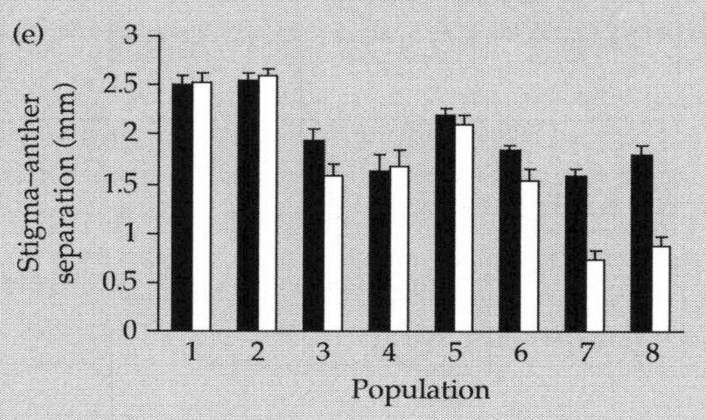

typical *C. repandum* subsp. *repandum*, *C. balearicum*, flowers produced by artificial hybridization among *C. repandum* and *C. balearicum* and flowers of other *C. repandum* subspecies (Plate 1). Analysis of genetic diversity using the technique of AFLP (Table 3.1) has shown that all plants on the limestone massif (whether they resemble *C. repandum*, *C. balearicum*, or hybrids) combine the diversity of *C. repandum* and *C. balearicum* (Box 3.1). This combination of genetic markers, specific to one or other of the parental species, tips the balance towards the idea that *C. balearicum* has been historically present on Corsica and that the populations on limestone represent secondary contact between a relict population of *C. balearicum* and local *C. repandum* subsp. *repandum*. Hybridization between these species in a single area of Corsica appears to have reproduced a remarkable array of morphological recombinant types whose variation encompasses the floral variability of the entire subgenus.

Another result of the genetic analyses is that all of the plants in the hybrid site studied show evidence of genetic introgression between the two species and none have a genetic constituency that falls within the range of variation present in allopatric *C. balearicum*. This illustrates the swamping role of gene flow in peripheral and marginal populations (Antonovics 1968; García-Ramos and Kirkpatrick 1997). All of the *C. balearicum* present in these populations are thus introgressed forms. However, many plants have a floral morphology which is identical to that of *C. balearicum*, others are like *C. repandum*, while some others have a combination of traits that resemble one or other of the two parental species. The maintenance of a floral morphology akin to *C. balearicum*, in the face of strong genetic introgression from *C. repandum*, strongly suggests a role for natural selection. Without wishing to be too speculative, I would suggest that temporal variability in the pollination environment, perhaps due to the absence of pollinators or years with a early summer drought and a need for rapid flowering and fruiting, may favour the persistence of a highly selfing strategy in these sites. Without some selective advantage, it is hard to understand what maintains the floral phenotype of *C. balearicum* in these sites given the levels of introgression observed.

Finally, the fact that *C. balearicum* has persisted on this limestone outcrop located on an otherwise 'granite island' (Fig. I.3) implies that initial divergence occurred in geographically peripheral and ecologically marginal populations on an unusual soil type. Populations in this zone of Corsica are very small, the number of plants barely exceed several hundred and occur in a small area compared to populations of *C. repandum* elsewhere on Corsica, where populations are usually much larger in terms of number and spatial extent. Differentiation may have been facilitated by developmental modifications since the newly opened flowers of *C. repandum* are similar in size and floral design to *C. balearicum* (Fig. 3.9). An abrupt developmental change or some form of instability in development could thus have triggered the evolution of the floral phenotype of *C. balearicum*. In contact with the parental species, plants which have persisted on limestone after the isolation of Corsica have, despite genetic introgression, maintained the floral phenotypes of *C. balearicum*. In contrast, molecular divergence has evolved in allopatry as *C. balearicum* has increased its distribution. Although geographic isolation may thus have contributed to genetic differentiation in this progenitor-derivative species pair, which otherwise forms a hybrid swarm in contact zones, I would argue that this *Cyclamen* story provides evidence for the local speciation model discussed at the beginning of this section in which speciation is initiated in local populations. If this interpretation was true then initial divergence must have been ancient, pre-dating the tectonic activity which isolated Corsica from the Balearic islands and southern France (Chapter 1).

Several other examples of ecological specialization in narrow endemic species relative to widespread congeners are known in Mediterranean plants (Box 3.2). The occurrence of edaphic adaptation in geographically peripheral and ecologically marginal populations may thus have been an important feature of the evolutionary process that has given rise to the plethora of narrow endemic species in the Mediterranean. However, in none of these cases is there evidence for adaptive differentiation, that is, that ecological differentiation contributes to fitness. To understand more fully the

Figure 3.9 From left to right, 3-day old flower of *C. repandum* subsp. *repandum*, newly opened flower of *C. repandum* subsp. *repandum*, and a 3-day old flower of *C. balearicum*.

role of ecological differentiation in local speciation will require that adaptive variation be identified.

3.4.4 Random differentiation in an archipelago system

In the Aegean Sea the multitudinous islands of different size and ecology and the history of island isolation have created a patchwork system in which many species groups may have diversified. In this part of the Mediterranean, small population systems, with marked among-population isolation, are a characteristic feature of chasmophytic species that inhabit cliffs and crevices, particularly on hard limestone (Chapter 2). Studies of two genera in this region, *Erysimum* and *Nigella*, illustrate random differentiation in such small population systems.

The different diploid taxa of *Erysimum* sect *Cheiranthus* which occur in the sheltered parts of cliffs in the Aegean region show a classic schizoendemic distribution pattern (Fig. 3.10). In this group a range of population characteristics interact with biological traits to facilitate the occurrence of genetic drift (Snogerup 1967).

1. 50% of populations have less than 50 individuals and taxa are self-compatible.
2. The cliff habitat isolates populations across the landscape. Since seeds are large with no adaptations for long-distance dispersal, isolation can hardly be countered by gene flow.
3. High seed germination rates and lack of dormancy prevent the establishment of seed banks that would normally buffer the effects of genetic drift.

The different taxa of *Erysimum* also show marked differences in their ecological distribution, *Erysimum corinthium* occurs primarily in maritime cliffs, *Erysimum senoneri* subsp. *senoneri* occurs inland at low altitude whereas *E. senoneri* subsp. *amorginum* is found only in cliffs above 500 m elevation, *Erysimum naxense* and *Erysimum rhodium* occur at intermediate elevations (between 500–800 m and 200–600 m on Naxos and Rhodes, respectively). So selection for ecological specialization may have played a role in population differentiation in the different parts of the range of this group.

The different taxa most probably evolved following an initial break up of continental areas and island isolation during the Pliocene. The amount of morphological differentiation that occurs among the different taxa closely parallels the history of isolation among their current distributions. For example, *E. naxense* (endemic to Naxos) is morphologically more similar to the different subspecies of *E. senoneri* than to other species. Successive land-bridge connections across this area during glacial maxima in

Box 3.2 Examples of ecological differentiation among closely related species in the Mediterranean

Comparative studies of endemic and widespread species in the Mediterranean (Chapter 2), the analysis of endemism in Mediterranean South Africa and patterns of habitat variation in *Cyclamen* mentioned in this chapter have revealed a consistent pattern of ecological differentiation among closely related species. This pattern suggests a possible role of ecological differentiation in marginal populations for species divergence. There are several other examples of ecological differentiation which can be used to strengthen this claim.

1. In western Mediterranean *Senecio* (Fig. 3.1) two species, *Senecio gallicus* and *Senecio petraeus* show a clear progenitor–derivative relationship (Comes and Abbott 2001). *S. petraeus* is endemic to a small calcareous mountain range in southern Spain whereas *S. gallicus* has a distribution which covers the Iberian peninsula and stretches into southern France.
2. The narrow endemic *Saxifraga cochlearis* shows ecological specialization relative to its widespread progenitor *Saxifraga paniculata* (Conti *et al.* 1999).
3. Despite broadly similar ecological requirements, three heathland *Erica* species which occur on nutrient-poor acidic sandy soils in southern Spain and Morocco show significant differences in soil chemistry and shading (Ojeda *et al.* 2000*b*).
4. *Antirrhinum lopesianum* only known from one population in Spain and Portugal occurs on serpentine soils whereas related *Antirrhinum mollissimum* and *Antirrhinum microphyllum* have a more typical rupicolous habitat (Mateu-Andrés 1999).
5. Several genera contain closely related calcifuge and calcicole species which may have diverged in relation to substrate, e.g. *Pinguicula* (Contandriopoulos 1962) and two sclerophyllous oaks: *Quercus calliprinos* on calcareous soils and the endemic *Quercus alnifolia* on ultrabasic rocks on Cyprus (Barbero *et al.* 1992).
6. The six subspecies of *Pinus nigra*, which have an almost completely vicariant distribution (Fig. 2.1; Barbero *et al.* 1998), all occur in the humid and/or subhumid bioclimatic zones in the supra- or montane-Mediterranean belts. The different subspecies also differ in the range of substrates they occupy (Quézel and Médail 2003). The subspecies with the most generalist substrate preferences are the most widespread.

the Pleistocene may have delayed the differentiation of this taxon on Naxos, where *E. senoneri* does not occur. This looks suspiciously like another example of the splitting off of an endemic species with a restricted distribution within the range of a more widespread species, as a result of random drift in small populations. Several populations show evidence of variation in morphology and a reduced fertility which could result from genetic instability in small inbred populations. In the more widespread *Erysimum candicum* on Crete and *E. corinthium* in Greece populations are more uniform, suggesting that historical and contemporary gene flow on larger areas of land have prevented the differentiation of populations in these taxa and their diversification into new taxa.

The confinement to mostly maritime limestone cliffs, the taxonomic isolation of many of the species, and the pronounced local differentiation all concord with the idea that the contemporary cliff flora is a relictual version of a more extensive flora that inhabited the coasts of Crete and the Aegean islands in the Pliocene. The geographic setting and the historical framework for the patterns of isolation and differentiation in this group of plants has made Snogerup's (1967) work well known, but we still do not know how much of the variation is truly genetically based. Quantitative genetic studies and a molecular phylogeographic study of *Erysimum* in and around the Aegean would be most useful here.

In his studies of the genus *Nigella*, and in particular the *Nigella arvensis* complex, Strid (1969,

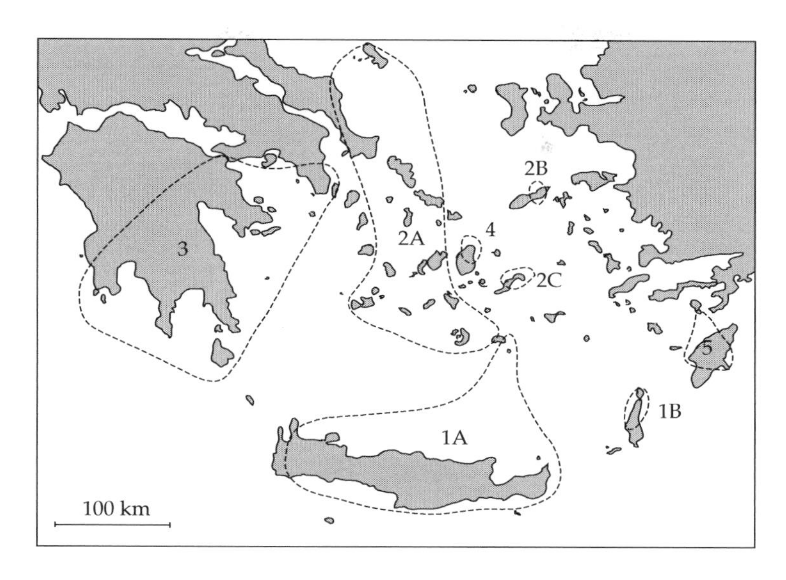

Figure 3.10 Distribution of the different taxa in *Erysimum* sect. *Cheiranthus* in the Aegean region. (1) *E. candicum*: (A) subsp. *candicum* and (B) subsp. *carpathum*. (2) *E. senoneri*: (A) subsp. *senoneri*, (B) subsp. *icarium*, and (C) subsp. *amorginum*. (3) *E. corinthium*, (4) *E. naxense*, and (5) *E. rhodium* (redrawn from Snogerup 1967).

1970, 1972) provides a parallel illustration of differentiation in a small population system in the Aegean islands, which he argued to be the result of allopatric differentiation as a result of genetic drift. He provided a comprehensive account of distinct morphological variation among the different taxa which make up the *N. arvensis* complex (Fig. 3.11(a)). On different islands in the Aegean Sea, morphological variation is highly discontinuous among the different taxa which differ strikingly in growth habit and a range of morphological characters. Across this archipelago there is thus a mosaic of non-overlapping distributions of morphologically distinct species and subspecies on different islands (Fig. 3.11(a)). These sharp morphological discontinuities among island populations stand in stark contrast to more clinal morphological variation among continental populations of two subspecies of *N. arvensis* (Fig. 3.11(b)). In current work on the nuclear and cpDNA sequence variation in 60 populations of the *N. arvensis* complex in the same region, H. Bitkau and H.P. Comes (University of Mainz, unpublished data) have analysed in detail the hypothesis that fragmentation during the Pleistocene and genetic drift in small isolated populations are responsible for the pattern of differentiation. Nuclear DNA sequences suggest a recent diversification at 1.7–2 Ma. For cpDNA variation, they have found that >80% of variation is due to differences among island populations, indicating that cytoplasmic gene flow is too low to prevent differentiation due to drift in small isolated fragments. Diversity was markedly higher among island populations than among continental populations and there was no evidence for isolation due to distance. In fact all the analyses made by these workers concord with the hypothesis proposed by Arne Strid: allopatric fragmentation during periods of climate change and isolation of small populations have set the scene for random differentiation due to genetic drift. Patterns of morphological differentiation in *Ranunculus* species on the Aegean islands, where 'almost every population could be regarded as a separate taxon' (Dahlgren and Svensson 1994: 268) further illustrate the extent of morphological variation that can be observed in this archipelago. Studies of the distribution patterns of other species in this region provide further evidence that patterns of differentiation result from allopatric fragmentation and that contemporary patterns are relicts of

(a)

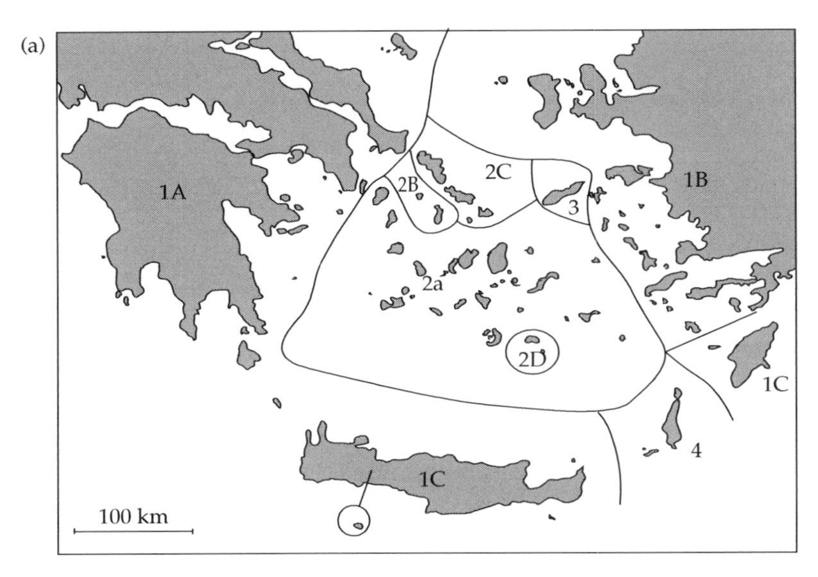

(b)

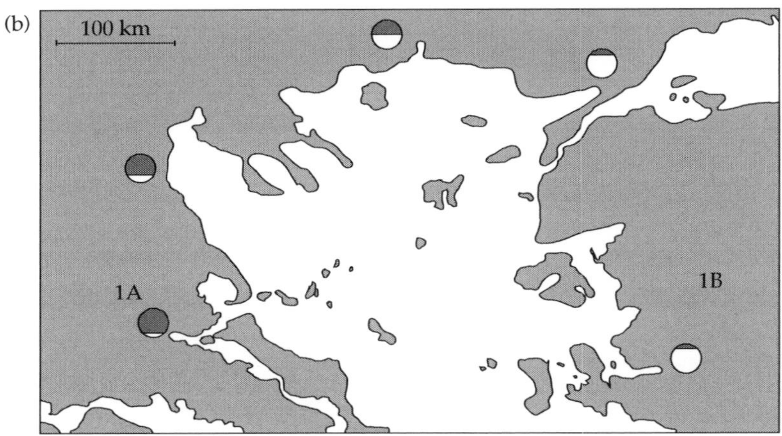

Figure 3.11 (a) The distribution of different taxa in the *N. arvensis* complex in the Aegean. (1) *N. arvensis*: (A) subsp. *aristata,* (B) subsp. *glauca*, and (C) subsp. *brevifolia*. (2) *N. degenii*: (A) subsp. *degenii*, (B) subsp. *jenny*, (C) subsp. *barbro*, and (D) subsp. *minor*. (3) *N. icarica*, and (4) *N. carpatha*. (b) Clinal variation in morphology among continental populations of two subspecies of *N. arvensis* in the northern Aegean region (redrawn from Strid 1972).

the once continuous distribution patterns (Greuter 1979).

One member of the *N. arvensis* complex, *Nigella degenii,* shows striking morphological variation among populations on the different islands where it occurs, to the extent that different subspecies have been recognized in the different parts of its range. This is in marked contrast to *Nigella doerfleri* which is morphologically more uniform across the Cyclades, Crete, and Andikithira. Whereas most *Nigella* species have traits which suggest predominant outcrossing (although most are thought to be self-compatible), *N. doerfleri* is one of only two species in the Aegean region with characteristics typical of a selfing species (Table 3.3). In addition, this species bears remnants of floral structures present in the outcrossing members of the genus (e.g. nectar producing petals). The morphological reduction,

Table 3.3 Biological traits of two *Nigella* species in the Aegean (from Strid 1969)

	Nigella degenii	*Nigella doerfleri*
Flowering time	May–July	Mid-April–May
Habitat	Large, mesic islands	Also on low-lying arid islands
Pollen production	High	Low
Seed fertility	Variable	High
Flowers	Large and highly coloured	Smaller and less coloured
Morphology	Highly variable among islands	Relatively uniform among islands

rapid development, and arid habitat of *N. doerfleri* fit the general pattern for the evolution of selfing. Based on the offspring of controlled crosses, Andersson (1997) suggested that pleiotropic relationships between floral and leaf traits may have facilitated a reduction in floral morphology of isolated populations in arid habitats in association with the evolution of inbreeding. The more uniform morphology of *N. doerfleri* across its range is indicative that original isolation and evolution was a highly localized phenomenon and that this species has subsequently expanded its range across the Aegean (either along ancient land-bridge connections or by dispersal across water barriers in recent history). A reliance on selfing may have allowed for rapid reproduction on small arid islands and provided reproductive isolation on the larger mesic islands where contact with other taxa is possible.

Finally, crosses among *N. arvensis* subsp. *aristata* and subsp. *glauca* are sterile, indicating that geographic isolation is associated with the evolution of sterility barriers (Strid 1970). Since this species has a fairly continuous distribution with areas of mixed populations between the subspecies, sterility of hybrid types may have facilitated the evolution of different forms in the extreme parts of the species range. In contrast, isolated subspecies on different islands produce fertile hybrids when crossed in cultivation (as do geographically isolated *Cyclamen* and *Centaurea* discussed above). In continental areas, reproductive isolation via hybrid sterility may thus be a key component of taxonomic divergence by virtue of its role in the maintenance of different forms in the absence of geographic barriers to gene flow (another example is shown in Box 3.3).

3.5 Hybridization and chromosome evolution

Common in plants, rare in animals, hybridization and polyploidy are key elements in plant evolution (Stebbins 1950; Levin 1983; Abbott 1992; Thompson and Lumaret 1992; Arnold 1997; Rieseberg *et al.* 2003). Hybridization and polyploidy can occur independently or in concert. The incorporation of two diploid genomes into a hybrid species which has a complete set of chromosomes from each parental species, the process of allopolyploidy, is a primary force in plant evolution. Hybridization in the absence of chromosome doubling, can have manifold effects on plant performance and evolution hence its recognition in the titles of some of the classic papers as being an 'evolutionary stimulus' (Anderson and Stebbins 1954) or a 'source of variation for adaptation to new environments' (Lewontin and Birch 1966) which more recently has been confirmed in the paper by Rieseberg *et al.* (2003), whose title reveals that 'major ecological transitions . . . [are] . . . facilitated by hybridization'. Finally, polyploidy can occur within a single species (autopolyploidy) in the absence of hybridization via the production of unreduced gametes (Bretagnolle and Thompson 1995). The role of these two processes in plant evolution took some time to be accepted in the evolutionary literature. The development and application of molecular tools in the last two decades of the twentieth century has however allowed evolutionary biologists to detect more and more clear examples of autopolyploidy and also to illustrate the hybrid origin of new species in the absence of chromosome doubling (homoploid diploid speciation), mostly based on comparative analyses of phylogenetic datasets based on nuclear and cytoplasmic genes (Rieseberg 1997). In this section, I provide illustrations of the role of hybridization and polyploidy in the evolution of diversity in Mediterranean plants. I tackle hybridization first, using a series of examples to illustrate its role in geographic differentiation, ecological adaptation, and speciation.

Box 3.3 Reproductive isolation and divergence in *Campanula dichotoma* (redrawn from Nyman 1991)

Campanula dichotoma contains a range of geographic morphological variation, shown here as different symbols, which some authors variously classify as endemic species or subspecies (Nyman 1991). The distribution patterns of the different morphological variants cross contemporary geographic barriers to dispersal, providing an illustration of the historical connections from the south-western Balearic islands, southern Spain, Morocco across North Africa to Sicily and southern Italy.

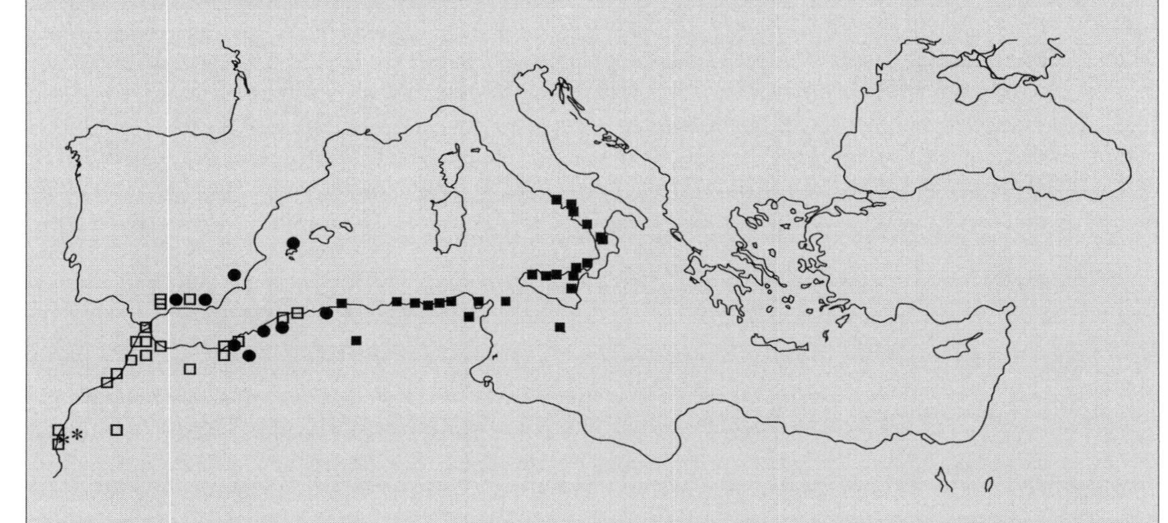

The distribution pattern of the four variants are almost non-overlapping in areas where there are no major geographic discontinuities, indicative of reproductive isolation. Although crosses among them produce seeds, the latter have poor viability. There are thus post-zygotic barriers to hybridization, which may facilitate divergence. The variants differ in floral traits and their ability to self-pollinate suggests that selfing may have facilitated differentiation, reproductive isolation, and their ability to colonize new areas.

The complex and dynamic nature of the history of the Mediterranean region means that fragmentation, contraction, and expansion of species ranges have repeatedly occurred. In this context, hybridization and polyploidy play prominent roles in plant evolution. A series of recent talks at the IOPB meeting on 'Plant Evolution in Mediterranean Climate Zones' in Valencia illustrate the repeated occurrence of hybridization in many genera in the Mediterranean, for example, *Limonium*, *Helichrysum*, *Phlomis*, *Armeria*, *Orchis*, etc. In some groups, hybridization may in fact be rife. In the genus *Thymus*, 60 hybrids have been reported in the Iberian peninsula where a total of 35 species occur (Morales 2002). Of the six possible hybrid combinations among four species of *Rhododendron* which have sympatric localities in Turkey, five showed evidence of hybrid formation in the wild (Milne *et al.* 1999) These authors illustrate that ecological differentiation favours reproductive isolation. This introduces an issue which I will emphasize in this section, genetic changes associated with hybridization and polyploidy provide the stimulus and ability for ecological adaptation.

3.5.1 Hybridization and speciation in *Senecio*

Molecular studies of the genus *Senecio* illustrate the role of hybridization in species divergence and the ecological success of derivative species.

Senecio flavus and *S. glaucus* have parapatric distributions in North Africa and the south-east corner of the Mediterranean region. A study of cpDNA and nuclear DNA diversity in section *Senecio* (in which both species occur) illustrates reticulation due to hybridization (Comes and Abbott 1999*a*,*b*; 2001). In a nutshell, although RAPD nuclear markers produce a phylogenetic tree which concords with the morphological classification of the section, cpDNA haplotypes and sequence variation in the ITS nuclear gene of *S. flavus* are more akin to those of *S. glaucus* than to the other species that are thought to be its closest relatives (Fig. 3.1). The latter result is strong evidence for hybridization and subsequent introgression of genes from *S. glaucus* into *S. flavus*. The combined use of different markers shows that the introgression of cpDNA haplotypes and sequence variation in the ITS nuclear region are not accompanied by introgression of the rest of the genome. Hybridization between these two species is also thought to have occurred early in the Pleistocene when climatic changes may have caused distribution changes which in turn brought the two species into reproductive contact. Hybridization, accompanied by chromosome doubling, is thought to have produced a third (polyploid) species *Senecio mohavensis* sometime in the Late Pliocene or Early Pleistocene (Coleman *et al.* 2001). The latter now occurs as subsp. *breviflorus* in the eastern Mediterranean and the arid areas of south-west Asia and as subsp. *mohavensis* in south-west North America. During the Pleistocene, *S. mohavensis* is thought to have dispersed to North America by long distance dispersal, perhaps in association with bird migration rather than wind dispersal (Coleman *et al.* 2003). The historical occurrence of hybridization and diversification has thus been modulated by climatic changes which have enabled contact and then isolated taxa in disjunct areas. In addition, recent hybridization may have provided the genetic material that has enabled *S. flavus* to diversify and expand its range.

A second example from the genus *Senecio* illustrates how hybridization can produce variant types which may produce a new species where reproductive isolation from the parents occurs. The case in point involves *Senecio squalidus*, a ruderal species in Great Britain which has colonized Britain since its introduction to the Oxford Botanical Gardens from seed sources collected on Mt Etna (Sicily). There is now molecular evidence that the seed source used in the introduction originated from hybridization between two Sicilian endemic species, *Senecio aethnensis* and *Senecio chrysanthemifolius* (Abbott *et al.* 2000, 2002). Studies of morphological and isozyme variation in cultivation provide two lines of evidence for this claim. First, *S. squalidus* closely resembles plants derived from hybrid swarms of the two endemic species present on Mt Etna. Second, *S. squalidus* also contains genes present in pure populations of the two endemic 'parental' species. The evidence thus points to a hybrid origin of a diploid species (all three species have the same diploid chromosome number of $2n = 20$), and thus a new example of homoploid hybrid speciation (see Rieseberg 1997 for the few known examples of this phenomenon). In this case, speciation in the natural range of the different taxa has not produced a new species probably because of a lack of reproductive isolation. In contrast, introduction of hybrid material to the Oxford Botanical Gardens in the eighteenth century has allowed for stabilization of hybrid material and speciation to occur in allopatry. So although the first step in the process of speciation occurred in the wild, via hybridization, the next and necessary step of stabilization and reproductive isolation only occurred in a geographically isolated site. The fact that neither of the two parental species have spread following their introduction to Britain suggests that the hybridization has conferred an ability to colonize a new environment, very different to that in which the parents occur on Mt Etna.

3.5.2 Geographic differentiation via hybridization in *Armeria*

The genus *Armeria* has a holarctic distribution, with a centre of diversity in the Iberian peninsula where ~60% of the ~120 mostly diploid species occur

(Fuertes Aguilar *et al.* 1999). Phylogenetic analyses of ITS nuclear gene sequence data for 55 samples in the Iberian peninsula (35 species and several samples of geographically disjunct subspecies of the morphologically variable *Armeria villosa*) is indicative of a significant role for reticulation in tree structure (Fuertes Aguilar *et al.* 1999). First, the different subspecies of *A. villosa* do not always group together within a single clade. Second, the composition of the five main clades detected in the analyses is more congruent with geographic distribution than their predicted systematic relationships. This pattern could be produced by lineage sorting in a highly polymorphic ancestral species. This is unlikely in *Armeria*, since it would require an extremely variable ancestral species. Lineage sorting would also require a loss of polymorphism (or lack of sampling) within accessions in the different clades. These two conditions make lineage sorting an unlikely cause of the separation of different subspecies of *A. villosa*.

A more plausible interpretation for the polyphyly in the gene tree, and thus the apparition of the morphologically different and geographically separated subspecies of *A. villosa*, is reticulation due to hybridization between a widespread ancestral subspecies of *A. villosa* and geographically contiguous populations of other species. In *Armeria* many taxa are cross-compatible, hence the formation and establishment of hybrids is a distinct possibility where contact occurs between different species. What is more, the morphology, ecology, and geographic distribution of the different subspecies in relation to the sympatric species with which there may have been hybridization, support the claim that the different subspecies have originated *via* hybridization (Fuertes Aguilar *et al.* 1999). These authors interpret their data by suggesting that *A. villosa* has 'captured' genetic polymorphism from sympatric taxa by introgression. In the case of *A. villosa*, failure to develop reproductive barriers with other species may have allowed hybridization with local sympatric species to promote geographic differentiation within the widespread species and in so doing stimulate adaptation to local ecological conditions.

In the Sierra Nevada, three *Armeria* taxa which share cpDNA haplotypes, *A. villosa*, *A. splendens*, and *A. filicaulis*, currently occur in distinct altitudinal belts (Gutiérrez Larena *et al.* 2002). The most plausible explanation for the haplotype sharing is that there have been repeated cycles of distributional changes associated with climatic warming and cooling during the Quaternary glaciations which initially allowed the three taxa (perhaps independently) to colonize the region. Climatic oscillations probably caused range extensions and contractions; allowing contact and subsequently isolating introgressed populations. Once again, the action of hybridization in plant evolution has been modulated by historical changes in climate which caused distribution changes and the possibility of reproductive contact among previously isolated but closely related taxa.

3.5.3 Hybridization at the polyploid level

The combination of two complete but different genomes within a single individual via hybridization and chromosome doubling is the process of allopolyploidy. Unlike many autopolyploids, allopolyploids are almost immediately recognized as new species due to their frequent morphological dissimilarity and their almost immediate reproductive isolation from parental diploids. The genetic diversity that this process introduces at the individual level, is thought to be critical for the spread of allopolyploid plants (Thompson and Lumaret 1992). In addition, allopolyploids, with independent origins, can hybridize among each other. Such hybridization may reshuffle available genetic variation and further contribute to evolution and speciation. Two examples of this phenomenon, whose domesticated relatives have become very well-known cultivated or horticultural plants, are the wild ancestors of cultivated wheats and garden peonies. These two groups illustrate how hybridization at the tetraploid level has played a decisive role in the establishment, persistence, and evolution of polyploid plants.

Wild wheats occur in two genera, *Aegilops* and *Triticum*, which occur wild in the eastern Mediterranean and the Irano-Turanian floristic province (Harlan and Zohary 1966). These genera contain six groups of diploids and three clusters of polyploids based on their genomic group. In the polyploids, the main 'pivotal' genome is identical to one chromosome set at the diploid level. Wild species within a cluster differ as a result of the additional genome they contain, which is not identical to one of the diploid ancestors. Polyploids thus show a rather peculiar situation in that they contain one 'unaltered' genome, common to a group of species, and one modified genome. This situation may have been critical for their evolutionary success which appears to have been determined by hybridization among different polyploids (Zohary and Feldman 1962; Zohary 1965).

Diploid wheats represent species with a sound taxonomic base. Each species shows limited variation in vegetative traits and characteristic spike morphology and seed dispersal strategy and can readily be classified into one of six genome groups. Species in different groups are isolated by sterility barriers. Diploids also have fairly specialized ecological requirements. At the diploid level, one thus observes distinctive morphology and habitat specialization. A very different situation occurs in the polyploids, which show marked variation in morphology within species and overlap among species, such that a series of morphological intermediates make species delimitation difficult. Some polyploids exhibit combinations of traits observed in different diploid groups. The differences in dispersal strategies observed at the diploid level do not occur among polyploids, and polyploid species in different clusters are not fully intersterile. As a result, the genome clusters resemble aggregates or species complexes.

In wild wheats, polyploids tend to be weedy with large overlaps in geographical and ecological distributions. Colonization of disturbed open habitats would appear to be closely related to hybridization among different polyploids, perhaps as a result of range changes in response to climatic variations and habitat availability during the Quaternary. The common genome in each cluster may have served as a buffer during the hybridization process and a source of the pivotal genomes, which Zohary (1965: 417) describes as the 'main evolutionary themes for the polyploid clusters'. The presence of a second genome has allowed for the introgression and differential modification that has evolved during the spread of the polyploids, a process also apparent during domestication (see Chapter 6). It is in this modified part of the genome that new chromosomal combinations, that is, 'the material for variations on the three basic themes' (Zohary 1965: 417), has evolved. The recent rise and spread of the wild polyploid wheats may thus have been favoured by the juxtaposition of two genome complements.

The Mediterranean is an important centre of distribution and diversity of wild peonies. In addition to a group of narrow patro-endemic diploids often restricted to islands, for example, *Paeonia rhodia* (Rhodes), *Paeonia clusii* (Crete and Karpathos), and *Paeonia cambessedesii* (Balearic islands) and continental diploids with more widespread distributions, for example, *Paeonia broteroi* (fairly widespread across the Iberian peninsula), a number of tetraploids also occur in the Mediterranean region, for example, *Paeonia coriacea* (southern Spain, Corsica, Sardinia, and North Africa), *Paeonia officinalis* (south and central Europe), and *Paeonia mascula* subsp. *arietina* (south-east Europe to Turkey). Most of the polyploids have more widespread distribution than the diploids, indicative that the genetic variability inherent in the polyploids may have favoured post-glacial re-colonization in the Pleistocene. The polyploids show much evidence of reticulation in phylogenetic analyses and discordance between nuclear and cpDNA gene trees (Sang *et al.* 1997; Ferguson and Sang 2001), two results that provide strong evidence that hybridization has played an important role in the colonization potential of polyploid peonies.

An intriguing and novel feature of this hybridization is that it has involved hybrid speciation at the polyploid level. For example, *P. officinalis* has evolved after hybridization between two allopolyploids, one in the group containing allopolyploid *Paeonia arietina* and one related to *Paeonia peregrina* (Ferguson and Sang 2001). The genetic configuration of this species in terms of similar copies of particular genomes (Ferguson and Sang 2001) may have facilitated meiotic chromosome pairing and thus the initial establishment of hybrid populations. As Ferguson and Sang (2001: 3918) point out, 'the integration of genes from multiple genomes is likely to have provided the variability necessary for the colonization of new, more wide-ranging habitats'. Such integration of multiple genomes has occurred in the absence of constraints associated with further chromosome doubling. This example also illustrates how current distributions can bear little resemblance to historical distributions, species now present only in Asia must have coexisted at some time (probably in the late Tertiary) with European Mediterranean species for the proposed hybridization to have occurred.

Increased colonization capacity may thus be strongly linked to hybridization among diploids to produce allopolyploids or among independently produced tetraploids (an idea championed by Stebbins 1985). The polyploid complex *Dactylis glomerata* illustrates this theme (Lumaret 1988). In this species nine diploid subspecies have disjunct endemic distributions within different parts of the Mediterranean region: for example, subsp. *ibizensis* on the Balearic islands, subsp. *juncinella* at high altitude in the Sierra Nevada and subsp. *reichenbachii* in northern Italy. This diversification at the diploid level has probably occurred since the end of the Tertiary. Diversification of diploids has been followed by independent production of different polyploids in the different parts of the species range. Although the derivative polyploids have roughly the same distribution pattern as the diploids they are not schizoendemic taxa, since they have independent origins from already differentiated diploid subspecies (Chapter 2). *D. glomerata* also contains three main tetraploid types that have cosmopolitan distributions across temperate or Mediterranean Europe and on the Macaronesian islands. These taxa have spread across an open landscape since the last glaciation, often in association with human activities which have spread their seeds actively for the creation of pastures for grazing. The three widespread tetraploids show distinct patterns of morphological variants in the different parts of their range indicative of hybridization with local tetraploid subspecies (Lumaret and Barrientos 1990). Crossing studies have shown that fertile hybrids can be obtained from crosses among several different diploid subspecies and among different polyploid subspecies (Lumaret 1988). Such hybridization may have been a key element in the successful spread of the now cosmopolitan tetraploids as a result of the incorporation of locally adapted genes from the endemic taxa (as in *A. villosa* discussed above). Likewise, distribution patterns in *Cruciata* species and in *Knautia dipsacifolia* also illustrate, to cite Ehrendorfer (1980: 49), how 'neopolyploids which combine genomes from different diploids (or . . . polyploids) often surpass their ancestors in variation, adaptability and invasion potential in new habitats and virgin areas'.

3.5.4 Reproductive isolation and the maintenance of genetic integrity in parental stocks

The above examples illustrate that hybridization can be frequent and have important effects on evolutionary potential. This raises the question of how species maintain their genetic integrity where their distribution overlaps with other closely related species. Reproductive isolation can be favoured by a number of traits which either by preventing pollen exchange among species (e.g. ecological differences in flowering phenology, pollinators, and habitat or if one species is highly selfing) or by preventing the formation of viable offspring (e.g. cross incompatibility and hybrid inviability). Depending on the precise situation, different factors are likely to be involved, ecology and pollination being particularly important.

Pre- and post-zygotic isolation in Mediterranean orchids

Orchid flowers are often thought to have highly specialized interactions with particular pollinators, which may contribute to reproductive isolation among coexisting species. In Mediterranean *Ophrys*, different species can have distinctly shaped and coloured flowers which produce a different bouquet of monoterpenes in their floral fragrance, traits which may differentially influence pollinator attraction and which should limit pollinator visitation to congeners. Different species often differentially attract male Hymenoptera, often specific to particular species, hence the possibility that odour, in combination with the shape and colour of the flowers, contributes to limit hybridization and maintain the integrity of different species in sympatry (Paulus and Gack 1990). Hybridization has nevertheless been detected among sympatric species, for example, between *Ophrys lutea* and *O. fusca* in North Africa (Stebbins and Ferlan 1956). In the group of closely related species that make up the *O. sphegodes* group, different species often flower simultaneously in sympatry in southern France and Italy. Soliva and Widmer (2003) have shown that although genetic differentiation among geographically distant populations of the same species was lower than differentiation among sympatric populations of different species, the strength of genetic differentiation among species was lower than that usually observed for different orchid species, probably as a result of some gene flow among species in sympatry. So, although in this group the sexual deceit pollination system may be less specific than thought, perhaps as a result of occasional mistakes by pollinators, the fairly specific nature of the *Ophrys*-pollinator interaction may be a key element of pre-zygotic reproductive isolation among taxa and thus a potentially important feature of orchid speciation (Paulus and Gack 1990). As I mention later in this chapter, post-zygotic factors such as chromosome divergence may also be important here.

Ecological isolation in Senecio

Population-level analyses of genetic structure in two morphologically distinct, highly outcrossing *Senecio* species whose flowering times and geographic distributions slightly overlap to form a zone of secondary contact in the eastern Mediterranean provides instructive information concerning the maintenance of species differences in the face of potential hybridization and introgression. Whereas *S. glaucus* is typically found in sandy dunes and arid regions of the steppic Irano-Turanian floristic zone or the Saharo-Arabian desert zone of the Near East, *S. vernalis* is very common in anthropogenic and disturbed habitats of the mesic-Mediterranean zone (Zohary 1973). Zones of potential hybridization occur in the eastern Mediterranean where there is clear evidence for introgression of cytoplasmic genes, although this has not impacted on the phenotypic morphological integrity of the two species (Comes and Abbott 1999*b*). The cpDNA introgression may thus be sporadic and historical rather than a common feature of contemporary populations, whose ecological isolation may limit such introgression and maintain species integrity.

3.5.5 The evolutionary significance of autopolyploidy

Polyploidy, even in the absence of hybridization among taxa, is a major process of plant evolution which allows the different stages of speciation, that is, genetic origins and population establishment and persistence, to be observed and studied experimentally (Thompson and Lumaret 1992). The occurrence of closely related taxa of different ploidy level is also an integral component of Mediterranean endemism (Verlaque and Contandriopoulos 1990). Finally, polyploidy provides fascinating model systems for the study of how taxa maintain their genetic integrity in areas where polyploids come into secondary contact with their related diploids. In plants, differentiation and diversification in association with polyploidy can occur following hybridization among closely related species (see above) or by chromosome doubling in a single taxon (autopolyploidy), usually as a result of fusion between gametes with an unreduced chromosome number (Bretagnolle and Thompson 1995).

A key question here concerns how newly formed polyploids establish and spread. A requirement for the establishment of any new derivative population is that it attains some form of reproductive isolation from its progenitors. A difference in chromosome number is, of course, very important here since hybrids may be immediately checked as a result of chromosomal imbalance. In addition, a doubling of the chromosome number, even in the absence of hybridization, can have manifold consequences since the increased number of gene copies at individual loci allows for new genetic combinations which provide a template for future evolution (Levin 1983). Such genetic changes may allow for polyploids to tolerate and colonize a wider range of ecological conditions by two mechanisms. First, individual plants may have increased capacity to modify their phenotype (i.e. show phenotypic plasticity) in different environments. Alternatively, the higher genetic variability may open new avenues of adaptive evolution in different environments. To understand these issues requires comparative study of the distribution of diploids and polyploids in order to quantify the processes which (a) favour the coexistence of diploids and derivative polyploids (both in primary and secondary contact) and (b) enable the spread of polyploids. The successful establishment and persistence of new polyploid variants may nevertheless require subsequent mutation, regularization of chromosome pairing, inactivation of duplicate loci, or hybridization among related but independently produced polyploids to further facilitate persistence (Stebbins 1980).

In the Mediterranean, polyploidy, with or without hybridization has been of prime importance in plant evolution. The fact that previous authors have been able to classify endemic species on the basis of ploidy variation (see Chapter 2) indicates the importance of ploidy variation in the flora and its endemic complement of species. Two important trends should be distinguished here (Verlaque and Contandriopoulos 1990). First, polyploidy has played a major role in the diversification of new (apo-) endemic Mediterranean taxa. Second, the flora is home to a large number of (patro-) endemic diploid species which represent the progenitor taxa of polyploids that have spread to occupy more wide-ranging distributions. It is thus not surprising that diploids and polyploids do not show consistent differences in geographical distribution (Ehrendorfer 1980; Stebbins and Dawe 1987).

3.5.6 Polyploidy and ecological differentiation

A common feature of distribution patterns in polyploid groups concerns the occurrence of segregated distributions of different ploidy levels. What this means is that polyploid complexes can be subdivided into 'single cytotype areas' over large parts of their geographic range. These single cytotype areas are thought to be maintained by the exclusion of the minority cytotype which incurs a reproductive disadvantage when it is rare (Levin 1975). This occurs because a rare cytotype, for example, a new polyploid in an otherwise diploid population, will primarily be pollinated by, and its pollen lost on the stigmas of, the majority cytotype. Such inter-cytotype mating may come at a significant cost due to hybrid inviability associated with triploid formation (e.g. see van Dijk *et al.* 1992).

Triploid inviability is common to many polyploid complexes, and may lead to the selection of reproductive isolation where the two cytotypes are in reproductive contact. In the polyploid complex *D. glomerata* crosses among diploids and polyploids are for the most part incompatible (Lumaret 1988). The offspring that are produced are either triploid or tetraploid. In the multiple contact zones which occur among diploids and polyploids of this species around the Mediterranean very few triploid hybrids have been found (Borrill and Lindner 1971; Lumaret 1988; Lumaret and Barrientos 1990). The absence of triploids in most areas is due to a combination of pre-zygotic isolation caused by differences in flowering time (Lumaret and Barreintos 1990; Bretagnolle and Thompson 1996) and habitat differentiation (Lumaret *et al.* 1987) and post-zygotic isolation caused by the abortion of embryos or lack of successful fertilization as a result of the unbalanced chromosome numbers in a diploid × tetraploid cross. The frequent absence of any triploids in *D. glomerata* contact zones is indicative that direct formation of tetraploids and crosses among diploids and tetraploids occur via the production of unreduced gametes in the diploids (Borill and Linder 1971; Bretagnolle and Thompson 1995). An exception to this trend concerns the occurrence of relatively large numbers of triploids in a mixed diploid–tetraploid population in Israel (Zohary and Nur 1959). Such triploids produced tetraploid offspring, indicating that triploids may serve as a bridge for gene flow from the diploid to tetraploid level. This 'triploid bridge' facilitates the incorporation of genetic variability into the polyploid and thus enhances their probability of persistence and spread. There is also strong evidence of gene flow among locally present ploidy levels via the 'triploid bridge' in *Ornithogalum umbellatum* (Box 3.4). Such gene flow can limit morphological differentiation and thus maintain genetic

Box 3.4 Variation in vegetative propagation and gene flow among the different ploidy levels of *Ornithogalum umbellatum* (from Moret 1991; Moret and Favereau 1991; Moret *et al.* 1991; Moret and Galland 1992)

Ornithogalum umbellatum contains diploids ($2x = 18$), triploids ($2x = 27$), tetraploids ($2x = 36$), pentaploids ($2x = 45$), and hexaploids ($2x = 54$). Diploids occur mostly in natural mountain areas close to the coast in the western Mediterranean, whereas higher ploidy levels occur to the north and in non-Mediterranean temperate Europe where they are common in disturbed habitats.

Polyploids above the triploid level rely on clonality and asexual reproduction, in which a combination of two strategies—vegetative propagation and apomixis—assure population persistence. Diploids are more dependent on normal sexual reproduction. In diploid populations which do practice vegetative propagation (production of sister bulbs), the precise modality is different to that of tetraploids (which propagate by bulbils).

The surprisingly high pollen and seed fertility of triploids, which occur in a number of disjunct populations, illustrates that triploidy can provide an effective means of gene transfer among ploidy levels. In fact, diploid and triploid populations in the Mediterranean region are morphologically more similar to local polyploids than to other disjunct populations of the same ploidy level.

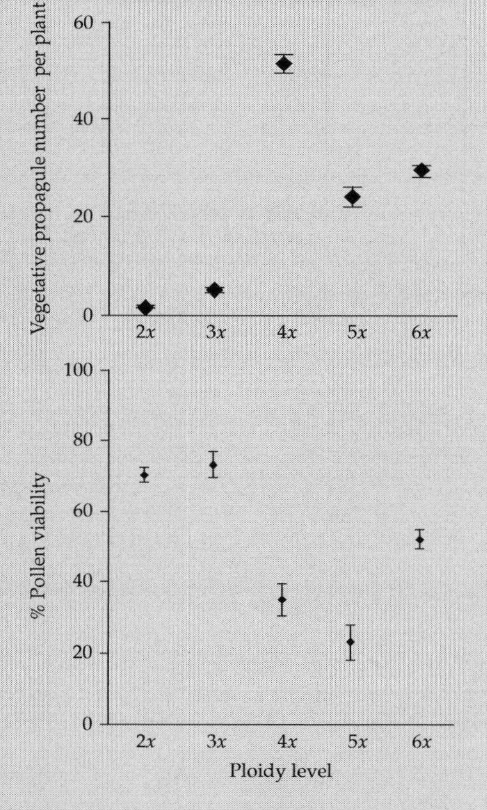

There is thus strong evidence that gene flow across the 'triploid bridge' maintains genetic cohesion among ploidy levels and introduces important genetic variation into the polyploids. The different ploidy levels can thus be considered as a single species complex in which spatial isolation is more important than ploidy level in the evolution of morphological differentiation.

cohesion in a polyploid complex as well as permitting the transfer of genetic variability among ploidy levels.

Unless a polyploid is reproductively isolated from its parental diploids, minority cytotype disadvantage may cause the elimination of one or other of the different cytotypes. An important question thus concerns whether polyploids differ from diploids in terms of their ecological amplitude. A response to this question is a first step towards understanding the conditions which regulate polyploid establishment and persistence and whether colonization ability in polyploids is determined by a greater capacity to buffer environmental variation (phenotypic plasticity) or more the result of adaptive differentiation to spatial heterogeneity of selection pressures. Two pieces of evidence suggest that in a given geographical region, polyploids do not generally have a wider ecological amplitude than diploids.

The first line of evidence comes from a study of the ecological distribution of endemic species on Corsica by Contandriopoulos (1962) who estimated the altitudinal distribution (in five altitudinal belts: littoral, lowlands <600 m, uplands 600–1,200 m, sub-alpine 1,200–1,800 m and alpine >1,800 m) of more than 100 diploid and tetraploid endemic species. If one compares the distribution of diploid and tetraploid species in her list (it is possible to do so for 29 polyploids and 65 diploids) one finds that both diploids and tetraploids occur in a surprisingly similar number of altitudinal belts (mean = 2.08 for diploids and 2.07 for polyploids). Second, in a study of the ecological amplitude of species in the flora of the Pyrenees, Petit and Thompson (1999) found no significant variation in the ecological amplitude of diploids and polyploids. This is not to say however that different ploidy levels occur in the same types of habitat. In fact, Petit and Thompson (1999) showed that the presence of polyploidy in a taxonomic group is closely associated with taxonomic diversity, which itself is closely correlated with the ecological range of the taxonomic group. This result suggests that successful establishment of polyploids, and thus taxonomic diversity of a given group, is facilitated by ecological divergence from progenitor diploids and adaptive differentiation in polyploids, and not the capacity to colonize a wider range of environments. Several other studies support this idea.

In their classic work on endemism and speciation in California, Stebbins and Major (1965) showed that patro- and apo-endemic species differ in their ecological requirements. Whereas patro-endemics are most frequent in the Central Coast Region, particularly where summer fogs maintain humidity, and at mid-elevation where an old stable giant redwood forest flora has persisted, apo-endemics are more frequent in areas with abrupt climatic gradients between two distinct areas (where ancestral diploids may have occurred). The frequency of recently derived endemics is thus closely correlated with climatic and ecological heterogeneity, which may favour the establishment of hybrid derivatives and the ecological segregation of autopolypoids.

Ehrendorfer (1980) suggested that in many groups diploids are more common in stable habitats of permanent climax communities whereas polyploids are often found in more open, disturbed successional communities, which often involve habitats that became available for colonization after the Pleistocene glaciations. This increased colonization capacity may be strongly linked to hybridization among diploids to produce allopolyploids (see examples above). In the genus *Pinguicula*, the widespread species in Europe are those with a high ploidy level, whereas diploids tend to have restricted ranges, for example, the patro-endemic *Pinguicula corsica* is endemic to the island from which it gets its name. In this genus, the different species and ploidy levels differ clearly in their occupation of limestone or acidic substrates (Contandriopoulos 1962). A similar pattern of European-wide distribution of polyploids and fragmented and smaller distribution ranges of diploids in Mediterranean Europe can be seen in many other plant groups, for example, *Plantago media* (van Dijk *et al.* 1992), *Rumex acetosella* (den Nijs 1983; den Nijs *et al.* 1985), *Cardamine pratensis* (Lihová *et al.* 2003), *D. glomerata* (Lumaret 1988), and *Arrhenatherum elatius* (Petit and Thompson 1997).

To quantify the capacity of diploids and tetraploids to buffer environmental habitat variation and/or adapt to variation in irradiance levels, Petit and Thompson (1997) studied the response of diploid and tetraploid populations of *A. elatius* to artificial shading. They found that tetraploids have greater vegetative size and flower production than diploids, as reported in other species, for example, *D. glomerata* (Bretagnolle and Thompson 2001) and *Anemone palmata* (Médail *et al.* 2002). Tetraploids did not however have a greater ability to buffer environmental variation than diploids. The plasticity of morphological traits, flowering phenology, and fertility were similar in the two cytotypes. In contrast, tetraploid populations showed greater differentiation among woodland and open habitat populations than diploid populations, indicative of local adaptation. In subsequent work (Petit and Thompson 1998) the direction of selection in these two types of habitat was reported to concord with the patterns of morphological variation, that is, taller plants favoured in open habitats. In this complex it would thus appear that adaptive differentiation among tetraploid populations has been an integral component of the spread of the polyploid cytotype into different environments.

A final example which illustrates the role of ecological differentiation in polyploid evolution comes from the classic work on annual *Mercurialis annua* in the western Mediterranean by Durand (1963). In this highly diverse species the three elements of my introductory triptych, geological history, climate change, and human activities, have jointly shaped differentiation and distribution in a single species in relation to ecological and genetic processes acting on plant diversification (Box 3.5).

The argument I thus propose is that the divergence and evolutionary success of polyploids is closely related to their ability to adapt to new ecological conditions. The new gene combinations present in polyploids, and the reshuffling of such variation by natural selection, are likely to be key ingredients here. The extreme rarity of mixed cytotype populations, even in overlapping regions, suggests the strong selective importance of local ecological conditions in this process.

3.5.7 Polyploidy and the evolution of uniparental reproduction

The potential for uniparental reproduction, either by a greater reliance on selfing or vegetative reproduction, is a reproductive strategy that may have contributed to the reproductive isolation of polyploids and their colonization capacity in new areas (Thompson and Lumaret 1992). Different groups in the Mediterranean illustrate these trends.

1. In *Senecio* sect. *Senecio*, diploids primarily have long well-developed outer ray florets which are female in addition to the central hermaphrodite disc florets. All examined diploids appear to be self-incompatible (R.J. Abbott, University of St Andrews, personal communication). In contrast, in tetraploids and hexaploids the outer female ray florets are absent or reduced in size (Alexander 1979) and they are self-compatible.
2. An improved capacity for vegetative reproduction may also have allowed many polyploids to establish in new, often disturbed, environments not exploited by their diploid progenitors. Examples of this increased reliance on vegetative propagation can be seen in *Tulipa oculus-solis* in the eastern Mediterranean (Horovitz *et al.* 1972) and in different *Ornithogalum* species in the eastern (Kushnir *et al.* 1977) and western (Moret and Favereau 1991) Mediterranean (Box 3.4).

In *Mercurialis annua* the trend towards increased selfing is also apparent, but with no direct link to colonization capacity. Diploids in temperate Europe are dioecious whereas the diverse Mediterranean polyploids are primarily monoecious and self-compatible and thus subject to some degree of inbreeding (Durand 1963). The variation in ploidy and habitat in this complex is thus accompanied by the evolution of a more inbreeding mating system. This may explain why polyploid populations show greater amounts of population differentiation for morphological traits than do diploid populations in this species (where morphological variation is clinal in nature).

3.5.8 The combination of auto and allopolyploidy

Western Mediterranean *Ornithogalum* illustrate how the combination of autopolyploidy and allopolyploidy contributes to diversification within a group

Box 3.5 The distribution limits of the different cytotypes of *Mercurialis annua* (redrawn from Durand 1963)

In *M. annua* distribution patterns of different ploidy levels coincide with bioclimatic regions. First, diploids (2*x*) have a mostly temperate distribution (note their presence across northern Spain which does not have a Mediterranean climate), with a meridional Mediterranean extension through Italy and Sicily to northern Tunisia. This points to a Mediterranean refugia for this taxon in Calabria, Sicily, and Tunisia, during the Quaternary glaciation cycles. In Tunisia, diploids are limited to humid and subhumid Mediterranean belts in the north.

In contrast, polyploids only occur in the Mediterranean.

In the Iberian peninsula, hexaploids occur around the Mediterranean coastal zones in the thermo-Mediterranean belt. Cold winters probably restrict the distribution of hexaploids to these coastal areas. In Tunisia hexaploids occur in the arid bioclimatic regions and octoploids in semi-arid conditions.

In Morocco only tetraploids and hexaploids are present. The latter have a more widespread distribution than in Spain (in a range of bioclimatic zones), while tetraploids are confined to semi-arid Atlantic coastal zone.

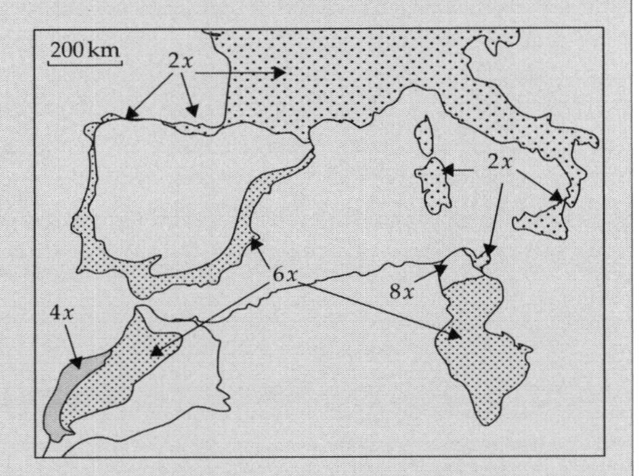

As Durand (1963: 616) points out (my translation), 'the adaptation of the different cytotypes to different bioclimatic zones is not the only factor contributing to their contemporary patterns of distribution'. In addition to contemporary climate, distribution patterns have been strongly moulded and constrained by the geological history of land connections, as the historical refugia zone through Italy and Sicily to Tunisia illustrates. Finally, all the different cytotypes grow in disturbed habitats (e.g. cultivated fields, gardens, road-sides, and waste-grounds). Hence the role of human activities in their dispersal, and their limits to dispersal, cannot be ignored. In this group, human activities may have favoured the expansion of diploids rather than polyploids.

of closely related species (Moret and Galland 1992). Polyploids of *Ornithogalum* which occur in the Iberian peninsula and North Africa have a very different morphology from those further north, which show almost exclusive reliance on sexual reproduction and are allopolyploid ($2x = 52$). These polyploids are thought to represent a distinct species which originated from hybridization among two ancestral diploids, one of which would have been in the *O. umbellatum* complex (see Box 3.4). In the mountains of Spain and North Africa, allopolyploidy has produced a lineage which occurs in both primary and secondary habitats over a wider and often more stressful range of conditions than diploid *O. umbellatum* in the same region. The occurrences of distinct geographic variants on either side of the Straits of Gibraltar indicate that the origins of this allopolyploid probably date to the Late Tertiary. This combination of autopolyploidy and allopolyploidy also occurs in several other Mediterranean plant groups, for example, *Thymus* (Morales 2002), *Mercurialis* (Durand 1963), *Campanula* (Geslot 1983; Contandriopoulos 1984), *Scilla* (Parker *et al.* 1991; Vaughan *et al.* 1997), and *Triticum* (Zohary and Feldman 1962). The different ways of being polyploid can thus greatly contribute to the diversification of large Mediterranean genera.

The conclusion to be drawn is that the different but closely interconnected processes of my introductory triptych, in relation with the genetic changes associated with polyploidy and hybridization, particularly introgression among different polyploids, have been crucial elements in the evolutionary dynamics of many species complexes in the Mediterranean flora. The genetic changes which accompany polyploidy and hybridization stimulate and allow for adaptive evolution in relation to local ecological conditions. Cytotype differentiation can represent a key element of within species differentiation in many taxa and thus may play a central role in the evolution of widespread distributions. An intriguing feature of this diversification is that it frequently occurs without major changes in morphological traits and in several of the above-cited studies one cannot be 100% sure of the ploidy level of plants in the field without recourse to chromosome counts or flow cytometry analysis of DNA content. In addition, in many groups, polyploids have had multiple origins in different parts of the range of the progenitor diploid. Adaptation to novel conditions, geographic barriers to dispersal, and the spread of different taxa in association with human-induced habitat changes have all contributed to patterns of diversification. Understanding plant diversity in such groups requires study of the joint roles of such processes in creating and maintaining diversity and thus a recourse to diverse methods in botany, ecology, and genetics, with a strong flavour of natural history. It was this wise combination of approaches that allowed G.L. Stebbins to do so much for our understanding of plant evolutionary biology.

3.5.9 Karyotype differentiation and evolution

In addition to polyploidy, there are two other major types of chromosome variation and evolution which can influence population differentiation and species divergence: quantitative variation in chromosome number and changes in structure and organization associated with breakage and reunion of chromosomes (Jones 1970; Stebbins 1971; Jones 1995; Rieseberg 2001).

Quantitative variation in chromosome number. This is frequent among closely related species and even within species. Many examples of this phenomenon have been reported in Mediterranean plants, some of which involve quantitative variation without any ploidy level variation while others involve chromosome loss and/or addition subsequent to polyploidization (Table 3.4). In the latter case, an amazing range of chromosome numbers can be observed within a given species or group of closely related species. One way in which such chromosome increments occur is associated with the presence of 'B' chromosomes (Jones 1995), in other cases chromosome number may decline via aneuploidy.

Table 3.4 Some examples of chromosome number variation within and between closely related Mediterranean plants

Species	Chromosome series (2*n*)	Reference
Cyclamen	20, 22, 30, 34 (68), 48, 84, and 96	Grey Wilson (1997)
Nonea	14, 16, 18, 20, 28, 30, 32, 40, 44, 60	Luque (1995)
Lithodora	16, 26, 28, 32, 35, 40	Luque and Valdés (1984)
Lupinus	32, 36, 38, 40, 42, 50, and 52	Plitmann (1981)
Leucojum	14, 16, 18, 22	Contandriopoulos (1962)
Phlomis	20, 22, 24	Azizian and Cutler (1982)
Reichardia	14, 16, 18	Siljak-Yakovlev (1996)
Cardamine pratensis	16, 24, 20, 32, 40, 46, 48, 55	Lihová *et al.* (2003)
Erysimum grandiflorum	14, 26, 34, 38, 40, 48	Kupfer (1981)

Variation in karyotype structure. This stems mostly from chromosome breakage and reunion which can cause deficiencies by elimination of segments, duplication via the integration of segments from homologous chromosomes, inversion within chromosomes, and translocation of segments among chromosomes. As a result of such changes, karyotype variation among populations or closely related taxa may occur in the absence of any variation in chromosome number, providing a new source of genetic variation for evolution and genetic isolation. In some cases such rearrangements may accompany a progressive reduction in chromosome number, for example, in species of *Reichardia* (Siljak-Yakovlev 1996). The Mediterranean flora provides numerous examples of how such variation may occur within and among closely related taxa, some of which provide evidence of the potential importance of the fixation of chromosomal variants for population differentiation. In some cases chromosomal rearrangements are not associated with any observable morphological differences, in others they occur among taxa that also differ in morphology.

Two examples illustrate how variation in karyotype can be accompanied by observable morphological differentiation (Verlaque *et al.* 1991).

1. *Scabiosa cretica* shows variation in karyotype organization but no variation in chromosome number, among populations in the different parts of its range. In this species, variation in karyotype organization among populations on different islands is accompanied by significant differences in overall plant morphology.
2. *Delphinium requienii* on the Hyères islands off the south-east coast of France shows marked differentiation in the symmetry of its chromosome complement compared to its disjunct 'sister species' *Delphinium pictum*, endemic to Corsica, Sardinia, and Majorca. These two species are however very similar.

Within species variability in chromosome structure and organization may occur in the absence of any observable morphological differences, as in many autopolyploid complexes where diploids and polyploids are often indistinguishable on the basis of morphological traits. Two geophytes in the Liliaceae provide pertinent examples of population differentiation in chromosomal structure within species that have widespread geographic distribution and similar morphology around the Mediterranean.

1. In *Muscari comosum* different inversion heterozygosities have been detected on different islands in the Aegean (Bentzer and Ellmer 1975) and among populations in continental Spain (Ruiz-Rejon and Oliver 1981). Bentzer and Ellmer (1975) interpret the pattern of fixation of different chromosomal variants on different islands as being the result of genetic drift. In contrast, the finding of similar patterns and types of variation in Spain, and their comparison with allozyme variability do not concord with this hypothesis, leading Ruiz-Rejon and Oliver (1981: 407) to the conclusion that 'although genetic drift cannot be rejected, we think that the interaction between adaptive and historical factors, related to the colonization process, have played the major role in determining the geographical distribution of the two coupled stable chromosome polymorphisms in *Muscari comosum*'.
2. In *Scilla autumnalis*, a widespread species around the Mediterranean, studies of variation in chromosome complement by Parker *et al.* (1991) and Vaughan *et al.* (1997) have detected 10 distinct chromosome races in the different parts of the range of this species (Fig. 3.12). The variation includes diploids ($2n = 14$), tetraploids ($2n = 28$), and hexaploids ($2n = 42$) plus changes in DNA content and chromosomal rearrangements. There are five haploid genome sets: A, B^7, B^5, B^*, D, which have given rise to a range of diploids and polyploids. The diploids occur in different parts of the distribution of this species around the Mediterranean, four on the island of Crete (Fig. 3.12). Three of the four variants on Crete were not detected elsewhere. Some variants are widespread, others endemic. Once again, it is the polyploid variants which have spread north to colonize temperate Europe while diploids have remained endemic to localized parts of the range in the Mediterranean. Northward migration would appear to have occurred from a Balkan refugia in

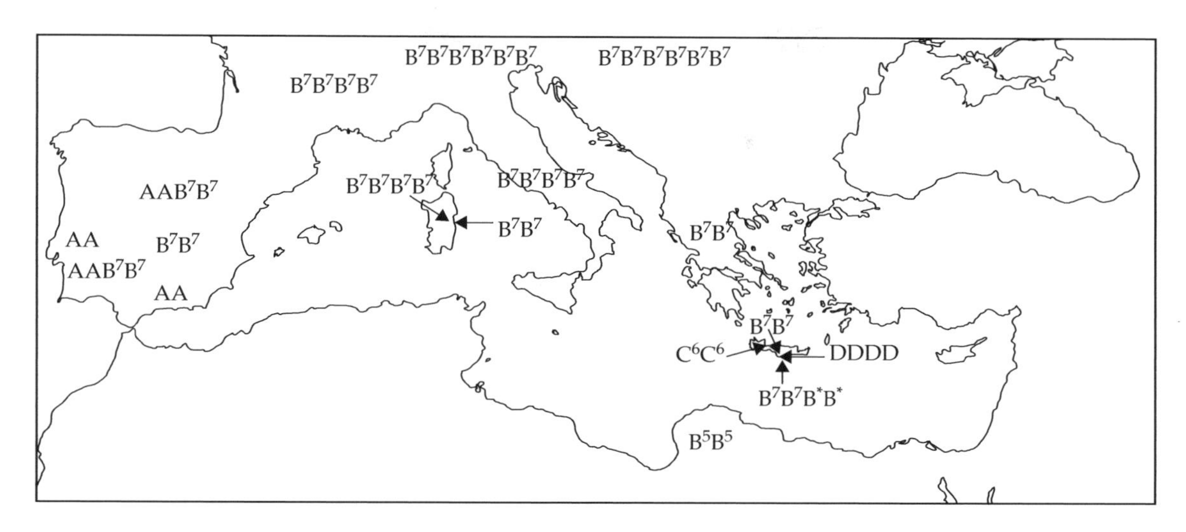

Figure 3.12 Geographical distribution of the chromosome variants of *Scilla autumnalis*. The $B^7B^7B^7B^7$ karyotype also occurs in Britain as does an additional type ($AAB^7B^7B^7B^7$) (redrawn from Vaughan *et al.* 1997).

this species. The morphology of the different 'races' is very similar and they are able to cross fertilize in the glasshouse (Vaughan *et al.* 1997), indicating that only small parts of the genome have diverged in allopatry.

Closely related endemic species may also show subtle differences in chromosome structure. In three endemic *Lilium* species in the mountains of the Mediterranean, yellow-flowered *Lilium pyrenaicum* from the Pyrenees and Cantabrian mountains, red-flowered *Lilium poponium* endemic to the Maritime Alps, and yellow/orange-flowered *Lilium carniolicum* which is endemic to the Balkans and the south-east Alps, Siljak-Yakovlev *et al.* (2003) have shown marked differences in the number and position of secondary constrictions and particular gene loci, and overall genome size (Fig. 3.13). Similar types of variation in karyotype have been documented in different groups with different life histories and in different parts of the Mediterranean, for example, *Vicia* in the Eastern Mediterranean (Zohary and Plitmann 1979), *Lactuca* in the Iberian peninsula (Mejías 1993), *Asphodelus* in the western Mediterranean (Díaz Lifante 1996), and the disjunct endemic Cedrus in different Mediterranean mountains (Bou Dagher-Kharrat *et al.* 2001). In the latter, differences in chromosome size are associated with variation in life-history among closely related species, as Stebbins (1950) first proposed. Different species vary in habitat occupation in relation to degree of aridity, suggesting a role for chromosome rearrangements in plant adaptation and evolution, as mentioned earlier in this chapter (see Klein 1970). Fixation of chromosomal variants in the different parts of the range of a widespread species or in disjunct populations may thus be a key process in the divergence of endemic species. Variation in chromosome number and structure can be of critical importance for the delimitation of new taxa, even in the absence of morphological variation, and may thus be a key ingredient of plant evolution in the Mediterranean.

Differences in karyotype also may play a role in the evolution of crossing barriers and reproductive isolation. An elegant demonstration of this has been provided for sympatric Mediterranean *Orchis* by Cozzolino *et al.* (2004). As mentioned earlier in this chapter, selection for pre-zygotic isolation and species-specific pollination may strongly contribute to reproductive isolation and diversification of orchids. However, some Mediterranean

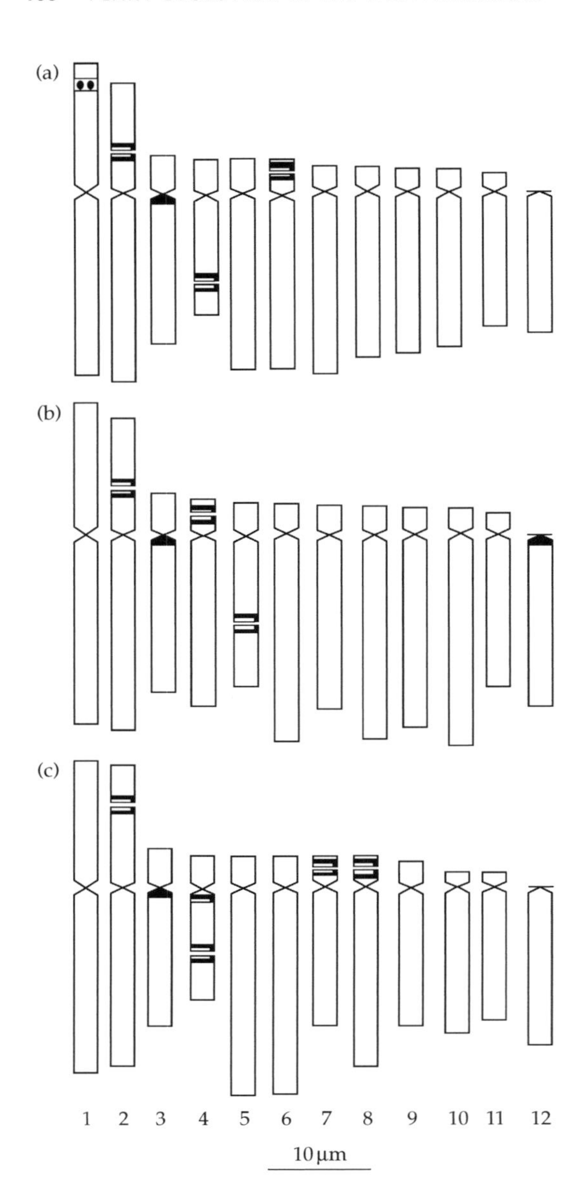

Figure 3.13 Schematic illustration of differences in the position of secondary constrictions (cut in the chromosome), banding patterns for chromomycin A3 (■), and the position of different ribosomal DNA loci, that is, 18S (□) and 5S (△) on each of the 12 chromosomes in the karyotype of three endemic *Lilium* species: (a) *L. pyrenaicum*, (b) *L. poponium*, and (c) *L. carniolicum*. Reproduced with permission from Siljak-Yakovlev *et al.* (2003).

orchids share pollinators, but maintain integrity in sympatry. The above authors show that pairs of species in sympatry which also share pollinators have significantly more divergence in karyotype structure than species pairs which have specific pollinators. Their interpretation is that post-zygotic reproductive barriers which result from karyotype divergence may be an important causal factor in orchid diversification in the Mediterranean. There is some evidence that such karyotype divergence may occur in peripheral isolates of widespread taxa (D'Emerico *et al.* 2002), another possible case of local population differentiation that may ultimately lead to speciation.

The importance of chromosomal rearrangements in divergence and speciation has received a fair amount of attention since early work which proposed that speciation can occur when populations become fixed for chromosomal rearrangements that reduce fitness when heterozygous (Lewis 1966; Stebbins 1971). In a recent review, Rieseberg (2001) discusses the various arguments developed in the literature which cast doubt on this idea. For this author, chromosomal rearrangements may reduce gene flow more often through their effects on recombination rates than through any affect on fitness. The potential for reduced recombination offered by chromosomal rearrangements increases the possibility of speciation where gene flow can occur, for example, in cases of sympatric speciation or as a newly derived peripheral 'neospecies' expands its range and encounters populations a progenitor species (see also Levin 2000). Rieseberg (2001), argues the need for additional work, now possible with the development of powerful molecular techniques and their application to plant genetics, to improve our understanding of the role of chromosomal variation in plant speciation. Some of the above examples would provide ideal model systems for such work.

3.6 Conclusions

In this chapter, I argue that by integrating species and population-level approaches a more precise understanding of the role of different microevolutionary processes, ecological factors, and historical effects in the evolution of new species can be reached. Evolutionary diversification of species-rich Mediterranean groups is a complicated problem due to the complex history of the region and the diversity of the evolutionary processes involved. The work I have presented in this chapter illustrates repeatedly how genetic discontinuities, within and among species, have arisen on the template laid out by the geological history of the Mediterranean Basin and the contractions and expansions of species during climatic change and oscillation. A similar tale can be told for other Mediterranean-climate regions such as California (Calsbeek *et al.* 2003), where climatic aridity became marked at ~3 Ma causing sharp ecological gradients across the region, and South Africa (Richardson *et al.* 2001). Geological history has thus left its mark not only on species distributions (previous chapter) but also the differentiation and evolution of new species. The patterns of diversification have however varied in time and thus differ for different groups of species (Vargas 2003). A characteristic feature of Mediterranean plant endemism is that many genera show distinct centres of species diversity in either the Iberian peninsula or in the Balkans, Greece, and Anatolia (Chapter 1). In addition to having been a refuge during periods of glacial maxima, such areas have also been important sites for the diversification of many genera. Future climatic change will likely cause more shifts in distribution and create more cases of endemism for us to delight upon around the shores of *Mare Nostrum*.

An integral component of endemism and speciation is chromosome evolution and hybridization. That the genetic changes which accompany polyploidy and hybridization stimulate and provide the genetic capacity for ecological differentiation and adaptation is now clearly demonstrated (Rieseberg *et al.* 2003) and the examples I discuss lend support to this idea. Other forms of chromosome variation may also be common in the Mediterranean flora and thus play an important role in diversification. However, in the absence of work linking chromosome rearrangements to natural selection in the wild, it is impossible to fully identify any adaptive function of such variation. So, despite the fact that Lewis (1966), nearly 40 years ago, and more recently Rieseberg (2001) have argued for an important role of chromosome rearrangements in plant speciation, this role has yet to be confirmed within the Mediterranean flora.

Several of the examples discussed in this chapter illustrate the potential evolutionary significance of local speciation in geographically peripheral or ecologically marginal populations of widespread taxa. Instructive for our understanding of evolution, these examples also illustrate the immense value of marginal populations of widespread species for the conservation of evolutionary potential. The existence of such populations should thus be recognized in the elaboration of priorities for conservation. Somewhere at the limits to the distribution of widespread species, new taxa are evolving that will become the rare endemics of future conservation efforts. The case of one of the prettiest *Cyclamen* species I have seen, *C. repandum* and its four allopatric subspecies, and its highly selfing derivative *C. balearicum*, is particularly enlightening here. Where some botanists may clamour for a range of disjunct species others may limit the distinction to two species, one with several subspecies. Understanding the precise nature of variation among and within such taxa is central to our delimitation of taxa in such groups and to our understanding of their evolution. The example of *Cyclamen* is one which repeats itself in many groups of Mediterranean plants. Yet there have been few studies which compare levels of quantitative variation in order to assess which traits may have been important in divergence, and where. In the case of *C. repandum*, it would appear that divergence of subspecies is ongoing, and that the variation which occurs within subspecies is that which ultimately causes differences among taxa.

Finally, I have discussed examples which provide evidence that ecological factors have been critical for differentiation and divergence and the maintenance of different taxa in geographic proximity. Although differentiation in response to ecological conditions is greatly facilitated by geographic isolation, in some cases divergence may proceed in local populations, particularly small populations in geographically peripheral and ecologically marginal sites. The interplay of factors which promote divergence (selection and drift) and those which homogenize gene frequencies in the landscape (pollen and seed movements) are critical to such divergence. They are also fundamental to the development of patterns of within-species differentiation in relation to the spatial configuration of habitats in Mediterranean mosaic landscapes, where ecological factors vary dramatically over short distances. This is the subject of Chapters 4 and 5.

CHAPTER 4

Trait variation, adaptation, and dispersal in the Mediterranean mosaic

> Adaptation, insofar as such a concept existed, could no longer be considered a static condition, a product of a creative past, and became instead a continuing dynamic process. Organisms are doomed to extinction unless they change continuously in order to keep step with the constantly changing physical and biotic environment. Such changes are ubiquitous, since climates change, competitiors invade the area, predators become extinct, food sources fluctuate; indeed hardly any component of the environment remains constant. When this was finally realized, adaptation became a scientific problem.
>
> E. Mayr (1982: 483–484)

4.1 Ecological constraints and adaptation in the Mediterranean

Evolutionary theory takes as one of its simplest and most fundamental premises that natural selection shapes patterns of trait variation in a way which maximizes the fitness of individuals in different environments. Fitness can be maximized in different environments in two ways.

- Traits may show genetically based adaptive differentiation, that is, different genotypes in distinct environments.
- Phenotypic plasticity may enable individual genotypes to alter their phenotype and thus adopt an adaptive phenotype in different environments.

Distinguishing among these solutions to environmental variation is a central theme in evolutionary ecology. In the Mediterranean region, microclimate (particularly rainfall and drought stress), geology, soils and a long history of human activities have fashioned a mosaic of ecological conditions, often on highly localized scales. Such localized spatial heterogeneity of ecological factors is ideal for the study of plasticity and adaptive differentiation in relation to levels of gene flow across the landscape.

The Mediterranean climate is characterized by strong seasonality which involves (a) the association of a drought period when temperatures are at their hottest and (b) a cool (and cold in many areas) moist period. The summer drought can limit growth, flowering, and fruiting, and is a major cause of seedling mortality, as illustrated by studies on *Nerium oleander* (J. Herrera 1991*a*), *Helianthemum squamatum* (Escudero *et al.* 1999), *Lavandula latifolia* (C.M. Herrera 2000*a*), *Olea europaea* (Rey and Alcántara 2000), *Frangula alnus* (Hampe and Arroyo 2002), *Thymus vulgaris* (my own data), and *Rhamnus ludovici-salvatoris* (Traveset *et al.* 2003). Summer drought may thus impose a severe demographic constraint. Cold temperatures in winter may also limit growth and cause mortality in many regions of the Mediterranean. The most favourable periods for growth are thus the autumn and spring. But even then, the unpredictability of rainfall and its occurrence in rapid bursts of intense rainfall can impose constraints on germination, establishment, and growth. Rainfall during the growing period may vary enormously from one year to another, and this variation can have marked effects on species composition (e.g. Figueroa and Davy 1991), growth, and flowering. Variability in the timing and amounts of rainfall can thus greatly impact population dynamics.

Initial work on how plant traits represent an adaptive response to the Mediterranean climate was mostly oriented towards the study of the adaptive significance of sclerophyllous vegetation in the different Mediterranean parts of the world (Box 4.1). There are in fact two main strategies of coping with drought stress: tolerance and avoidance. To tolerate drought requires functional attributes that allow for the maintenance of cell functions and plant persistence in the absence of the necessary moisture for normal growth and development. To avoid drought, plants may adopt a strategy of summer dormancy. As I will discuss in this chapter, both the tolerance and avoidance strategy are well illustrated in the Mediterranean flora. This co-occurrence of different strategies is well known in the other 'Mediterranean' parts of the world. Take, for example, the plant communities of Southern California, where drought-deciduous coastal sage communities and evergreen chaparral 'occur intimately associated in an intricate mosaic pattern throughout large areas of the landscape' (Harrison *et al.* 1971: 868).

An important point to make from the outset is that moisture limitation can allow variation in other environmental factors, such as diversity in topography, aspect, and nutrient availability, to exert an important controlling effect on plant distribution and population ecology. Indeed, mosaic vegetation in the Mediterranean is often linked to variation in soils, which are generally poor in nutrients and highly variable in space. Two integral ingredients of this variation are depth and moisture retention capacity. Topography can also contribute to this mosaic, as the differential distribution of species and vegetation belts on south-facing and north-facing slopes illustrates. For a well documented and illustrated description of such vegetation patterns, I refer the reader to Gamisan's floristic inventory of the island of Corsica (1991). The striking nature of such vegetation differences on north- and south-facing slopes in southern France is such that local languages evolved words (now used in the botanical and ecological literature) to describe the baked south-facing slopes (*'adret'*) and cooler, shady north-facing slopes (*'hubac'*). Differences in water supply may be of primary importance for the occurrence of vegetation patterns on such slopes (Pigott and Pigott 1993), with an additional limitation imposed by the drying effect of some of the strong winds which impose themselves in the Mediterranean region (e.g. see Barbero and Quézel 1975). Elsewhere it has been reported that even long-lived tree species on north- and south-facing slopes can show marked genetic differentiation (Linhart and Grant 1996), hence the distribution patterns on 'hubac' and 'adret' expositions are not without consequence for plant evolution. Several other examples of distribution patterns of individual species (Thuiller *et al.* 2003) and patterns of vegetation structure (Zavala *et al.* 2000) which are determined at least in part by aridity are known in the Mediterranean landscape.

In addition, human activities have greatly modified the selection pressures acting in local populations and the spatial organization of natural habitats in the Mediterranean landscape. A major feature of such variation involves not just the fragmentation and disturbance of natural habitats but also, following the abandonment of human activities, the colonization of open areas (pastures, cultivated fields, old mine sites) and the successional development of forest vegetation. The variable stages of such succession are often all locally present, even over very short distances. Population differentiation in such landscapes will ultimately depend on rates and distances of dispersal across the landscape, any spatial heterogeneity in dispersal patterns, whether selection pressures are strong enough to override the homogenizing effects on gene flow, and the capacity of plants to establish populations and persist in a heterogeneous landscape. Let us not forget, if selection is strong enough, even in the presence of gene flow, adaptive variation may rapidly evolve.

My investigation of variation and adaptation of functional traits in Mediterranean plants in this chapter is structured around three main themes.

1. I explore variation in functional traits and phenology in relation to the classical themes of aridity and nutrient stress in Mediterranean communities. The contemporary climate is a recent phenomenon. The association of trait combinations in many species no doubt evolved in the region prior to the onset of a Mediterranean-climate regime, as the presence of a large group of Pre-Pliocene sclerophyllous, unisexual, fleshy-fruited species illustrates. I would argue that the occurrence of sclerophylly in the Mediterranean is an issue which has largely overshadowed other important aspects of plant variation and adaptation, which I will bring more to the fore in this chapter as I examine evidence for variation in precise functional attributes. My standpoint is that to observe plant evolution in relation to climate requires analysis of trait variation within and among closely related species in relation to ecological stress and environmental variation.

2. I examine the constraints and selection pressures acting in plant populations during the colonization of open habitats following the abandonment of human activities and the successional establishment of forest vegetation. I stress the roles of dispersal limitation and spatial variation in dispersal patterns and regeneration capacity in relation to habitat heterogeneity for the evolution of population differentiation.

3. Finally, I take up the theme of why there are so many aromatic plants in the Mediterranean flora. My focus here is on variation among closely related species and among populations of individual species in their dominant secondary metabolites and whether such variation is adaptive.

Compared to previous chapters, the focus has now shifted from problems of diversity to problems of adaptation.

4.2 Summer drought and nutrient stress: functional traits and their variability

4.2.1 Constraints and functional attributes

Vegetation–climate relationships are one of the foundations of plant biogeography and provide the basis for the classification of Mediterranean forests and other vegetation of the region (Quézel and Médail 2003; Chapter 1). As our climate continues to evolve, enormous efforts are being channelled into attempts to predict how future climate change will affect the distributions of species and communities. To make realistic predictions requires an understanding of how contemporary trait variation and species distributions have been shaped by historical climatic variation.

The dramatic changes in vegetation across the xeric Mediterranean–desert transition on the eastern fringes of the Mediterranean illustrate the importance of constraints on plant growth associated with increasing aridity (Kutiel *et al.* 1995). Since the beginning of the twentieth century, ecologists have been fascinated by the similarities of vegetation structure and function in the different Mediterranean-type ecosystems of the world and the possibility that such patterns represent independent convergence of traits as species adapt to the Mediterranean summer drought. Much of this interest has been directed towards discerning the adaptive nature of the evergreen and sclerophyllous habit of Mediterranean shrubs and trees (Box 4.1).

In the Mediterranean, soil moisture becomes decreasingly available during the drought period, even for deep-rooted sclerophyllous species, which do not show a physiologically latent phase during the summer drought. Their vegetative activity is fairly constant throughout the year with seasonal peaks in photosynthetic activity associated with bud-opening and shoot growth (Gratani *et al.* 1992). As the progressive drought sets in, stomatal control is probably the most effective means of regulating water loss due to transpiration (Mooney and Dunn 1970*a*; Joffre *et al.* 1999). During late summer, water stress can cause cell turgor and water potential values to drop to extremely low values which may cause major physiological dysfunction. Stomatal closure may provide an effective control on such stress by allowing plants to increase water use efficiency during summer drought (Cowan and Farquhar 1977), although this may be less than predicted if transpired water affects air humidity in and around the canopy (Tenhunen *et al.* 1990). Seasonal changes in leaf physiological efficiency may also be of prime importance in controlling water loss during the summer drought (Tenhunen *et al.* 1990; Gratani *et al.* 1992). Summer is also characterized by a high air temperature, a vapour pressure deficit, and high solar irradiance. Stomatal closure in response to water stress will also cause reduced gas exchange and thus reduce photosynthesis. Last but not least,

Box 4.1 Sclerophyllous vegetation in the different Mediterranean regions of the world

Sclerophyllous trees and shrubs are a characteristic feature of Mediterranean vegetation. There has thus been much interest in the adaptive significance of sclerophylly and the possible evolutionary convergence of form and function in the five geographically disjunct Mediterranean-type climate regions of the world.

In the Mediterranean, the (winter) deciduous strategy has traditionally been thought to be disadvantageous due to a short photosynthentically active period which coincides with summer drought. As Mooney and Dunn (1970*b*) argued, the cost of maintaining evergreen leaves that can withstand drought stress and deter herbivory is less than that of producing a completely new foliage each year. The combination of leaf traits associated with sclerophylly (i.e. a relatively low photosynthetic capacity, a high proportion of stored carbon, a low leaf-nitrogen concentration, a low surface to volume ratio, thick cell walls, and a thick rigid cuticle) may facilitate tolerance of negative turgor pressure under water stress. Sclerophyllous species also show a much higher degree of stomatal regulation than several deciduous species in the Mediterranean (Duhme and Hinckley 1992). Finally, the nutrient acquisition and allocation strategy of sclerophyllous species may permit growth in a nutrient-poor soil and their chemical defence may assure long leaf lifetime. Many authors have considered such traits, and the sclerophyllous evergreen habit, to represent an adaptation to life in a Mediterranean climate, as a couple of citations illustrate.

- 'Thus, the environmental conditions are such that evergreenness is an appropriate strategy in Mediterranean-type climates' (Mooney and Dunn 1970*b*: 298).
- 'The evergreen shrub seems to be the morphological life-form best adapted to Mediterranean stress conditions with seasonally restricted growth' (Seufert *et al.* 1995: 352).

Drought resistance of sclerophyllous species may arise from an ability to recover from damage to xylem conductance rather than a faster recovery from higher rates of water loss. Sclerophyllous species cannot however tolerate overly severe drought stress, and are excluded from extremely arid areas. Sclerophylly in Mediterranean plants may thus be effective only if drought stress does not cause a reduction in transpiration which provokes irrevocable damage to xylem water conductance as a result of cavitation. If drought stress is too intense or too long it may prevent recovery of cavitated xylem, and there will be little advantage to being sclerophyllous. In a comparison of drought resistance in two sclerophyllous (*Viburnum tinus* and *Ilex aquifolium*) and non-sclerophyllous (*Hedera helix* and *Sambucus nigra*) species, Salleo *et al.* (1997) showed that sclerophyllous species do not recover from water loss and stress more rapidly but do recover more completely from xylem cavitation. In their study of the mosaic distribution of coastal sage and chaparral vegetation in California, Harrison *et al.* (1971) showed that although evergreen chaparral plants are less sensitive to long periods of summer drought by virtue of their low rates of transpiration, their low photosynthetic activity during the favourable season puts them at a disadvantage. Hence in extremely arid areas, a summer-deciduous drought avoiding strategy, as adopted in the coastal sage community, may be a more beneficial strategy by virtue of the higher rates of assimilation of the species in this community during the favourable season for growth. In contrast, in areas which benefit from a slight increase in soil moisture availability during summer, the xeromorphic morphology and greater water economy of the chaparral vegetation can allow them to maintain photosynthetic activity during the dry period. The mosaic of chaparral and coastal sage communities in California, with the latter dominating the most xeric sites fits this idea and that of Kummerow (1973: 166) for whom sclerophyllous species represent 'moderate xeromorphic plants'.

In fact, despite broad similarities, and probable convergence of traits linked to overall form and function, there are important differences in patterns of divergence in form and function of sclerophyllous species in different Mediterranean-climate regions (Naveh 1967; Specht 1969*a*,*b*; Mooney and Dunn 1970*b*; Raven 1971; Parsons 1976; Cody and Mooney 1978;

Cowling and Campbell 1980; Shmida 1981; Milewski 1983; Cowling and Witkowski 1994). Historical and contemporary differences in local climate and the importance of different fire regimes and soil nutrient status are likely to be the most important causes of such differences (Joffre and Rambal 2002). Indeed, a recent review by Rambal (2001) illustrates that leaf photosythetic performance of Mediterranean woody species does not generally differ from that of species in other biomes and that drought tolerance and stomatal sensitivity are not consistently related to rooting depth, resource availability, or the degree of seasonal water stress. In fact, the maintenance of a canopy of drought tolerant sclerophyllous leaves through the summer may be more closely associated with nutrient stress rather than drought stress per se (Loveless 1961, 1962; Rundel 1988, 1995; Specht and Rundel 1990; Joffre *et al.* 1999). Overall, growth rates will be slow, except during marked seasonal flushes during periods of water availability, and thus allocation to maintenance costs and a longer leaf longevity could be favoured. Indeed, levels of annual dry matter production of evergreen and deciduous species can be equivalent in nutrient-poor sites, whereas in nutrient-rich sites, deciduous species have a greater productivity, hence, some authors go so far as to consider sclerophylly as a simple epiphenomenon of phosphorous and other nutrient deficiency (Loveless 1961, 1962; Rundel 1988).

Having long-lived leaves with a low photosynthetic activity and high carbon concentrations may come at a cost if herbivory is frequent. The low photosynthetic capacity means that the replenishment of stored carbon reserves, lost as a result of herbivory, will be slow. Leaves must therefore be defended against herbivory and have a long lifetime in order for their construction costs to be recovered. The leathery and thick cuticle of sclerophyllous leaves, which are not consistently more expensive to produce than leaves of deciduous species (Merino *et al.* 1982), and which often contain various monoterpenes and tanins as significant proportions of fixed carbon, illustrate that traits associated with sclerophylly introduce benefits other than coping with a summer drought. Indeed, Herms and Mattson (1992: 305–306) conclude that 'the balance of evidence suggests that low nutritive quality, high concentrations of secondary metabolism, and tough, sclerophyllous foliage interact to provide formidable barriers to herbivory in resource poor environments'.

Finally, many sclerophyllous species (which have been the object of the majority of ecophysiological studies) evolved from a tropical and temperate sclerophyllous vegetation present during the Tertiary, that is, prior to the onset of the seasonal Mediterranean climate. Such species did not evolve sclerophylly in response to the onset of a Mediterranean-type climate. In a detailed analysis of the character syndromes of the woody vegetation of southern Spain, C.M. Herrera (1992*a*) revealed how character associations in contemporary Mediterranean woody plants in Spain (notably the occurrence of sclerophylly in association with unisexual, uncoloured flowers and large vertebrate-dispersed seeds and fruits) are not observed in a dataset limited to those genera that have only been present since the onset of a Mediterranean-climatic regime. Verdú *et al.* (2003) produced similar results for woody plants in the Mediterranean parts of the Iberian peninsula, California, and Chile. The similarities of vegetation in different Mediterranean-climate ecosystems are thus strongly linked to the long-term persistence of Tertiary lineages which evolved under a subtropical climate. The general conclusion is thus that although sclerophylly may provide certain advantages in a Mediterranean setting, it pre-dates the Mediterranean climate and, at least in terms of its basic structural features, is not an evolved adaptation to the highly seasonal Mediterranean climate (Seddon 1974; Axelrod 1975; C.M. Herrera 1992*a*; Box 1997; Joffre and Rambal 2002). The similar traits one observes in the different regions have previously been interpreted in relation to convergence in association with the independent onset of Mediterranean climates in different parts of the world (Cody and Mooney 1978). Traits such as sclerophylly were present prior to the onset of the Mediterranean-climate regime in the different Mediterranean regions and thus do not represent convergent evolution (Verdú *et al.* 2003). This is a clear-cut illustration of how ecological patterns can result from historical processes associated with the dynamics of regional taxonomic assemblages and the persistence of historical lineages.

a reduction in transpiration can cause a modification of the energy balance of the leaf, which may result in overheating, a problem that can be exacerbated by high solar irradiance and temperature. If excessive, such overheating can cause photo-inhibitory damage. Several constraints can thus simultaneously limit plant growth and development.

Mediterranean plants show a diverse array of morphological and physiological attributes which help accommodate drought stress. Joffre *et al.* (1999) have classified these according to the timescale of drought stress.

1. Seasonal variation in water availability may be buffered by reduced leaf size. Low leaf surface area is common in Mediterranean shrubs, as the plethora of Mediterranean Lamiaceae (e.g. *Satureja*, *Thymus*, and *Rosmarinus*) illustrate. At xeric sites in California, species with small leaves contribute to ~80% of vegetation cover. In some Lamiaceae, individual plants may vary leaf size among seasons—producing small leaves in summer and large leaves in winter (Margaris 1976). During a season, as the water supply diminishes, plants may favour the growth of deeply penetrating roots, and soil below ~2 m contributes an increasing fraction of total plant water uptake. For example, in a study of *Quercus coccifera* Rambal (1984) showed that this contribution may be as much as ~25% of total uptake in severe drought periods.
2. Diurnal variation in water availability can be countered by the regulation of stomatal activity. For instance, stomatal closure can allow for rapid reductions in rates of transpiration and thus a more conservative water-use. Such stomatal closure may be progressive, for example, in *Quercus* species (Acherar and Rambal 1992) and *N. oleander* (Gollan *et al.* 1985) or, as in *Pinus halepensis*, more dependent on a threshold water stress (Joffre *et al.* 1999).

Plants which do not escape the summer drought via dormancy must also tolerate high solar irradiance and thus cope with excess intercepted solar radiation at the time when carbon assimilation is limited by stomatal closure and photosynthesis by water stress. Light energy may thus exceed that required for carbon fixation and cause damage or dysfunction in the photosynthetic system. A range of different physiological mechanisms may permit the dissipation of such excess absorbed energy, for example, decreased chlorophyll content (Kyparissis *et al.* 1995) and the production of volatile oils (Section 4.5).

A range of additional functional attributes may enable deciduous species to persist in a Mediterranean setting, alongside their evergreen relatives. Deciduous trees, neglected by many early evaluations of plant–water relations in the Mediterranean flora, are an important component of Mediterranean woodlands. Indeed, one of the major climax oak species in the western Mediterranean, *Quercus pubescens*, is deciduous. This species does not avoid summer drought by early leaf senescence as do some deciduous oaks in California (Griffin 1973). In *Q. pubescens* leaves appear in spring and begin to senesce just prior to the autumn rains. Individual trees thus renew their entire foliar tissue and must maintain sufficient resources to survive the drought period and then renew their leaves. Photosynthetic assimilation prior to the physiological stress triggered by drought stress is thus essential.

Measures of water potential made on *Q. pubescens* during the hot and dry summer of 1994 in southern France (when March–August rainfall was only one-third of the mean rainfall for that period in the study site), showed that water losses lead to a gradual and severe decrease (similar to that shown by the evergreen *Quercus ilex* at the same site) in predawn water potential (Damesin and Rambal 1995). The similarity of low water potential values in the deciduous and evergreen species and the fact that they do not decrease below a certain level suggests that even the deciduous species regulates its physiological activity in order to tolerate extreme aridity. In this species, conservative water use is promoted by progressive stomatal closure as water potential drops and by daytime reductions in net gas exchange (measured as the stomatal conductance to water vapour and net CO_2 assimilation rate), preventing irreversible cell dehydration. In addition, the above authors found no evidence of photodestruction; inhibition of the photosynthetic apparatus due to high solar irradiance and temperature was only temporary. Deciduous species can thus cope with drought stress and have a lower area-based leaf construction cost. Although *Q. pubescens* does not have a higher photosynthetic capacity than *Q. ilex* (to offset a shorter period of photosynthetic activity), the proportion of nitrogen in fallen leaves relative to mature leaves on the tree is only 44%, compared to 78% in fallen leaves of *Q. ilex* at the same site (Damesin *et al.* 1998). Hence the deciduous species may more efficiently remove nitrogen from senescing leaves. These studies thus illustrate the importance of studying a range of trait functions that may be involved in coping with drought stress; no single factor allows individuals to withstand the extreme drought conditions that occur in some years.

The precise nature of plant adaptation to drought stress may also vary among scleropyllous species. For example, sclerophyllous Mediterranean plants show marked variation in the thickness of the cuticle and the structure of their outer cell walls and their mechanisms of drought resistance (Kummerow 1973). Comparisons of the physiological strategy of drought resistance in three species with roughly equal leaf dry weight/surface area ratios by Lo Gullo and Salleo (1988) showed differences in the physiology of leaf–water relations. *Olea europaea* shows a 'drought tolerant' strategy (with a large diurnal osmotic stress in the warmest hours of the day late in summer), *Ceratonia siliqua* continues extracting enough water such that turgor and transpiration are little affected by drought ('water-spending' strategy), and *Laurus nobilis* had intermediate values of leaf conductance and high leaf relative water content (a more 'drought avoidance' strategy) due to low losses of water and rapid recovery of leaf water content.

4.2.2 Drought and nutrients: adaptation in the mosaic

Whether Mediterranean plants show intraspecific genetic variability in their capacity to withstand and/or respond to the summer drought, particularly in relation to the functional attributes discussed above, has unfortunately been a neglected issue. There are nevertheless some examples of population variation in relation to climatic and soil variation in the Mediterranean region.

In a study of leaf morphology and structure of *Arbutus unedo* and *Pistacia lentiscus* at two sites on Sardinia, Gravano *et al.* (2000) reported marked variation in structural leaf traits in relation to local climate. Although these authors did not detect differences in overall morphology (leaf area, dry weight, and weight/area ratio) they did find an increased tannin content in *A. unedo* and a thicker leaf cuticle in *P. lentiscus* in an arid site compared to a more mesic habitat. Both these traits could occasion a greater resistance to water stress in the arid site. In *Dactylis glomerata*, plants from populations in the Mediterranean region (relative to populations from temperate Europe) show better turgor maintenance under low water potential which facilitates survival of the summer drought (Roy 1981) and an enhanced ability to accumulate water-soluble carbohydrate which permits rapid regrowth in the autumn (Volaire 1995). Likewise, the leaf anatomy of *Satureja horvatii* populations suggests a xerophytic ecotype in lowland Mediterranean habitats (280 m) when compared with plants from the oro-Mediterranean mountain conditions (1540 m elevation) in the Balkans (Todorovíc and Stevanovíc 1994).

It is not only leaf morphology which can show variation among populations in relation to contrasting ecological conditions associated with variation in climatic conditions. For example, *Narcissus triandrus*, a widespread and locally abundant geophyte in the central, north-western, and northern parts of the Iberian peninsula, shows continuous geographic variation in stature and floral morphology across its range (Barrett *et al.* 2004). In central Spain, plants bear 1–2 pale-lemon flowers per inflorescence, have small narrow leaves, and are generally small in stature. In contrast, in the north-western part of the Iberian peninsula, that is, outside of the Mediterranean-climate region, plants are wide-leaved, taller, and bear larger inflorescences with up to 10 creamy white flowers, what Blanchard (1990: 112) calls 'an impressive plant'. Flowers are significantly larger in size in the non-Mediterranean parts of the range of this species. The overall change in phenotype thus accompanies an ecological gradient from Mediterranean to non-Mediterranean conditions. It would be fascinating to test by manipulative transplant experiments the extent to which such patterns of variation, that is, the small-leaved and small-flowered plants of central Spain, represent adaptive variation in Mediterranean conditions.

In *L. latifolia,* a common herb in lowland and mid-elevation open habitats in the western Mediterranean, maternal plants differ both in their inherent capacity to produce seeds, which produce seedlings when sown into natural sites in southern Spain (C.M. Herrera 2000*a,b*). This variation is primarily related to a differential ability to tolerate and survive the summer drought. In addition there exists an inverse relationship between seed production and seed and seedling viability, hence, adaption to drought will be strongly influenced by a trade-off between fecundity and viability. Increased fitness associated with increased flower number will be compromised by lower viability of the offspring from plants with many flowers, probably due to among-flower self-pollination and inbreeding depression (reduced viability of offspring produced by selfing compared to outcrossing). The precise analysis of the different components of variation in maternal fitness represents an essential element in our understanding of fitness variation in natural plant populations.

To examine the idea that aridity may cause greater levels of population differentiation, as a result of greater isolation among populations (Stebbins 1952), Comes and Abbott (1999*b*) compared the spatial organization of genetic variation in populations of two closely related annual, outcrossing *Senecio* species, *Senecio vernalis* and *S. glaucus.* In the eastern Mediterranean *S. vernalis* is common in mesic habitats in the Mediterranean-climate zone of Israel whereas *S. glaucus* is primarily a species of maritime sands and semi-desert environments. Based on comparisons of genetic differentiation for nuclear and cytoplasmic genes, these authors found no evidence for greater genetic differentiation in arid habitats. Volis *et al.* (2002*a,b*) found a similar lack of difference in the organization of genetic variation at allozyme loci and for a range of phenotypic traits among desert and Mediterranean populations of wild barley (*Hordeum spontaneum*) in Israel. Hence, the hypothesis of greater levels of population differentiation in arid environments does not apply to these species. The lack of higher levels of differentiation in the deserts which fringe the Mediterranean probably stems from the fact that Mediterranean habitats are themselves highly heterogeneous.

Reproductive effort may vary in relation to aridity. For example, several studies illustrate increased allocation of resources to reproduction in more arid habitats. This trend has been reported in desert populations of *Erucaria hispanica* (Brassicaceae) and two grasses *Bromus fasciculatus* and *Brachypodium distachyon* relative to Mediterranean populations in Israel (Aronson *et al.* 1993; Boaz *et al.* 1994) and in *Bromus erectus* populations from sites with dry, stony, fersialitic soils in sites with mild winters compared to populations on deeper moister soils that are less stony in sites with colder winter temperatures (Ehlers and Thompson 2004*b*). The higher proportional investment in reproduction in more arid sites may result from a trade-off with allocation to vegetative biomass and competitive ability or a competition–colonization trade-off if seedling survival is low in arid sites. Whatever the cause, evidence from several species suggests that increasing aridity is associated with the adoption of a life-history strategy which may be adaptive in unpredictable environments (see Box 4.2).

The soil environment may also be an important component of trait variation among plant populations. In his classic work on the genetic differentiation of *Anthoxanthum odoratum* populations grown on the different soils of the Park Grass Experiment, Snaydon (1970) revealed a mosaic of *A. odoratum* populations locally adapted to soil heterogeneity. Differentiation among populations had evolved rapidly (in plots less than 40 years old) and over very short distances (plots were sometimes <30 m apart). In the Mediterranean mosaic, substrate and soil conditions (Box 4.3) can vary markedly among and within sites, creating a mosaic of variable selection pressures over very localized scales. Such variation may have a strong effect on spatial and temporal patterns of seedling recruitment and thus the dynamics of natural plant populations.

A striking example of localized variation in female reproductive success has been documented by Albert *et al.* (2001) for a narrow endemic species

Box 4.2 Temporary marshes: diversity and adaptation in relation to spatial and temporary variation in ecological conditions

The alternation of a cool moist period and summer drought is pushed to the extreme in wetland areas which dry out in summer. Such temporary marshes (called vernal pools in California) occur in localized depressions with no natural outlet, show striking variation in the relative length of flooding (which is nevertheless long enough to allow the development of hygromorphic (water saturated) soils and aquatic vegetation), and show strong gradients in environmental conditions both in space (among marshes and from the centre to the peripheral parts of a given pool) and time (among years and during a given year in relation to the alternation of flooding and drying). In the Mediterranean Basin, temporary marshes also show immense variation in size, (Grillas and Roché 1997). Temporary marshes are not exclusive to Mediterranean-climate regions and provide ideal systems for the study of ecology and evolution in relation to spatio-temporal habitat variation. Three main features of these habitats are relevant to the study of plant evolution.

First, temporary marshes show striking spatial variation in ecological conditions, both among and within marshes. In the temporary marshes of the Mediterranean Basin, species composition varies spatially as a result of the interaction between life-history traits and strong salinity gradients, both in space and time, which are closely related to the date of flooding, its duration, and the depth of water (Grillas *et al.* 1991; Grillas and Roché 1997). Position in a marsh is of particular importance here. Towards the periphery of the pool, drying is more frequent and severe and the environment more variable. Hence plants may often germinate and grow in shallower or drier conditions, and may experience a more variable environment and in some cases less intraspecific but higher interspecific competition in the peripheral parts of a pool relative to the centre of a pool. In addition, the granulometry of the sediment, temperature, irradiance, and salinity may all vary along a gradient from the centre to the periphery of a marsh. This spatial variation can create divergent selection pressures on plants of a given species at different places in a pool (Linhart 1988). Most of this trait variation fits the idea that spatio-temporal environmental gradients are strong enough to counter local gene flow and produce striking patterns of within-population adaptive variation.

Second, temporary pools are well known for their species diversity, e.g. in California (Holland and Jain 1981), Chile (Bliss *et al.* 1998), and the Mediterranean Basin (Barbero *et al.* 1982; Grillas and Roché 1997; Quézel 1998). Like the vernal pools of California (Zedler 1990; Bliss and Zedler 1998), temporary marshes in the Mediterranean Basin are characterized by a large number of annual plants with a rapid life-cycle (Barbero *et al.* 1982) and a highly sensitive germination strategy which reduces the probability of germination at an unfavourable time and maintains a large seed bank (Bonis *et al.* 1995). This bet-hedging strategy is typical of plants in a temporally variable environment. These authors argue that this seed bank (which can reach tens to hundreds of thousands of seeds per square metre in the topsoil) has a 'storage effect' which magnifies the effect of favourable years relative to poor years. In conjunction with variation in reproductive strategies, this seed bank may contribute to the persistence of a species-rich community. The interaction between species traits and environmental gradients in space and time can thus act to maintain species diversity in these communities, in a way similar to that proposed by lottery models (Chesson and Warner 1981).

Finally, temporary marshes in several areas of the Mediterranean Basin are currently threatened and in many areas in danger of being lost from the landscape. Close to the coast, some are prime sites for drainage to provide a rich soil for farming or yet more space for tourism. Others are threatened by invasion of shrub species and their sustainability will require the implementation of management techniques to counter such encroachment (Médail *et al.* 1998; Rhazi *et al.* 2004). A sound scientific understanding of these marshes is thus important, not just for those of us interested in plant evolution, but also for the establishment of conservation plans (Grillas and Roché 1997).

Box 4.3 Major types of soil in the Mediterranean region

Mediterranean soils are primarily of two main types (Lossaint 1973; Zohary 1973; Bottner 1982):

1. Brown forest soils (fairly shallow clay-loams) which develop under various woody vegetation around the Mediterranean and which differ from those in temperate regions by their frequent high concentrations of Ca^{2+}, Mg^{2+}, and Na^{+} ions in mineral–nutrient cycles, the persistence of carbonates in the soil and in which the seasonal drought has a strong effect on the development of humus and litter decomposition.
2. Iron-rich *terra rossa* which is derived from hard or dolomitic limestone and is characteristic of garrigue and maquis vegetation particularly in subhumid and humid Mediterranean bioclimates (see Chapter 1).

In addition, one can observe carbonate-rich soils with a superficial crust that develop on limestone in arid and semi-arid bioclimates, rendizina with a dark surface horizon that sharply contrasts with the parent rock (usually softer limestones, chalks, and marly limestones) and alluvial soils in the river plains and serpentine soils in some sites.
Soils in the Mediterranean region are often thin and stony with strong local affinities to the parent rock, which undergoes relatively slow chemical weathering (Paskoff 1973). They tend to occur in one of three nutritional states: (a) moderately leached with a well balanced (albeit fairly low) nutrient status (particularly nitrogen and phosphorous), (b) strongly leached and nutrient poor, and (c) calcium rich (high pH) with micronutrients in relatively high concentrations that can render phosphorous insoluble (Specht and Moll 1983). All in all, soils are often oxidized and nutrient poor, a key feature being the paucity of available nitrogen and phosphorous.

of *Erodium* which occurs in a single locality in the Lozoya Valley in central Spain. At this locality, plants occur on either rocky or lithosol microhabitats on an isolated dolomitic outcrop in a landscape dominated by extensive siliceous formations. The rock micro-habitat is characterized by chasmophytes and the lithosol micro-habitat by species more typical of meadows and pastures. Plants are fairly evenly distributed in these two microhabitats which are themselves equally represented in the habitat island. In this mosaic of different communities, plant reproductive success is highly variable: seed production varies among years in the lithosol microhabitat (where seed production is very high in favourable years) and is low but stable in rocky microhabitats. As a result, in favourable years high seed production can occur (in the lithosol habitat) while in less favourable conditions persistence will be favoured (rock habitat). This species thus occurs in a mosaic of microhabitats with different dynamics, another example of highly localized metapopulation function, which in this case is closely related to local ecological conditions (see Chapter 2).

In a long-term field experiment on patterns of seedling recruitment in *L. latifolia* in the Sierra de Cazorla in southern Spain, C.M. Herrera (2002) showed how spatial patterns of the soil environment affect both the probability of seedling emergence and the survival of seedlings over the six years of his study. Although all the soils at the site originate from calcareous or dolomitic limestone, they show marked spatial differences in texture, organic matter content, and nutrient concentrations over small distances. Emergence was closely related to textural features of the soil (associated with drainage), while survival depended more closely on soil fertility. The patchiness of seedling establishment (and adult plant distribution) in this species may thus be closely related to how the regeneration niche is differentially affected by spatial heterogeneity of the soil environment. Later stages of the life cycle, for example, flowering probability and reproductive success, may also have different ecological correlates.

All these studies illustrate patterns of morphological and functional trait variation in relation to contrasting ecological conditions associated with climatic and/or edaphic variation. More examples will come later in this chapter as I discuss other features of spatial habitat variation in the Mediterranean mosaic. It should however be realized that most of the examples concern patterns of variation which will require experimental validation. Manipulative studies using transplanted material among sites which quantify performance variation in relation to environmental conditions would be extremely valuable here.

Another informative way of detecting patterns of adaptive trait variation is to compare estimates of population variation based on molecular markers (which quantify variation which may have evolved as a result of both random and/or adaptive causes) with variation in quantitative traits (morphology, phenology, and physiology) related to plant fitness, that is, trait variation that is likely to result from the action of natural selection. If molecular markers vary little compared to traits related to fitness then one can suspect variation in selection gradients among sites to be an important cause of trait variation among populations. A brief review of the literature shows that different species show very different patterns.

By comparing variation at polymorphic allozyme loci with variation in trunk morphology, growth and survival for 19 native populations of the maritime pine, *Pinus pinaster*, mostly in the Iberian peninsula, González-Martínez *et al.* (2002) found that quantitative traits showed higher differentiation than allozymes. Variation in the three quantitative traits was closely associated with local patterns of precipitation, temperature, and soil type, indicative of local adaptation to variation in these features of the environment. In particular the authors detected a latitudinal cline in survival rates which could reflect climatic effects on survival. Despite the long-lived nature of this tree, populations of maritime pine show evidence for adaptation to highly diverse ecological conditions, adaptation that has evolved since the last glaciation. In the annual legume, *Medicago truncatula*, Bonnin *et al.* (1996) reported that populations are more differentiated for morphological traits than marker loci, suggesting that quantitative characters have undergone strong divergent selection in different sites, among which gene flow is restricted. An evaluation of the ecological factors creating such divergent selection was not, however, examined.

In studies of genetic variation and differentiation among natural populations of the wild relatives of two cultivated crops, wild barley (*H. spontaneum*) and wild emmer wheat (*Triticum dicoccoides*), Nevo *et al.* (1983, 1988*a*) reported significant variation in allozyme frequencies in relation to soil type. In this area, brown basaltic soils produced during the Pleistocene and terra rossa soils on Middle Eocene hard limestone vary over small distances (transects were 100 m long). Similar patterns of population differentiation were detected in relation to highly localized and geographic variation in the climate (Nevo *et al.* 1979, 1982, 1988*b*). Selection gradients in the Mediterranean mosaic could thus be acting on the allozyme markers. In particular, heterozygosity was often positively correlated with increasing aridity and unpredictable rainfall, indicative that aridity stress represents a major selective cause of genetic differentiation. Morphological variation among populations supports this idea (Nevo *et al.* 1984*a*,*b*). Although Volis *et al.* (2002*a*) found low levels of genetic differentiation between desert and Mediterranean populations of wild barley (*H. spontaneum*), they detected significant differentiation in phenotypic traits such as awn morphology, plant size (but surprisingly not leaf size which was more variable among populations within the two regions), and spikelet biomass. Reciprocal transplanting of seeds and seedlings among these two types of habitat produced strong evidence for local adaptation to the two types of environment (Volis *et al.* 2002*b*). As mentioned in other species above, plants from Mediterranean habitats have a higher competitive ability (vegetative vigour) and lower reproductive effort than those from desert habitats.

Within-site variation in climatic and soil conditions can develop as a result of modifications to the local environment by dominant species. Such effects may alter the balance of competitive interactions among species and even facilitate the establishment of later-successional species (Connell and Slatyer 1977; Bertness and Callaway 1994). For instance, individual trees are well known to modify the local soil environment on which they and other species may grow (e.g. Boettcher and Kalisz 1990), shading may impede the reproduction of gap species but provide a more favourable microclimate for their regeneration, and roots may compete with other species for nutrients and water. Concomitantly, increased litter decomposition may create a nutrient-rich environment, favourable to the growth of associated species. This is important because in the Mediterranean litter decomposition may be slow due to the coriaceous leaf structure of many species, their high concentration of secondary compounds, and low moisture content of soils (Arianoutsou and Radea 2000). Two examples from the western Mediterranean illustrate these processes and how species not only respond to environmental variation but also create spatial heterogeneity for other species (Box 4.4).

4.3 The phenology of flowering and fruiting

4.3.1 Summer dormancy

Many species survive the summer drought by avoiding it. Such avoidance can be achieved by an annual life history, very prevalent in the Mediterranean flora and in the Mediterranean-climate region of California (Shmida 1981) or, in perennial plants, by summer dormancy. This dormancy involves the dieback of above-ground parts and a resting period of highly reduced physiological activity.

A well-known form of summer dormancy in perennial plants is the geophyte life form, in which all the above-ground parts of the plant are shed and the plant remains dormant with a perennating bud on a subterranean storage organ (tuber, bulb, corm, etc.). Perennial geophytes are an emblematic component of the Mediterranean flora (Shmida 1981) not only in monocotyledons (*Narcissus*, *Asphodelus*, *Iris*, *Muscari*, *Crocus*, *Tulipa*, *Ophrys*, etc.) but also in some dicotyledons (*Cyclamen* is a well-known example).

Summer dormancy may be triggered by one or a combination of several environmental factors which co-vary as summer develops: daylength, drought, and temperature. The relative importance of these factors is not constant among species. Whereas in some species temperature plays a critical role in the onset of dormancy, for example, in *Anemone coronaria* (Ben-Hod *et al.* 1988), daylength, which may be further stimulated by high temperatures, is the primary cue in other species such as *Poa bulbosa* (Ofir and Kigel 1999). High concentrations of abscisic acid in leaf tissue accompany this photoperiodic induction (Ofir and Kigel 1998).

In the Mediterranean, autumn is an important period of photosynthetic activity and may thus be an appropriate moment to flower. Autumn flowering enables rapid deployment of flowers after the summer drought when few other species are in flower and competition for pollinators may be reduced (Shmida and Dafni 1989) and there is probably less chance of pollen loss or stigma clogging due to heterospecific pollen transfer. There are however several constraints on flowering in the autumn. First, flowering occurs in the rainy season. Contact with rain may inhibit pollen germination and reduce performance, physically damage the flower, and dilute nectar. Shmida and Dafni (1989) and Dafni (1996) discuss how the several floral traits, such as timing of flower opening, pendulous flowers, pollen germination inhibitors, and hydrophobic floral parts, may be adaptive in such conditions. Second, flowering may have to be rapid once rains begin. Finally, pollinators are less abundant than in spring (Shmida and Dafni 1989).

Mediterranean geophytes provide an example of how to exploit the autumnal pollination niche. The subterranean storage organ not only provides an effective dormant period during the summer drought, but also provides a supply of nutrients for rapid growth at the beginning of the rainy season. Rapid development after the summer drought is important because the exact timing of the period favourable for photosynthetic activity and growth is unpredictable and the length of this period variable, and quite short in some years. Once the geophyte bulb attains a given size, resource stockage may exceed that required for flowering. This accumulation of stored nutrients is likely to be very important for flowering in autumnal flowering geophytes (see below). Although most Mediterranean geophytes flower in spring, there is a secondary

Box 4.4 Local modification of ecological conditions by dominant species

The savannah-like dehesa landscape (photo kindly supplied by R. Joffre) which covers ~55,000 km^2 of the southern Iberian peninsula has a low density (40/50 per hectare) of oak trees (mostly *Q. ilex* and/or *Q. suber*) managed for acorn, cork, and cereal production (Joffre et al. 1999).

Species number and vegetation structure differ in the open areas and under the canopy (Marañon 1986; Joffre *et al.* 1988): for example, the mean number of species under the oak canopy (17 species/4 m^2) is significantly less than that on the canopy edge (32 species/4 m^2) and in open grassland >5 m distant (27 species/4 m^2). These floristic differences may arise because soils under the canopy have a greater water-retention capacity (Joffre and Rambal 1993) and are richer in nutrients and organic matter due to litter decomposition (factors which are likely to promote the dominance of a small number of species) and/or the shelter trees provide for domestic livestock (Escudero 1985). In this ancient agroforestry system, natural selection pressures may vary sharply and repeatedly over small distances; a fascinating mosaic for the study of local adaptation.

The woody shrub *Thymus vulgaris* is associated with a significant modification of soil properties in garrigue habitats in southern France where this species is a common and abundant feature of local communities. Under thyme plants, soils have large amounts of leaf litter due to the shedding of leaves in late summer. Soil from under thyme plants has a significantly higher organic matter content compared to soils <5 m distant but in patches of grass, and associated grass species show higher biomass on soil collected from under thyme plants relative to soil from elsewhere in a given site (Ehlers and Thompson 2004*b*).

peak in autumn. In fact, perennial geophytes can comprise 80% of the autumn-flowering species in a given Mediterranean community (Shmida and Dafni 1989). Two phenological strategies are closely associated with this bimodal flowering phenology (Dafni *et al.* 1981*a,b*).

1. In species with 'synanthous' leaves peak flowering occurs when the leaves are fully developed. The dormant period is followed by a period of leaf growth and resource acquisition and storage followed by flowering and fruit production. Nearly all spring-flowering geophytes have this strategy.
2. In species with 'hysteranthous' leaves flowering is uncoupled from leafing and occurs primarily in the autumn or in winter. The first event after dormancy is the appearance of flowers, and only once flowering passes its peak, do leaves appear.

Hysteranthy may have evolved from synanthy in autumnal flowering species in response to the unpredictability of rainfall and the need to flower rapidly after the onset of the first autumn rains (Dafni *et al.* 1981*a,b*). This strategy may be more prevalent in arid conditions, indicating that flowering is in response to temperature and/or daylength and not the onset of autumn rains. Although in most spring-flowering Mediterranean geophytes, flowering time occurs later with increasing altitude (e.g. Arroyo 1990*a,b*), the opposite is true for autumnal-flowering species. For example, on Mt Hermon in Israel increasing elevation is correlated with later flowering in spring-flowering *Hyacinthus orientalis* and earlier flowering in autumn-flowering *Crocus olbanus* (Dafni *et al.* 1981*b*). On Corsica, two congeneric *Cyclamen* species (spring-flowering *Cyclamen repandum* and autumn-flowering *C. hederifolium*) which co-occur over an altitudinal gradient of ~1,000 m, show a similar pattern.

Cyclamen illustrates the different aspects of flower, leaf, and fruiting phenology in Mediterranean geophytes. The first point to make here is that, whatever the time of year there is always at least one species in flower. From the autumnal drifts of *C. hederifolium* flowers in the low mountain forests across the northern shores of the Mediterranean, through the winter appearance of *C. coum* and *C. persicum* in the east to the spring appearance of the most western Mediterranean species such as *C. repandum*, there is always a *Cyclamen* to be seen in flower during the cooler moister season of the Mediterranean climate. To see a species in flower during the summer drought, you have to leave the Mediterranean and head to the Alps, where *C. purpurascens* completes the annual cycle of flowering in the summer. This variation in flowering phenology among species, combined with the thrill of seeing a host of pink cyclamens in flower in the litter of a Mediterranean forest, is enough to make any biologist want to study the evolution of floral traits in this genus.

The genus *Cyclamen* has ~20 species, mostly distributed around the Mediterranean Sea and the Black Sea, plus one species in central Europe (*C. purpurascens*) and one in Somalia (*C. somalense*). Once a *Cyclamen* flower is fertilized, the single-flowered stem coils or bends downwards to place the capsules at soil level where they mature and open for seeds to be dispersed by ants (Hildebrand 1898; Affre *et al.* 1995). Whereas flowers are easy to see, sometimes in spectacular carpets on the forest floor, fruits are extremely difficult to find in the wild. Hierarchial analyses of variation of flowering time show that flowering time in this genus is primarily determined by subgenus membership (Box 4.5). This result indicates that major phenological shifts (between the two main peaks of autumn and spring flowering) occurred prior to the diversification of species in each subgenus. The divergence of ancestral taxa thus represented an important stage in the diversification of phenological strategies in this genus. Since then, phenology has been relatively conserved as species have diverged within different lineages, except in the subgenus Gyrophoebe in the eastern Mediterranean. This phylogenetic component to flowering phenology has been detected in several other studies (e.g. Johnson 1992; Ollerton and Lack 1992). In *Cyclamen*, the combination of leafing and flowering phenology shows a gradient of variation from strictly hysteranthous species, for example, *C. africanum*, *C. hederifolium*, and *C. rohlfsianum*, to strictly synanthous species, for example, *C. persicum*, *C. coum*, and *C. repandum*, while in some species leaves appear before the end of flowering (Debussche *et al.* 2004). After flowering, the long period of leaf production in autumnal flowering species may allow for a greater accumulation of stored reserves which in turn may enable rapid autumn flowering prior to leaf production (autumn-flowering species tend to have a much large maximum corm size).

Hysteranthous species also occur in the geophytes of other Mediterranean-climate regions (Ornduff 1969; Dafni *et al.* 1981*b*; Johnson 1992). Hence, there is evidence for convergent evolution of this strategy in different Mediterranean-climate regions. Geophytes which are active throughout the year tend to occur in the tropics or (as in *C. purpurascens*) in non-Mediterranean temperate climates which do not experience a summer drought. This suggests that the abundance of geophytes in a Mediterranean-type climate has been favoured by the onset of this climate since the Pliocene. As aridity becomes more important, hysteranthous leaf phenology may have been selected for (Dafni *et al.* 1981*a,b*). Unfortunately the genetic basis and degree of plasticity of this trait is unknown.

4.3.2 General patterns of phenology in the Mediterranean

Asymmetric bimodality of flowering phenology with a sharp peak in spring and a secondary peak in the autumn is a conspicuous feature of Mediterranean plant communities (Box 4.6; Dafni and O'Toole 1994). The summer drought thus imposes a severe constraint on phenology, and flowering time, as illustrated by several studies of intraspecific variation in the timing of germination, growth, and reproduction.

1. In *Quercus ithaburensis* in the eastern Mediterranean, one can observe variation in leaf phenology; some trees are deciduous and others almost evergreen due to a very short duration of leaflessness (Ne'eman 1993). Both phenotypic plasticity and genetic variation appear to influence this variation in phenology, which may be in a transition stage from a deciduous to an evergreen strategy.
2. The phenology of annual grass species populations from California and the Mediterranean Basin have a phenology which corresponds with a longer and earlier summer drought in California (Jackson and Roy 1986).
3. Mediterranean populations of the annual composite *Crepis sancta* flower three weeks earlier than populations from Northern France when grown in a common environment (Imbert *et al.* 1999*a*). This difference is despite the fact that Mediterranean populations germinate later, after the onset of autumnal rains.
4. The innate dormancy of *Senecio vulgaris* seeds from Mediterranean populations (an inhibitor in the seed coat may prevent the synthesis or liberation of gibberellins required for germination) over a wide range of temperatures relative to seeds from British populations, which germinate (~80%) immediately at 20°C, allows for a winter annual life cycle in a Mediterranean climate (Ren and Abbott 1991).

In the coastal shrublands of southern Spain, species which flower outside of the spring peak tend to occur in more mesic sites (J. Herrera 1986, 1991*a*). Arroyo (1990*a,b*) reported a similar but less marked pattern. The similarity of species phenology and their patterns of interannual variability suggest that climatic constraints are more important than phylogenetic relationships in determining flowering time (Petanidou *et al.* 1995). This may be particularly important for high-mountain species, for example, *Hormathophylla spinosa* in the Sierra Nevada where the extremely short and dry growing period imposes a severe constraint on flowering phenology (Gómez 1993).

As in *Cyclamen*, the constraint can also be of a phylogenetic nature. In various Tertiary endemics (see Chapter 2) flowering phenology no doubt evolved prior to the onset of the Mediterranean climate. For example, the early summer flowering of *N. oleander* (J. Herrera 1991*a*) and the long flowering period (November to June) of *Cneorum tricoccon* (Traveset 1995) suggest a historical constraint on flowering phenology which is a highly conserved trait in such species. The reproductive failure of *Ruscus aculeatus* has also been attributed to a historical constraint, this species persisting in a pollination environment very different from its original habitat in the Late Tertiary (Martínez-Pallé and Aronne 2000). In contrast, the flowering of the relict populations of *F. alnus* in southern Spain is significantly earlier than in populations in temperate Europe (Hampe 2004).

Box 4.5 The phenology of *Cyclamen* species

Morphological and molecular phylogenetic analyses suggest that the genus *Cyclamen* can be be subdivided into four subgenera. Debussche *et al.* (2004) quantified the phenology of leafing, flowering (grey bars), and fruit maturity (F) in 17 species in controlled conditions under a Mediterranean-climatic regime (column numbers refer to months). H—hysteranthous species (flowering prior to leaf emergence), S—synanthous species (flowers and leaves at the same time). Figure drawn from data in Debussche *et al.* (2004), photos kindly supplied by Max Debussche.

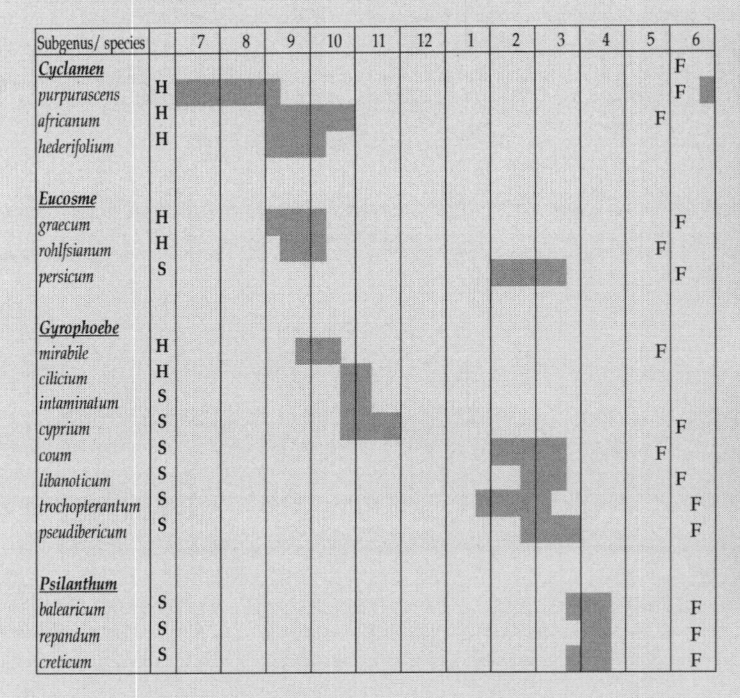

Hysteranthous
C. hederifolium

Synanthous
C. pseudibericum

All studied species are summer dormant except for the non-Mediterranean *Cyclamen purpurascens* which occurs under a continental climate and continues growth in the summer. Caught between the constraints of cold winters and summer droughts, species in the the spring-flowering subgenus *Psilanthum* have a very short period of photosynthetic activity, and thus a high capacity for resource acquisition (i.e. high values of specific leaf area—a trait correlated with photosynthetic activity.

Despite the strong seasonal differences in flowering among species in different *Cyclamen* subgenera, there is a striking lack of variation in the timing of seed release. Even species such as the spring-flowering synanthous *Cyclamen repandum* and the autumn-flowering hysteranthous *Cyclamen hederifolium* which grow together in the wild show almost identical periods of seed release. The length of seed maturation thus varies from 2–3 months in *C. repandum* to 10 months in *C. purpurascens* and the seed release period is independent of the flowering period and closely linked to the end of vegetation activity prior to the onset of summer drought.

Box 4.6 Flowering phenology in Mediterranean plant communities

(a) Total number of species which flower in a particular period, with the mean number per month in parentheses.

Community and region	Spring	Summer	Autumn	Winter	Reference
Greek phrygana	27 (9)	2 (1)	5 (1.7)	0 (0)	Petanidou and Vokou (1990)
Mediterranean coastal scrub community	19 (~6)	3 (1.5)	5 (1.7)	3 (<1)	J. Herrera (1986, 1988)
Mediterranean grassland	14 (~5)	2 (1)	1 (<1)	0 (0)	Bosch *et al.* (1997)

(b) Number of species in flower during each month over three years in the Greek phrygana. Drawn from data for 133 species in Petanidou *et al.* (1995).

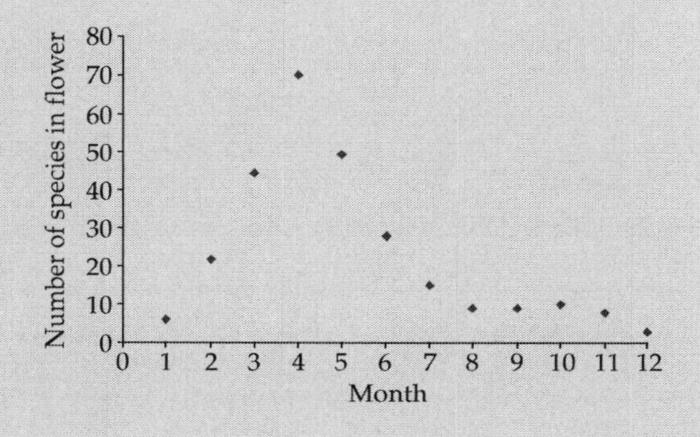

In southern Spain, the flowering period is very short and flowers that develop in late spring lose a large proportion of ovules to desiccation. Contemporary climate thus imposes a strong constraint on reproductive phenology and fruiting in this taxon.

In a study of the summer flowering shrub *L. latifolia* which flowers in mid-summer, that is, when fewer other species are in flower, C.M. Herrera (1992*c*) reported that although early flowering plants had higher seed production than later flowering plants, this result was not due to variation in water availability. This suggests that summer flowering is associated with some form of inherent constraint that may indirectly cause the pattern of temporal variation in seed production. In other species, extended flowering may allow reproductive failure in some years to be compensated by reproduction during an extended period of flowering, for example, in *Lobularia maritima* in coastal scrubland in north-east Spain (Picó 2000; Picó and Retana 2001).

The spring peak of flowering produces a large supply of flowers which in some situations may create severe competition for local pollinators. The result is a large investment in reward (nectar and pollen) and attractiveness (large floral displays, colourful flowers) in spring-flowering species (Cohen and Shmida 1993). In contrast, outside of the spring peak, there are fewer species in flower, but also fewer insects in flight, and it has been predicted that one should observe lower investment in reward (Shmida and Dafni 1989). A trend from high-rewarding spring flowers to less-rewarding summer flowers is seen in the Lamiaceae in Israel where there is a decline in flower size (Shmida and Dukas 1990) and a parallel reduction in nectar reward (Petanidou *et al.* 2000) during the spring and early summer. Larger flowers and nectar reward in spring may enhance competitiveness in terms of pollinator attraction. Flowers may nevertheless remain showy outside of the spring to attract the few pollinators that are present at this time, what has been termed 'discovery advertisement' (Shmida and Dafni 1989; Cohen and Shmida 1993). Long-distance attraction of potential pollinators may thus be at a premium in autumn-flowering species when pollinators and flowers are rare. Experimental analysis of these ideas and trends would be useful.

4.3.3 Variation in fleshy-fruit phenology and morphology

Seed dispersal by animals brings a mutual benefit: animal dispersers obtain food and plants disperse their offspring, hopefully towards a suitable site for germination and establishment. In the contemporary Mediterranean flora, vertebrate-dispersed plants are abundant (C.M. Herrera 1995*a*) and, in contrast to temperate plant communities, local groups tend to be taxonomically diverse at the family level. This is primarily because of the persistence of 1–2 species in several genera and families in the tropical element of the contemporary Mediterranean flora, for example, *Jasminum*, *Phillyrea*, *Olea*, and *Osyris* to name but a few (Fig. 4.1).

The abundance and nutritional characteristics of fleshy-fruited species makes them an important element of biodiversity functional relationships in the Mediterranean (C.M. Herrera 1984*a*,*b*; Debussche 1988; Rey 1995). Many fleshy-fruited species in the Mediterranean have peak fruit maturity when avian dispersers are most abundant (C.M. Herrera 1984*a*,*b*; Debussche and Isenmann 1992); but whether this is due to selection by dispersal agents and not just a consequence of resource limitation imposed by the dry summer has been questioned (Herrera 1985). Although rates of fruit removal are often very high, they may vary among and within species (Jordano 1987; Thébaud and Debussche 1992; C.M. Herrera 1995*c*) and many species show variation in size and shape of fruits which may be differentially consumed by frugivores, for example, *Prunus mahaleb* (Jordano 1995) and *O. europaea* (Rey *et al.* 1997). Hence the question: is phenology and morphology adaptive in relation to vertebrate dispersers?

Several studies have attempted to evaluate whether variation in fruiting phenology and fruit morphology represent a response to selection by avian dispersers in Mediterranean fleshy-fruited plants. In a 12-year study of fruit dispersal in 12 species (9 genera) in the Sierra de Cazorla in south-east Spain, C.M. Herrera (1998) revealed a striking independence of fruit production and bird abundance over several years. This was despite a more than 10-fold annual variation in fruit production by all the studied species and marked variation in the composition of the fruit spectrum available to dispersers. Except for a positive relationship between *Phillyrea latifolia* and *Sylvia atricapilla*, which was primarily due to a greater abundance of birds in the two mast-fruiting years of this shrub, there was no correlated variation between fruit availability and either fruits in the diet of avian dispersers or disperser abundance. Bird abundance was more closely related to November temperatures than fruit availability. Variation in fruit production most probably arises as a result of resource availability exceeding a sufficient level for their allocation to reproduction and high levels of fruit production. The study by C.M. Herrera (1998) and previous work by Jordano (1987) indicate that the primary limiting factor here is likely to be water availability. The absence of a consistent relationship between fruit traits and frugivory and any significant variation among populations in the interaction between frugivory, plant traits, and reproductive success has also been reported for *Juniperus communis* in the Mediterranean mountains of south-east Spain (García *et al.* 2001). In contrast, in ant-dispersed species it has been reported that habitat type, by its effect on disperser communities, may cause plant–disperser interactions to adopt distinct evolutionary trajectories in different populations (Garrido *et al.* 2002).

Another limitation on adaptive variation in fruit traits stems from the finding that, with just a few exceptions (C.M. Herrera 1984*b*), fleshy-fruited plant species in the Mediterranean tend to have their seeds dispersed by either a diverse array of bird species (C.M. Herrera 1984*a*,*b*; Jordano 1987, 1989;

Figure 4.1 Examples of fleshy-fruited species present (or with ancestral species present) in the Mediterranean region prior to the onset of the Mediterranean climate: (a) *Phillyrea latifolia*, (b) *Olea europaea*, (c) *Osyris quadripartita*, (d) *Jasminum fruticans*, (e) *Rhamnus alaternus*, and (f) *Pistacia lentiscus*. Photos kindly supplied by C.M. Herrera ((a)–(c)) and M. Debussche and B.O. Krüsi ((e)–(f)).

Debussche and Isenmann 1989; Guitián Fuentes and Sanchez 1992) or combinations of birds with either rodents, for example, *Juniperis communis* (García *et al.* 2001), or ants, for example, *Rhamnus alaternus* (Aronne and Wilcock 1994). Spatio-temporal heterogeneity in the composition and relative abundance of dispersal agents across the distribution of a given plant species may limit the opportunity for adaptive evolution in fruit traits (C.M. Herrera 1995*c*).

Other factors may further contribute to reduce the selective impact of dispersal agents on fruit traits (Box 4.7). First, although variation in traits may influence rates of dispersal, the effect on final reproductive output, and fitness, may be minimal, as a result of the complexity of interactions that may occur and the overwhelming importance of total crop size. Second, fruit number may be significantly reduced by predispersal abortion and predation of fruits (Jordano 1989) which reduce the contribution of variation in crop size to reproductive success.

In the hemiparasitic dioecious woody shrub *Osyris quadripartita* (Fig. 4.1), common in warm shrublands in the western Mediterranean, a rather surprising temporal pattern of fruit production can be observed (C.M. Herrera 1984*c*, 1988*a*). In populations of this species in southern Spain, females produce ripe fruits throughout the year, despite the fact that individual plants only flower once a year. Due to the gradual ripening of fruits from a single flowering bout, fruits resulting from the previous year's flowering ripen as new fruits develop. The only seasonality is in the fraction of plants ripening fruits which peaks in late-autumn/early winter. In *Osyris*, the persistence of a continuous phenological strategy through the summer drought, may have been facilitated by the 'generalist' hemiparasitic nature of this species, which permits additional nutrient and water uptake (C.M. Herrera 1984*c*, 1988*b*).

This long period of growth, flowering, and fruiting, with extensive temporal overlap in such activities, is typical of species in a tropical climate and is not related to potential selection pressures imposed by contemporary dispersal agents. As I discussed in Chapter 1, many fleshy-fruited woody species (e.g. species in the Oleaceae, Anacardiaceae, Santalaceae, Rhamnaceae, Myrtaceae, Cneoraceae) in the Mediterranean flora, originated prior to the onset of a Mediterranean seasonal climate and form a fairly distinct Pre-Pliocene group. Species in these groups were thus present when conditions were subtropical in this region. Of woody shrub and tree genera now present in southern Spain that were present prior to the Pliocene 94–95% have fleshy fruits, whereas only 5–6% of genera that have immigrated since the onset of a Mediterranean-climate regime have fleshy fruits (C.M. Herrera 1992*a*; 1995*a*). Although, many species disappeared from the region as the climate evolved, several have persisted. In the Late Tertiary, vegetation extinctions would have been paralleled by the extinction of animals, particularly large birds capable of dispersing fruits which were present prior to the Pliocene onset of the Mediterranean climate (Mourer-Chauviré 1989). Hence many plant species may have persisted following the loss of their seed dispersal agents, hence their dispersal mode may not be adaptive with respect to contemporary dispersal agents and the Mediterranean climate. Nevertheless, since the Pliocene, fleshy-fruited genera have gone extinct at a lower rate than non-fleshy-fruited genera, suggesting that their high frequency in the flora is not just due to phylogenetic inertia.

The tropical element also has a majority of species with small often unisexual flowers and fleshy fruits (C.M. Herrera 1982*a*; Aronne and Wilcock 1994; Quézel and Médail 2003), a syndrome which is certainly not limited to the Mediterranean. Although it has been predicted that such dioecious species should show greater absolute maternal investment in seed dispersal compared to hermaphrodites (Givnish 1980), C.M. Herrera (1982*a*) found no evidence of this in a study of dispersal-related maternal reproductive investment in 73 species in southern Spain. Three points are important here: (a) selection pressures due to dispersal agents may be complex and highly variable in space and time; (b) the significance of crop size for fitness is limited since many seeds are dispersed into unfavourable habitats; (c) many dioecious shrubs in the Mediterranean have probably shown little modification over a long period of time. This association of characters is thus primarily due to phylogenetic constraints and sorting processes, having evolved long before the Mediterranean-climatic regime set in (Box 4.1).

In the tropical relict, *Jasminum fruticans* (Fig. 4.1; Plate 2), fleshy fruits are also ripened gradually such that many are mature only in late winter (Puech 1986). Late ripening in such species is associated with a lower pulp biomass and smaller fruits (Puech 1986; C.M. Herrera 1988*a*). Species which have a ripening period which occurs in late autumn and early winter also have different fruit characteristics (noticeably a decrease in water and an increase in lipid content in late-ripening fruits) compared to those which ripen their fruits in summer and early autumn (C.M. Herrera 1982*b*; Debussche *et al.* 1987). Such variation is unrelated to variation in avian dispersers. The co-occurrence of peak fruit availability and the abundance of avian dispersers is more the result of climatic constraints, pest pressure, and opportunistic behaviour of frugivores (Debussche and Isenmann 1992; C.M. Herrera 1995*c*) than the result of selective pressures exerted by dispersal agents. Hence, it would appear that dispersal agents impose only weak selective pressures on the fruiting phenology of fleshy-fruited shrubs in the Mediterranean region, due to the complexity of the interaction between single plant species and their multiple dispersal agents. Several studies suggest that a similar conclusion can be drawn for variation in the size and shape of fleshy fruits in the Mediterranean (Box 4.7). There is thus a fairly large body of evidence which attests to a lack of direct interaction and little evidence of any adaptive variation in fruiting phenology and morphology in relation to avian dispersal agents. However, as I

Box 4.7 Fleshy fruits: morphology and vertebrate dispersal

At the present time there is a paucity of evidence of adaptive variation in fruit morphology in relation to avian dispersal agents in Mediterranean fleshy-fruited species.

First, C.M. Herrera (1988*a*) reported significant variation among individuals of *Osyris quadripartita* in fruit size and production over four years of study. However, the patterns of variation were highly variable over a single year and among years due to differences in rainfall events in different years (once again the unpredictability of rainfall events is important). Although fruit characteristics influence dispersal rate, they only have a minor contribution (0.2%) to realized reproductive output in a population, due to the overriding contribution of total fruit production. For these two reasons, variation in fruit morphology does not translate into fitness differences and thus may have little selective value. This primary importance of absolute numbers of fruits produced has been documented in other Mediterranean herbs and shrubs, for example, *Lavandula latifolia* (C.M. Herrera 1991) and *Prunus mahaleb* (Jordano and Schupp 2000). More fecund plants may thus disperse more propagules despite variation in fruit size or relative rates of seed set.

Second, in comparative analyses (correcting for statistical non-independence associated with phylogenetic relationships) of 117 fleshy-fruited species (from 35 families) in the Iberian peninsula, C.M. Herrera (1992*b*) found that fruit shape variation in Mediterranean vertebrate-dispersed plants does not support the prediction that groups of plants with dispersal agents of contrasting sizes and fruit handling capabilities have differently shaped fruits (in terms of length–width relationships). Instead, allometry and genus- and species-specific variation in fruit shape are more important than dispersal agent in explaining the patterns of variation. Nested analyses of variance showed a significant phylogenetic component to variation in fruit size and shape. Diversification in fruit morphology occurred at a level above the species level and closely related species share similar traits. The combined effects of allometry and taxonomic relationships rule out adaptive explanations of variation in such traits.

Third, any potential direct effect of fugivorous birds on plant traits that enhance fruit removal may be masked or buffered due to the impact of other ecological parameters. One of the most important of these is herbivory and pre-dispersal seed predation. In a study of the effects of such factors on fruit production and removal in *Olea europaea* in southern Spain, Jordano (1987) reported that although individual fruit yield was positively correlated with fruit removal by dispersers, plants with high yield were massively infested by *Dacus oleae* (a Tephritide fly), which reduced the contribution of trees to total fruits removed in a population.

Nevertheless, in species whose distribution extends beyond the bounds of the Mediterranean fruit traits have been reported to vary between the Mediterranean and temperate regions, perhaps in relation to the local climate and variation in disperser presence and activity (Hampe and Bairlein 2000; Hampe 2003). Hence, it is possible that colonization and/or persistence in the Mediterranean has to some degree impacted on particular dispersal-related traits in some species.

will discuss later, dispersal patterns show marked spatial patterns in the Mediterranean mosaic, with potential consequences for the evolution of genetic differentiation among populations.

4.4 Dispersal and establishment: the template of local differentiation

4.4.1 Landscape change: process and consequences

Landscapes are changing around the Mediterranean. Although vegetation patterns and species distributions have frequently changed in relation to climatic variations, even since the initial onset of a Mediterranean-type climate in the Pliocene (see Chapter 1), contemporary rates of change are probably unprecedented in Mediterranean history. Instead of gradual climatic oscillations, Mediterranean vegetation is now faced with the rapid, extensive, and often brutal effects of human activities. Near the sea, wetlands and semi-natural coastal and lowland habitats are rapidly being destroyed and fragmented by urbanization and tourist resort development, causing major loss of biodiversity and creating severe problems for the viability of natural populations. Inland, a different story can be told. Throughout the European backcountry of the Mediterranean region, forests are spreading due to agricultural decline and rural depopulation (Fig. 4.2). Some famous Mediterranean landscapes are changing, as woody species colonize the rolling hills of Tuscany, the limestone grasslands near Roquefort, and the mountains of Crete. The contours of the Mediterranean habitat mosaic are thus changing rapidly.

The reforestation of many inland areas around the northern rim of the Mediterranean is a natural ecological process, but one that has been set in motion by important socioeconomic changes that began in the late nineteenth and early twentieth centuries (Lepart and Debussche 1992). These changes caused traditional agricultural practices, pastoralism, and forest activities to become economically unattractive and in many areas unprofitable (e.g. Marty *et al.* 2003). Too much labour and not enough profit, traditional farming has modified its structure and in many areas been quite simply abandoned. The result was, and continues to be, rural depopulation. For example, in the Languedoc-Roussillon region of southern France the number of inhabitants in rural areas decreased 2–3 fold over the period 1936–90 (Cheylan 1990). Towards the coast, human populations remained stable during this period, and since the 1960s have begun to grow, at rates faster than anywhere else in France. A similar story occurs in many areas of the Mediterranean. On Crete, major landscape changes in the second half of the twentieth century involve tourist development along the north coast (where human populations have remained stable) and forest spread and closure of previously open woodland in inland areas, where the mountain village populations have declined by more than one-third (Arianoutsou 2001). Hand in hand with these changes in human populations, the landscape has changed as woody plants have spread to form woodland and forest on previously open landscapes. This forest spread is extremely rapid in some areas (Fig. 4.3).

On the northern shores of the Mediterranean there is a distinct difference in the degree of human impacts on natural ecosystems which depends on location: whereas human impacts are increasing near the coast, further inland, particularly in low mountain areas, they have actually diminished. In this section, I will focus on the ecological processes and consequences of natural forest spread. This discussion is motivated by the question: what are the implications of such landscape change for plant population ecology and differentiation?

The principal land-use changes associated with modified human impact and rural depopulation include the abandonment and dramatic decline of terrace agriculture on hillsides, a shift towards intensive smaller-scale farming with a decline in extensive grazing practices over large expanses of lowland and upland plains and reduced exploitation of woodlands for glass-making, charcoal, and even firewood collection (Lepart and Debussche 1992). Striking examples of the rapidity of forest spread associated with these changes can be seen in the Mediterranean backcountry (Figs. 4.2 and 4.3), for example, in the Cévennes (Debussche *et al.* 1999) and Maritime Alps (Barbero *et al.* 1990) in

Figure 4.2 Some examples of landscapes currently experiencing natural spread of woody vegetation and forest closure in the Mediterranean: (a) the spread and establishment of *Pinus brutia* woodland on Crete, (b) the spread of deciduous oak (*Quercus pubescens*) on the Causse du Larzac (southern France), (c) the spread of *Buxus sempervirens* on the Causse du Larzac in southern France (photo kindly supplied by P. Marty).

Figure 4.3 The rapidity of natural spread of woody vegetation (mostly evergreen oaks) and forest closure in the Basses Cévennes area of southern France: (a) a postcard view in 1929 and (b) the same view in 1992 (reproduced with permission from Lepart *et al.* 1996).

southern France and in mainland Greece and Crete (Arianoutsou 2001; Grove and Rackham 2001). In the Languedoc-Roussillon region of southern France mentioned above, forest cover has more than doubled since the beginning of the twentieth century, increasing from 400,000 ha to ~975,000 ha (Agence Méditerranéenne de l'Environnement 2000). An important element of this reforestation involves the return of evergreen oak woodland and the spread of pioneer pine trees. For example, *P. halepensis*, which covered ~36,000 ha of Mediterranean France at the end of the nineteenth century, now occupies more than 200,000 ha of this region (Barbero and Quézel 1990). The Mediterranean mosaic is thus subject to important changes: woodland interspersed with patchy open habitats and localized intensive cultivation is becoming a typical landscape in many areas.

Forest spread may have significant ecological and evolutionary consequences. First, local ecological conditions are modified, creating new constraints and selection pressures for plants which arrive at a site. Second, the spatial configuration of habitats in the landscape has been dramatically modified. This means that dispersal of seeds and pollen, and thus gene flow, among sites are subject to marked modifications. Populations of woodland plants isolated from one another for several centuries may now be increasing in size and may come back into close contact with possibilities for crosses among previously isolated populations. In contrast, populations of species that occur in open habitats or gaps in the forest may decrease in size and become more isolated from another if they cannot survive shading and forest spread. The implications for genetic differentiation in these two groups of species are thus very different.

4.4.2 Dispersal and establishment: colonization processes and population differentiation

Colonization and population establishment depend on rates of arrival at a site (and thus dispersal-related traits and rare episodes of long-distance dispersal) and the establishment of a local population from the initial colonists (i.e. the regeneration of seedlings in a novel environment). Because plant mortality tends to be high at the seedling stage in many plant populations (see references above) there should be strong selection on dispersal-related traits that consistently favour the chances of putting a seed into a site favourable for seed germination, seedling establishment, and seedling survival. Recognized as a key element in the maintenance of diversity in plant communities (Grubb 1977), the regeneration niche is a key element in the population ecology of individual species. In addition, in recent years there has been increased recognition of the important role played by dispersal limitation in the patterns of abundance and distribution of individual species and community structure (Hubbell 2001). Evidence from studies of different Mediterranean systems illustrate the combined role of spatial patterns of dispersal and its restricted nature and ecological factors influencing regeneration for the dynamics of natural populations in the Mediterranean mosaic landscape. The point that is important here is that the interplay of factors affecting dispersal and establishment during the colonization process can create sharp spatial patterns in natural populations.

4.4.3 Spatial patterns of dispersal across open habitats

The rate and spatial extent of establishment of vertebrate-dispersed woody plants in open landscapes in the Mediterranean, like elsewhere, are closely dependent on the spatial distribution of seed sources. Studies of the several fleshy-fruited and wind-dispersed woody plants shows that most dispersal occurs close to maternal trees, that is, over a few metres or tens of metres, for example, *Buxus sempervirens* and *Fraxinus angustifolia* in Mediterranean old-fields (Debussche and Lepart 1992), mast-fruiting vertebrate-dispersed *P. latifolia* in mid-elevation scrubland and forest in southern Spain (C.M. Herrera *et al.* 1994), wind-dispersed *P. halepensis* on the lower slopes of Mt Carmel in Israel (Nathan *et al.* 2000), several fleshy-fruited pioneer plants in early-succession old-fields (Debussche *et al.* 1985; Debussche and Isenmann 1994), and *O. europaea* subsp. *europaea* var. *sylvestris* in scrubland and grazed sites in southern Spain (Alcántara *et al.* 2000). In several of these studies the distribution pattern of seeds in the seed rain was very similar (particularly in open scrubland) to the spatial pattern of seedling establishment. The spatial aggregation of seedfall around maternal plants and the coupled pattern of seed rain and established seedlings provide strong evidence that dispersal limitation will create spatial pattern, with a potentially 'lasting impact on ... population dynamics" (C.M. Herrera *et al.* 1994: 315).

An important component of this spatial pattern involves nucleation around initial colonists. For many Mediterranean fleshy-fruited species, the presence of other tree and shrub species in a more or less open landscape promote nucleation as birds use conspecific trees and other species as perching points (Debussche *et al.* 1985; Debussche and Lepart 1992; Debussche and Isenmann 1994; C.M. Herrera *et al.* 1994; Verdú and Garciá-Fayos 1996; Alcántara *et al.* 1997, 2000; Garciá *et al.* 2000; Traveset *et al.* 2003). For example, Debussche *et al.* (1985) showed that although 15% of the seeds of 14 fleshy-fruited species in Mediterranean old-fields are dropped within 20 m of a seed bearer, noticeable peaks in the seed shadow occur around perching places (isolated trees). In wind- and animal-dispersed species, dispersal limitation is thus not completely random in the landscape. A pattern of nucleation can also be observed in the colonization of open grassland by wind-dispersed species, for example, *Pinus sylvestris* which is actively colonizing peri-Mediterranean grasslands (Box 4.8).

Dispersal patterns can be closely related to microhabitat variation in the Mediterranean mosaic, particularly in vertebrate-dispersed plants. The first point to note here is that the mechanical and chemical action of the dispersal agent has little importance for overall germination capacity in many Mediterranean species (Debussche 1985, 1988; Izhaki and Safriel 1990). Frugivory may however promote dispersal into favourable sites for germination. Alcántara *et al.* (2000) integrated the patterns of seed fall both in the seed shadow (under trees) and the seed rain (population level) to examine the relative contribution of spatial, microhabitat, and tree characteristics for seed dispersal from wild olive trees (*O. europaea* subsp. *europaea* var. *sylvestris*) at two study sites in the Sierra Sur de Jaén in southern Spain. They found that differences among frugivores in sizes of seeds dispersed and foraging movements created a pattern of seed dispersal in which different microhabitats received different distributions of seed sizes. The spatial distribution of seeds was found to depend on a complex interaction between the fruit traits which attract frugivores, distance to the source tree, and microhabitat. The latter was associated with high seed densities in patches of shrub species with dense foliage (*Phillyrea latifolia, Pistacia lentiscus*, and *Q. coccifera*), low seed densities under species with open foliage (*Pistacia terebinthus* and *Rhamnus lycioides*) and low seed dispersal into open spaces.

A low rate of seed dispersal by frugivores into open spaces is commonly observed in fleshy-fruited species in Mediterranean habitats (C.M. Herrera and Jordano 1981; Izhaki *et al.* 1991; Debussche and Isenmann 1994; C.M. Herrera *et al.* 1994). Debussche *et al.* (1985) showed that a complex vegetation structure can favour dispersal over longer distances: the establishment of seedlings >100 m from any potential source illustrating this. In a study of bird-dispersed *P. mahaleb* in southern Spain, Jordano and Schupp (2000) detected a highly heterogeneous spatial pattern of seed dispersal. In this species the seed shadow was determined by a complex interaction between the foraging patterns of a suite of frugivores that differ in their habitat preferences and the precise distribution of different habitat patches in the landscape. Covered patches received seed dispersed by 7–11 bird species whereas seed dispersal into open habitats was by only 1–2 species. As Jordano and Schupp (2000) suggest, one would predict greater genetic variability among *P. mahaleb* propagules within the covered habitat patches. In addition, reduced and spatially aggregated seedfall patterns in more open habitats may favour differentiation among patches (each from potentially different seed sources but with little variation within patches).

4.4.4 Spatial heterogeneity and regeneration

Contemporary reforestation of many areas is an ecological change involving primarily native species and is thus a natural example of secondary succession. Vegetation succession has fascinated ecologists ever since the science of ecology shifted its emphasis from describing patterns to understanding their causes. In their classic paper which focused attention on the mechanisms which allow and favour the transition from pioneer communities to climax forest, Connell and Slatyer (1977) outlined three models of vegetation succession:

- Facilitation: pioneer species modify the environment in a way which makes ecological conditions suitable for the establishment of later successional species. Without such modifications the latter cannot establish at a site
- Tolerance: late successional species establish independently of pioneer species
- Inhibition: establishment of late successional species only occurs as a result of disturbance or mortality of pioneer and early succession species.

Box 4.8 Dispersal, nucleation, and colonization of limestone grasslands by *Pinus sylvestris* in southern France

An example of the role of nucleation for the colonization of open areas by wind-dispersed species can be seen in pine trees colonizing upland limestone grasslands in southern France. In this area, the spread of *P. sylvestris* since the beginning of the twentieth century (when cereal cultivation began to decline and milk production became based on intensive, specialized methods) illustrates dispersal mediated patterns of spatial aggregation (Debain 2003; Debain *et al.* 2003*a*,*b*). Dispersal across the plateau from the western forested part of the plateau eastwards onto the open 'Causse' has occurred by occasional episodes of long distance dispersal and establishment of isolated trees, followed by local colonization and nucleation around these pioneer trees. Isolated trees produce up to four times as many cones as dominant trees in groups and more than ten times the number of cones produced by subordinate trees in dense stands. Long-distance dispersal and nucleation rapidly increase the speed of colonization in this landscape due to the creation of new focal points (or 'nascent foci') of colonization, distant from the primary source populations, which become secondary source populations for subsequent spread (see also Moody and Mack 1988; Hengeveld 1989; Cain *et al.* 2000). The expansion of species such as the pine trees colonizing the Causse Méjean, is also favoured by their reproductive and seed biology, that is, early juvenile reproduction, a short interval between large seed crops and the production of many small winged seeds: that is, a suite of traits which favours dispersal and rapid population establishment (Barbero *et al.* 1990; Rejmánek and Richardson 1996). This expansionist strategy (*sensu* Barbero *et al.* 1990) can also be observed in *Pinus halepensis* colonizing abandoned agricultural fields in the plains of southern France, Spain, Croatia, and north-west Italy (Quézel and Médail 2003).

These models, although quite simple when compared to the complex reality of the process, stimulated ecologists to identify mechanisms of interaction and their role in what is a central process in plant population ecology. Since the publication of the above paper it has become clear that to understand the process of succession requires explicit consideration of how interspecific interactions occur in relation to abiotic environmental conditions, particularly soil water and nutrient levels and light, and the role of diverse trade-offs in colonization, community structure, competitive ability, and allocation to different traits (Tilman 1990; Garnier *et al.* 2004).

A range of different modes and mechanisms of succession no doubt occur in the succession of vegetation after agricultural abandon in the Mediterranean (Escarré *et al.* 1983; Lepart and Escarré 1983; Debussche *et al.* 1985, 1996; Debussche and Lepart 1992; Zavala *et al.* 2000). In fact, the relative importance of different processes may show a shifting balance between positive and negative interactions in relation to the life-cycle stage of individual plants, local environment (in particular abiotic stress), and successional stage (Pugnaire and Luque 2001). Studies of plant–plant interactions in successional old-fields, semi-arid grasslands, and woody plant colonization of upland pastures provide some illustrations of this dynamic balance between competition and facilitation in Mediterranean communities.

First, in old-field populations that have developed following the abandonment of agricultural activities near Montpellier in southern France, there is a well-documented sequence of vegetation succession over time (Escarré *et al.* 1983). Following a period of initial dominance by annuals, perennial Lamiaceae become important components of the vegetation within ~10 years. This is followed by the establishment of a community containing dominant perennial grasses (20–50 years), in which establish several woody species. Forest closure occurs after

50 years or more. In the early parts of this succession, Sans *et al.* (1998) reported that facilitation is important for regeneration of the short-lived herbaceous species *Picris hieracioides* whereas nutrient competition is an important determinant of persistence. So although neighbouring plants positively promote seedling recruitment (perhaps via an improvement of microclimatic conditions), they have a strong negative affect on later growth and reproduction.

Second, field experiments in two semi-arid grassland communities in south-east Spain by Maestre *et al.* (2003) revealed how the bunch grass *Stipa tenacissima* consistently facilitates the establishment of the shrub *P. lentiscus* due to a positive effect on soil moisture, fertility, and microclimate despite strong below-ground competition between these two species. The magnitude of this facilitation effect varied significantly in relation to small-scale environmental variation, being greatest in stressful conditions.

Third, experimental work on the Causse du Larzac, an upland limestone plateau in southern France, where *Buxus sempervirens*, *Juniperus communis*, and *Q. pubescens* are rapidly spreading across the landscape (Fig. 4.2), illustrates how the establishment and persistence of woody species depends on the balance of positive and negative interactions (Rousset and Lepart 2000). In this area *Buxus* has spread rapidly due to it no longer being cut by farmers (it was formerly used in stables and sheep pens and for soil enrichment). *Quercus* seedlings are 50 times more abundant under the canopy of *Buxus* and *Juniperus* (particularly on north-facing side of such shrubs) than in open grassland. Experimentally planted acorns of *Q. pubescens* under pioneer *Buxus* and *Juniperus* showed significantly higher germination and reduced seedling mortality than in open areas, or under shrubs where the canopy had been cut. Microclimate and reduced grazing caused these results. Although seedling growth was less under *Buxus* than under *Juniperus* and in open areas, once the canopy was overtopped such effects disappeared. Facilitation due to reduced summer drought stress and grazing can thus offset the negative effects of competition. Important here is that different results were obtained under *Buxus* and *Juniperus*, hence the spatial heterogeneity of the environment in which oak seeds germinate, that is, the precise distribution and identity of other species, may have an important effect on the pathway and speed of woody plant colonization and succession.

The works of Rousset and Lepart (1999, 2000) illustrate two important points. First, facilitation is not limited to early successional stages where stressful conditions commonly occur. Even the establishment of late succession woody species, albeit in a hostile environment where drought and grazing limit seedling recruitment, also depends on positive interactions. Indeed, a positive balance in species interactions, due to the mediation of improved soil conditions (particularly reduced nutrient and water stress), may be a primary force regulating regeneration in semi-arid communities (Pugnaire *et al.* 2004). Second, the extensive spread of woody species across the Causses limestone plateaux of southern France is more due to the abandonment of cultivation than to reduced grazing pressure. Although seedling recruitment is strongly limited by sheep grazing, which is responsible for 2/3 of seedling mortality of *B. sempervirens*, grazing did not have a strong effect on plant growth. Grazing controls the spread of woody species but does not completely prevent it. Of course, a combination of intense herbivory and drought stress (in addition to limited seed dispersal) can restrict establishment and survival of herbaceous species in open areas. For example, in disturbed sites in south-west Spain, the winter annual *Geranium purpureum* has a localized distribution that is almost invariably spatially aggregated under dominant *Juniperus* trees (J. Herrera 1991*b*).

Microhabitat specific patterns of seed germination, seedling survival and growth, and reproduction are also known to occur in fleshy-fruited species whose seed rain is non-random (C.M. Herrera *et al.* 1994; Jordano and Herrera 1995). Indeed, in some species there may be a strong spatial discordance between seed rain and seedling establishment, indicating the importance of ecological constraints on regeneration. In *Rhamnus ludovici-salvatoris* (endemic to the Balearic islands), Traveset *et al.* (2003) found that whereas most seeds

Plate 1 Floral diversity and habitats of *Cyclamen* subgenus *Psilanthum* (see chapter 3). (a)–(e) The different subspecies of *C. repandum*: (a) *C. repandum* subsp. *repandum*, (b) and (c) *C. repandum* subsp. *peloponnesiacum* (varieties *peloponnesiacum* and *vividum* are depicted respectively although all intermediates can be observed in the wild), (d) *C. repandum* subsp. *rhodense*, (e) *C. repandum* subsp. *creticum*. (f)–(i) Examples of habitats: (f) mid-elevation *Pinus maritima* forest on the island of Corsica (*C. repandum subsp. repandum*), (g) woodland and garrigues in the Peloponnese penisula (*C. repandum subsp peloponnesiacum*), (h) mid-elevation *Pinus brutia* forest on Crete (*C. repandum* subsp. *creticum*)—*C. repandum* subsp. *rhodense* can also be observed in this type of habitat on the island of Rhodes, (i) and (j) *C. balearicum* habitat (evergreen oak woodland in a limestone gorge in southern France) and flowers respectively. (k)–(n) The range of floral diversity (k)–(m) observed in populations on limestone on Corsica (n) where both *C. repandum* subsp. *repandum* occurs and a relict population of *C. balearicum* may have persisted.

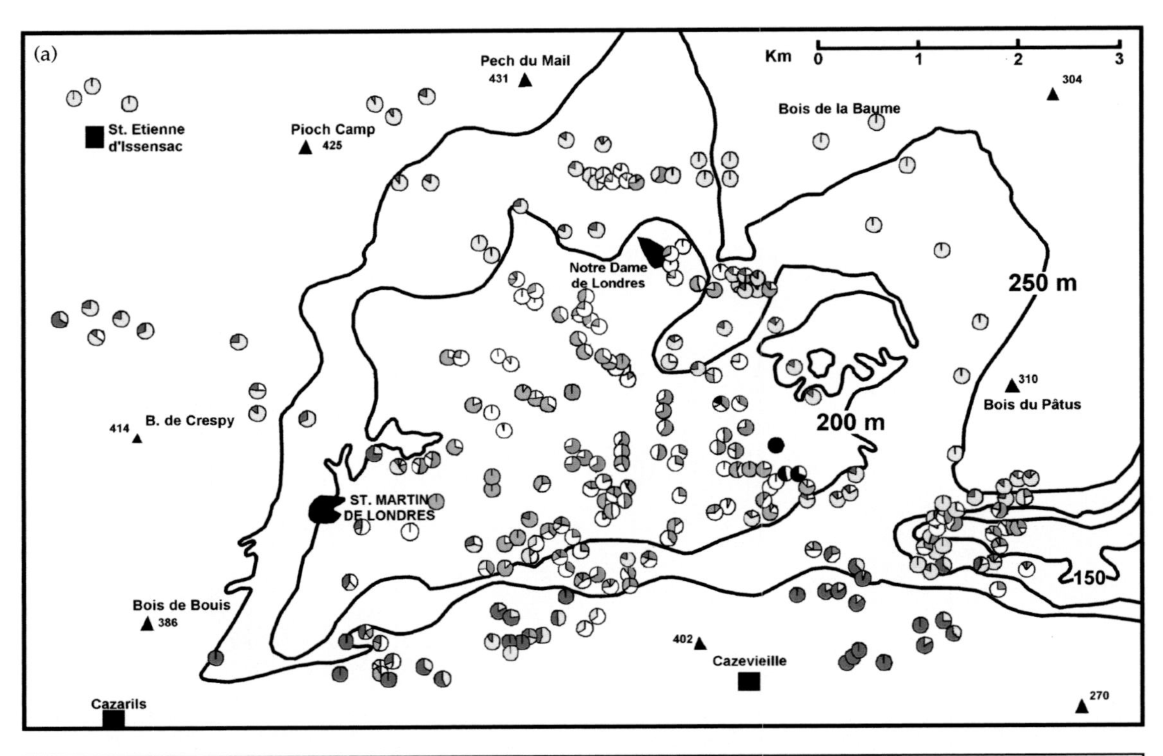

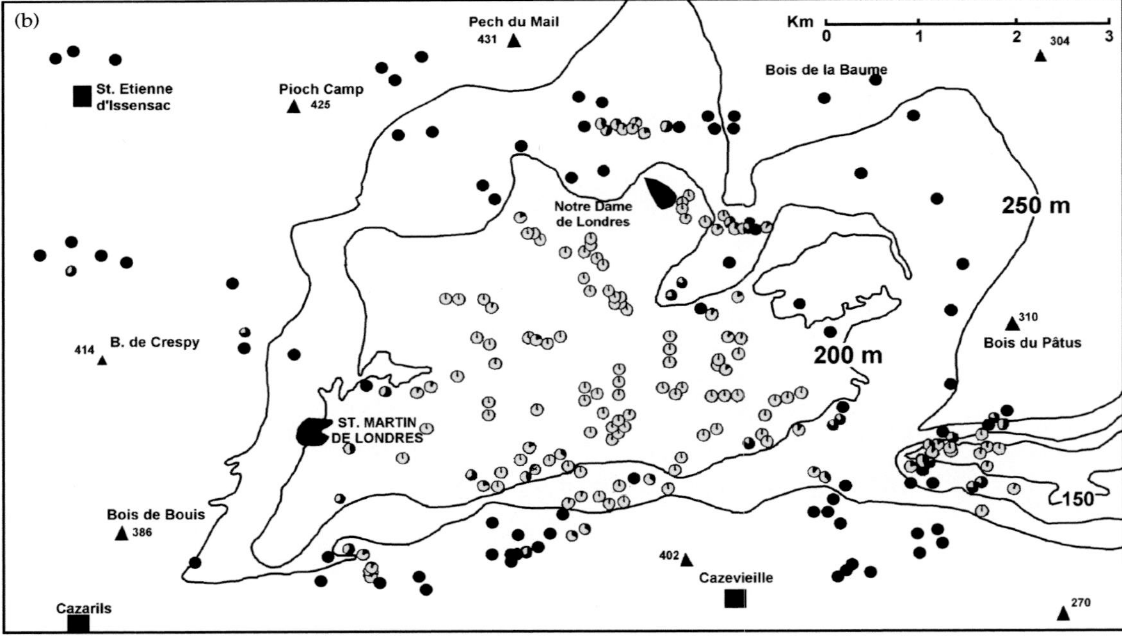

Plate 2 The spatial distribution of *Thymus vulgaris* chemotypes in and around the St Martin-de-Londres basin ~25 km north of Montpellier in southern France (see chapter 4). (a) The relative proportion of the different monoterpenes in a bulk sample of the oil at each site, with each colour representing one of the six monoterpenes which characterize the six chemotypes: geraniol (black), α-terpineol (blue), thuyanol (green), linalool (white), carvacrol (red) and thymol (yellow); (b) the relative proportion of the oil composed of phenolic (black) and non-pehnolic (yellow) monoterpenes at each site. Redrawn with permission from Thompson (2002) and based on original data collected by Vernet *et al.* (1997*a*,*b*) and Gouyon *et al.* (1986).

Plate 3 Examples of species discussed in Chapter 5 in relation to the role of pollinator mediated selection on reproductive trait variation in Mediterranean plants. (a) *Aquilegia viscosa*, (b) *Aquilegia vulgaris*, (c) *Jasminum fruticans*, (d) *Erysimum mediohispanicum*, (e) *Paeonia broteroi*, (f) *Lavandula latifolia*, (g) *Viola cazorlensis*, (h) *Hormathophylla spinosa*. Photos kindly supplied by C. Herrera (e)–(g), J. Gómez (d) and (h) and G. Debussche (a) and (b).

Plate 4 *Narcissus* species which show the three types of style-length polymorphisms discussed in Chapter 5. (a) *N. assoanus*, (b) *N. dubius*, (c) *N. papyraceus*, and (d) *N. tazetta* are all species with a stigma-height dimorphism; (e) *N. albimarginatus* is distylous and (f) *N. triandrus* is tristylous. Photos kindly supplied by J. Arroyo (c) and (e) and S.C.H. Barrett (f).

are dispersed to sites under conspecifics, both in open and woodland habitats, patterns of seedling distribution are not the same in the two habitats. In a *Quercus ilex* forest, most established seedlings were found under the oak trees (despite the fact that most seeds were dispersed to sites under conspecific trees) whereas in an open field, juveniles are more evenly distributed across different microhabitats. In this endemic species, population dynamics are more limited by seed dispersal than the availability of microsites in both habitats. In addition, the ecological processes acting on seedling recruitment and survival vary markedly between microhabitats. Spatial patterns of recruitment thus vary markedly in space and cannot simply be predicted from seed dispersal patterns in this species. In species with a secondary means of dispersal, such as by water in species that occur in riparian habitats (e.g. Thébaud and Debussche 1991; Hampe 2004), a spatial decoupling of wind or vertebrate-dispersed seed rain and seedling distribution may be frequent.

In other cases, nucleation due to dispersal patterns can be further enhanced by a positive (nurse) effect of adults on seedling establishment due to reduced herbivory, improved microclimate due to the buffering effect of litter on evapotranspiration and temperature extremes, improved soil nutrient and organic matter content, less compact soil structure and reduced aridity. In *P. latifolia* in the Sierra de Cazorla of southern Spain seedling survival varies significantly among microhabitats in scrubland but not in forest due to the role of facilitation played by shrubs in the former habitat type (C.M. Herrera *et al.* 1994). The role of stress avoidance in facilitation is well known (Bertness and Callaway 1994) hence facilitation, and consequent nucleation, may be key processes in the spatial organization of plant populations in Mediterranean habitats where abiotic stress is common (Koechlin *et al.* 1986).

This section illustrates the diversity of ecological factors influencing dispersal and regeneration, and thus the mechanisms of succession, in the Mediterranean mosaic landscape. The close fit between seed rain and seedling distribution of woody species in old-fields supports the tolerance model of succession while the frequent installation of fleshy-fruited species under pioneer shrubs and trees is indicative of a model of facilitation via the presence of perching posts which facilitate both arrival at a site and seedling establishment due to improved local environmental conditions and reduced herbivory. The relative importance of different processes may vary in relation to the life-cycle stage of individual plants, the local environment (in particular abiotic stress), and the successional stage of the vegetation. When conspecifics are scattered in an open landscape, spatial aggregation may be intense. However, a recruitment deficit may actually occur close to seed bearers due to competition, either with the maternal plant or conspecific seedlings when at high density. Indeed in several species, high seedling density under conspecifics has been found to be associated with higher rates of seedling mortality (Debussche and Lepart 1992; C.M. Herrera *et al.* 1994; Traveset *et al.* 2003), probably because of intraspecific competition, although localized patterns of herbivory and parasitism could also contribute. In contrast, a more continuous shrub cover may extend the tail of the seed shadow and potentially reduce intraspecific competition and thus increase local recruitment. The dynamics of such populations will thus be closely linked to the spatial structure of the vegetation within and among populations, which may set the stage for differentiation to evolve.

4.4.5 Consequences of reforestation for landscape pattern and biodiversity

Contemporary reforestation is a natural process which shows distinct spatial heterogeneity of dominant species and speed among regions (Lepart and Debussche 1992). For example, in southern France, old-fields are primarily colonized by vertebrate-dispersed fleshy-fruited species (e.g. in *Pistacia*, *Rhamnus*, *Juniperus*, and *Phillyrea*), many areas of lowland garrigues have become dominated by *P. halepensis*, in riparian areas a multispecies forest with *Populus* and *Alnus* has developed and upland limestone plateaux are being actively colonized by *B. sempervirens*, *J. communis*, *Q. pubescens* and *P. sylvestris* in many areas, or exotic *Pinus nigra* close to early twentieth century plantations. This regional specificity of forest spread has developed in response to three principal causes (Debussche *et al.* 1999): (a) differences in the proximity of an existing pool of tree species for reforestation and herbaceous understorey species, (b) local ecological conditions and topography, and (c) the spatial organization of habitats in the landscape.

The importance of proximity of source populations is illustrated by the spectacular disappearance of terrace agriculture, where the small size and difficult access of terraces rapidly made them unsuitable and unprofitable for mechanized agriculture (Lepart and Debussche 1992; Grove and Rackham 2001). Terraces by definition occur on hill sides, that is, close to remnant areas of forest in less accessible and less hospitable areas (on crests, ridges and steep slopes, and closer to summits). This proximity has allowed for the rapid arrival of tree species onto abandoned terraces, whose deep soils and open flat surface promotes the rapid reconstitution of forest. In many areas, oaks have directly colonized such areas, without a temporary stage of colonization by pioneer species (Debussche *et al.* 1999). In the absence of disturbance, succession leads primarily to deciduous oak dominance, whereas repeated disturbance leads to the development of more open sclerophyllous vegetation (Tatoni *et al.* 1994). The colonization of burnt areas by *P. halepensis* closely depends on the proximity of seed bearing trees near the burnt site (Abbas *et al.* 1984). Dispersal limitation may thus slow re-colonization after a large fire. The spontaneous spread of species from plantations of human origin also illustrates the importance of the proximity of source populations. Examples of this phenomenon (see Chapter 6) include *P. pinaster* in the Cévennes and the Maures massif in Provence (Mazurek and Romane 1986), *Fraxinus ornus* in riparian vegetation along the Hérault river (Thébaud and Debussche 1991) and *P. nigra* on the Causse Méjean (Lepart *et al.* 2001). In addition to the spread of some exotic species, there are other ecological consequences of rapid reforestation.

First, in some areas, the development of fairly natural and diverse understorey vegetation may be rapid, but in other areas very slow due to their isolation from areas where a forest cover was maintained during periods of extensive traditional agricultural processes. In addition, species in understorey herbaceous vegetation are not renowned for their rapid colonization capacity. A study of temporal changes in the vegetation of *Q. pubescens* woodland after the abandonment of coppicing and grazing illustrates that although the vegetation of previously exploited and grazed forest has become similar to that of undisturbed forest, the floristic diversity may never regain levels similar to those of undisturbed woodlands (Debussche *et al.* 2001). The rarity of such undisturbed woodland in the landscape and the lack of mechanisms favouring long-distance dispersal in many forest plant species are probably the causes. So even though a natural forest cover may be re-establishing itself in many areas, its biodiversity, in terms of herbaceous species richness, may be low, relative to the species diversity in open habitats (e.g. the species rich open grassland and semi-rupicolous communities of southern France) and the original primeval forests that probably occurred in these areas. Plantations of exotic species clearly have significantly negative effects on species diversity, notably a decline in the presence of endemic species, as reported for heathlands in southern Spain (Andrés and Ojeda 2002).

Second, although natural reforestation and other landscape changes associated with alterations in human land-use patterns may have little effect on natural populations of many protected species, due to their rupicolous and chasmophytic nature Chapter 2, some protected species, in addition to those in coastal habitats, are clearly threatened by contemporary landscape change. Species which require gaps and open habitats for flowering are a prime candidate. An example in point concerns native peonies such as *Paeonia mascula* and *Paeonia officinalis* which give a bright splash of colour to forest gaps during their flowering period. Populations of both species (in Israel and southern France, respectively) are threatened by rapid habitat change and forest closure. Both species show significantly reduced flowering in dense shade compared to open clearings and forest gaps (Andrieu 2002; Ne'eman 2003), a pattern also reported for peonies in the Californian chaparral (Keeley 1991). For such long-lived perennial species, the speed of forest closure in some areas is so fast that plants now in the forest include many which established themselves in an open habitat. Population persistence will thus depend on how different components of regeneration (germination, seedling establishment, and survival) and reproduction (flowering, fruiting, and seed dispersal) are affected by forest spread.

Finally, a decrease in the grassland–shrubland mosaic landscape as a result of reduced grazing pressure and cultivation may in some areas be associated with a loss of biodiversity. For example, in a comparison of evergreen oak woodland, grassland, shrubland, and grass-land–shrubland mosaic vegetation, Verdú *et al.* (2000) reported that species richness is maximum in the grassland–shrubland mosaic landscape. A return to the forest may thus have a negative consequence for plant biodiversity.

4.4.6 Trait variation in relation to succession and reforestation

Populations of a single species which occur in different successional stages provide model systems for the study of trait variation as environmental conditions change. Yet, there have been surprisingly few direct studies of trait variation along successional gradients and the few studies that have been done provide little evidence for adaptive variation, population persistence along such gradients being greatly favoured by phenotypic plasticity (Gray 1993; Thompson *et al.* 1993). The Mediterranean mosaic of old-fields of variable age and contrasting open areas and woodland provides an ideal landscape context for such work.

Two studies based on reciprocal transplant experiments have provided consistent support for the idea that trait variation along successional gradients is primarily a result of phenotypic plasticity. Using a pioneer (3 year old) population and a 40 yr old population of the biennial *P. hieracioides*, Sans *et al.* (1998) showed that different populations in the succession show a similar response to the selection pressures in the two habitats. Likewise, experiments with the colonist annual *Crepis sancta* in three Mediterranean old-fields that differ in age since abandonment (6, 10, and 15 years) by Imbert *et al.* (1999*b*) showed only very limited evidence for local adaptation in life-history traits related to survival, growth, and reproduction. In the Mediterranean mosaic of neighbouring populations which differ greatly in age and ecology, populations may frequently colonize ecological conditions different from their source populations, which, in conjunction with the temporary nature of pioneer populations may limit the evolution of adaptive variation. In addition, within-habitat spatial heterogeneity of local conditions may mask any selective differences among sites. The achene heteromorphism in this species is likely to be important here. Peripheral achenes have a poor dispersal capacity in space but produce more competitive offspring that the smaller more easily dispersed central achenes (Imbert *et al.* 1997). Hence, within individual variation in seed traits, an excellent expression of phenotypic plasticity, may provide a highly flexible adaptive strategy in a mosaic of successional old-fields where persistence of annual plants depends on both high dispersal rates and good competitive ability.

The role of functional trait variation in the successional dynamics of Mediterranean old-fields has been analysed by Garnier *et al.* (2004). These authors reported that early succession habitats are dominated by fast-growing plant species with high specific leaf area (the ratio of water-saturated leaf area to leaf dry mass which is positively correlated with photosynthetic activity) and leaf nitrogen content but low leaf dry matter content (the ratio of leaf dry mass to water-saturated fresh mass) and tend to have high rates of resource processing per unit biomass. In late succession habitats the reverse is the case. These differences in traits among successional stages point to the replacement of fast-growing species which acquire external resources rapidly in early stages by more slowly-growing species which efficiently conserve resources as succession proceeds. The fact that these relationships are maintained when only the traits of the two dominant species are considered strongly suggests that modifications to community dynamics and structure depend more on species traits than on species number. These results reflect what Herms and Mattson (1992) term a balance between growth and differentiation in which growth-dominated plants invest a high proportion of their resources into resource acquisition and high relative growth rates, while differentiation dominated plants primarily invest resources into the non-growth processes and structures which facilitate efficient use of resources, and consequent low relative growth rate.

Rapid reforestation in the Mediterranean means that plants have to cope with deep shade in addition to summer drought. Sack and Grubb (2002) subjected a range of species, including two species common in Mediterranean forests, *Viburnum tinus* and *Hedera helix*, to this combination of constraints. For these species they did not find, as trade-off theory would predict, a greater impact of drought stress when plants were also subject to deep shade. This is surprising since it is difficult for a plant to specialize both in shoot traits that favour irradiance capture and root function to avoid and/or tolerate drought. The result suggests that some traits must allow for both reduced demand for irradiance and water simultaneously; for example, reduced size and low respiration rates, that is, the presence of long-lived leaves with low specific leaf area. In the polyploid complex *Arrhenatherum elatius* (see Chapter 3), Petit and Thompson (1997, 1998) reported marked differentiation in morphological traits among woodland and open habitat populations of the tetraploid cytotype which concords with the direction of selection in such habitats (taller plants favoured in open habitats). These authors suggest that adaptive differentiation among tetraploid populations in open and woodland habitats in the Mediterranean mosaic has been an integral component of the spread and persistence of the polyploid cytotype in a wider range of environments than the diploid progenitor.

A fascinating example of plant response to variation in irradiance occurs in the carnivorous endemic *Pinguicula vallisneriifolia*. In this species variation in light availability influences investment in carnivory in natural situations. This species inhabits damp rocky habitats over a range of light environments in south-east Spain (Zamora *et al.* 1996). In an experimental study of transplanted individuals, Zamora *et al.* (1998) reported a decreasing gradient in performance from sunny environments to deep shade over a distance of a few metres. However prey availability is greater in damp shady habitats than in open habitats where plants capture significantly fewer insects (Zamora 1995). The decline in performance was manifest even when plants received additional prey, hence carnivory did not compensate for diminished photosynthesis. In this species, plants in open habitats bear curled leaves (probably to reduce water loss) with higher levels of mucilage secretion than plants in deep shade which bear more flattened leaves to maximize photon capture. The latter may only capture smaller insects but in such habitats insect prey is more plentiful than in open areas (Zamora 1995; Zamora *et al.* 1998). These patterns illustrate once again the complexity of factors acting on plant response to nutrient uptake (and in this case prey availability and capture), water loss, and irradiance in a Mediterranean context. The favoured habitat of this species is transitional habitats between open areas (where there is a surplus of light but little prey and potential water stress) and deep shaded areas (where water and food are available, but additional prey cannot compensate for lack of light). Hence, along the environmental gradient on which this species occurs, the limiting factor changes. The message is clear, to study plant response and adaptation requires appreciation of responses to diverse environmental variables.

4.4.7 Forest fires: successional dynamics and trait adaptation

Fires were an integral component of the Mediterranean landscape long before humans learnt their use (Chapter 1) and may thus have influenced trait evolution. Mediterranean fires may impact on plant population ecology and evolution in two main ways. As a disturbance, fires open the environment, creating spatio-temporal heterogeneity of the environment and new opportunities for colonization and succession. As Barbero *et al.* (1987*a*: 48) commented 'the fire-shaped landscape perfectly corresponds to the definition of the mosaic community'. Second, as a selective agent, fire may shape the evolution of traits that allow species to adapt to fires.

The dynamics of vegetation after fire has been well documented in Mediterranean communities (Trabaud and Lepart 1980; Trabaud 1987, 1992).

Immediately following the passage of a fire, species diversity remains low, reaching a peak within three years after the passage of a fire. After about five years, species diversity declines to a level equivalent to that prior to the fire. A key point here is that the composition of many plant communities returns to its pre-fire state fairly quickly (Casal 1987; Trabaud 1987). For example, garrigue communities in southern France studied by Trabaud and Lepart (1980) recovered more than 75% of their original vegetation within 10 years after a fire. After fire, one thus observes the return of previously present species, rather than a process of species turnover which is more characteristic of secondary succession.

Fire frequency can affect pattern and diversity in Mediterranean plant communities. Very high fire frequencies create a patchwork vegetation structure with a low species diversity dominated by resprouting species which rapidly propagate into a dense cover, for example, *Q. coccifera* in the western Mediterranean. In north-east Spain, sites with a high fire recurrence have recently shifted species composition from domination by *P. halepensis* to domination by evergreen oaks because the fire frequency is shorter than the time (15–20 years) for *P. halepensis* individuals to produce seeds and accumulate a seed storage capacity for self-replacement (Pausas 2001). In contrast, as shown by Trabaud and Galtié (1996) working in the Massif des Aspres in southern France, low fire frequency causes greater spatial heterogeneity in the landscape and greater diversity in plant communities, with the establishment from seed of numerous pioneer woody species such as *Thymus*, *Lavandula*, and *Cistus*, which develop rapidly. Such seeder species may however be eliminated at very low fire frequency in the absence of resprouting species if serotinous species are present. In the absence of any disturbance such as fire, evergreen oaks such as *Q. ilex* will replace *P. halepensis* by virtue of its enhanced shade tolerance (Zavala *et al.* 2000). These patterns illustrate that the predicted dominance of seeders at high levels of disturbance (in this case high fire frequency) does not apply to fire-prone ecosystems of the Mediterranean which contain species with traits that facilitate resprouting. In fact, in such systems, if fires are too frequent, biodiversity may be threatened (Moreno *et al.* 1998), hence the frequency of fires may have a major effect on the mosaic pattern of vegetation and species diversity. This effect of fire frequency on mosaic structure of vegetation has also been discussed in other Mediterranean-climate ecosystems (e.g. Zedler *et al.* 1983).

Although it is difficult to assign the status of a true pyrophyte (i.e. a species whose propagation or reproduction is stimulated by fire or which resist fires) there are a number of traits which illustrate adaptation to fire. These have been classified in different ways by different authors (Kunholtz-Lordat 1958; Naveh 1975; Grove and Rackham 2001). Here, I identify five main strategies:

Resistance. Traits such as the thick insulating and protective bark of trees such as *Quercus suber* (Pausas 1997), *Q. ithaburensis* (Naveh 1975), or *Pinus brutia* (Eron 1987) provide a means of surviving a fire and thus regeneration. The flammability of many pines, sclerophyllous shrubs and trees, and aromatic species may also contribute to this resistance by hastening the passage of a fire. In *P. halepensis* the loose fallen needles act as fuel which allows ground fires to spread rapidly.

Resprouting. Many plants may have their aboveground parts burnt to ground level without major damage to the roots, allowing them to resprout after fire: e.g. *Q. coccifera* and *Q. lusitanica* in the western Mediterranean, and *Q. ithaburensis* in the eastern Mediterranean, and species of *Phillyrea*, *Rhamnus*, *Arbutus*, and *Erica* (Naveh 1975; Trabaud and Lepart 1980; Barbero *et al.* 1987*a*; Kummerow *et al.* 1990; Ojeda *et al.* 1996*b*). Critical to this resprouting strategy is the presence of burls, root crowns, and rootstocks, particularly swollen lignotubers rich in starch at the base of the stem. As shown for *Erica* species of the Cape floristic region, plants with a resprouting strategy build-up starch reserves in the roots, even in young seedlings which provide a stock of rapidly usable reserves for growth after a disturbance such as fire (Bell and Ojeda 1999; Verdaguer and Ojeda 2002). The resprouting strategy is favoured by high fire frequency (Barbero *et al.* 1987*a*) and severe summer drought (Ojeda 1998).

Seed and seedling biology. In situations where fire frequency is reduced, a high reproductive output and a suite of seed traits may be favoured, for example, small seeds, with hard coats and a fairly long persistence in the soil seed bank, or the fire-stimulated germination and rapid seedling establishment observed in *Cistus* in the western Mediterranean (Thanos and Georghiou 1988; Trabaud and Oustric 1989; Roy and Sonié 1992). In *Cistus*, some species produce two types of seeds, some which germinate without fire and others, with a hard seed coat that germinate after treatment at high temperatures (Thanos and Georghiou 1988). Seedlings of species with a 'seeder' response to fire also show a better drought resistance compared to seedlings in species which resprout and produce few seedlings after fire (Ojeda 1998). Another seed trait which may favour resistance to fire is trypanocarpy, the capacity of seeds to bury themselves in the soil. Such seed movements are often associated with the presence of a hygroscopic awn, for example, in *Stipa tortilis* in fire-prone steppe and arid vegetation of the eastern Mediterranean (Naveh 1975). In other Mediterranean-climate regions, the adaptive nature of germination strategies (e.g. in response to heat and charred wood, or following the release of nitrogen and smoke) is firmly established (e.g. Keeley *et al.* 1985; Thanos and Rundel 1995). The persistence of such seeder species depends on their ability to produce seeds between two fires, seed survival during a fire, and enhancement of regeneration and recruitment as a result of a fire passage.

Dispersal and re-colonization. Traits which favour dispersal and re-colonization may allow species which lack resistance to rapidly re-colonize burnt areas. In *P. halepensis* the light, wind-dispersed seeds favour dispersal to open areas and have favoured the spread of this species in burnt areas (Achérar *et al.* 1984; Barbero *et al.* 1987*b*).

Serotiny. This is the capacity to maintain old unopened cones for several years in a 'canopy-stored' seed bank and release seeds only after the passage of a fire. It is present in just a few species in the Mediterranean flora, for example, *P. halepensis* and *P. brutia*, *Cupressus sempervirens*, and *Tetraclinis articulata* (Quézel and Médail 2003). In *P. brutia* serotiny may favour re-colonization of burnt sites (Eron 1987) and in *P. halepensis* young trees require a higher temperature for seed release than do older trees (Tapias *et al.* 2001). This age-related plasticity may be adaptive. However, critical analysis of this phenomenon has shown that a large amount of seed release (~60%) may occur in the absence of fire, suggesting that this species is only partially serotinous and that this trait may have evolved in response to other environmental factors (Nathan *et al.* 1999). Serotiny is more common in other Mediterranean-type ecosystems than in the Mediterranean flora.

Interpretation of the adaptive significance of such traits requires a certain prudence. The different strategies which favour resistance to, or re-colonization after, fire, also favour tolerance of summer drought and other environmental stress (Naveh 1975). Hence, it would be unwise to conclude, on the basis of present knowledge, on the adaptive nature of such traits solely in terms of a response to selection pressures imposed by fire. As Pausas (2001) argues, it is necessary to consider a range of traits (age at maturity, serotiny, seed dormancy and germination, and survival in shaded conditions) and fire regimes in order to make biologically realistic predictions of the relative fitness of different strategies.

Different combinations of traits may evolve in systems with different fire histories. Such variation could help explain the relative lack of serotinous species in the Mediterranean flora relative to other Mediterranean-climate areas. Pausas *et al.* (2004) developed a fourfold classification of species in relation to the basic fire-response strategies (i.e. resprouting ability and propagule persistence): resprouters (R+), non-resprouters (R−), propagule persisters (P+) or propagule non-persisters (P−). They then examined how combinations of supposed fire-response traits vary in different regions and the extent to which different strategies co-occur with other traits. Their study showed marked variation in the relative abundance of each of the four types in different Mediterranean ecosystems. For example, most resprouters in the Mediterranean flora do not persist as propagules, that is, they are R+ P−, while in California, resprouters are evenly distributed among propagule persisters (R+ P+) and propagule non-persisters (R+ P−). The two basic response traits were found to be closely correlated with other traits, for example, resprouters are longer-lived and allocate more resources to storage and basal buds. However, not all resprouters may respond in a similar way to fire because of different combinations of associated traits in different regions, that is, differences in stored seed banks and modes of dispersal. Hence attempts to predict population-level changes and the relative fitness of seeder and resprouting species require integration of the various combinations of traits and fire regimes in different regions.

One should thus be cautious in any interpretation of the adaptive significance of the above traits in relation to fire. Several of these traits may have evolved in association with other selection pressures, perhaps prior to the onset of a Mediterranean-climate regime. They may thus represent an example of pre-adaptation in Mediterranean fire-prone communities. Two studies back up this claim. First, resprouting is a widespread trait in plants of shrubland in the mountains of Mexico (Lloret *et al.* 1999). The marked morphological and floristic similarity of this 'mexical' shrubland to Mediterranean-type vegetation is strong evidence that it represents a relict of the Tertiary flora which has persisted either under (in California) or in the absence of (Mexico) a Mediterranean-type climate. The prevalence of resprouters characterized by the presence of a lignotuber or burl in the mexical shrubland community strongly suggests that resprouting evolved to cope with disturbance other than fire, that is, prior to the onset of the Mediterranean climate in California. Second, in a phylogenetically controlled analysis of the correlated and independent occurrence of resprouting capacity and propagule-persistence ability in plants of the Iberian peninsula, Pausas and Verdú (2004) detected a significant negative association between the two response strategies, that is, R+ P− and R− P+ were more frequent than expected by chance. In addition, their results do not suggest a correlated evolution of these two traits in R+ P+ species as a result of selection on one trait. In their dataset, resprouting capacity (R+) preceded propagule persistence (P+) in species which combine these traits. Their study also confirmed the co-occurrence of longevity with resprouting capacity and the trend for juveniles of resprouters to grow more slowly than seeders (see also Verdú 2000). These results are consistent with the idea that allocation trade-offs cause the negative relationship between the capacity for resprouting and propagule persistence. The fact that resprouting precedes propagule persistence strongly points to an origin for the former trait in the older lineages present in the Tertiary flora, that is, prior to the onset of the Mediterranean-climate regime. Resprouting capacity thus appears to have been part of the Pre-Pliocene suite of traits that has not evolved in response to environmental changes associated with the onset of a Mediterranean climate.

4.4.8 Colonization and adaptation on abandoned mines

Since the early 1960s, plant colonization on mine spoils has provided a clear illustration of how strong natural selection can be and how quickly plant populations can adapt to local conditions (McNeilly and Antonovics 1968; MacNair 1987). These studies illustrated that plants from soils polluted with heavy metals such as zinc, lead, cadmium, copper, and nickel can show a greater capacity for growth and survival in the presence of such elements at high doses than plants of the same species from adjacent unpolluted sites. What fascinated biologists at the time was that even though gene flow was occurring from unpolluted sites into the polluted population, this was not enough to counter genetic differentiation between populations on the different soils. The clear message was that selection can be strong enough to produce adaptation, even when gene flow occurs at significant rates. To reduce the 'harmful' effects of breeding with 'unadapted' individuals from different populations, populations on polluted sites were also found to have a greater chance of self-fertilization than those on unpolluted soils. In the absence of heavy metals however, plants from polluted sites were less vigorous than those from unpolluted sites, they have a cost associated with their resistance in terms of a fitness reduction in the absence of the metals. As a result, the frequency of resistance genes should decline away from polluted sites. More recently, there has been much interest in the precise physiological and genetic mechanisms which allow plants to (a) adapt to what are highly toxic molecules (MacNair 1987, 1993) and (b) hyper-accumulate heavy metals in their aerial parts (Escarré *et al.* 2000) .

Abandoned mines litter the back country of several Mediterranean regions where they provide replicated sites for experimental analysis of the traits which facilitate colonization and adaptation of new environments. Rates of gene flow among populations can now be assessed with powerful genetic markers that were not available to evolutionary biologists in North Wales in the early 1960s. How plants cope with several compounds at high doses can also be evaluated now. This is important because tolerance of one metal does not go hand in hand with tolerance of others, the genes for resistance to different compounds can be different (MacNair 1987, 1993). Adaptation to one metal may thus be compromised by genetic adaptation to another metal. Finally, in the case of mines in the Mediterranean region let us not forget that the toxic environment is made even more stressful by summer drought, hence tolerance may be even more difficult to achieve.

Work on *Thlaspi caerulescens,* well studied for its metal tolerance and ability to colonize polluted soils in temperate Europe illustrate some of these issues, as well as being of particular interest to restoration ecologists and landscape project managers by virtue of its capacity to accumulate high concentrations of heavy metals in its tissues (Box 4.9). Despite ecotypic differentiation between the two types of population, the genes which confer tolerance are not limited to contaminated sites and plants from contaminated and uncontaminated sites show no clear pattern of isozyme differentiation (Dubois *et al.* 2003). Either the two ecotypes have evolved recently or there is ongoing gene flow among the two types of population.

Plants of *T. caerulescens* also have the ability to accumulate high concentrations of the heavy metals in their aerial parts, although this accumulation does occur in the same way for different metals. Whereas plants from metallicolous populations accumulate cadmium, those from non-metallicolous populations accumulate zinc (Escarré *et al.* 2000). This ability to hyperaccumulate toxic compounds is particularly interesting for the restoration of such ecosystems and also for studies of the genetic basis of physiological traits that allow plants to colonize severe environments. Based on controlled crossing experiments, Frérot *et al.* (2003) found that zinc accumulation was significantly higher in the offspring of crosses among non-metallicolous parents than in the offspring of both crosses among metallicolous parents and crosses between the two ecotypes (the latter having intermediate capacities for zinc uptake). There is thus heritable variation for zinc uptake and accumulation capacity in the studied populations, which may involve a simple monogenic control with two alleles. A homozygote recessive for the locus involved would be non-metallicolous, a single dominant gene producing intermediate zinc uptake and a dominant homozygote blocking uptake and accumulation capacity. Such a simple genetic control may facilitate the evolution of hyperaccumulation capacity, both in the wild and in breeding programmes designed for phytoremediation and restoration purposes.

4.5 Variation and adaptation in aromatic plants

4.5.1 Abundance and diversity of aromatic plants in the Mediterranean

Where else in the world do plants smell so good? Although most plants produce secondary compounds, their presence is enhanced in the Mediterranean where one of the most striking features of many plant communities is their fragrance. One of the most characteristic and abundant types of volatile compounds produced by Mediterranean plants are the monoterpenes produced in essential oils. Monoterpenes have a 10-carbon, 16-hydrogen structure which may take the form of acyclic and cyclic non-aromatic molecules as well as truly aromatic types. They are ubiquitous in higher plants, and accumulate to significant levels in ~50 angiosperm families (Seufert *et al.* 1995). To put a number on essential oil presence in the Mediterranean flora, Ross and Sombrero (1991) recompiled data in Guenther's (1948–52) classic work on the presence of essential oils in the angiosperms. In a total of 153 genera (from 49 families) that have species which accumulate

Box 4.9 Re-colonization and adaptation to heavy metals on abandoned mines in southern France (photos and graph kindly supplied by H. Frérot)

In southern France, *Thalaspi caerulescens* and *Anthyllis vulneraria* occur in 'metallicolous' (M) populations on a range of abandoned mine sites where soils are contaminated with zinc, lead, and cadmium and in 'non-metallicolous' (NM) populations in the surrounding landscape. Metallicolous populations show greater tolerance of polluted soils (based on the ratio of aerial biomass on contaminated/uncontaminated soils) than non-metallicolous populations (Escarré *et al.* 2000). This tolerance is however 'constitutive', that is, present (albeit at lower levels) in plants in non-metallicolous populations. In fact, in *T. caerulescens* most plants can survive on contaminated soils regardless of their origin, unlike in other species such as *A. vulneraria* where populations from polluted sites have a greater tolerance. In *T. caerulescens*, however, growth and reproduction on contaminated soil is severely reduced in plants from non-metallicolous populations.

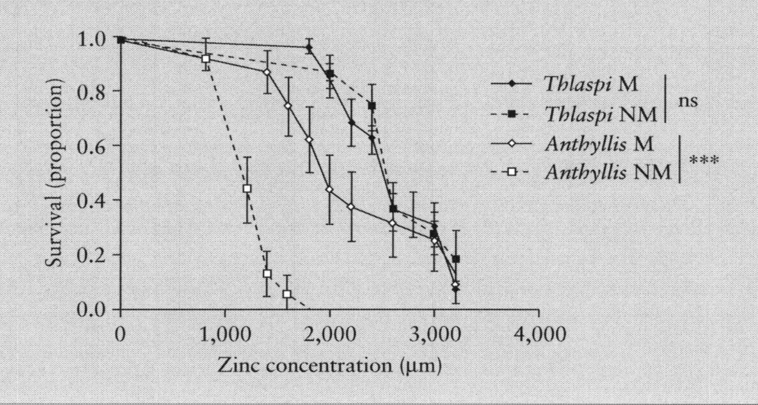

essential oils, 58 occur only in Mediterranean-type ecosystems, 32 in Mediterranean and other regions, 42 are tropical, 17 temperate, and 4 are cosmopolitan. Hence, 90 of the 153 genera occur in a Mediterranean-type ecosystem. Although, this exploration lacks a rigorous comparative analysis of numbers of available genera for study in each region and the necessary correction for phylogenetic relatedness and effects of sorting processes (the large number of genera may be linked to high rates of diversification in a few families such as the Lamiaceae, Apiaceae, and Asteraceae which are rich in secondary compounds), the abundance of aromatic plants is clear.

Essential oils and resins which have a volatile oil component are typical of perennial, evergreen, xeromorphic shrubs, and are, not surprisingly, less abundant in fast growing and drought avoiding species, particularly annuals. In many species the biosynthesis of volatile oils represents a sizeable fraction of resources in terms of the amount of fixed carbon used up in their production (Ross and Sombrero 1991). For example, in *Q. ilex*, monoterpenes represent a significant feature of carbon cycling (Staudt *et al.* 2001).

Volatile oils may contain toxic byproducts of primary metabolite synthesis and thus require storage in specialized tissues where they do not disrupt normal cell function (Table 4.1). Glandular hairs and trichomes on the leaf surface are a common form of such storage. Glandular hairs can be of two main types: peltate and capitate, which differ in structure and function (Werker *et al.* 1985). Peltate hairs consist of a basal cell, a short, wide stalk cell, and an almost circular head of secretory cells. Capitate hairs differ by having a uni or bicellular head. In most species (for an unknown reason *Rosmarinus officinalis* is an exception) the hairs can be seen on both leaf surfaces, on calyces and on green stems.

Monoterpenes are ideal compounds for the study of how ecological interactions are mediated by secondary compounds for a number of reasons (see also Lerdeau *et al.* 1994): their biosynthetic pathways are fairly well known, they can play multiple ecological roles due to their effects on biological processes in the surrounding environment, they represent an important investment of fixed carbon and may thus be costly to produce, and they show variation in production among closely related species and among populations of individual species. Although the material of this section differs somewhat from that of previous sections, the questions are simple variants on the common theme of the chapter: what is the function of such compounds and is intraspecific variation adaptive in a Mediterranean context?

Table 4.1 Storage organs and their volatile oils in some common Mediterranean genera

Storage organ	Family	Genus	Compounds
Resin canals	Coniferae	*Pinus*	Oleoresins
		Picea	
		Juniperus	
	Anacardiaceae	*Pistacia*	
		Cotinus	
External glandular hairs	Cistaceae	*Cistus*	Aromatic resins
		Helianthemum	
Glandular trichomes	Lamiaceae	*Salvia*	Monoterpenes
		Thymus	
	Cannabaceae	*Humulus*	
		Cannabis	
	Myricaceae	*Myrica*	
Oil cavities	Rutaceae	*Citrus*	
	Myrtaceae	*Eucalyptus*	
Schizogenous oil ducts	Umbelliferae	*Coriandrum*	
	Asteraceae	*Carthamus*	

To understand why volatile essential oils become accumulated in specialized organs in many Mediterranean plants requires recognition of why they are produced in the first place. Two view points can be read on this issue. First, volatile oils represent secondary compounds. In other words they represent the byproducts of phytohormone, phytosterol, and carotenoid production (Banthorpe *et al.* 1972), although their accumulation may then be of adaptive significance in relation to several features of the abiotic and biotic environment. Alternatively, some authors (e.g. Seigler and Price 1976; Croteau 1987) have proposed that such compounds are not simple waste products because they are often catabolized in a highly specific and ordered fashion and may thus play a primary role in physiological processes. Although I recognize the possibility of this primary function, I will often refer to volatile essential oils, simply for convenience, as secondary compounds. They nevertheless have several important ecological functions.

4.5.2 Ecological functions of secondary compounds: biotic interactions

A principal function of chemical compounds in plants is their role in the mediation of biotic interactions, in particular (and there is an immense literature on these subjects) defence against parasites, predators, and herbivores, allelopathic effects on associated species which reduce competition and pollinator attraction.

Following the claim by Fraenkel (1959) that defence against herbivores and parasites represents the veritable *'raison d'être'* of secondary compounds, Ehrlich and Raven (1964) argued for the idea that novel chemical compounds allow plants to escape such attacks. They discussed how the coevolutionary response by predators and parasites, which in turn evolve to counter new defence mechanisms, push defensive systems to continually evolve if they are to be effective. The result of this is radiation, both for the plants and their enemies. There is now a solid body of literature on the importance of secondary compounds in defence against herbivores (Herms and Mattson 1992).

There are many precise examples from Mediterranean ecosystems of how herbivory and parasitism can have a lasting and sometimes severe impact on population dynamics (Thébaud *et al.* 1996; Escarré *et al.* 1999) and community ecology (Noy-Meir *et al.* 1989; Sternberg *et al.* 2000), how variation in pest infestation may be linked to particular plant traits (e.g. Alonso and Herrera 1996), and how variation in pest populations may closely parallel patterns of the host (Burban *et al.* 1999). However, there is surprisingly little published information on the role of monoterpenes in defence against herbivory. The role of conifer resins and monoterpene alcohols in defence against insect and small mammal attack due to their impact on digestive processes is well known from studies elsewhere (Seufert *et al.* 1995) and studies on *T. vulgaris* by Linhart and Thompson (1995, 1999) in controlled conditions provide evidence that potential herbivores may respond to variation in monoterpene composition in Mediterranean species. Several Mediterranean species have high concentrations of secondary compounds in their fruits (e.g. Milesi *et al.* 2001), suggesting a role in seed protection. But what is lacking, as I write this book, is direct evidence from a natural situation of monoterpene-mediated defence in a Mediterranean plant. Evidence of a role for chemical defence is variable, in some studies increased phenolic content of leaves across species has been shown to be associated with reduced browsing (Massei *et al.* 2000) while in others variation in the phenolic content of the leaves was not associated with differences in insect attack (Glyphis and Puttick 1989).

The presence of secondary compounds may mediate or influence community structure where they affect the germination, growth, and survival of associated plant species. The role of such 'allelopathy' as an ecological process acting to modulate competitive interactions and structure plant communities has been a hotly debated subject. In the early 1960s observations of bare zones around some aromatic shrubs in Californian chaparral, combined with experiments in controlled conditions, led to the claim that allelopathic effects of dominant shrub species such as *Adenostoma fasciculatum* suppress the germination and growth of associated species and inhibit the germination of conspecific seeds (e.g. McPherson and Muller 1969). Such observations were used to develop the idea that allelopathic interactions are a key feature of plant–plant interactions and community organization in such ecosystems. At about the same time, Whittaker and Feeney (1971) emphasized the role of chemical compounds in the mediation of interspecific interactions. However convincing experiments in natural conditions have rarely if ever been performed and as a result the experimental results and their applicability to natural systems are far from being clear cut (Rice 1979; Inderjit and Dakshini 1995). In the case of *A. fasciculatum* it was later found that the concentration of phenolics in the soil under adult plants was insufficient to prevent germination and growth of associated species, soil toxicity being more the result of phytotoxins produced by the soil microbes under the dominant shrubs (Kaminsky 1981).

In the Mediterranean, an allelopathic role of essential oils has often been proposed. For example, Barbero *et al.* (1987*a*: 45) stated that 'Mediterranean communities characterized by *Thymus vulgaris, Dorycnium suffruticosum, Staehelina dubia,* and *Lavandula latifolia* are not very diversified ... because of allelopathic reactions'. Although direct proof that this is true in the wild remains to be furnished, there have been some encouraging studies of several Lamiaceae and the Cistaceae.

1. In the eastern Mediterranean, individuals of the aromatic shrub *Coridothymus capitatus* are often encircled by a ring in which annual plants are absent (Katz *et al.* 1987). These authors showed that germination of two annuals (*Plantago psyllium* and *Erucaria hispanica*) is suppressed both by leachates and the volatile oil extracted from shoots in laboratory conditions and when seeds are hand planted into bare rings around plants of *C. capitatus* in the field. They also note that plants vary in their capacity to inhibit the establishment of seedlings of the two annuals.
2. The essential oil and leaf litter of *T. vulgaris* inhibits the germination of an associated species *Brachypodium phoenicoides*, a potentially important competitor species in garrigue communities where wild thyme occurs in southern France (Tarayre *et al.* 1995).
3. Vokou and Margaris (1982) showed a strong inhibitive effect of high concentrations of the oils of *Thymus (Coridothymus) capitatus, Satureja thymbra, Teucrium polium,* and *Rosmarinus officinalis* on two cultivated species but only a weak effect on three wild annual species (*Astragalus hamosus, Hymenocarpus circinatus,* and *Medicago minima*) which grow in the same communities as the species used as a chemical source.
4. Vokou *et al.* (2003) have shown potential strong effects of the monoterpenes present in many Mediterranean Lamiaceae on seedling growth of cultivated plants.
5. Finally, Robles *et al.* (1999) have reported potential allelopathic and auto-allelopathic effects of *Cistus albidus*.

It is known that monoterpenes are soluble and that they can be transported to the soil surface by precipitation (e.g. Hyder *et al.* 2002) but whether concentrations in the topsoil are ecologically significant remains to be precisely determined. The addition of essential oils obtained from aromatic plants (*S. thymbra* and *Thymus (Coridothymus) capitatus*) can stimulate soil respiration and cause an increased organic matter content of soils and at the same time diminish fungal spore germination in controlled conditions (Vokou and Margaris 1984, 1988). Increased respiration is the result from increased numbers and activity of the bacterial community in the soil which may use the oils as a substrate for growth. Recent work by Ehlers and Thompson (2004*b*) and Y.B. Linhart (University of Colorado, unpublished data), who grew seedlings of a variety of species which occur in Mediterranean garrigue communities on soils collected in the field, suggest that monoterpenes leached from the aromatic plant *T. vulgaris* are indeed ecologically significant, with beneficial and negative effects on seedling growth, depending on the species and the monoterpene. The search for the ecological significance of allelopathy is thus an important issue in the Mediterranean, particularly in open garrigue-type vegetation where community dynamics may be strongly influenced by aridity and nutrient stress and competition among low woody and herbaceous vegetation. The task ahead will be to identify exactly the role of allelopathy in the functioning of plant–plant interactions in natural situations.

The release of monoterpenes during litter decomposition, or their presence in calyces and fruits which enclose the seed, may also influence the germination of conspecific seeds. In his seminal paper on plant regeneration, Grubb (1977) discussed the idea, and the lack of critical evidence for it, that regeneration of seedlings beneath their parents is less than under a comparable degree of shade due to leaching of chemical compounds. McPherson and Muller (1969) were among the first to claim a role of allelochemicals in the dormancy of *A. fasciculatum* seeds in the California chaparral until fires consume the adult plant. More recently, however, Keeley (1991) suggested that the direct effects of fire (e.g. charred wood and smoke) and other ecological constraints may be more important determinants of germination. The leaching of volatile oils which inhibit early germination could in fact represent an adaptive response to such irregular germination cues (Augspurger 1979; Keeley 1991; Fox 1995). In a Mediterranean ecosystem the first bout of heavy rain in late summer is often followed by a subsequent period of drought stress. Germination may thus be more successful if it is timed to occur when more consistent autumnal rains occur. There is in fact diverse evidence for climatically adapted germination strategies in Mediterranean plants?

- Mediterranean populations of *D. glomerata* set seed early in the spring but only drop their seeds in the fall when rainfall becomes more reliable, whereas populations in Atlantic climate areas of France set seed later in spring and release their seeds in the summer (Knight 1973).
- Margaris and Vokou (1982) illustrate how, for a number of aromatic species in the phrygana of Greece which produce seeds in late spring/early summer, seeds remain in the soil and because they are (p. 456) 'equipped with after-ripening dormancy' seeds are 'protected from untimely germination'.
- In *C. capitatus*, germination is significantly slower in the presence of calyces (which are the unit of dispersal and which contain the essential oil) than when seeds are germinated alone, an effect relieved by leaching of the essential oil over time (Thanos *et al.* 1995). In this species the strong inhibitory effect of pure oils is lost once seeds are transferred and washed and leaf litter has no effect on the germination of conspecific seeds (Vokou and Margaris 1986).
- In *T. vulgaris* the presence of litter or pure oils causes a ~50% inhibition of early seed germination, and such effects may gradually wear off (Tarayre *et al.* 1995).

Volatile oils may also play a role in pollinator attraction (Williams *et al.* 1983; Ackerman *et al.* 1997). For volatile compounds to have an attractive potential they must be associated with a modification of the reward, otherwise the insect will remain indifferent to any fragrance and its variation. Several examples from the Mediterranean flora provide insights into the role of fragrance variation in pollination biology.

In some species vegetative tissues may emit volatile compounds that attract pollinators. In the Mediterranean dwarf palm, *Chamaerops humilis* (Dufaÿ *et al.* 2003) the leaves can attract a specific pollinating weevil due to fragrances emitted by fairly large structures at the sinuses of the palmate leaf. Of course the leaves themselves provide no reward to the insect, which in the process of pollination also lays its eggs in the rachis of male plants. In *Majorana syriaca* (Beker *et al.* 1989) honey bees react differently to and discriminate between leaves and inflorescences that vary in their ratio of carvacrol and thymol. It is the vegetative parts which emit the volatile oils which facilitate long-distance attraction to the plant. The close range signal, once a bee is close to the plant, is then provided by the flower.

Floral fragrance is easy to appreciate in *Narcissus* which are well known for their sweet smell. The volatile floral fragrance of nine *Narcissus* species in southern Spain, which cover seven sections of the genus, includes no less than 84 volatile compounds (Dobson *et al.* 1997). Some of these (such as trans-β-ocimene) are well known in cultivated forms, others represent a whole new range of natural product diversity. These authors recognized three main 'fragrance types': fatty acid derivatives, isoprenoids, and benzenoids (without trans-β-ocimene) or isoprenoids and benzenoids (with trans-β-ocimene). The basic organization of biosynthetic pathways is common to all species, hence, fragrance differences are most probably due to differences in the regulation of enzyme activity rather than to the presence or absence of specific enzymes. *Hypecoum* species in the eastern Mediterranean show marked variability in their floral emissions (Dahl *et al.* 1990).

In *Ophrys* orchids, individual plants have a floral fragrance containing many compounds and several species show evidence of intra and interspecific variation in fragrance composition. For example, *Ophrys lutea* has a fragrance with farnesol and geraniol or a mixture of the two whereas *Ophrys fusca* is a mixture of farnesolic, cironellalic, and other alcohols (Kullenbeg 1976). In *Ophrys* orchids pollinated by male wasps, Bergstrom (1978) showed how their fragrance is (a) similar to that of female insects in terms of the presence of a certain number of compounds and (b) associated with the vital needs of the pollinating bees in terms of their requirements for mating, feeding, and nesting. In Mediterranean *Ophrys* the low frequency of pollinator visitation to many species is likely to sharpen the interaction between attractive traits and pollinators since traits which favour the chance of visitation and also the specificity of the interaction should be under strong selection. Indeed, Ayasse *et al.* (2000) have shown that in *Ophrys sphegodes* (a common species in Mediterranean garrigue) biologically active components of the floral fragrance show less variation among plants than do non-active compounds, a result which suggests pollinator mediated selection on the composition of the bouquet.

Volatile oils may thus be critical elements of mutualistic plant–pollinator interactions, both in highly specialized interactions such as those between orchids and their pollinators and dwarf palms and their specialist pollinator (see Chapter 5) and in species where the interaction is less species-specific, such as in *Narcissus* and various Lamiaceae.

4.5.3 Ecological functions of secondary compounds: abiotic interactions

The diversity and abundance of aromatic plants has stimulated much discussion of the ecological function of volatile oils in relation to abiotic environmental constraints in the Mediterranean region (Margaris and Vokou 1982; Ross and Sombrero 1991; Seufert *et al.* 1995; Thompson *et al.* 1998; Joffre *et al.* 1999). This literature illustrates the diversity of abiotic roles that volatile oil accumulation may play in Mediterranean plants.

First, the presence and accumulation of essential oils may improve tolerance of water constraints and high solar radiation. In a Mediterranean-climate, plants receive high levels of intercepted solar radiation at times when photosynthetic capacity and growth are limited and carbon assimilation is restricted by water stress and high temperatures. As discussed earlier in this chapter, during the summer drought, the absorption of light energy may be in excess of that required for carbon fixation, with potential damage to the photosystem. Different physiological regulatory mechanisms may help dissipate the excess of absorbed energy without any damage to cell function and it has been suggested that the production and accumulation of essential oils could serve this function (Ross and Sombrero 1991; Joffre *et al.* 1999). During the summer drought, photosynthates may be shunted into secondary metabolic pathways, minimizing their accumulation and potential negative effects on the photosynthetic system. The fact that different aromatic Lamiaceae in Greece (Vokou and Margaris 1986) and Israel (Basker and Putievsky 1978) have a significantly higher yield of essential oils in summer than in spring and autumn supports this idea. In *Ruta graveolens*, a pungent smelling herb in dry rocky areas in lowland garrigue-type habitats, the yield of furanocoumarins is strikingly higher (10-fold) in the fruits than in stems, leaves, or roots, indicative of a seasonal increase in production during summer (Milesi *et al.* 2001). This could also represent variation which enables enhanced protection of fruits and seeds from predators and pests. However, such studies do not precisely control for seasonal variation in other traits, such as leaf surface area, which may decrease in summer (Ross and Sombrero 1991), hence, more controlled investigation will be necessary for any adaptive significance to be unravelled.

Second, glandular trichomes and hairs containing volatile oils on the leaf surface may enhance tolerance of high leaf surface temperatures and reduce excessive water loss by trapping air which helps buffer high temperature and increases resistance of the laminar boundary, allowing for a reduction in transpirational water loss (Audus and Cheetham 1940). Volatile oils may also have a cooling effect and initiate stomatal closure and the emission of isoprene and monoterpenes may provide thermal protection for leaves subject to high temperature and potential damage of cell and membrane functions (Sharkey and Singsaas 1995).

Third, secondary compounds may be reconverted and re-utilized after their release. In *Mentha piperata* in California neo-menthyl glucoside produced by leaves is transported to roots and rhizomes where it is converted into other lipid-like metabolites (Croteau 1987). According to this author (p. 951), several other species provide... 'compelling evidence that monoterpenes can be degraded to metabolites that are reutilized in lipid biosynthesis in the developing root or rhizome or can conceivably be further oxidized in energy production'.

Finally, essential oils may play a role in enzyme maintenance during summer when metabolism and growth are depressed. It has been hypothesized (Banthorpe *et al.* 1972) that the biosynthesis of essential oils could maintain the appropriate enzyme systems in a state which could allow the rapid reactivation of the metabolic system once favourable conditions (the onset of which is often unpredictable) for rapid regrowth occur.

Precise experimental tests of these hypotheses remain to be done in order to fully evaluate their pertinence in a Mediterranean setting.

4.5.4 Secondary compound production: are costs reduced in a Mediterranean context?

To fully understand the function and adaptive nature of secondary compound production requires a formal understanding of the costs of manufacture, the benefits of their production, and their multiple roles. The production of secondary compounds requires the use of fixed carbon and the differentiation of structures for their storage requires that resources be diverted from growth, into other functions. This resource allocation may represent an important use of carbon and other resources, hence the accumulation of secondary compounds may come at a cost in terms of reduced growth. The 'trade-off' this implies may have fundamental ecological and evolutionary consequences. Indeed, as Herms and Mattson (1992) describe, the production of secondary metabolites in high concentrations can represent something of a dilemma to plants, and to grow or to defend is a question that is often posed. In their in-depth review of the theory of resource allocation to defence, the above authors outline two main hypotheses.

The carbon/nutrient (C/N) balance hypothesis. This hypothesis predicts that the C/N ratio of a plant should be positively correlated with the concentration of carbon-based secondary compounds (e.g. monoterpenes) and inversely correlated with concentrations of nitrogen-based compounds. This hypothesis relies on the assumption that moderate nutrient deficiency limits growth more than photosynthesis, leading to an increase in C/N ratio of the plant. Excess carbon is then allocated to the production of secondary metabolites. Evidence for this idea comes from a study of four woody species in different genera in southern France by Glyphis and Puttick (1989). These authors detected a negative relationship between foliar nitrogen concentration and leaf-phenolics among sites. In sites where plants had a lower phenolic content they also had a lower leaf surface area and lower concentrations of phosphorus.

The growth-differentiation (G/D) balance hypothesis. This hypothesis is similar, but slightly more comprehensive than the previous hypothesis. It takes as its basic premise the idea that there is a physiological trade-off between growth and differentiation (e.g. the production of structures necessary for secondary compound storage). Unlike the C/N balance hypothesis, the G/D balance hypothesis integrates the role of developmental constraints and external agents and thus provides a more complete picture of the environmental constraints influencing the production and defence function of secondary compounds. This hypothesis predicts that more photosynthates are available for the production of carbon-based secondary compounds in nutrient-poor environments. However, theoretical models which incorporate a feedback effect of plants on resource supply indicate that the simple prediction (based on resource availability) of high investment in antiherbivore defence in resource-poor environments may not always be valid (Loreau and Mazancourt 1999).

Although there is evidence in some species for a cost to secondary compound production, trade-offs in terms of reduced growth due to allocation to carbon-based defence compounds may not always occur. This may be particularly the case in the Mediterranean where water shortage and nutrient deficiency limit growth. The opportunity cost (in terms of not being able to grow because resources have been allocated to defensive compounds) of secondary metabolism will thus be low (Coley *et al.* 1985). In Mediterranean habitats plant growth is often limited by nutrients, hence growth may decline more rapidly than photosynthetic activity, creating an excess carbon above the requirements for growth. In such conditions woody plants may accumulate carbon-based compounds which lack nitrogen and phosphorous. Such carbon input to secondary compounds may thus be at little resource cost because even if more resources were available, growth cannot be increased. Such a reduced resource-based cost to growth (in terms of the use of fixed carbon for secondary compound production) is a distinct possibility for Mediterranean plants which may continue photosynthesis in the summer when environmental constraints prevent growth.

In addition, some species may shift allocation from one type of defence to another depending on attack rates and the resource cost of accumulating essential oils may be buffered by their multiple roles. As outlined by Herms and Mattson (1992), in resource-limited environments, the production and storage of essential oils which act as a deterrent to herbivores and parasites may also contribute to the competitive ability of a plant or its tolerance of environmental constraints and thus further enhance fitness. In other words, defence may not be as costly as some models for the evolution of chemical defence, assume. A dual role of reduced herbivory and reduced competition could thus improve plant fitness in a resource poor Mediterranean environment, without incurring major costs. In some (perhaps many) species, storage organs such as glandular hairs are already well developed and abundant on very young seedlings (references in Herms and Mattson 1992). Indeed in *T. vulgaris* it suffices to touch the first leaves as they open to realize that the glandular hairs are rapidly filled with monoterpenes. Such precocious differentiation of storage sites suggests the importance of essential oils in the defence of very young seedlings and illustrates that growth and differentiation can occur both rapidly and simultaneously.

Finally, to my knowledge, the exact nature of the relation among resource availability, cell differentiation related to secondary compound production, and plant growth, and thus the true nature of the cost of such compounds, has not been precisely identified in any Mediterranean plant. What I suggest here is that a combination of multiple roles and an environment where aridity and nutrient stress reduce the effective (or opportunity) cost of producing monoterpenes provide a possible explanation for the abundance of species with volatile oils in the Mediterranean flora.

4.5.5 Variation in chemical composition among closely related species

Closely related plant species in the Mediterranean flora often show striking differences in the blend of monoterpenes they produce. Species of the genus *Lavandula*, aromatic species *'par excellence'*, provide a convenient place to start. The species cultivated in the high plains of Provence since the end of the nineteenth century, *Lavandula angustifolia* (or true lavender), is probably the most well known of any of the different species and commercial varieties. Its rich oil, coveted by the perfume industry for over 100 years, and its vivid purple bouquet have made it an unmistakable emblem of the Provence landscape. This species grows wild above ~500 m in the western Mediterranean, and produces an essential oil rich in linalool and linalyle acetate. Below 600 m elevation, its congener *L. latifolia* is common on limestone in natural garrigue formations. Its oil is quite different, being composed mostly of 1,8-cineole, borneone, camphor, and linalool. At similar elevations on acidic schists and granite, *Lavandula stoechas* has a different oil, which is rich in borneol. The ecological and geographic segregation of these species, as well as some hybridization (to produce the high-yielding sterile lavindin hybrid), and their chemical diversity, make them a promising model for the study of plant evolution, a work which would surely interest the flourishing lavender industry.

Less well known, but by no means less significant, is the variation among Mediterranean oak species in the blend of terpenes they emit; some species are predominantly isoprene emitters, others predominantly monoterpene emitters (Loreto *et al.* 1998; Cisky and Seufert 1999). Instead of being stored in specialized secretory structures, as is the case for most aromatic oils, or induced when plants are damaged, the volatile organics emitted by oak trees are constitutively produced but not stored.

Synthesized from recently fixed carbon, they are almost immediately emitted at rates dependent on light and temperature conditions. Although Loreto *et al.* (1998) suggest that the variation among species may be of little ecological and adaptive significance, it being an ancestral character that may have served multiple functions, further work would be worthwhile on this issue.

Finally, the composition of the essential oil may be a distinguishing feature of endemic species. For example, *Ruta corsica*, endemic to Corsica and Sardinia has an essential oil characterized by a predominance of alkyl acetates whereas congeneric and more widespread species in the western Mediterranean have high concentrations of 2-ketones (Bertrand *et al.* 2003).

4.5.6 Variation in chemical composition within species

Variation in secondary metabolite formation can also occur within species, either as qualitative variation, which allows the recognition of two or more forms (or 'chemotypes') or as quantitative variation when individuals produce a different blend of similar monoterpenes.

The presence of qualitative intraspecific variation in chemical composition, which suggests the occurrence of distinct chemotypes in natural populations, has been documented in several groups of Mediterranean plants.

- Different species of *Thymus* (Stahl-Biskup 2002; Table 4.2), notably the six chemotypes of *T. vulgaris* in southern France (Thompson 2002) which I discuss in detail below.
- The common mint, *Mentha spicata* which has four distinct chemotypes in Greece (Kokkini and Vokou 1989).
- *Withania somnifera* which has been reported to have three chemotypes in Israel (Abraham *et al.* 1968).
- *Origanum vulgare* subsp. *hirtum* which has plants whose oil is dominated by either thymol or carvacrol in Greece (Vokou *et al.* 1993).
- Individual plants of *Coridothymus capitatus* and *Majorana syriaca* in the eastern Mediterranean contain either high amounts of thymol and small concentrations of carvacrol or vice versa (Ravid and Putievski 1983).

Quantitative variation in the monoterpene composition of essential oils may be common in many Mediterranean aromatic species, as the following examples illustrate.

1. In *Q. ilex*, a sample of 146 trees from natural sources in southern France revealed the existence of three main chemical profiles based on five main monoterpenes; α-pinene, β-pinene, sabinene, limonene, and myrcene (Staudt *et al.* 2001). All plants emitted at least a small amount of each of these molecules, but in distinctly different proportions. Three groups can be identified: a pinene type, dominated by two types of pinene (with sabine and myrcene as secondary components and limonene as a minor component) occurred in 71% of sampled trees; 31% of the trees were of the limonene type; and 8% emitted a monoterpene blend dominated by myrcene. Lack of environmental or seasonal variation suggests that such variation is genetically based in *Q. ilex*.
2. In *P. lentiscus* on Corsica, limonene and pinene types have been described with similar levels of relative abundance in sampled trees (Castola *et al.* 2000).
3. In the widespread *P. halepensis*, resin composition is highly variable among geographic localities (Schiller and Grunwald 1987).
4. In *R. officinalis*, the major monoterpene is variable, with plants having an oil dominated by one of four main molecules—camphor, eucalyptol, verbenone, or α-pinene (Granger *et al.* 1973).
5. The two different varieties of *Cistus ladanifer* in southern France, var *albiflorus* and var. *maculata*, show significant variation in the blend of monoterpenes in their essential oil (Robles *et al.* 2003).

The first step towards understanding the evolutionary significance of intraspecific variation is to document its genetic basis. In *T. vulgaris* six well-defined chemotypes have been identified in southern France (Granger and Passet 1973; Vernet *et al.* 1977*a*,*b*, 1986).

Table 4.2 Dominant monoterpenes and potential chemotype differentiation (based on a review by Stahl-Biskup 2002) in different species of the genus *Thymus* in the western Mediterranean

Section	Species	G	A	B	U	18C	CA	L	C	T
Mastichina	*T. albicans*					*		*		
	T. mastichina					*		*		
Micantes	*T. caespititius*		*		*				*	
Piperella	*T. piperella*								*	*
Pseudothymbra	*T. funkii*					*				
	T. moroderi					*	*			
	T. membranaceus					*	*			
	T. longiflorus				*	*	*			
	T. villosus	*					*	*		
	T. antoninae					*				
Thymus	*T. camphoratus*			*		*				
	T. carnosus			*	*			*		
	T. hyemalis							*	*	*
	T. orospedanus					*	*		*	
	T. baeticus	*		*	*	*		*	*	*
	T. vulgaris	*	*		*	*		*	*	*
	T. loscosii					*				
	T. zygis									
	subsp. *gracilis*	*						*	*	*
	subsp. *sylvestris*	*	*			*		*	*	*
	subsp. *zygis*		*					*	*	
	T. serpylloides	*	*		*	*		*	*	
	T. hirtus								*	
Hyphodromi	*T. bracteatus*							*		
	T. leptophyllus							*		
Serpyllum	*T. wilkomii*		*		*			*		
	T. pulegioides									*
	T. nitens	*	*						*	*
	T. herba-barona								*	*

Notes: G: geraniol, A: α-terpineol, B: borneol, U: thuyanol, 18C: 1,8-cineole, CA: camphor, L: linalool, C: carvacrol, T: thymol.

These chemotypes can be distinguished on the basis of their dominant monoterpene (present in glandular trichomes on the leaves, stems, and calyces), that is, geraniol (G), α-terpineol (A), thuyanol-4 with terpinen-4-ol (U), linalool (L), carvacrol (C), and thymol (T). The six monoterpenes are all produced from geranyl pyrophosphate in a series of changes in configuration and hydroxylation and thus have fairly similar molecular structures, but with a major distinction between phenolic and non-phenolic chemotypes (Box 4.10) whose importance will become apparent later. A large series of crossing experiments and progeny analysis first identified that the underlying genetic control involves 5 (or 6) loci (Vernet *et al.* 1986). A plant with a dominant *G* allele at one or other of two loci will have the geraniol phenotype, regardless of whether it has dominant or recessive alleles at the other loci. If the plant is homozygous recessive at the G loci and has at least one dominant *A* allele then it will have the α-terpineol phenotype. The sequence continues in the order (*G, A, U, L, C*) until all loci are

Box 4.10 Biosynthesis, dominant molecule and genotype of the six chemotypes of *Thymus vulgaris* in southern France

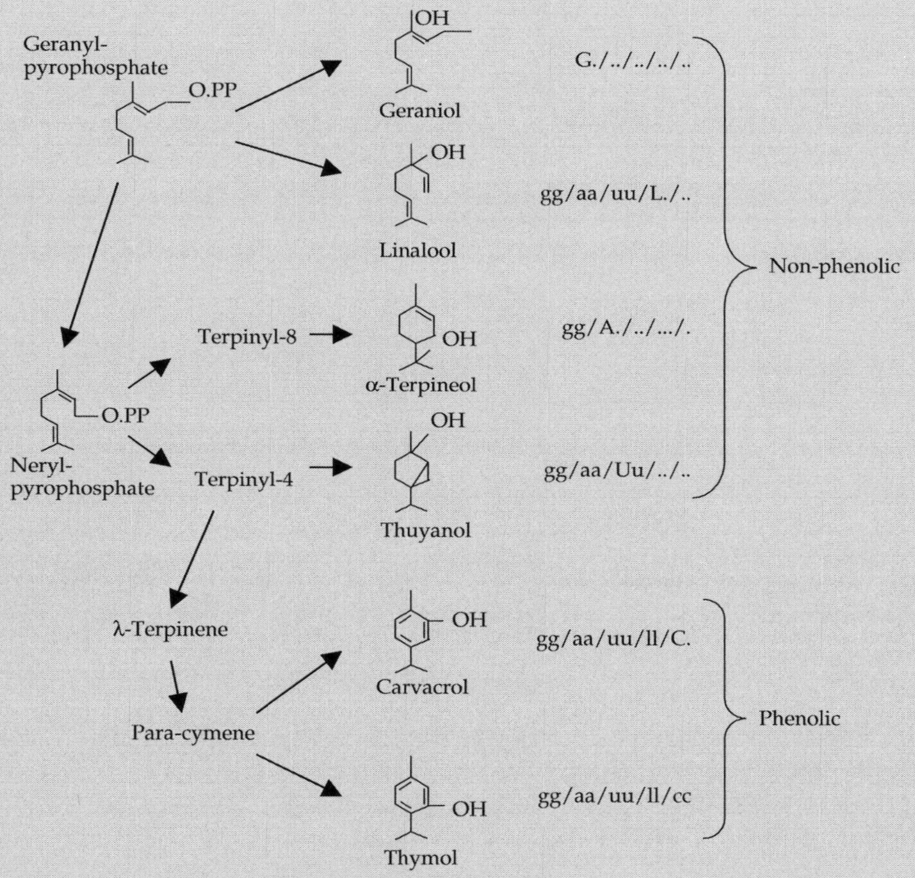

Geranyl-pyrophosphate
O.PP
OH
Geraniol
G./../../../..
OH
Linalool
gg/aa/uu/L./..
Non-phenolic
Terpinyl-8
OH
α-Terpineol
gg/A./../.../.
O.PP
Neryl-pyrophosphate
OH
Terpinyl-4
Thuyanol
gg/aa/Uu/../..
λ-Terpinene
OH
Carvacrol
gg/aa/uu/ll/C.
Para-cymene
Phenolic
OH
Thymol
gg/aa/uu/ll/cc

Plants with a phenolic chemotype have the characteristic thyme odour which makes them readily distinguishable from the four non-phenolic types. Another chemotype easy to distinguish in the field is geraniol, which has a lemon fragrance. All these chemotypes have a single dominant monoterpene, except the thuyanol chemotype whose oil is dominated by a combination of primarily thuyanol-4 and terpinene-4-ol and also contains non-negligable amounts of myrcenol-8 and linalool (Granger and Passet 1973; Thompson *et al.* 2003*b*).

homozygous recessive, which produces the thymol chemotype.

This mode of inheritance may be slightly more complicated than just a series of epistatic interactions (Thompson *et al.* 2003*b*). These authors obtained samples of each chemotype from two different population contexts: (a) where a chemotype is the most abundant chemotype ('home' sites) and (b) populations where the same chemotype is rare or occurs as part of mixed-chemotype populations ('away' sites). For five chemotypes, plants sampled at away sites had a significantly lower proportion of their dominant monoterpene in the oil than plants at a home site (Fig. 4.4). For nearly all plants at 'away' sites, the decline in proportion of the characteristic dominant monoterpene is correlated with a significant increase in a secondary component of the oil, which is the monoterpene present in the locally abundant chemotype. So, when linalool plants were sampled in populations dominated by the thuyanol chemotype, the decrease in linalool in their oil was accompanied by an increase in thuyanol, whereas if carvacrol was predominant then the linalool plants had a high proportion of carvacrol. A similar pattern was observed for geraniol plants, which had a high proportion of linalool in their oil at sites where the linalool chemotype was locally abundant, for α-terpineol plants in a population dominated by the thuyanol and linalool chemotypes, and carvacrol plants in a population dominated by the thymol chemotype. The thymol chemotype showed in almost all cases a decline in thymol associated with an increase in the composition of its precursors, and thus not the presence of one of the other characteristic monoterpenes. The latter result is not surprising since thymol is at the end of the chain.

The decline in the percentage composition of the dominant monoterpene was thus associated with a significant increase in the presence of a monoterpene produced by a dominant allele at a locus that would normally not be expressed in the chemotype (e.g. an *L* allele in a geraniol plant). The presence of a 'secondary' dominant gene may thus not be completely switched off by the dominant gene prior in the chain, allowing for a limited production of a secondary monoterpene by plants of a given chemotype. In fact, rather than being due to epistasis, the genetic control of monoterpene production may be a simple consequence of gene dosage effects along a chain of loci. Under this hypothesis, one would expect an increases in the production of linalool across a range of genotypes of the geraniol chemotype: from GG/ll (no linalool), to GG/Ll, Gg/Ll, and Gg/LL. Modifier genes could also allow for the production of secondary monoterpenes. Distinguishing the precise cause will necessitate more detailed genetic investigations of secondary compound production in this species.

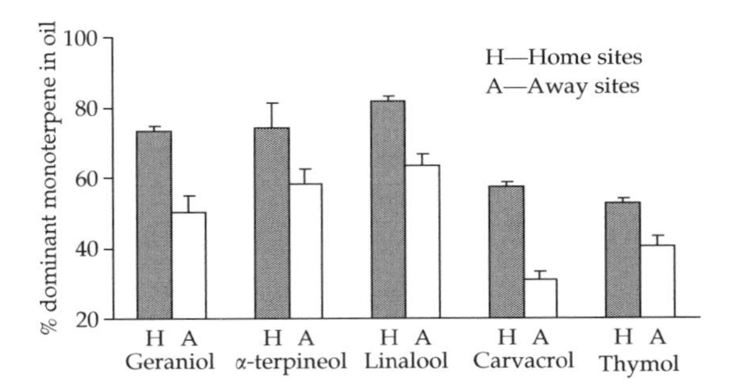

Figure 4.4 Mean percentage (±SE) of the dominant monoterpene in the essential oil of five *T. vulgaris* chemotypes in southern France at 'home' sites where that chemotype is the majority type (closed bars) and 'away' sites where that chemotype is rare or in a mixed-chemotype population (open bars). (Reproduced with permission from Thompson *et al.* 2003*b*).

In *M. spicata* in Greece (Kokkini and Vokou 1989) four chemotypes occur; one dominated by linalool (65–75% of the essential oil), one with variable amounts of carvone and di-hydrocarvone, one with either piperitone oxide or piperiteonene oxide depending on the site where plants are sampled, and one with a blend of menthone (18–45%), isomenthone (3–15%), and which may or may not have some pulegone (up to 31%). These types fit nicely onto the probably pathway of monoterpene synthesis, which would produce a first branch to linalool, a second branch to carvone (and di-hydrocarvone), and a subsequent series through piperitonene (and piperiteonene), pulegone and finally menthone (Murray and Lincoln 1970). The chemotypic segregation of interlinked compounds in a chain of synthesis with branches strongly suggests that dominant genes regulate

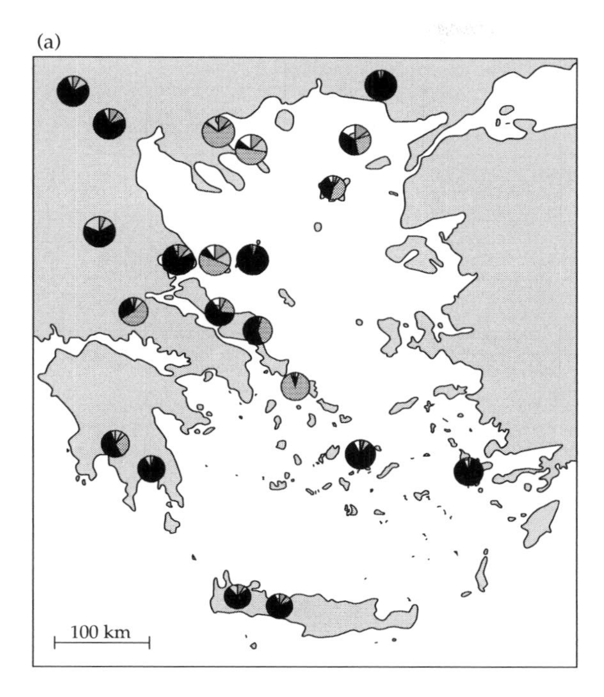

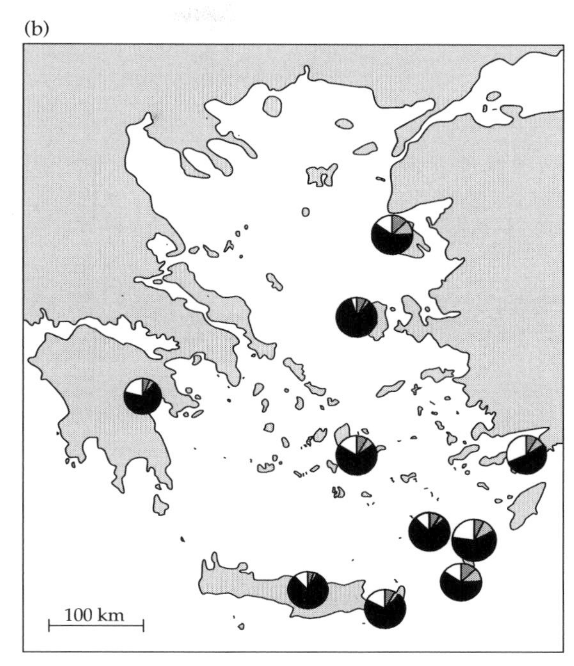

Figure 4.5 Geographic variation in the percentage of carvacrol (black sector) and thymol (dotted sector) in the essential oil of populations (each circle represents a population) of (a) *O. vulgare* subsp. *hirtum* and (b) *O. onites* (drawn from data in Vokou *et al.* 1988, 1993). The lightly shaded and open sectors of each circle represent precursors and their compounds in the oil.

enzyme activity at points in the chain which suppress activity further down the chain. In addition, several species show spatial variation in chemotype abundance, inviting investigation of the adaptive significance of variation in secondary compound production.

The genus *Origanum* provides an indication that local climate is closely related to population variation in monoterpene production. In a study of 23 populations of *Origanum vulgare* subsp. *hirtum* across Greece and onto the Aegean islands, Vokou *et al.* (1993) found that the local climate, in terms of thermal efficiency, had a significant effect on essential oil yield, which is higher at low altitude and decreases in cooler sites and at high elevation. Although there is no consistent effect on composition, the sum of the two phenolic monoterpenes, and even the sum of the four compounds in the phenolic pathway (thymol, carvacrol, and their two precursors para-cymene and λ-terpenene) were positively related to thermal efficiency at a site, and thus favoured in warmer climates. In addition the southern and eastern populations, that is, those with the highest thermal efficiency (milder winters and drier hot summers) tend to produce an oil composed of a relatively higher total content of phenolic monoterpenes and tend to be pure carvacrol (Fig. 4.5). The potential influence of climate on this pattern is supported by work on other closely related taxa:

1. In *O. vulgare* subsp. *vulgare* and subsp. *viridulum,* which have more northerly distributions, essential oil yield is lower than in subsp. *hirtum* (Kokkini *et al.* 1991).
2. In *Origanum onites,* which has a more southerly and easterly distribution, all plants have carvacrol as the dominant component of their oil (Vokou *et al.* 1988; Fig. 4.5).
3. In Portugal, the phenolic content of the oil of *O. vulgare* increases in a southerly direction (Carmo *et al.* 1989).

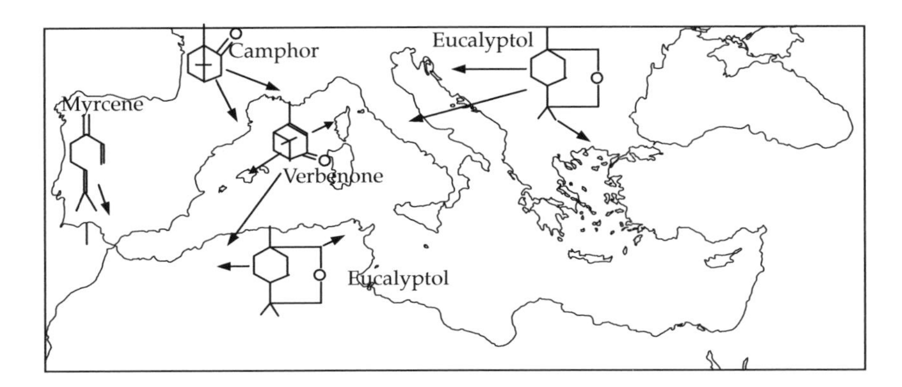

Figure 4.6 Geographic variation in the principal chemotype of *R. officinalis* in the western Mediterranean. Redrawn from Granger *et al.* (1973).

A study of 509 populations of 109 different aromatic labiates taxa in the flora of Greece by Kokkini *et al.* (1989) further supports this pattern. These authors identified three categories of essential oil production: oil-rich taxa (>2 ml/100 g dry weight) restricted to low elevation (<300 m); low-yielding taxa (<0.5%) at all elevations (sea level to 2,000 m); and intermediate taxa which show a gradual decline in frequency of populations with altitude. Within the high-yielding taxa there is a gradual decline in yield with altitude. There is thus a dominant trend for oil-rich plants, particularly those with phenolic monoterpenes, to occur in the hotter and drier Mediterranean communities.

In *Rosmarinus officinalis* the relative proportion of one of 4–5 major monoterpenes—all of which occur in most plants—varies geographically (Granger *et al.* 1973). In Spain and southern France, the oil is rich in camphor, in the eastern and southern (i.e. hotter and drier) parts of the range, eucalyptol is the dominant component of the oil, in the middle of the western Mediterranean and on Corsica, the oil is almost exclusively composed of verbenone (Fig. 4.6). Unfortunately, I know of no study which has explored the ecological and genetic basis to this pattern of differentiation.

Chemotype differentiation is probably a general feature of adaptive genetic diversity in western Mediterranean *Thymus*. Species of thyme are one of the most characteristic plants of the open garrigue vegetation of southern France and the tomillares of Spain, where they produce carpets of pink flowers in April and May and shrivel to produce greyish small bushes in late August. A true Mediterranean plant, if ever there was one. Several species in different sections of the genus show evidence of chemotype variation (Table 4.2). For example in *T. zygis* three allopatric subspecies show chemical variation in relation to latitude: the most southern subsp. *gracilis* has thymol, carvacrol, and linalool chemotypes; the more widespread subsp. *sylvestris* in central Spain has thymol, carvacrol, linalool, α-terpineol, 1,8-cineole, and geraniol chemotypes; and the most northern subsp. *zygis* has carvacrol, linalool, α-terpineol, and geraniol chemotypes. The trend is thus towards non-phenolic oils in more temperate sites with colder winters. A similar spatial pattern can be observed in *T. vulgaris* in southern France, where phenolic chemotypes (carvacrol and thymol) dominate populations in hot dry sites close to the Mediterranean Sea and non-phenolic chemotypes (geraniol, α-terpineol, thuyanol-4 and linalool) are abundant further inland, particularly above 400 m elevation, that is, in wetter, cooler climates (Granger and Passet 1973; Thompson 2002).

In *T. vulgaris* a highly localized pattern of spatial differentiation occurs in and around the St Martin-de-Londres Basin in southern France (Plate 2). Inside the basin a temperature inversion during winter causes the accumulation of cold moist air against the imposing north face of the Pic St Loup. In winter, temperatures are often several degrees lower (extreme freezing temperatures may reach −20°C in the basin) than in the hills that surround the basin where soils are more stony, shallower and drier, and different in terms of carbon and nitrogen mineralization when compared with those inside the basin (Billès *et al.* 1971, 1975; Gouyon *et al.* 1986). Cold winter temperatures probably prevent the colonization of the basin by *P. halepensis*, and differences in soil create a mosaic for the distribution of deciduous oaks on deeper soils and evergreen oaks in more rocky areas (see photograph in Box 4.11). This climatic and soil gradient, which occurs on a localized mosaic of geological variation (see Introduction) has been found to be correlated with differences in the distribution of different chemotypes over a very short distance. Whereas phenolic chemotypes predominate over large areas around the rims of the basin above 250 m elevation, below 200 m elevation inside the basin, where populations are often fragmented by agricultural land-use, there is a mosaic of smaller thyme populations where non-phenolic chemotypes are most abundant (Plate 2).

Since geraniol is produced by the dominant allele of the locus at the start of the chain, and is thus the only gene that cannot 'hide' behind other genes which cause a particular phenotype to be expressed, the dominant *G* allele may be more prone to loss during episodes of colonization and extinction. This chemotype is the rarest chemotype in the region. Hence, some of the variation in frequency among populations could have a stochastic cause due to random loss of genes and fixation of others. This is however, hardly likely to cause the striking spatial segregation of phenolic and non-phenolic chemotypes in different ecological conditions. Indeed, there is growing evidence for the role of selection in the maintenance of spatial chemotype differentiation in this region.

First, in 336 populations analysed by Gouyon *et al.* (1986) ~20% contained only a single chemotype and 50% contained a mixture of two chemotypes, another 20% had three chemotypes, and few populations had more than three chemotypes (Fig. 4.7). Although populations with two chemotypes are the most common, those with mixtures of phenolic and non-phenolic chemotypes are rare and, in general, populations containing both phenolic

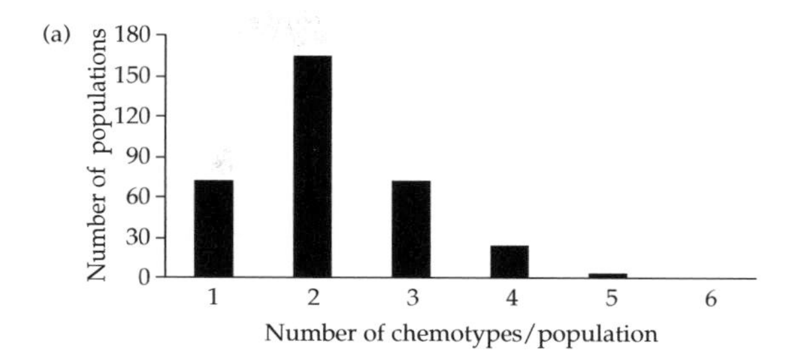

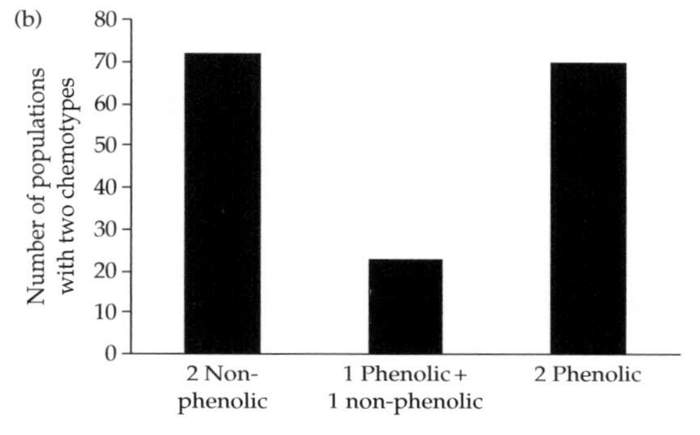

Figure 4.7 Chemotype diversity in natural populations of *T. vulgaris* in southern France: (a) the number of populations with different numbers of chemotypes in an overall sample of ~330 populations; (b) the number of populations with two chemotypes that have either two phenolic chemotypes, two non-phenolic chemotypes, or one phenolic and one non-phenolic chemotype. Reproduced with permission from Thompson (2002) and based on original data collected by Vernet *et al.* (1997*a*,*b*) and Gouyon *et al.* (1986).

and non-phenolic chemotypes (i.e. a combination of those with 2, 3, 4, or 5 chemotypes) are rarer than expected (Fig. 4.7). Most populations which contain appreciable percentages of both phenolic and non-phenolic chemotypes occur in the elevation transition between 200 m and 250 m (Plate 2). The sharp spatial separation of phenolic and non-phenolic chemotypes, despite the large number of mixed chemotype populations, is strong evidence that natural selection eliminates one or the other, depending on the spatial location of the site.

Second, the comparison of isozyme variation with chemotype variation indicates that allele frequencies based on isozymes show little genetic differentiation relative to those coding for the monoterpenes (Box 4.11). Most pollination of thyme is by honey bees which fly mostly between adjacent plants

Box 4.11 The frequency of chemotypes, and isozyme variation at two loci (PGM-2 and GOT-1) along a transect in southern France across the St Martin-de-Londres Basin

In the basin, left of dotted line, non-phenolic chemotypes predominate, whereas over the shoulder of the Pic-St Loup and onto its south-facing slope only phenolic chemotypes occur. Isozyme data from Tarayre and Thompson (1997), chemotype data from Thompson *et al.* (2003*b*). Each circle represents the frequency of alleles or chemotypes in a given population.

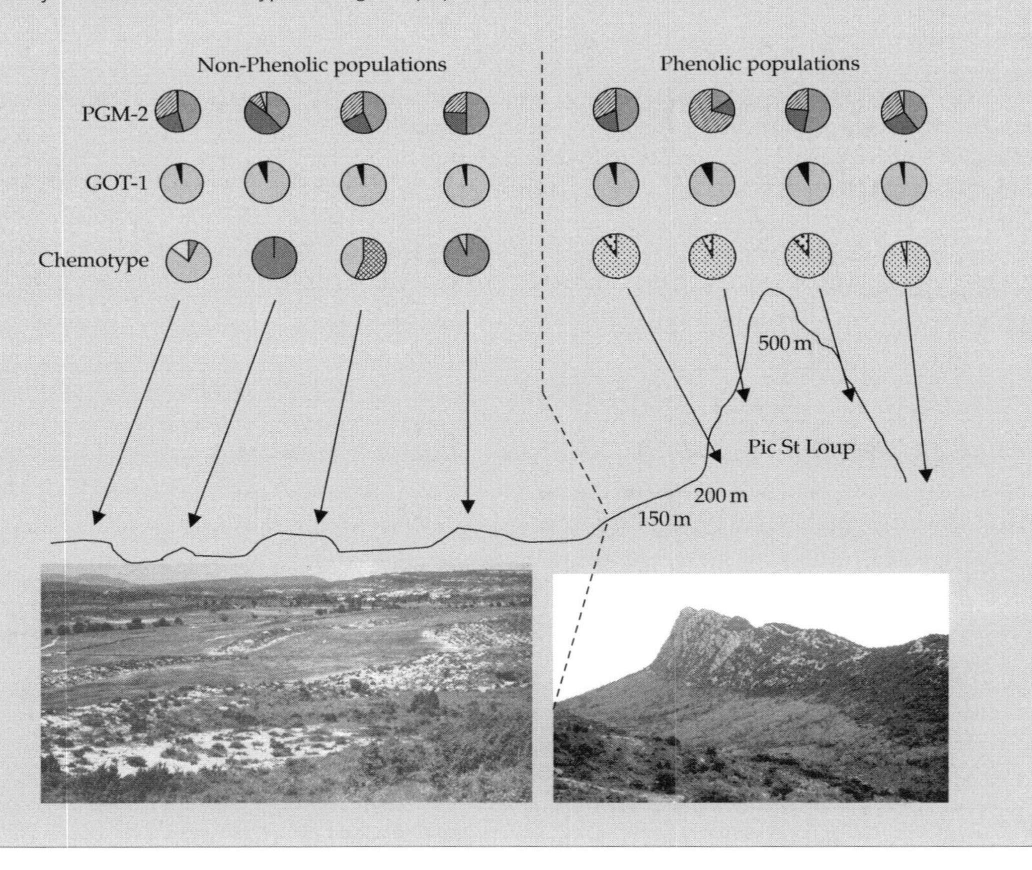

(Brabant *et al.*, 1980) but can transport pollen over longer distances and even between populations (B. Vaissière, INRA Avignon, personal communication). In addition several butterfly species which travel long distances between plants create a potential for pollen flow among populations. Tarayre *et al.* (1997) show that gene flow for pollen is much greater than that via seed, attesting to at least occasional among-population pollinator movements. In the face of such pollen movement, selection on chemotype genes must be strong.

Third, transplant experiments I have done with several colleagues have given us a closer insight into the nature of local chemotype adaptation in this region.

1. In seedlings transplanted among eight pairs of phenolic and non-phenolic populations reciprocally transplanted in and around the St Martin-de-Londres Basin (Plate 2), seedlings from non-phenolic sites show a marked reduction in survival compared to seedlings of phenolic parents in sites where the latter naturally occur. This reduced survival appears to be due to poor resistance of soil conditions. In the non-phenolic sites, there is a tendency for seedlings from the carvacrol population to have reduced survival compared to non-phenolic seedlings.

2. Transplantation of replicated cuttings from 15 clones of each of the six chemotypes into an experimental garden in Montpellier (where phenolic chemotypes—primarily carvacrol—would naturally be the most dominant form) and near the village of Mévouillon at 800 m elevation in the hills north of the Mt Ventoux in Provence (where only non-phenolic chemotypes are present in natural populations) show that the survival and biomass of phenolic and non-phenolic chemotypes is maximized in their original environment.

3. In a study of the second generation offspring from two phenolic and two non-phenolic populations grown in the experimental garden in Montpellier, Thompson *et al.* (2004) have shown (Fig. 4.8) that size and survival of offspring from maternal plants from phenolic populations, particularly those from the nearby carvacrol population, are significantly greater than those from maternal plants sampled in non-phenolic populations. An interesting twist to this study is the finding that vigour depended not only on the maternal origin of the offspring but also on the origin of the pollen donor: offspring size being significantly greater if the pollen donor originated from a local phenolic population (Fig. 4.8).

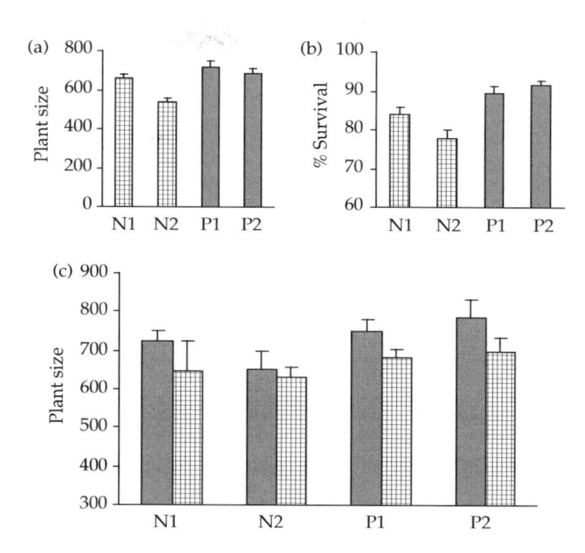

Figure 4.8 Performance of *Thymus vulgaris* offspring grown in a common garden at a site where phenolic, primarily carvacrol, chemotypes would normally be present: (a) plant size at reproduction, (b) four-year survival rates in maternal offspring from two non-phenolic (N1 and N2, hatched bars) and two phenolic (P1 and P2, filled bars) populations, (c) size at reproduction in maternal offspring (as above) with pollen donors from phenolic (filled bars) or non-phenolic (hatched bars) sites (redrawn from Thompson *et al.* 2004).

Although these studies provide convincing evidence for monoterpene-mediated local adaptation and that natural selection contributes to the pattern of spatial chemotype differentiation in thyme, they do not identify the precise ecological cause(s) of population differentiation. Based on the distribution pattern one would predict that soil and localized climatic variation are two main factors that either directly or indirectly (perhaps because of correlated variation in herbivores and parasites) contribute to spatial differentiation. There is some evidence that this may be the case.

In controlled conditions Couvet (1982) found that the non-phenolic α-termineol chemotype is significantly less resistant to drought and hot temperature stress than the carvacrol, thymol, and linalool chemotypes. Likewise, Pomente (1987) reported that seedlings of phenolic (thymol and carvacrol) plants had a better tolerance of drought stress than those of non-phenolic plants. Phenolic types may thus be better adapted to drier soils. Investigations are currently underway comparing the growth and survival of different chemotypes on the different soils collected from different sites where they occur.

Phenolic chemotypes, particularly carvacrol, are absent from sites which experience extreme subfreezing temperatures. Although Varinard (1983) found no effect of freezing on the survival of seedlings of the carvacrol, thymol, and thuyanol chemotypes in controlled conditions, Amiot *et al.* (unpublished manuscript) have found that carvacrol and thymol plants are more sensitive to freezing below −10°C than non-phenolic plants, particularly early in winter. Transplanted out of the Mediterranean into very cold winter climates, the germination and survival of offspring from phenolic plants is significantly less than those from non-phenolic chemotypes (Y.B. Linhart, University of Colorado, unpublished data). Phenolic chemotypes of thyme may be less resistant to freezing, perhaps due to a greater toxicity of the phenolic molecules which, following freezing and the rupture of cell membranes, cause mortality during harsh winters. Why plants with a carvacrol phenotype may be less resistant to colder sites than those with a thymol chemotype is not known. The ecological segregation of carvacrol and thymol may in fact be related to a better tolerance of heat and/or drought in the latter, since I have found that survival of transplants of the carvacrol chemotype has exceeded not only that of non-phenolic transplants but also those of the thymol chemotype in very hot and dry sites during the prolonged and very hot summer of 2003. Climate effects appear to exert a strong selective pressure on genetic variation in this species.

Spatial differentiation of thyme chemotypes may also be closely related to herbivory and parasite attack, as illustrated by several studies in controlled conditions (Gouyon *et al.* 1983; Linhart and Thompson 1995, 1999). Linhart and Thompson (1995) showed that snails (*Helix aspersa*) have a preference for non-phenolics, particularly the plants with a linalool chemotype, and a marked distaste for the two phenolics. What is more, snails fed on a diet of exclusively carvacrol plants lose weight. An interesting twist to this tale is the possibility that when plants with a genotype that should produce a linalool phenotype are at the young seedling stage, their leaves may not have a linalool phenotype but a phenolic phenotype, only developing their 'correct' phenotype after the young seedling stage. Linhart and Thompson (1995) suggest that such developmental plasticity could allow the genotype which is the least repellent to snails (linalool) to 'hide' behind a less palatable phenotype (thymol or carvacrol) during early seedling development.

Attempts to observe fitness effects of such snail herbivory in the field have however been inconclusive, even on small transplants, defying further interpretation of the above studies. In contrast, the occurrence of a specialized parasite, the gall-forming Dipteran fly *Janetiella thymicola*, can be frequently observed in the wild (Box 4.12). Observations of the occurrence of galls induced by oviposition and larval development in the area around the St Martin-de-Londres Basin have shown that rates of infestation show spatial aggregation in relation to chemotype (Box 4.12). Gall-formation is rare on plants in populations dominated by the phenolic chemotypes and higher in non-phenolic populations, with a striking peak in populations where the geraniol chemotype is abundant. In a mixed population of geraniol and linalool, infestation rates were significantly higher on geraniol plants. It could thus be that the rarity of the geraniol chemotype in the region where this work has been done (Plate 2) is at least in part due to its susceptibility to such parasitism. Where galls develop, they incorporate a large proportion of leaf biomass on a stem and thus may have a fitness cost in terms of resource acquisition and reproduction. This is currently being investigated alongside the issue of whether adults can choose oviposition sites (chemotypes) or whether chemotypes differ in their toxicity for larval development. Examples of genetic variation in resistance to gall-forming insects has been documented elsewhere, for example, on *Salix* (Fritz 1990) and *Solidago* (Horner and Abrahamson 1992). The case of thyme and its specialized fly may well provide another intriguing model system for this theme.

Studies of a range of different potential herbivores and different microorganisms show that variation in chemical defence among chemotypes depends on the external agent and no one chemotype provides the best defence across the spectrum of potential herbivores and pathogens, although in general the phenolic chemotypes do tend to be more deterrent than non-phenolic chemotypes (Linhart and Thompson 1999). In addition to their resistance to snail herbivory and parasite attack phenolic chemotypes of *T. vulgaris* have a stronger antibacterial effect than non-phenolic chemotypes (Simeon de Bouchberg *et al.* 1976), a pattern which can be seen in comparisons of different species which have either phenolic or non-phenolic chemotypes (Tantaoui-Elaraki *et al.* 1993). However due to variation in deterrence of the spectrum of potential antagonistic organisms, any spatio-temporal variation in the abundance of different herbivores, parasites, and pathogens could contribute to spatial variation in

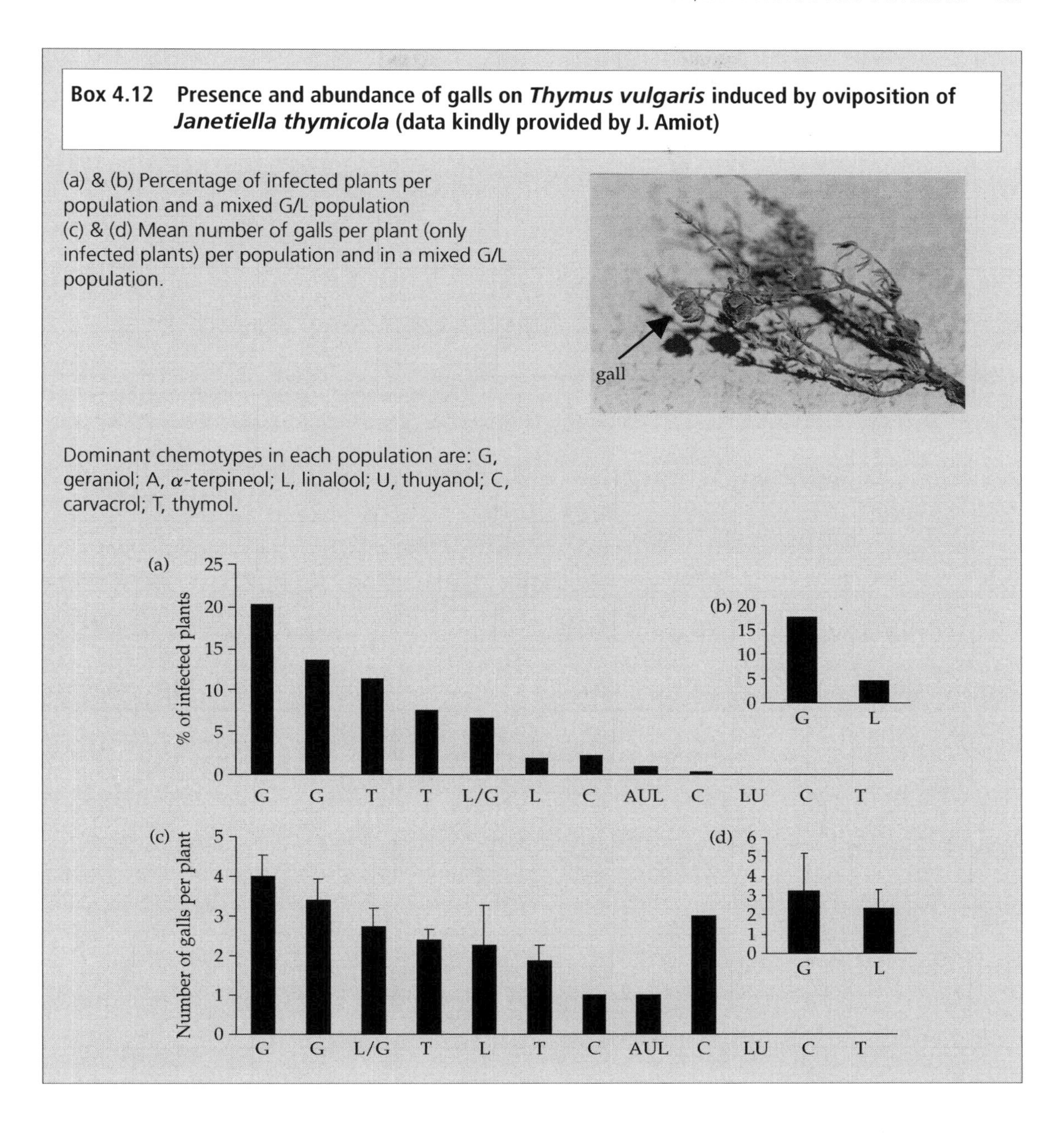

the relative abundance of chemotypes and the maintenance of the chemotype polymorphism (Linhart and Thompson 1999).

These different facets of the biotic environment may interact with spatial variation in the abiotic environment to influence chemotype abundance. In controlled conditions in an experimental garden Pomente (1987) reported that the thuyanol chemotype is particularly favoured in humid conditions, and that the presence of an associated grass was correlated with a decrease in the survival of thyme plants in conditions of drought stress. This effect of grass presence was not an effect of competition, but rather because the grass maintained a more humid environment in which slugs sheltered and subsequently caused a greater mortality of thyme plants. The thuyanol chemotype grew best in humid conditions and is also the chemotype most deterrent to slugs (Gouyon *et al.* 1983). The response to different ecological factors may thus go hand in hand, complicating our efforts to understand causal relationships in the maintenance of this complex genetic polymorphism.

Chemotype variation in thyme may also influence interactions with co-occurring plant species. Experimental work in controlled conditions shows how the litter and soils under different chemotypes can affect the regeneration and growth of associated species. Phenolic leaf litter or soils from under phenolic plants of *T. vulgaris* and other species significantly reduce germination and growth (Vokou and Margaris 1982; Tarayre *et al.* 1995; Karamanoli *et al.* 2000; Ehlers and Thompson 2004*b*) relative to those from non-phenolic plants. An interesting question raised by some of the work by Despina Vokou and her colleagues working in the phrygana of Greece is whether associated species may show tolerance to the presence of the compounds in the soil. Ehlers and Thompson (2004*b*) have shown (a) that *Bromus erectus* seedlings and adult plants have greater biomass on soils collected from thyme patches dominated by non-phenolic chemotyes than on soil collected from patches of phenolic chemotypes and (b) that seedling and adult biomass of *B. erectus* plants from non-phenolic thyme sites are on average greater on soil from thyme patches in the original non-phenolic site than brome from a phenolic site. The reciprocal is also true on soil from phenolic sites. This pattern of 'home site advantage' is detected on soil from patches of thyme but not on soil from the same sites but in patches with no thyme plants, indicative that the growth response of *B. erectus* to topsoil heterogeneity is closely related to modifications associated with the local thyme chemotype. Species in communities where aromatic plants are abundant may thus adapt to high concentrations of secondary metabolites in the soil or other modifications of the soil environment linked to the decomposition of litter rich in such compounds. Leaf litter may thus represent not only a critical component of mineral nutrient cycling, which affects the composition and function of local communities and their successional development, but also a component of spatial heterogeneity of the local environment, and thus a cause of divergent selection pressures.

Spatial heterorgeneity in the soil environment may thus develop as a result of genetic differentiation among populations of common species. To understand how species interactions shape the balance between facilitation and competition in Mediterranean garrigue communities may thus require an appreciation of both the processes which shape adaptive variation at the population level alongside those which regulate species interactions. To understand the dynamics of biodiversity in such systems requires a close link between population and community ecology.

4.5.7 More than just essential oils

I have focused this section on a small subset of the arsenal of chemical compounds that plants contain and employ for diverse purposes. Such roles are of course far from being limited to essential oils. Closely related Mediterranean species can also differ significantly in secondary compound profile for other types of compound. To finish this chapter, let us consider two examples.

First, western Mediterranean species of *Ruta* which show variation in essential oil composition (see above) also show marked differences in the relative composition of four furanocoumarins (Milesi *et al.* 2001). Whereas psoralen (46%) and xanthotoxin (43%) are predominant in *Ruta angustifolia*, bergapten (67%) is the main compound in *Ruta chalepensis* and xanthotoxin is the single most important furanocoumarin in *Ruta montana*. In *Ruta graveolens*, the species with the highest total yield of furanocoumarins, a blend of three main furanocoumarins was detected—bergapten (40%), psoralen (33%), and xanthotoxin (25%). The high yield and blend of different compounds in *R. graveolens* are no doubt associated with the use of this species in aroma-therapy, to clean bee hives (when used with thyme and rosemary) and in traditional medicine (San Miguel 2003). Aqueous extracts from *R. graveolens* have a significant fungistatic activity (Oliva *et al.* 1999) and can inhibit seed germination of cultivated and weedy species (Aliotta *et al.* 1994). The elegant work by Berenbaum *et al.* (1991) illustrating the adaptive significance of furanocoumarin variation, suggests that the natural substances in Mediterranean *Ruta* may play a significant ecological role and may have contributed to the diversification of different (morphologically similar) species in different parts of the range of this group. The functional and evolutionary significance of secondary compound variation in this genus thus awaits attention.

Second, *Ferula communis* shows chemotype differentiation in terms of its hydroxycoumarin composition, with one chemotype being highly toxic to livestock in Morocco and on Sardinia. On the island of Sardinia, Marchi *et al.* (2003) have reported that genetic differentiation for enzyme loci is significantly greater among populations of the two chemotypes than among populations of geographically distant populations of the same chemotype. This pattern of gene flow suggests reproductive isolation among chemotypes which may facilitate chemotype differentiation on the island.

4.6 Conclusions

The Mediterranean mosaic is changing, and so are the plant populations it contains. Of prime importance here is the way human activities currently modify and structure the Mediterranean mosaic. Such effects are ongoing and require consideration and integration, not just as an external agent, but as a key ecological factor influencing patterns of dispersal and colonization and the functioning of natural plant populations. The evidence abounds for a strong impact of spatial habitat heterogeneity on patterns of dispersal and regeneration, two critical events for the development of spatial differentiation among populations. More work which integrates both demographic, dispersal and genetic data within the context of ecological habitat heterogeneity would be particularly worthwhile here. The message here is that to understand plant evolution in the contemporary Mediterranean requires a full appreciation of how ecological variation creates differences in selection pressures in different habitats and how the dramatic contemporary changes in the spatial configuration of these habitats associated with human activities modifies the spatial pattern of seed dispersal and thus the potential for genetic differentiation.

That plants have not had time to adapt to the Mediterranean climatic regime is to my mind not a generally acceptable idea. When selection is strong, plants evolve rapidly and there are many examples of highly localized genetic differentiation within plant populations, even long-lived trees (Linhart and Grant 1996). Of course, many phenological and growth strategies such as sclerophylly represent nothing more than the 'ghost of selection past', or 'ecological phantoms' since they clearly evolved prior to the onset of the Mediterranean climate (Herrera 1992*a*). In this context, Grove and Rackham (2001: 45) argue that ...

> European ecology is dominated by environmental change rather than evolution. Most environments have not existed long in evolutionary terms. Plants, except perhaps for annuals which breed every year, have had to make the best of environments into which accidents of climatic and geological history have thrust them, rather than becoming adapted to some specific environmental niche. ... Plants have lived with the present climate for barely longer than they have lived with major human activity. They have yet to become fully adapted to it.

In this chapter I have attempted to balance this argument with illustrations of how diverse functional attributes and phenological strategies vary in relation to the complexity of the climatic constraints and selection pressures acting in Mediterranean habitats. Most of the evidence for adaptive variation concerns geographic and regional patterns of trait variation and variability among closely related sclerophyllous species in the precise mechanisms they use to resist water stress. Although limited in extent, this evidence can be used as a basis for manipulative experimental investigation of intraspecific variability in traits related to aridity, nutrients, and other features of the landscape mosaic. Less emphasis should be given to the occurrence of trait associations and syndromes, such as sclerophylly *per se* and more experimental research on intraspecific variation is necessary for us to conclude on whether species in the contemporary flora show evidence for climatic adaptation. Such experimental work would also be of value for the development of realistic model predictions for biodiversity changes as the climate continues to change in the Mediterranean. Species which occur on the fringes of the Mediterranean or which cover the complete climatic gradient from Mediterranean to non-Mediterranean-climatic regimes would be particularly useful model systems here (see also Chapter 6). My viewpoint is that more precise analysis, such as the transplant experiments described in the section on aromatic plants, could well show that localized adaptation to climatic and other ecological features of the Mediterranean scene is common in many plants, and not just annuals.

In a resource-limited environment such as the Mediterranean, several factors may favour the investment of large amounts of fixed carbon in the metabolism of secondary compounds. The most important of these concerns selection pressures associated with herbivory, parasitism, and competition and the potential contribution of secondary metabolites to multiple functions. The take home message here is that the multiplicity of ecological roles that secondary compounds play may furnish them with a low opportunity cost, and thus favour their presence in the Mediterranean flora. In addition, the production of many secondary compounds shows genetically based variation, which may thus be a key feature of diversification. Such intraspecific variation in secondary metabolite concentrations occurs in other Mediterranean floras (Mabry and Difeo 1973) and is thus of general significance.

The emission of volatile organic compounds may be of global significance since they may rival the emission of methane and other compounds of anthropogenic origin. The abundance and diversity of aromatic plants in the Mediterranean flora make the region an important source of such emissions. Since CO_2 levels likely influence volatile oil production it is clear that research on the dynamics of landscape change at the level of within-species variation in plant performance combined with close analysis of genetic variation in volatile oil production represent important themes to help broaden the debate on the ecological consequences of climate change. The productivity of Mediterranean woody species can be greatly influenced by levels of CO_2 (Hättenschwiler *et al.* 1997; Rathgeber *et al.* 1999), hence such selection pressures may also change as the climate alters.

Finally, landscapes are rapidly changing in the Mediterranean. First, fire regimes are far from being constant, and as the climate alters, landscapes may vary dramatically in the future as a result of massive fires and future changes in the fire regime. Second, the loss of natural coastal habitat on the northern shores and woodland in North Africa, and the progression of woodlands in the back country of the European Mediterranean countries is also causing the spatial structure of the Mediterranean mosaic to change dramatically. Selection pressures on plant populations and opportunities for gene flow are thus also being rapidly altered. An important component of this landscape change concerns the processes involved in plant–plant and plant–animal interactions which may set the scene for population differentiation during dispersal and regeneration. As we have seen, facilitation may promote the establishment of woody species due to the stressful effects of soil aridity in Mediterranean communities and species may respond to genetic variation in associated species, in a way which may alter the outcome of interspecific interactions. To further understand these ecological processes, and the evolutionary significance of species interactions, will require close contact between the disciplines of population and community ecology.

CHAPTER 5

Variation and evolution of reproductive traits in the Mediterranean mosaic

Through the pioneering efforts of Darwin . . . the naturalists of the last quarter of the nineteenth century were given good reason to believe that cross-pollination was beneficial to most flowering plants and that there were, as a result, distinct advantages to the plants in the attraction of potential pollinators. As a result of this stimulus, the naturalists searched for, and satisfied themselves that they had found, adaptive advantages in every morphological and behavioural feature of flowers and inflorescences.

H.G. Baker (1961: 64)

5.1 Reproductive trait variation: the meeting of ecology and genetics

To understand plant evolution, one must know how a plant reproduces. The study of plant reproduction includes two central themes in evolutionary ecology: the balance between sexual and asexual reproduction and the evolution of the mating system (i.e. who mates with who, and how often). The evolution of the mating system has long attracted the attention of evolutionary biologists because it determines the transmission of genes and thus has a crucial effect on the levels of genetic variability in a population. Many plants interact with animals which serve as vectors for pollination and seed dispersal, others depend on wind or water. Pollen transfer and seed dispersal are the means by which genes migrate among populations, hence, understanding the ecological processes of pollination and seed dispersal is critical for our understanding of plant population differentiation and evolution. What is more, diversification in floral morphology can contribute to reproductive isolation and speciation. The study of plant reproduction is where diversity and adaptation meet.

The subject matter of this chapter is the population ecology and evolution of reproductive traits in the spatially heterogeneous Mediterranean environment. For plant species which interact with animals for their pollination, reproduction and fitness depend on an efficient interaction. The Darwinian approach to evolution views floral traits in animal-pollinated plants as adaptations to pollinators. However, the flowers of many species are visited by large numbers of insect species. Any spatial and temporal variation in the abundance and composition of pollinator assemblages, when coupled with differences in pollinator efficiency or in their response to floral trait variation, may create variation in selection pressures which may condition the adaptive responses of flowers to their pollinators. In addition, attracting pollinators may come at a cost if herbivores, predators, and parasites are also attracted. This cost may further limit pollinator-mediated selection and adaptation. In plants, many species are sexually monomorphic but show fairly continuous quantitative variation in floral traits and/or gender. Others show polymorphic variation in their reproductive system, that is, the occurrence of two or more mating types (or morphs) in a population. Understanding the evolution and maintenance of such polymorphisms has been a central theme in evolutionary biology ever since Darwin. A fundamental feature of many species with sexual polymorphism is that the relative frequency of the different morphs may vary across the landscape due to a complex interaction between the ecological and genetic factors acting on sex expression, pollination, and the population dynamics of the species. In many cases, the stability of such polymorphisms is closely linked to both the precise functional interactions between plants and their pollinators and the population ecology of the plant.

In this chapter, I discuss three of these issues with reference to studies of reproductive trait variation and evolution in Mediterranean mosaic habitats.

- Spatial and temporal pollinator diversity and its effects on floral trait variation and evolution.
- The interactive effects of pollination and herbivory on plant fitness.
- Sexual polymorphisms and their variation among populations.

5.2 Specialization and generalization in a mosaic pollination environment

5.2.1 Specialized interactions in a seasonal climate

Plant species that rely on animals for pollen transfer may evolve specialized floral traits that facilitate their interaction with particular classes of pollinators (Darwin 1877; Grant and Grant 1965; Stebbins 1970). The evolution of specialized floral morphology in relation to interactions between floral traits and pollination efficiency (a theme I will pick up on later in this chapter in my discussion of style–length polymorphisms) can be illustrated by reference to the stability of species-specific plant–pollinator mutualisms, in which the plant benefits from pollination by a single species of insect which itself is specific to the plant, often because it lays its eggs in plant tissues and thus assures its reproduction. Two Mediterranean plants, the common fig and the dwarf palm, illustrate several aspects of the stability of such mutualisms in the highly seasonal Mediterranean climate. They are species that originated from the Tertiary flora present in the Mediterranean region prior to the onset of the Mediterranean climate, the mutualisms have thus persisted despite the important climatic oscillations and distributional shifts the plants and their pollinators have experienced over the last 2 million years.

The genus *Ficus* is well known for its tight mutualistic interaction with particular pollinators (Janzen 1979; Anstett *et al.* 1997). A typically tropical genus, figs find their way into this book by virtue of a single species, the dioecious common fig *Ficus carica*. The sole pollinator of this species is the agaonid fig wasp *Blastophaga psenes*. The inflorescence of the fig occurs inside an almost-closed urn-shaped syconium, which has a small entrance called an ostiole which is covered by several bracts. The inner wall of the syconium is lined with uni-ovulate female and in some cases male flowers. The mutualistic interaction involves a cycle of events (Valdeyron and Lloyd 1979; Kjellberg *et al.* 1987). The wasp enters a syconium when female flowers are receptive (prior to male flower maturity) and lays its eggs in the ovules. Some flowers (with a long style) produce seeds (if pollinated) others (usually with a shorter style relative to the length of the wasp ovipositor) produce a new generation of fig wasps which are adult when male flowers produce pollen. The wasps actually mate within the syconium and it is the female wasps, usually bearing pollen, which emerge and then search for a receptive fig for the next round of oviposition. *F. carica* is functionally dioecious because on some trees all the female flowers in a syconium have short enough styles for the fig wasps to lay eggs, and they are thus functionally male. In contrast, other trees have only flowers with longer styles where oviposition is not possible and only seeds are produced, that is, the tree is female. Clearly, a fig wasp which enters a syconium on a female tree will have no reproductive success, it is as if it enters a trap, a deadly one.

One would thus predict strong selection on the wasp to avoid female figs. However, the sex-specific seasonal phenology of *F. carica* means that male and female syconia are almost never simultaneously available to the fig wasp. Thus asynchrony of attractivity among sexes prevents wasps from having a choice (Valdeyron and Lloyd 1979; Kjellberg *et al.* 1987). When the majority of wasps emerge from male syconia (thus bearing pollen), the only receptive syconia elsewhere (with no wasps inside them) are female. Wasps only live for ~2 days so they have no choice, enter a female synconia or perish. In fact, in experimentally manipulated arrays, the pollinators of *F. carica* show an ability to discriminate between male and female inflorescences (Anstett *et al.* 1998). The blend of volatile compounds emitted by receptive male and female syconia contains a very similar array of compounds (Gibernau *et al.* 1997; Grison-Pigé *et al.* 2001), however, the proportions of the different compounds in the blend differ among males and females (Grison-Pigé *et al.* 2001). Hence wasps can choose, if they are given a choice. Pollination of females is thus by deceit and the asynchronous phenology of males and females, in a highly seasonal environment, causes selection on strict sexual mimicry in this system to be weak, thereby promoting the stability of the mutualism.

On the northern shores of the western Mediterranean, the dioecious dwarf palm, *Chamaerops humilis* is pollinated by a single species of weevil (*Derelomus chamaeropsis*: Curculonidae), whose eggs develop and pupate in the rachises of male inflorescences and adults transfer pollen from males to females (Anstett 1999). So, in exchange for pollen dispersal, dwarf palms provide the weevils with shelter, egg-laying sites, and food, and the two partners have highly co-adapted life cycles and reproductive biology. Although weevils lay eggs in females as they pollinate, processes associated with fruit development prevent larval development (Dufaÿ and Anstett 2004), hence males assure the next generation of potential pollinators. Despite their lack of reproductive success, weevils continue to visit females, albeit perhaps less often. An important facet of this interaction concerns the chemical attraction of the weevils by volatile chemicals emitted by vegetative parts of the plant (Dufaÿ *et al.* 2003; Chapter 4). Although the blend composition of the volatile attractive fragrance varies among plants, and males may produce more scent, there is no significant variation in the composition of the fragrance emitted by females and males (Dufaÿ *et al.* 2004). Such similarity in attractive potential may prevent complete discrimination by the pollinator. Females thus benefit from pollination but do not provide a site for larval development, they thus 'cheat' their pollinator which may have reduced reproductive success as a result of visiting females (Dufaÿ and Anstett 2004).

In both these 'nursery pollination mutualisms' where the plant hosts larval development, an important feature of the plant species is that they are dioecious: female plants prevent egg laying or larval development of the pollinator, and thus avoid the major potential cost of such a mutualism. The costs of such larval development thus differ between males and females, primarily since the phenology of larval development is asynchronous with male function. Both species are single representatives of Pre-Pliocene palaeo-tropical ancestral lineages in the Mediterranean and thus pre-dated the evolution of a highly seasonal climate (Chapter 1). Although the tight mutualistic interaction pre-dates their existence under a Mediterranean-climate regime, these two species have nevertheless maintained a highly specific interaction with their pollinator. During the repeated oscillations of the climate since the Pliocene the pollinators have thus tracked distribution changes in the plant and the mutualism has persisted.

5.2.2 Generalization in Mediterranean plant communities

For a plant to evolve floral adaptations to a particular class of pollinators, it is necessary that different potential pollinators vary in the strength of their interactions with the plant (Schemske and Horvitz 1984). Such variation will be governed by two key elements of the interaction: the relative 'quantity' of visits by different insects and the 'quality' of the interaction in terms of rates and distances of pollen dispersal within and among plants (C.M. Herrera 1987, 1988*c*, 1989).

If a plant species relies on more than one species for pollination, and if each pollinator visits more than one plant species, generalized pollination systems may develop. Several factors favour the maintenance of generalist pollination systems and a network of plant–pollinator interactions in local communities (Waser *et al.* 1996; Olesen and Jordano 2002): (a) spatial or temporal variation in pollinator abundance which could cause pollen limitation of seed set or even plant extinction in a particular region if a preferred pollinator is absent; (b) similarities in floral rewards among co-occurring plant species; (c) if pollinators suffer energetic constraints on pollinator flight distances and a long lifetime of insects relative to the flowers in a community; (d) lack of variation in pollination efficiency, that is, a sort of functional equivalence. Waser *et al.* (1996) and Olesen and Jordano (2002) argue that these situations commonly exist in nature, to quote Waser *et al.* (1996: 1053) they may represent 'the rule rather than the exception'. However, spatial variability in pollinator abundance, if coupled with differences among pollinators in their pollination efficiency, could also create a spatial mosaic of more or less specialized interactions (Thompson 1994). In this case, different evolutionary trajectories in plant–animal interactions and different floral trait adaptations may evolve in a single species in different environments.

Several Mediterranean plant communities have generalized pollination systems, in which plant species are visited by a large number of insects, themselves unspecialized on a single plant species, for example, in coastal scrub community, at the Reserva Biológica Doñana in south-west Spain (J. Herrera 1988), open garrigues of southern France (my own data), herbaceous grassland in northern Spain (Bosch *et al.* 1997) and the phrygana ecosystem in Greece (Petanidou and Vokou 1990; Petanidou and Ellis 1993). The occurrence of generalized pollination systems does not mean that there is no observable pattern in communities, nor does it rule out the existence of a 'certain structure in the plant pollinator interactions' (Bosch *et al.* 1997: 590). The latter authors identified four functional groups of plants based on their main types of insect visitors, with plants with similar reward composition (in terms of relative dependence on pollen and nectar) having similar visitors. So, for example, Fabaceae tend to be visited primarily by bees, Cistaceae by Diptera and pollen collecting bees, and *Lonicera* by hawkmoths. Such patterns remain, however, coarse (J. Herrera 1988). There is also a degree of asymmetry in the plant–pollinator relations of such systems. Although individual plant species are visited by many insect species, individual species have a smaller number of plant targets. In the coastal scrubland of southern Spain, J. Herrera (1988) reported that whereas a single insect visited on average just two plant species, a given plant species was, on average, visited by 16 different insect species. The large number of rare insect visitors and the tendency to focus studies on fairly common plant species no doubt contribute to this asymmetry. Additionally, plants flowering at the same time tend to be visited by the same insect taxa, more so than plants with similar floral morphologies but which have different dates of maximum flowering (Petanidou and Vokou 1993).

At the plant community level the pollinator fauna can also change dramatically during the flowering season (J. Herrera 1988; Petanidou and Vokou 1990). In the Lamiaceae of the Greek phrygana, Petanidou and Voukou (1993) reported that plants which flower in late winter or early spring have a small number of flowers and are visited by few insect taxa while those which flower later in spring bear many flowers and are visited by a large number of different insects.

The major type of reward in Mediterranean flowering plants is pollen (Kugler 1977). Many Mediterranean plant species, in fact most animal-pollinated species are pollinated by wild bees. This predominance of bees as pollinators has been documented in Spain (J. Herrera 1987, 1988), Greece (Petanidou and Vokou 1990; Petanidou and Ellis 1993), and Israel (Potts *et al.* 2001). In southern Spain only 35% of a total of 122 plant species were considered nectariferous by J. Herrera (1985). In the phrygana ecosystem of Greece, Lepidoptera represent a small fraction of insect visitation which is primarily by pollen collecting bees (Petanidou and Vokou 1990). These authors interpret this pattern in relation to drought which can have a negative effect on nectar production (see also Potts *et al.* 2001). Hence, to quote J. Herrera (1985: 51) 'species that rely on a copious and constant production of pollen may have a higher or more constant . . . reproductive success'. This conclusion awaits experimental confirmation.

An extreme functional extension of this lack of nectar reward can be observed in the formation of completely rewardless flowers that rely on deception of insects for pollination (like in figs above). For example, the orchid flora of the Mediterranean region, in common with that in southern Australia, shows an important element of mimicry and non-rewarding flowers (Dafni and Bernhardt 1990). In *Orchis caspia* in the eastern Mediterranean, this deception involves more than one insect species and several other plant species (Dafni 1983). In different sites, *O. caspia* attracts pollinators from different rewarding species according to the abundance of the latter. In this case the deception relies on the occurrence of a generalized pollination system in which pollinators show little discrimination among plant species. In *Nerium oleander*, the lack of nectar is associated with a strategy of pollination by deceit since even the pollen is difficult to access due to the concealed position of the anthers in the flowers (J. Herrera 1991*a*). In the dioecious hemiparasitic shrub *Osyris alba*, females neither produce pollen, nor have any nectar reward, and are thus pollinated by deceit (Aronne *et al.* 1993). Such pollination by deceit requires that pollination be to some extent unspecialized. In nectarless *Cyclamen*, several species possess a buzz-pollination strategy where bumblebees actively vibrate flowers to extract pollen. In such species, the functional aspects of the plant–pollinator interaction are likely to be specialized.

Many Mediterranean plant species have diverse pollinator assemblages (Table 5.1). This includes species with papillonaceous flowers and those with fairly long and narrow tubular corollas which could in theory restrict visitation to particular types of insect (J. Herrera 1988; C.M. Herrera 1996; Bosch *et al.* 1997; Thompson 2001). In addition, some species may be pollinated by a combination of very different groups, for example, insects and ants in *Lobularia maritima* in south-west Spain (Gómez 2000) and lizards and insects in *Euphorbia dendroides* on the Balearic islands (Traveset and Sáez 1997). In addition, some species may combine both wind and animal pollination, for example, *Hormathophylla spinosa* in the Sierra Nevada (Gómez and

Table 5.1 Examples of insect-pollinated plant species with diverse pollinator assemblages in the Mediterranean region

Species	Family	Region	Pollinator assemblage	Pollinator variation
Cistus libanotis[a]	Cistaceae	Eastern Mediterranean	Diverse Hymenoptera, Coleoptera, and Diptera	Among sites
Three *Cistus* species[b]	Cistaceae	North-east Spain	Diverse Hymenoptera, Coleoptera, and Diptera	Not studied
Paeonia broteroi[c]	Paeoniaceae	Iberian peninsula	>25 insect species, mostly Hymenoptera	One species of bee is the principal pollinator
Asphodelus albus[d]	Liliaceae	Northern Spain	13 species of Hymenoptera, Diptera, and Lepidoptera	Among years
Satureja thymbra[e]	Lamiaceae	Greece	At least 34 bee species	Not studied
Lavandula latifolia[f]	Lamiaceae	Sierra de Cazorla, south-east Spain	~70 species of Hymenoptera, Diptera, and Lepidoptera	Among and within sites, and among and within years
Jasminum fruticans[g]	Oleaceae	Southern France and north-east Spain	>28 species of Hymenoptera, Diptera, and Lepidoptera	Among sites and among years
Lobularia maritima[h]	Cruciferae	North-east Spain	>50 species of Hymenoptera, Diptera, and Coleoptera	Not studied
Hormathophylla spinosa[i]	Cruciferae	Altitudinal gradient in the Sierra Nevada	~70 species (19 families in 5 orders)	Among 3 populations

[a] Talavera *et al.* (2001). [b] Bosch (1992). [c] Sánchez-Lafuente *et al.* (1999).
[d] Obeso (1992). [e] Potts *et al.* (2001). [f] C.M. Herrera (1988*c*).
[g] Thompson (2001). [h] Gómez (2000). [i] Gómez and Zamora (1999).

Zamora 1996) and *Urginea maritima* in coastal areas of the eastern Mediterranean (Dafni and Dukas 1986). Despite this diversity, in insect-pollinated species which have received detailed investigation, for example, *Lavandula latifolia*, (C.M. Herrera 1988*c*), *Jasminum fruticans* (Thompson 2001), and *H. spinosa* (Gómez and Zamora 1999), a small number of abundant pollinators dominate the pollinator assemblage which otherwise contains a large number of fairly rare floral visitors. For example, the 10 most abundant taxa represent ~80% of all visits to both *L. latifolia* and *J. fruticans*. These diverse pollinator assemblages thus show a highly skewed (log-normal) frequency distribution of species abundance and rates of visitation, indicative that a small number of strong interactions may exist within the buzzing array of visitor diversity. Evidence for specialized interactions in species with diverse pollinators remains however elusive, as the two examples in the next section illustrate.

5.2.3 The spatio-temporal pollination mosaic: consequences for floral trait variation and evolution

Detailed analyses of pollinator diversity in a six-year study of *L. latifolia* (Plate 3) in south-east Spain by C.M. Herrera (1987, 1988*c*, 1989, 1990*a*, 1995*a*) illustrates very clearly the spatial and temporal variation that may occur in the cortège of pollinators visiting a single plant species. Flowering at a time of the year when very few species are in flower (peak flowering is in July and August) *L. latifolia* is a target for ~70 insect species searching for pollen and nectar. Although most interactions involve a small number of species, there is no consistent relationship between abundance and visitation rates, since no pollinator was both abundant and an efficient forager, and the few species with the most visits occur in diverse taxonomic groups. What is more, pollinator species abundance and the composition of the pollinator assemblage showed significant temporal and spatial variation (Fig. 5.1).

1. Variation in time occurred among years, at different moments of the flowering period and at different times of the day. Hymenoptera were the dominant component of the pollinator assemblage in four years, Lepidoptera numerically dominated the assemblage in one year, and a more balanced assemblage occurred in the remaining year. Even the identity of species in the group of abundant visitors varied among years. Intensive observations revealed that the cortège of visitors showed strong seasonal dynamics during a single year and that pollinators do not all visit at the same time of the day. For example, *Anthophora* bees exhibited two peaks, one in mid-morning the other in mid-afternoon, a pattern observed in other species with different flowering ecologies, for example, *Petrocoptis grandiflora* (Navarro *et al.* 1993) and *Narcissus assoanus*.

2. Pollinator abundance and composition varied significantly in space both among populations, in relation to proximity to water, and within-sites, due to a mosaic of sun-shade patches. Although for some pollinators foraging behaviour varied among sun and shade patches, others principally visited plants in full sunshine. Hence, variation in the pollinator assemblage was highly localized in relation to microhabitat conditions. Although this pollination mosaic can be related to some environmental features and the population cycles of some insects, it also has a strong stochastic component.

For such variation to create a spatio-temporal mosaic in selection regimes on floral traits pollinators must vary in their activity and pollination efficiency in a way which has consequences for plant fitness. There should thus be some association between abundance, foraging efficiency, and pollination effectiveness (rates and distance of pollen transfer). In *L. latifolia* the quality and quantity of pollination do not go hand in hand (C.M. Herrera 1987). Pollinators differ in (a) the number of pollen grains they deposit on the stigma during flower visitation, (b) their visitation rates to male-phase and female-phase flowers (*L. latifolia* flowers are protandrous), and (c) patterns of flight distance among plants. For example, pollination by Lepidoptera may occasion greater rates of outcrossing as a result of their longer flight distances among plants. Hence, subtle differences in the composition of the pollinator assemblage in the landscape could

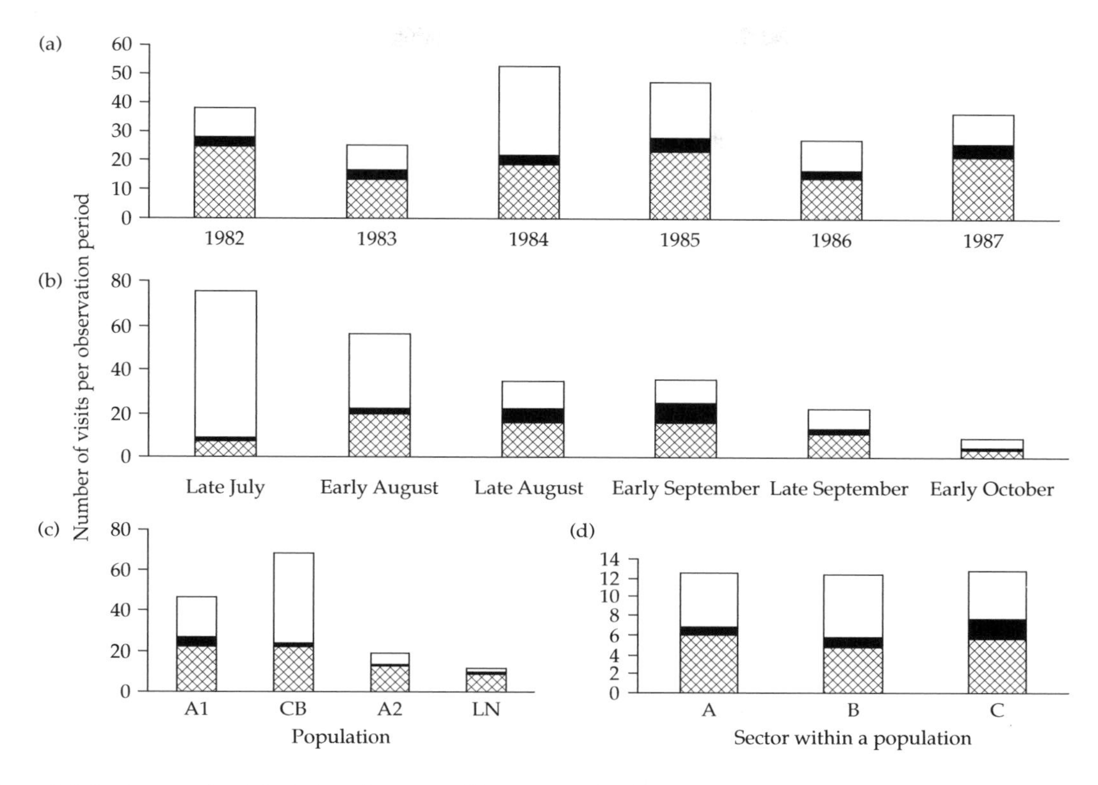

Figure 5.1 Spatio-temporal variation in the pollinator assemblage visiting *L. latifolia* at sites in south-east Spain. Pollinator composition can be seen in the relative portions of each bar and is based on the mean number of individuals per observation period by Hymenoptera (hatched portion of bar), Diptera (closed portion), and Leipdoptera (open portion). Graphs show (a) temporal variation among years, (b) temporal variation within a year in a single population, (c) spatial variation among populations, and (d) spatial variation among locations within a population. Data are number of visits per observation period (redrawn from C.M. Herrera 1988*c*).

have profound consequences for plant population dynamics (C.M. Herrera 2000*a*,*b*). Finally, manipulation of the corolla lobes of *L. latifolia* (one, both, or neither of the two corolla lips were modified), showed that the modification of one trait was without consequence for pollen removal and pollen deposition (C. Herrera 2001). Corolla integration is thus not closely related to selection pressures imposed by pollinators, and may stem more from genetic correlation among traits.

To sum up, the spatio-temporal pollination mosaic described for *L. latifolia* provides no evidence of a mosaic of specialized interactions, rather a diversification which may reduce the impact of pollinator-mediated selection on plant traits.

A second example of the spatio-temporal mosaic in the pollinator assemblage and its consequences for plant reproduction and floral evolution concerns *J. fruticans* (Table 5.1). Whereas *L. latifolia* occurs in the Lamiaceae, which is part of the Mediterranean element of the flora that has shown explosive radiation during the recent history of the Mediterranean, *J. fruticans* is a tropical relict in the Oleaceae (from the ancient Tertiary flora), and is the only native member of the genus in the western Mediterranean, where it occurs at the northern distribution limits of the genus. *J. fruticans* is a clonal shrub (Plate 3) common on scree slopes and around fields and vineyards in the western Mediterranean. It bears tubular yellow flowers in May when the garrigues of southern France is alive with potential pollinators. Its fleshy fruits are dispersed by birds and hence its frequent occurrence under other trees and shrubs or along old stone walls used as perching posts.

At least 28 insect species (Hymenoptera, Diptera, and Lepidoptera) in six main groups visit *J. fruticans* flowers, where they actively feed on nectar or collect pollen (Thompson 2001):

(1) short-tongued bees (mostly *Apis mellifera*: Hymenoptera);
(2) long-tongued bees (*Anthophora*, *Eucera*, and *Xylocopa*: Hymenoptera);
(3) bumblebees (*Bombus*: Hymenoptera);
(4) bombylid flies (*Bombylius*: Diptera);
(5) hawkmoths (*Macroglossum stellatarum* and *Hiemaris fuciformis:* Lepidoptera);
(6) several butterfly species (Lepidoptera).

As in *L. latifolia*, the composition of the pollinator assemblage varies significantly in space (among populations in southern France and northern Spain) and time (among years in a single population) (Fig. 5.2), and insect groups vary markedly in their foraging efficiency (Fig. 5.3). For example, whereas hawkmoths and bombylid flies rapidly visit a large proportion of open flowers on a given flowering stem, butterflies slowly visit a small proportion of flowers. Insect groups also differ in their behavioural response to variation in floral design (e.g. flower size) and floral display (numbers of open flowers). Depending on the insect, visitation rate was positively related to the number of flowers on a stem (short-tongued bees), the number of open flowers in a patch (bee flies and butterflies), corolla tube length (hawk moths), or corolla lobe length (butterflies). These different foraging responses reflect differences in the biology of the insects.

The divergent selection pressures that variation in pollinator visitation may create are all the more intriguing in *J. fruticans* because of its floral biology. This species is distylous (Box 5.1), a floral polymorphism in which the morphs differ in the sequence of heights at which stigmas and anthers are positioned within the flowers (see Section 5.5). The reciprocal positioning of stigmas and anthers in the different floral morphs promotes cross-pollination among morphs due to precise pollen positioning on pollinators and efficient pollen transfer among sex organs at each level in the flower

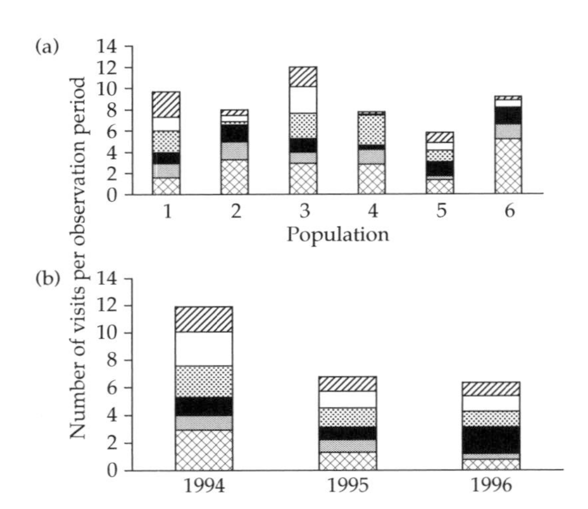

Figure 5.2 Spatio-temporal variation in the number of visits of different pollinators to *J. fruticans* (a) in six populations and (b) over three years in a single population. Data are number of visits per observation period. From the bottom of each bar: hatched portion—short-tongued bees, grey portion—long-tongued bees, black portion—bumblebees, dotted portion—bombylid flies, open portion—hawkmoths, diagonal portion—butterflies (drawn based on data in Thompson 2001).

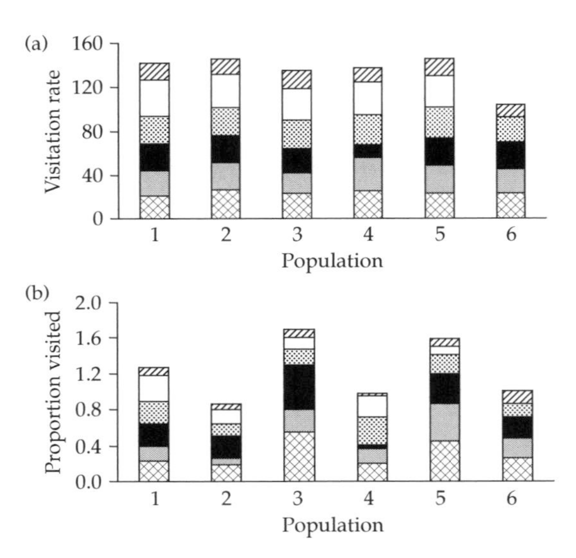

Figure 5.3 Variation in the foraging behaviour of different pollinators visiting *J. fruticans* (a) visitation rates to flowers in a given time and (b) proportion of flowers visited per stem during a visit (drawn from data in Thompson 2001). From the bottom of each bar: hatched portion—short-tongued bees, grey portion—long-tongued bees, black portion—bumblebees, dotted portion—bombylid flies, open portion—hawkmoths, diagonal portion—butterflies.

(see later in this chapter). One would thus predict strong selection on stigma and anther position. However, in *J. fruticans*, in contrast to many distylous species (see Thompson and Dommée 2000), there is marked quantitative variation in these floral traits (Box 5.1). Two aspects of this variation are particularly intriguing.

First, style length shows continuous variation, and in some short-styled plants (S-morph) the stigma is at the same level as the anthers (at the mouth of the corolla tube). Plants with this floral morphology have pollen grains that are (a) similar in size to those produced by plants of the S-morph, that is, significantly larger than pollen grains of long-styled plants (L-morph) and (b) incompatible on other plants of the S-morph. As in many distylous species, *J. fruticans* is self-incompatible and crosses among plants of a given morph are also incompatible. This occurrence of S-morph plants which lack any stigma–anther separation is an unusual feature for a distylous species, where the occurrence of such 'homostyly' is usually associated with the occurrence of a novel variant morphology and a breakdown in the heterostylous system (Section 5.5). This is not so in jasmine where homostylous flowers are part of the S-morph variability.

Second, many plants of the L-morph have a curled style, and the longer the style, the greater the number of plants with a curled style in a population (Box 5.1). Any advantage to having a longer than average style may be counterbalanced by costs associated with placing the stigma too far out of the flower. Indeed, as stigma exertion of the L-morph increases, I have observed a marked decline in the probability that a stigma will receive S-morph pollen and variation among insect groups in whether they cause pollination. There is also a significant correlation between maternal style length and seed and cotyledon size of young seedlings in the offspring of the L-morph (but not the S-morph). Such enhanced seedling vigour may result from some advantage to increased style length in the L-morph, perhaps via gametophytic selection (i.e. where more vigorous pollen outcompetes pollen of poorer viability for fertilization in flowers with longer than average styles) or correlated variation of other traits.

Why then do both morphs show variation in style length? We can see above that there may be divergent selection on style length in the L-morph. For the S-morph, however, it is unclear whether there is any cost or benefit associated with variation in style length. Quantitative variation in the S-morph may thus be neutral and maintained by a simple genetic correlation between style length in related individuals of the two morphs. In distylous species such as *J. fruticans* where each morph is only fertilized by the other morph, and because the genetic control of the polymorphism has a simple genetic basis (the L-morph is usually homozygous recessive '*ss*' and the S-morph heterozygous '*Ss*') each morph produces equal numbers of the two morphs in its offspring. This means that if style-length variation within a morph is due to the expression of quantitative genetic effects or modifier genes, then offspring of both L-morph and S-morph maternal parents will show both qualitative and quantitative style-length variability. In *J. fruticans*, Thompson and Dommée (2000) reported correlated variation in the mean style-length of the two morphs among populations.

In summary, the patterns of floral variation suggest that the spatio-temporal mosaic of variation in the pollinator cortège, coupled with variation in pollination efficiency of different groups, may create divergent selection on floral traits such as style length and stigma position, even within a floral morph. In the L-morph, having a longer than average style may provide an advantage (production of more vigorous offspring), however it is necessary to keep the stigma close to the mouth of the corolla in order to receive pollen. A long, but curled, style satisfies these conflicting selection pressures and may represent an 'adaptive compromise' to divergent selection pressures on floral design (see Armbruster 1996). In addition, a genetic correlation may maintain variation in both morphs. One last point here, in *J. fruticans* style-length variation occurs within the confines of a complex genetic polymorphism, which itself no doubt evolved in this genus (the tropical and cultivated flowers of this genus I have seen have a floral morphology which suggests distyly) prior to the Mediterranean-climate regime and thus in a very different pollination environment. Quantitative floral variation may thus have evolved in association with a fairly recent diversification of pollinators. Indeed, studies

Box 5.1 Floral biology of *Jasminum fruticans*.

Jasminium fruticans is a distylous species (Guitián *et al.* 1998; Thompson and Dommée 2000), that is, with two floral morphs

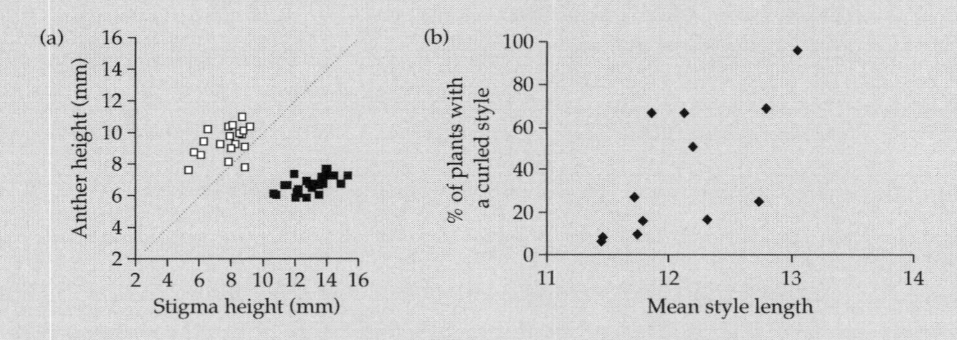

(a) Some short-styled plants (open squares) have stigmas adjacent to the anthers, while in long-styled plants all plants (filled squares) all plants have a marked stigma–anther separation and (b) many have a curled style. The degree of curling in 12 populations in southern France and Spain is positively related to the mean style length of the L-morph in a population.

of endemic *Viola cazorlensis* (C.M. Herrera 1990*b*,*c*; 1993; Plate 3) and *Arum italicum* (Méndez and Diaz 2001) in Spain produced no evidence that floral or inflorescence traits influence fitness.

Abiotic factors may also affect the functioning of plant–pollinator interactions (as the patch-dependent pollination of *L. latifolia* would suggest) and cause pollen limitation on fruit and seed production to occur and vary in space and time. An important factor here is climate unpredictability which can have a strong impact on pollinator activity. In *Narcissus longispathus*, which flowers early in spring when low temperatures may hinder bee activity and successful pollination, a favourable microclimate inside the large tubular flowers, a long floral lifetime (14-day-old flowers are still capable of high seed set), and the thermal biology of the principal bee pollinator (*Andrena bicolour* which raises thoracic temperature by basking in and on the corolla) are critical for pollen limitation to occur in this species (C.M. Herrera 1995*b*). In the Lamiaceae, nectar production is maximized at high temperature (Petanidou and Smets 1996) and species which flower late in spring have a higher nectar sugar concentration than earlier flowering species (Petanidou *et al.* 2000). In *N. assoanus* in southern France, local climate (late snow and heavy rain during peak flowering) reduces seed production to almost zero in upland populations on limestone plateau areas (>700 m) which contain tens of thousands of plants in some years. Indeed, several Mediterranean species show spatio-temporal heterogeneity in pollen limitation on seed production, for example, *N. assoanus* in France (Baker *et al.* 2000*a*) and dioecious *Rhamnus ludovici-salvatoris* on the Balearic Islands (Traveset *et al.* 2003). In *Helleborus foetidus*, pollinator visitation may be highly variable and infrequent because of early flowering in relation to pollinator abundance and/or climatic limitations (C.M. Herrera *et al.* 2001).

Several of the above examples suggest that pollinator-mediated selection may be rare. However, natural selection is notoriously difficult to detect. So in some ways the examples I discuss here illustrate more that pollinator-mediated selection may be very complex and that to study floral evolution requires precise, detailed and long-term evaluation of traits in relation to variation in diverse ecological parameters that create selection and constraints in natural situations. In perennial plants, establishing the precise role of traits in fitness variation and assessing the true contribution of single-year estimates of fertility is difficult due to demographic costs of reproduction and the complexity of untangling the role of different ecological factors (e.g. C.M. Herrera 1991). In addition, floral trait variation may be closely related to differential male fitness via an effect on pollen removal and transfer. So evidence for adaptation, and its absence, must be cautiously and carefully interpreted.

5.3 Attracting pollinators . . . but avoiding herbivores

In animal-pollinated plants, reproduction requires that pollinators be attracted to a flower. At the same time, however, plants must deter potential herbivores and seed predators, which may negate the benefits provided by mutualistic pollinators. Pollination and herbivory may rarely operate independently and their ecological effects and evolutionary significance may be closely interconnected, particularly when they affect the same traits (Strauss and Armbruster 1997). Indeed, attractive and defensive traits may be linked by function or because of phylogenetic constraints and some traits involved in defence may have subsequently evolved to attract pollinators (Armbruster *et al.* 1997). If herbivores affect traits which influence pollination, potential selection exerted by pollinators may be altered, masked, or rendered insignificant. Herbivory may thus alter the nature and strength of pollinator-mediated selection on floral traits and may exert quite different, even conflicting, selective pressures on reproductive traits. To fully understand how reproductive traits evolve thus requires that any non-additive (i.e. interaction) effects of herbivory (including predation of seeds and parasitism) and pollination be evaluated.

In the Mediterranean, insect diversity is high, hence the need to fully appreciate the combined effects of pollination and herbivory on reproductive success. In several Mediterranean species, pre-dispersal flower, fruit, and seed predation has been reported to have significant impact on fitness, and thus potentially constrain selection related to pollinators. In *Lavandula stoechas*, pre-dispersal seed predation by insects can account for a 31% reduction in seed set (J. Herrera 1991*c*). In *N. assoanus*, flower herbivory and fruit predation may cause up to 60% of plants in a given site in a given year to have no fruit production. In relict populations of *Frangula alnus* in southern Spain, floral herbivory strongly limits reproductive output (Hampe and Arroyo 2002; Hampe 2004). Finally, in five *Onopordum* species in Greece, 34 out of 46 populations have reduced seed production due to infestation by the weevil *Larinus latus* (Briese 2000). In endemic *Ebenus* on the Mediterranean coast of Egypt and Libya, ~90% of seeds are consumed or damaged by a bruchid beetle (Hegazy and Eesa 1991). Post-dispersal seed predation is also very important in several Mediterranean woody species (Hulme 1997; Quézel and Médail 2003).

The balance of positive affects related to pollination and the negative affects of herbivory have been studied in a number of Mediterranean plants. Studies of the interactive effects of herbivory and pollination on seed production in a population of *Paeonia broteroi* (Plate 3; C.M. Herrera 2000*c*) and the recruitment of seedlings by maternal plants of *H. foetidus* in two different regions (C.M. Herrera *et al.* 2002) have revealed such non-additive fitness consequences of exposure to herbivory and pollination with two important trends (depicted schematically in Fig. 5.4). First, pollination has a positive effect on seed production, seedling emergence, and recruitment only if herbivores are excluded. Second, the exclusion of pollinators renders the negative effect of herbivores on seedling recruitment non-significant. In other studies, the effects have been found to be additive in nature since herbivores exert a negative effect on performance even in the absence of pollinators, due to the capacity of the studied species to autonomously self-pollinate (e.g. Box 5.2).

In the montane herb *Erysimum mediohispanicum* (Plate 3), Gómez (2003) studied whether pollinator-mediated selection on floral traits is modified by the effects of insect and ungulate herbivory. This monocarpic perennial herb occurs from

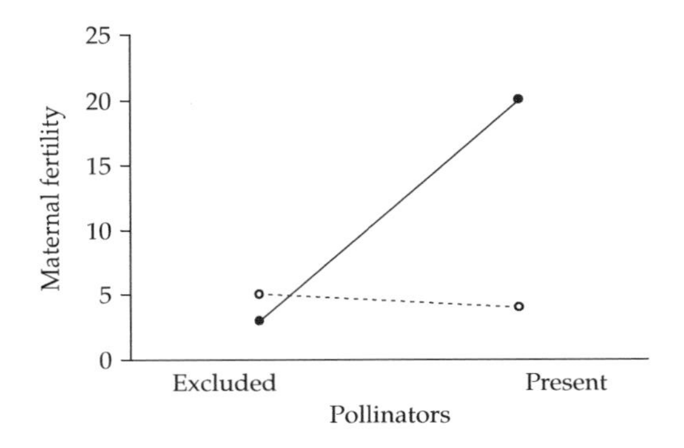

Figure 5.4 Schematic representation of the interactive effects of pollination and herbivory on maternal fitness (seed production, seedling emergence, and seedling recruitment) in *P. broteroi* and *H. foetidus* populations in Spain (based on C.M. Herrera 2000 and C.M. Herrera *et al.* 2002, respectively). Open symbols: herbivores present, closed sysmbols: herbivores excluded.

1,000 to 2,000 m above sea level in several mountain ranges in southern and eastern Spain. In the Sierra Nevada, plants grow as perennial rosettes for up to ~4 years, after which they produce many reproductive stems bearing up to several hundred bright-yellow hermaphroditic flowers. In the Sierra Nevada, *E. mediohispanicum* is primarily pollinated by the pollen collecting beetle *Meligethes maurus* (Nitulidae), has its flowers and fruits browsed by the Spanish Ibex (*Capra pyrenaica*), its sap sucked by several bugs (*Corimeris denticulate* and *Eurydema* subsp.), and its stems and fruits bored by weevils. Gómez (2003) reported two results indicative of potential pollinator mediated selection in the absence of ungulate herbivores (the major herbivores on this species). First, lifetime maternal reproductive success (easily measured since although plants are perennial they die after one bout of flowering) was positively correlated with pollinator visitation rates. Second, several traits influenced pollinator visitation rates, indicative of directional selection on these traits. For instance, plants with more flowers had higher absolute (total seed number per plant) and relative (seed number per fruit) seed production. Positive directional selection was also detected for flowering stalk height and two floral traits (petal length and inner diameter). In the presence of ungulate herbivory, selection on flower number was markedly reduced and selection on

Box 5.2 A comparative study of herbivory and pollination in natural populations of two congeneric *Aquilegia* in southern France (Plate 3) (graphs drawn from data in Lavergne *et al*. (2004a)

Aquilegia viscosa is endemic to the Languedoc Roussillon region of southern France and north-east Spain and its congener *A. vulgaris* is widespread across western Europe. In these two species floral predation (primarily Curculionideae larvae, which cause major damage to floral buds and prevent flowering, and various Diptera and Lepidoptera) and pollen limitation (pollination is essentially by bumblebees) thus combine to limit the maternal fertility of the endemic species but not the congeneric widespread species in the same region.

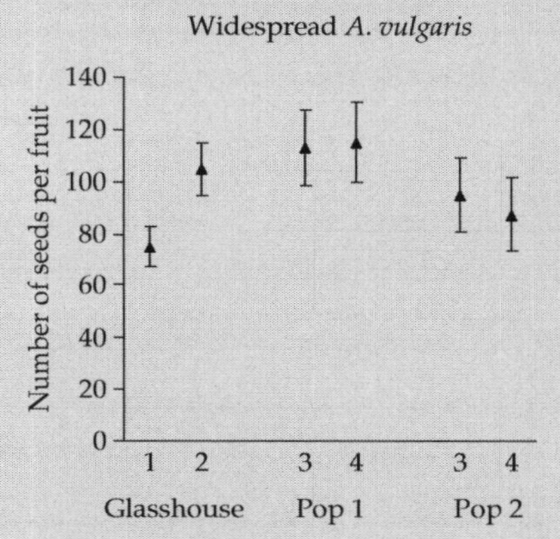

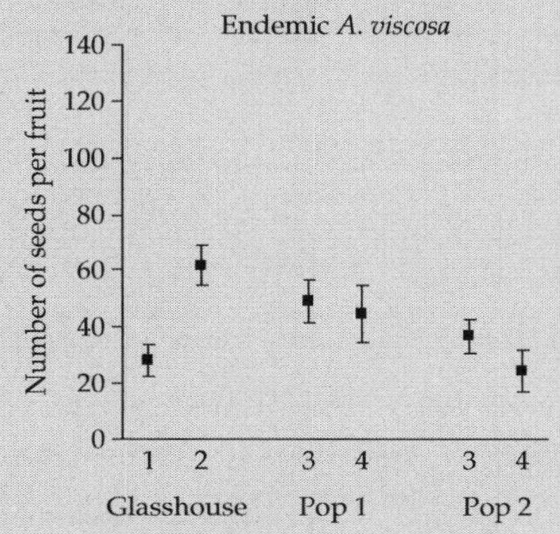

1. Absence of predators and pollinators
2. Absence of predators with outcross pollination
3. Predator exclusion and open to pollination
4. Open to predation and pollination

Herbivory significantly limits reproductive success (by up to 56%) in the endemic species but not in the widespread species. In unpredated flowers, seed set in the endemic species remains significantly less than that of the widespread species. Given that in the absence of both pollinators and herbivory (i.e. in the glasshouse) the two species show equivalent levels of autonomous self-pollination compared to seed set on outcrossing and because pollinator visitation rates are significantly higher in populations of the widespread species than in populations of the endemic species (where pollinators are extremely rare), it would appear that pollen limitation also constrains female fertility in the endemic species. The ability to attract pollinators, but avoid floral herbivores, both contribute to female fertility and perhaps influence the differences in distribution of the two species. In endemic *A. viscosa*, herbivory imposes a strong constraint on maternal fertility even in the absence of active pollination due to autonomous self-pollination which provides some reproductive assurance in the absence of pollinators. Herbivory may thus continue to limit maternal fertility even where pollinators are rare.

floral traits became non-existent, that is, as in the above studies, the detrimental effect of herbivory on maternal fecundity was so strong that it cancelled out any pollinator mediated selection on particular traits. In fact, whereas tall plants with many flowers are strongly favoured in the absences of herbivores, they are counter-selected when ungulate herbivores are present. These two opposing selective forces may limit the evolution of adaptive variation in flower number.

In the high mountain zone of the Sierra Nevada, the mass-flowering crucifer *Hormatophylla spinosa* (Plate 3) shows significant variation in traits related to how it interacts simultaneously with a number of antagonistic herbivores and seed predators and mutualistic pollinators (Gómez and Zamora 2000; Box 5.3). These authors found that only ungulates (mainly Spanish Ibex) and pollinators significantly affect plant fitness and the relative effects of these interacting organisms varied greatly among study sites—indicative of functional specialization within populations. In two populations, ungulate pressure was most intense and significantly reduced performance. In the population where this effect was most severe, plants had a significantly higher number of thorns. In contrast, where ungulates were less abundant, and thus caused less damage to plants, maternal fitness was closely correlated with pollinator visitation. As above, floral traits only significantly contributed to performance variation (via pollinator attraction) in sites where herbivore pressures were non-significant. So in this species, only a highly restricted sample of all the potential interacting organisms is likely to exert a selection pressure on plant traits in a given population.

Ultimately, plants with traits that enhance pollination and allow some form of defence against or escape from herbivory should be selected. However, the frequent occurrence of herbivory on reproductive structures of herbaceous Mediterranean plants suggests that such interactive effects may commonly affect plant reproduction in this region and perhaps elsewhere (C.M. Herrera 2000*c*; C.M. Herrera *et al.* 2002). Once again, the message is clear, to fully quantify and understand the role of pollinator-mediated selection requires appreciation of the range of ecological factors influencing reproductive success in natural environments.

5.4 Mating system and gender variation

5.4.1 Mating system variation and evolution

Most plant species are hermaphroditic (co-sexual), with flowers that contain both male (stamens) and female (pistils) sex organs. Co-sexuality is favoured in species with high resource allocation to attractive structures that are shared by male and female function, particularly if they suffer pollen limitation. Hermaphrodism nevertheless comes with a cost. This cost may be of a genetic nature if inbreeding occurs and progeny suffer reduced viability (inbreeding depression), or more ecological in nature if in self-incompatible species self-pollination causes reduced mating opportunities (i.e. self-interference).

Despite the prevalence of hermaphroditism, in flowering plants there is an immense diversity of mating strategies. Between the extremes of 'perfect' hermaphrodite flowers and unisexuality (dioecy) lie a wide range of intermediate types, the most prominent of which include:

- monoecy: male and female flowers on the same plant;
- andromonoecy: male and hermaphrodite flowers on the same plant;
- gynomonoecy: female and hermaphrodite flowers on the same plant;
- gynodioecy: the coexistence of female and hermaphrodite plants;
- androdioecy: the coexistence of male and hermaphrodite plants.

These different sexual systems combine variation among flowers/inflorescences on a single individual and among plants in a population.

Even within hermaphrodites there can be much variation in the mating system. First, some species may be almost completely outcrossing, while others may rely on selfing to assure seed production, and others may have a mixed mating system with intermediate, and often variable, rates of selfing and outcrossing (Barrett 2002*a*). Second, hermaphrodites may vary in their relative allocation of resources to male and female function as a result of environmental and ontogenetic causes (Lloyd and Bawa 1984).

The occurrence of variation in plant sexual systems entered early into the literature on plant evolution. There thus exists a long tradition of population genetic studies of the evolution of the mating system and the relative effects of selfing and outcrossing on genetic variation which has been paralleled by the growth of pollination ecology. In the last 15 years these two fields have gradually merged to produce a new synthesis in plant evolutionary biology whose central theme is the ecology and evolution of plant

Box 5.3 The selction mosaic of interactions in *Hormathophylla spinosa* positive impacts are represented by solid lines; negative impacts by broken lines (redrawn from Gómez and Zamora 2000), the bolder the line, the stronger the interaction

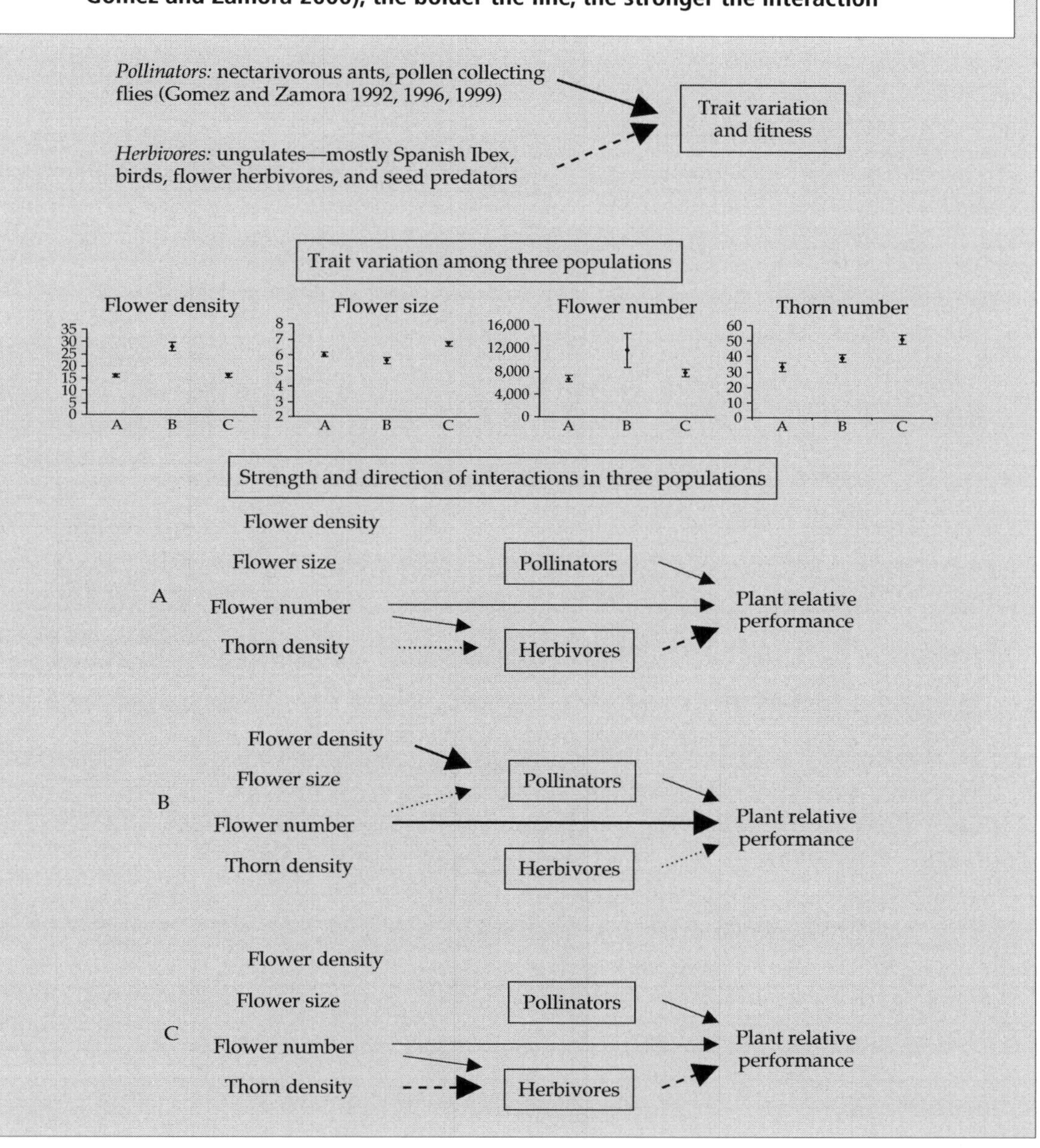

reproduction. The evolution of plant reproduction is thus not only where adaptation and diversity meet, but also where an integrative approach based on ecology and genetics is necessary. In addition, stochastic factors associated with population history and colonization dynamics can closely interact with natural selection (due to environmental differences in resource status and other ecological parameters) to shape patterns of variation and the evolution of genetic polymorphism.

5.4.2 From outcrossing to selfing

'It is well established that, in certain circumstances, outbreeding and the consequent promotion of genetical variation are adaptively advantageous while, in other circumstances, the genetical fixity and reliable seed formation made possible by inbreeding fit the needs of a population more appropriately' (Baker 1966: 349). Many species have thus 'abandoned' outcrossing for a highly autogamous mating strategy and there has been a great deal of interest in the causes and biological consequences of the evolutionary shift from outcrossing to selfing, all the more so because of the frequent occurrence of this change (Barrett 2002*a*).

The evolutionary shift from outcrossing to selfing depends on four main factors, two of them genetic in nature (gene transmission and inbreeding depression), the other two ecological (reproductive assurance and pollen wastage due to self-pollination). First, because inbreeding species transmit genes to their offspring by both female and male function whereas outcrossing plants only contribute one copy of their genes to each offspring, selfers have a genetic transmission advantage. However inbreeding is often accompanied by a reduced viability of the offspring. If such inbreeding depression is low then selfing is favoured, whereas high levels of inbreeding depression should act to maintain outcrossing (Husband and Schemske 1996). The expression of inbreeding depression can also depend on local ecological conditions. Since environmental stress is likely to have a greater than average impact on individuals with inherently low vigour, a stressful environment may increase the magnitude of inbreeding depression.

The possibility that inbreeding depression may vary among populations is particularly pertinent in the Mediterranean habitat mosaic where ecological conditions vary dramatically in a localized manner (Chapter 4). Studies of *Crepis sancta*, a ruderal annual common in old-fields, vineyards, and along roadsides, where it adds a vivid splash of yellow to the landscape in spring, illustrate how diverse ecological conditions can affect the evolution of the mating system. In a study of several populations from Mediterranean old-fields of different successional stage, both drought stress (Cheptou *et al.* 2000*a*) and increased competition (Cheptou *et al.* 2000*b*) have been reported to cause the magnitude of inbreeding depression to increase, creating a strong selection pressure against the evolution of inbreeding. This may underlie why, in this species, selfing is significant in young successional populations where competition is weak but not in older populations, which are almost exclusively outcrossing (Cheptou *et al.* 2002), and where competition from perennial grasses is important feature of the environment. Spatial heterogeneity of abiotic and biotic environmental conditions in the mosaic of Mediterranean old-fields may thus create the conditions for divergent selection on the mating system of pioneer plant species. The frequent episodes of population extinction and re-colonization that are part of the population ecology of such pioneer species provide another illustration of meta-population function in which environmental conditions vary among patches, in this case as a result of successional habitat variation. Indeed, Cheptou *et al.* (2002: 753) conclude that mating system variation in this species may result from a 'balance between the cost of outcrossing and inbreeding depression in a metapopulation context'.

Increased selfing rates and levels of inbreeding are frequently accompanied by reduced pollen–ovule ratio (i.e. reduced investment in male function) and smaller flowers with less pigmentation, shorter floral longevity, and a greater facility for autonomous self-pollination due to reduced stigma–anther separation in selfing species. This suite of floral traits has been reported to be a general feature of endemic species relative to their more widespread congeners in the western Mediterranean (Lavergne *et al.* 2004*b*) and among outcrossing *Lactuca viminea* subsp. *chondrilliflora* and selfing *L. viminea* subsp. *ramosissima* in the Iberian peninsula (Mejías 1994). Autonomous selfing in small flowers with stigmas close to the anthers can provide an advantage over outcrossing plants that rely on pollen vectors for successful pollination. This 'reproductive assurance' when pollinators are rare or absent represents a positive selection pressure on traits favouring selfing if pollinator servicing is consistently and dramatically low, especially in annual plants. Such reproductive assurance has often been proposed to represent an important feature of colonization ability and if you delve into the literature you will often see this idea referred to as 'Baker's Law' due to its origins in the early writings of the plant evolutionary ecologist Herbert Baker.

In the Mediterranean region, the colonization of new regions and isolation on islands may be associated with reduced pollinator visitation, providing a context for the evolution of selfing by virtue of the reproductive assurance it provides. There is evidence for this idea. First, *Anchusa crispa*, a species which is endemic to sand dunes on Corsica and Sardinia where it occurs in small and patchy populations, is capable of autonomous self-pollination due to the proximity of stigmas and anthers in its small blue flowers (Quilichini *et al.* 2001) and is homozygous at all studied isozyme loci (Quilichini *et al.* 2004), indicative of a highly inbred mating system. Second, western Mediterranean *Cyclamen*, where autonomous self-pollination in *Cyclamen balearicum* appears to have evolved at the range limits of the species in this genus and in habitats where pollination by bees and bumblebees is rare (Chapter 3). In other species traits allowing reproductive assurance have not evolved to replace the need for pollinators. *Medicago citrina*, which has a typical bee-flower, occurs on the Columbretes archipelago (between the Balearic islands and mainland Spain) and on some small islands in the Balearic archipelago, where bees are absent and seed production in natural sites is low compared to that after hand-pollination, indicative of pollen limitation (Pérez-Bañon *et al.* 2003). In the absence of bees, pollination on these islands is assured by occasional visitation by hoverflies (*Eristalis tenax*) and bowflies (*Calliphora vicina*). Nonetheless, the unusually long flowering period of this species (January–May) and the longevity of individual flowers may represent a response to the uncertain pollination environment in which this species occurs. In *H. foetidus* in the Iberian peninsula, northern populations, which have the greatest capacity for autonomous self-pollination also had the highest rates of pollinator visitation during two consecutive years. This result goes against the reproductive assurance hypothesis, suggesting that historical selection has moulded trait variation in a way which does not fit with the contemporary pollination environment (C.M. Herrera *et al.* 2001).

These examples from the Mediterranean flora serve as illustrations of the factors involved in the balance between outcrossing and selfing (I discussed the role of polyploidy in Chapter 3). The only one of the four main selective factors influencing the evolution of selfing that has not been studied in a Mediterranean plant is 'pollen discounting', which represents an ecological cost to self-pollination by reducing the amount of pollen available for outcrossing.

5.4.3 Sex and size in hermaphrodite plants

'The concept that the sex of a seed plant may be adaptively modified by its circumstances has a long history in the botanical literature' (Lloyd and Bawa 1984: 255). However, it is only since roughly the time that these authors made this statement that detailed attempts to understand the adaptive significance of gender variation have been made. The modular construction and flexibility in resource allocation to reproductive functions that characterize many flowering plants mean that gender variation may commonly occur, particularly in relation to size and resource status.

The theory of sex allocation in hermaphrodites requires that resources allocated to reproduction are fixed and partitioned between male and female function (Charnov 1982). As a result, allocation to one sexual function may be compromised by allocation to a different function, what is termed a trade-off. Hence, in unisexual species, where individual plants have either only a male or female function, a plant may compensate the absence of one function by increased allocation to the single function it performs. In hermaphrodites, trade-offs and compensatory effects may regulate gender expression and variation such that hermaphrodite plants do not allocate equally to male and female function. Indeed, it has been argued that individual hermaphrodite plants may rarely have a functional gender based on equal contributions to the next generation via male and female function (Lloyd and Bawa 1984). Ecological and ontogenetic factors may greatly contribute to such variation by affecting the resource status of individuals as they grow and age in natural populations. An important prediction here is that as hermaphrodite plants increase in size (due to age and/or resource availability) their functional gender will become biased towards female function, the so-called 'size-advantage hypothesis' (e.g. Wright and Barrett 1999).

A study of gender variation in relation to plant size in the Balearic endemic peony, *Paeonia cambessedesii* illustrates this bias towards femaleness with increasing plant size (Méndez and Traveset 2003). Although allocation to both male (number of stamens) and female (number of ovules and number of seeds) functions are positively correlated with plant size, seed production shows a stronger increase than stamen number, hence gender becomes female-biased with increasing plant size (at least in terms of size of individual ramets). In many plants, particularly those which have a subterranean storage organ with few flowers per stem (such as these peonies) flower size can be positively correlated with plant size due to the positive effect of resource status on resource allocation to different functions, as reported for Mediterranean *Narcissus* (Worley *et al.* 2000). Hence, gender variation at the plant level may also be manifest at the flower level. In *P. cambessedesii*, there is however a partial uncoupling of plant gender from flower size since there is no variation in gender expression with increasing flower size, despite the weak correlation between plant size and flower size. Hence, in this species, gender variation may have an adaptive function and is not simply the result of an ontogenetic constraint associated with differences in flower size. In other Mediterranean examples the relationship is more complex. For example, Méndez (1998) reported variation in floral sex ratio in *Arum* which followed a pattern of increasing femaleness of later produced inflorescences. This shift in gender was not correlated with inflorescence size or number since plants with a single inflorescence could be male, female, or even functionally sterile. In a subsequent study, Méndez (2001) found that an increase in inflorescence mass was related to a disproportionately greater increase in biomass allocation to male rather than female flowers. This trend did not however translate into a more male-biased functional gender (i.e. in terms of gamete production). Hence it is important to understand how changes in size relate to variation in the number and size of the flowers, and how size-dependent changes translate into functional modifications in gender.

A strong effect of resource status on plant size and thus the expression of size-dependent gender modification may mask the expression of trade-offs between the different sexual functions performed by a flower. In Mediterranean *Narcissus*, Worley *et al.* (2000) showed that although flower size and number show a significant negative relation, and thus evidence for a trade-off, among different species, no such trade-off was observed among plants in one species (*Narcissus dubius*) with a highly variable flower number per inflorescence. Several causes may limit the expression of this trade-off within species, for example, resource status and environmental variation may mask such a trade-off due to the positive relationship between vegetative (bulb) size and both flower size and number. In addition, high levels of genetic divergence in flower size may be necessary for trade-offs with flower number to become apparent.

Plant size is a critical element of the reproductive ecology of clonal plants, due to the spatial spread of individual clones. In plants capable of clonal growth, reproduction and regeneration strategies may vary in relation to spatial heterogeneity of Mediterranean habitats. For example, in *Cornus sanguinea* in three different habitats in southern France, Krüsi and Debussche (1988) reported that although a population in an abandoned olive grove showed significantly higher rates of flowering and fruiting compared to trees in forest interior or on forest edges, the percentage cover of this species was greater in the latter habitat. In the closed habitat, clone size was inferred to be three times that in the open habitat. Hence the reproductive strategy of this species is closely related to habitat conditions, with abundant sexual reproduction and seed dispersal in open habitats and reliance on clonal growth in closed forest habitats. This result has a distinct similarity to the reliance on vegetative persistence in *Paeonia* species subject to forest closure in many areas (discussed in Chapter 4).

Clonality facilitates the display of massive numbers of flowers. Although this floral display may be beneficial it is not without important reproductive consequences, both in animal- and wind-pollinated species, as the two following examples illustrate.

As I mentioned earlier in this chapter, in the insect-pollinated climbing shrub *J. fruticans*, the most important plant trait influencing pollinator visitation rate to a given stem is the number of open flowers in the surrounding patch (Thompson 2001). The bigger and more floriferous the clone, the more pollinators visit the plant. However, in large patches, many visits will be unsuccessful due to the self-incompatible nature of the species and the fact that the proportion of overall flowers visited on a stem (and thus in the patch) actually declines. So in a highly patchy mosaic environment, where individual clones occupy large patches, fruit and seed production, as well as pollen dispersal out of the patch (necessary for it to contribute to reproduction in this self-incompatible species) can be limited, despite enhanced pollinator visitation rates.

A size-dependent limitation on reproduction may be even more severe in wind-pollinated species, which do not have the benefit of increased pollinator attendance in large patches. An example of the negative effect of clone size on seed set has been reported in the self-incompatible clonal macrophyte *Scirpus maritimus* in the natural reserve of Roquehaute in southern France, where this species occurs in a dozen or so small and discrete temporary marshes (Charpentier *et al.* 2000). Each marsh occurs in a small depression of the basaltic plateau on which the reserve occurs and different marshes are separated by only tens of metres but by dense shrub vegetation. At this site, a study of 12 small marshes (which vary in surface area from ~10 to ~400 m^2) showed that seed production barely exceeds a single akene per spikelet in the different ponds. Pollen supplementation in five of these ponds revealed a significant increase in maternal fertility after among-pond pollination but not after within-pond pollination, indicative that local populations suffer from a deficit in outcross pollination. It would appear that the colonization of ponds is an almost one-off event, once a clone arrives in a pond it spreads to occupy all available space. As a result, there is a lack of compatible pollen to fertilize ovules. The landscape of distinct ponds is thus akin to a spatial mosaic of genetically distinct clones, each occupying the best part, or perhaps all, of a given patch, within which sexual reproduction is limited by an absence of compatible pollen. So even in a highly localized system of marshes with short spatial distances among pools, there is sufficient isolation to cause reduced fertility of this wind-pollinated plant.

5.4.4 Gender variation and sexual dimorphism: conceptual basis

Species which show gender polymorphism provide useful model systems for the study of the ecological and genetic factors influencing plant reproduction. Although less than 10% of flowering plant species contain unisexual individuals, the evolutionary transition from hermaphrodism to dioecy has repeatedly occurred in diverse families. The evolution of dioecy requires the establishment of sterility mutations in natural populations and that unisexual plants have a fitness advantage relative to hermaphrodites. This evolution is thought to occur by two main pathways (see recent discussion in Barrett 2002*a*) which differ in their intermediate stage: one which passes through a gynodioecious stage, the other via monoecy. The rare breeding system known as androdioecy is probably a reversion from dioecy back towards hermaphrodism. Four main factors govern the establishment and persistence of gender polymorphisms in flowering plants:

1. The fitness consequences of selfing versus outcrossing: the combination of significant selfing and inbreeding depression is of critical importance for the spread of sterility mutations, particularly male sterility, since together these two factors give rise to what is termed the 'outcrossing advantage'.

2. Trade-offs and resource allocation to male and female function: by specializing on a single sex function, plants may compensate for the absence of the opposing function and thus increase their fitness.

3. Frequency-dependent selection: the reproductive success of different morphs (or mating types) may depend on the relative frequency of other sexual phenotypes with which it can reproduce. Frequency-dependent selection usually involves a negative correlation between frequency and fitness (rare morph has higher fitness than abundant morphs) and thus creates conditions for the maintenance of a polymorphism.

4. The genetic basis of female and male sterility and the migration of these genes among populations.

5.4.5 Evolutionary transitions in gender: monoecy, dioecy, . . . , and androdioecy

The theme for this section concerns among population variation in gender dimorphism. In particular, I illustrate cases of geographic variation among closely related species and localized population variation within polymorphic species.

Variation in sexual systems is well documented in *Ecballium elaterium* (Costich and Galán 1988). These authors describe a dioecious variety *'dioicum'* which occurs in drier, hotter localities than a monoecious form *'elaterium'* (Fig. 5.5). This pattern of variation fits the theoretical prediction for the evolution towards unisexuality in stressful environments as a result of either high levels of inbreeding depression or selection on sexual specialization. In a transplant experiment encompassing the two different areas of Spain where the two varieties occur, Costich (1995) reported that the dioecious plants have a higher investment in either male function or female function compared to co-sexual plants. In a nutshell, males have a greater number of flowers and females have higher seed set than monoecious plants. Such gender specialization illustrates how unisexual individuals benefit from allocating resources to a single sex function, providing evidence of reproductive compensation. In drier sites, co-sexuals show a male-biased gender, indicating that selection on gender specialization may occur due to environmental stress, particularly aridity, which more strongly limits female function than male function (see Barrett *et al.* 1999 for a similar example

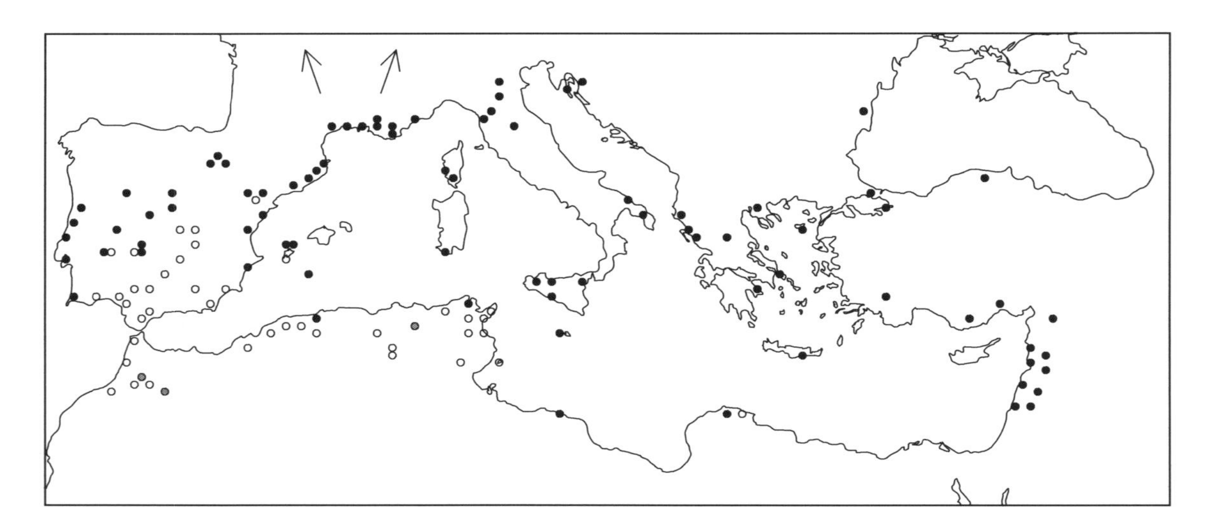

Figure 5.5 Distribution of dioecious (open circles) and monoecious (closed circles) populations of *E. elaterium* in the Mediterranean region (redrawn from Costich and Meagher 1992). Grey circles indicate mixed populations. Arrows indicate that monoecious populations continue north-west into non-Mediterranean France.

in the Mediterranean-climate region of south-west Australia). Drought stress may thus provide a strong selection pressure on plant reproductive strategies.

If environmental limitations were the only cause one would expect to see dioecy in at least some eastern Mediterranean sites. Pérez Chiscano (1985) argued that *E. elaterium* has spread westwards across the Mediterranean in two waves of migration, one in the north (*elaterium*) and one (*dioicum*) across the south to North Africa and subsequently into southern Spain. The distribution pattern (Fig. 5.5) suggests a principal role of land-bridge connections from Sicily to Tunisia in this colonization process and that geographic isolation following the loss of this connection may have been involved in the evolution of dioecy. The genetic analyses of Costich and Meagher (1992) support the idea of a secondary contact zone between dioecious and monoecious types as a result of independent colonization of the Iberian peninsula by the two varieties (and not the evolution of dioecy from monoecy in this zone). The spatial pattern of sexual system variation and ecological specialization should thus be interpreted within the context of the history of plant migration in this species.

Geographic variation in gender dimorphism occurs in the annual *Mercurialis annua* (Box 5.4) in which widespread diploids in temperate Europe are dioecious and the diverse Mediterranean polyploids are primarily monoecious and self-compatible, and thus subject to some degree of inbreeding (Durand 1963). This study contains an intriguing report of populations with male individuals alongside monoecious (hermaphrodite) individuals, that is, an illustration of androdioecy, a rare breeding system in plants. Rare, because the conditions for its maintenance are strict since males (to persist) must fertilize more than twice the number of ovules fertilized by hermaphrodites. This stringent condition is made even stricter if hermaphrodites are self-compatible. Charlesworth (1984) challenged previous descriptions of androdioecy, suggesting that in most species with a morphological expression of androdioecy, hermaphrodites lack a male function, rendering the species functionally dioecious. In other words, to more precisely evaluate plant sexual systems, functional features of reproduction are more informative than morphological traits. Notwithstanding, three examples from the Mediterranean illustrate that androdioecy does indeed occur, and that its maintenance may depend on a complex interaction between environmental and genetic effects on sexual resource allocation and the population ecology of the species.

First, Pannell (1997*a*) has confirmed that androdioecy occurs in populations of *M. annua* in southern Spain and Morocco. In this species, males have no seed production and hermaphrodites reproduce via both sexual functions, although a large majority of their pollen may be used in selfing. Males invest 10-fold more resources in pollen production than hermaphrodites and the relative frequencies of males (up to 30% of a population) shows marked variation among and within populations (Pannell 1997*b*). Finally, based on the analysis of progeny from a crossing experiment, Pannell (1997*c*) reported that the determination of sex expression in this species may be quite simple. Maleness appears to be determined by the presence of a dominant allele at a single nuclear locus, with hermaphrodites homozygous for that locus. This means that in the absence of a new mutation a population with no male individuals can only become androdioecious if males (or male genes) immigrate into the population. Pannell (1997*c*) also reported lability in the sex determination system of this species, males may switch to hermaphrodism in benign conditions. Sex expression in this species thus has both a genetic basis and an environmental component. Throughout this series of papers, Pannell developed two main arguments for the maintenance of androdioecy in *M. annua*. First, the frequent occurrence of colonization in an annual species will favour self-fertile hermaphrodites which provide reproductive assurance and a better guarantee on population establishment and persistence. Second, at higher densities in older populations, reproductive assurance will lose its primary advantage and the outcrossing nature of males and their high pollen production and greater facility for pollen export (their inflorescences are taller than those of male flowers on hermaphrodite plants) will favour their maintenance. The population ecology of the species

Box 5.4 Population variation in sex expression in hexaploid *Mercurialis annua* (redrawn from Durand (1963))

Although most populations in the Iberian peninsula and Morocco contain only monoecious plants (which occur in all studied populations), some populations, especially in Morocco, contain either completely male or completely female plants or monoecious plants with very few or reduced male flowers or very few female flowers.

In fact, the sexual system of a population can take one of several forms: monoecy is the most widespread sexual system, while some populations are almost dioecious, gynomonoecy, andromonoecy, androdioecy, and trioecy also occur.

Each circle gives the relative proportion of females (open), males (filled), monoecious (open dots), and various forms of monoecious plants with reduced female or male function (grey portion).

Such patterns suggest quantitative gender variation (Lloyd and Bawa 1984).

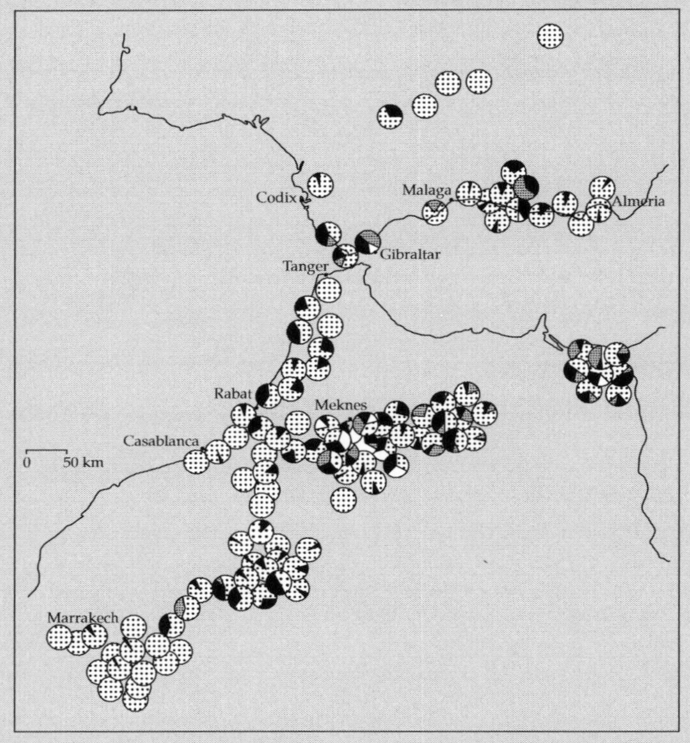

The only sexual system absent at the level of the population is gynodioecy, since whenever females occur in a population, males are also always present, indicative that when females are present they cause strong selection on male function, and thus the frequency of male plants (Lloyd 1976). Hexaploids occur in smaller and more isolated populations than their diploid (dioecious) relatives (Durand 1963), hence, inbreeding and periodic reductions in population size and isolation may have created the conditions for random deviations among populations in morph frequency. In *Ecballium elaterium*, which like *M. annua* occurs primarily in ruderal habitats disturbed by human activities in the Mediterranean region, monoecious populations are more inbred than dioecious populations (Costich and Meagher 1992).

and the spatial mosaic of available habitats may thus be critical for the evolution of the breeding system in the Mediterranean hexaploid cytotype of this species.

The reproductive traits of two Mediterranean Oleaceae also suggest the occurrence of androdioecy. The first of these is the wind-pollinated, fleshy-fruited shrub (sometimes a small tree) *Phillyrea angustifolia*, which is androdioecious in southern France (Lepart and Dommée 1992), south-east Spain and the Balearic islands (Traveset 1994*b*), and south-west Spain and Portugal (Pannell and Ojeda 2000). In southern France, Lepart and Dommée (1992) reported that crosses among hermaphrodites in one population produced no seed and the occurrence of 1 : 1 sex ratios (i.e. a much greater proportion of males than one would expect). These results indicate that some populations may be functionally dioecious. However in one population in southern France, Vassiliadis *et al.* (2002) demonstrated, using paternity analysis of progeny from open-pollinated mother plants, that pollen from hermaphrodites has a high rate of participation in the fertilization of ovules. Hermaphrodites also regularly set seed and are thus functionally hermaphrodite and this species is thus functionally androdioecious, even in sites where male frequencies attain ~50%. Elsewhere in the distribution, males are less common, usually at frequencies <30% (Traveset 1994*b*; Pannell and Ojeda 2000), in accordance with theoretical predictions. In this species, males may obtain a male-fertility advantage by flowering more regularly among years than do hermaphrodites (Pannell and Ojeda 2000). This is a typical result for dioecious species, where the resource cost of maternal investment in fruits and seeds reduces the frequency of flowering of females. Males of *P. angustifolia* may also have (an almost twofold) greater fertilization success in some conditions and hermaphrodites are self-incompatible (Vassiliadis *et al.* 2000*a*). The latter result is important because theoretical studies indicate that a genetic linkage between the self-incompatibility locus and genes controlling female sterility may allow for the spread of males (Vassiliadis *et al.* 2000*b*). This hypothesis and the precise nature of any male fitness advantage await rigorous experimental demonstration.

A second example of androdioecy in Mediterranean Oleaceae can be seen in the flowering ash *Fraxinus ornus*. In southern France, hermaphrodites of this species produce viable pollen in dehiscent anthers and can be self-compatible whereas males have no female function (Dommée *et al.* 1999). In terms of functional gamete production this species is thus androdioecious. However, as in *P. angustifolia* some populations have 1 : 1 sex ratios, a feature one would expect to see in functionally dioecious species (Charlesworth 1984). However, the analysis of physiological traits indicates that males are more drought tolerant than hermaphrodites and produce 1.6 times more inflorescences (Verdú 2004), there may thus be an advantage to being male. The male contribution of hermaphrodites to seed offspring nevertheless appears to be low, indicative that this species is cryptically dioecious (Verdú *et al.* 2004). Since many species are dioecious or polygamodioecious in the genus *Fraxinus*, androdioecy is probably an evolutionary reversion from dioecy.

To wrap up this section on gender polymorphism I will describe a rather unique case of sexual polymorphism which involves sex separation both among and within individuals. This occurs in the essentially Mediterranean genus *Thymelaea* (in which species are frequently narrow endemics to localized regions). In a systematic revision of the genus, Tan (1980) mentions a diversity of sexual systems in this genus. These range from hermaphrodism in a single species, the circum-Mediterranean *Thymelaea passerina*, through monoecy (in ~7 species) to dioecy (~15 species), with the possible occurrence of androdioecy, gynodioecy, and andromonoecy. As described by Dommée *et al.* (1988), the dioecious and monoecious species have different geographical distributions. Dioecy occurs in species in the Iberian peninsula and southern France, Turkey and the Middle East, and in the western part of North Africa, and monoecy along the northern shores of Morocco, Algeria, and Tunisia. In small areas on either side of the Straits of Gibraltar, one can observe species with both these sexual systems. *T. passerina*, the only hermaphrodite species in the genus, has a widespread circum-Mediterranean distribution which extends into the Irano-Turanian and Pontic floristic zones.

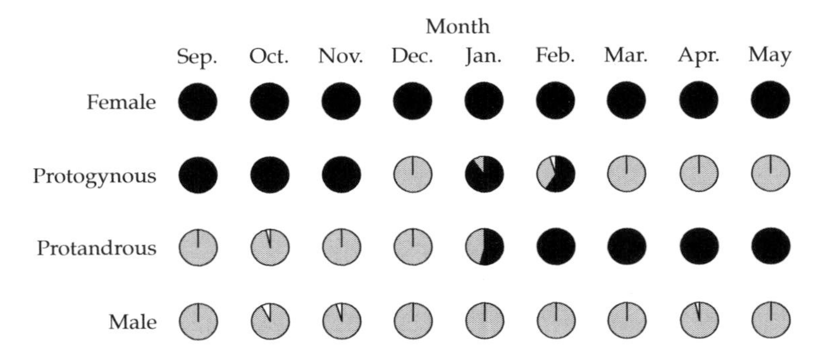

Figure 5.6 Sexual variation in *T. hirsuta* (redrawn from Dommée *et al.* 1988). Each circle contains the proportion of male (hatched sectors), female (black sectors) and rare hermaphrodite (open sectors) flowers on the four different sexual phenotypes of *T. hirsuta* in a population in southern France during the flowering period of this species.

An intriguing feature of dioecy and monoecy in this genus is that they co-occur in a complex polymorphism, often called trioecy, in at least one species, *Thymelaea hirsuta*. Long-term studies of plants in natural populations in southern France and offspring grown in garden conditions (Dommée *et al.* 1990, 1995) have shown that the sexual system involves four sexual morphs: males, females, and two types of monoecious plants which switch sex (Fig. 5.6). The latter are either protandrous at the level of the whole plant, that is, they bear only male flowers in the autumn then switch sex to produce female flowers in early spring, or protogynous (again at the whole plant level), bearing only female flowers in autumn and male flowers in spring. The switch, referred to elsewhere as a form of hetero-dichogamy, occurs during a short transition period when days start to get longer and warmer in January–February. At this time one can observe both male and female flowers on some plants as they switch sex. Dommée *et al.* (1990) report that plants of the protandrous morph have a greater seed production than those of the protogynous morph, indicative of female gender specialization in the former. There is thus a fairly strong trend towards functional dioecy in these populations; indeed on a given date, populations are almost always functionally dioecious. Gender polymorphism and lability are thus a constant and genetically determined feature of the sexual system of this species, which may represent a transitional stage in the evolution of strict dioecy.

In *T. hirsuta*, gender variation shows spatial variation: the proportion of the different sexual types showing marked variation among populations in southern France (Dommée *et al.* 1990) and Egypt (El-Keblawy *et al.* 1995), although in general males and protogynous plants (the dichogamous morph which may have a male-biased gender) tend to be more common than females and protandrous individuals in most populations. In the offspring of open-pollinated maternal parents, one can observe all four types in the offspring of each sex-morph although the combination of males and protogynous individuals is >60% of offspring sex ratios (Dommée *et al.* 1995). In this species, male-biased sex ratios may be favoured in more arid environments, where plants are smaller and males are larger than females (Ramadan *et al.* 1994; El-Keblawy *et al.* 1995). In this genus, resource limitation may cause spatial variation in gender and sexual specialization.

5.4.6 Sex ratio variation in gynodioecous species

Gynodioecy is characterized by the coexistence of females and hermaphrodites in natural populations. The repeated evolution of this polymorphism and its potential importance in the evolution of dioecy has stimulated a great deal of interest in the evolutionary stability of gynodioecy (Charlesworth 1999). Several Mediterranean plant species illustrate the occurrence of this polymorphism and some of its most characteristic functional features: sexual

Table 5.2 Examples of gynodioecy in Mediterranean plant species and the functional characteristics of the polymorphism

Species	Relative seed set	Flower size	Gender variation	% Females
Romulea bulbocodium[a]	Females > hermaphrodites	Hermaphrodites > females	Hermaphrodite seed set	Some variation present
Thymus vulgaris[b]	Females > hermaphrodites	Hermaphrodites > females	Hermaphrodite pollen production and seed set; degree of stamen abortion in females	6–93%
Silene vulgaris[c]	No difference	Hermaphrodites > females	Occurrence of male-sterile flowers on hermaphrodites	Not studied
Hirschfeldia incana[d]	Not studied	Hermaphrodites > females	Occurrence of male-sterile flowers on hermaphrodites	2–10%
Daphne laureola[e]	No differences	Hermaphrodites > females	No evidence of gender variation	21–56%
Echium vulgare[f]	Females > hermaphrodites	Hermaphrodites > females	Not studied	2–14%

[a] Moret *et al.* (1992). [b] Reviewed by Thompson (2002). [c] Dulberger and Horovitz (1984).
[d] Horovitz and Galil (1972). [e] Alonso and Herrera (2001). [f] My own data.

dimorphism in flower size, greater seed set by females, sex ratio variation among populations, and the marked variation in the functional gender of hermaphrodites relative to females (Table 5.2). In some species ecological factors may contribute to sex ratio variation, as predicted for other cases of sexual polymorphisms (see above). For example, in *Daphne laureola* in the Iberian peninsula, females show marked sex ratio variation among populations, with a negative correlation between female frequency and elevation (Alonso and Herrera 2001). These authors suggest that high female frequency may be favoured at low elevation by increased intensity of summer drought, hence, sex ratio variation may be mediated by ecological factors, in accordance with the theoretical predictions for increased sexual dimorphism in harsh conditions. A similar situation has been reported for other gynodioecious species in more desert environments that fringe the eastern Mediterranean (Wolfe and Shmida 1997).

In gynodioecious populations, females do not produce pollen and therefore do not transmit genes to the next generation via male function. Females must obtain some form of fitness advantage over hermaphrodites in order to persist. The traditional explanation for the presence of females and their maintenance invokes two main possibilities. First, because females do not produce pollen they do not incur self-pollination and can thus avoid the genetic costs associated with inbreeding depression. They may thus produce more offspring of greater vigour than hermaphrodites. Second, lack of pollen production means that females may have more resources available for seed production, hence they may 'compensate' for their lack of male function by increased seed production and greater progeny vigour due to increased seed provisioning. Females may thus produce a higher 'quantity' of better 'quality' offspring. Third, in many gynodioecious species, sex expression is determined by an interaction between cytoplasmic and nuclear genes. The maternally inherited cytoplasmic genes inhibit male function, causing the sexual phenotype of an individual plant to be female. The effects of these cytoplasmic genes are repressed by nuclear alleles at particular loci which restore male function, causing the individual to be hermaphrodite. The sexual phenotype of an individual will thus depend on the combination of its nuclear and cytoplasmic genome. The interaction between cytoplasmic and nuclear genes may not just affect qualitative sex expression but may also influence quantitative variation in performance of individuals of each sex (Delph and Mutikainen 2003). An important aspect of this potential interaction is that combinations of nuclear and cytoplasmic genes are likely to be reshuffled as populations go extinct and plants re-colonize new sites. Hence, meta-population ecology in a spatial mosaic of habitats may be a key factor influencing sex ratio variation in gynodioecious species, particularly pioneer species in the early stages of succession (Couvet *et al.* 1998).

Theoretical and empirical research over the last 40 years on gynodioecious wild thyme *Thymus vulgaris* (whose polymorphism for secondary compound production was discussed in Chapter 4) in the garrigue vegetation of southern France illustrates this theme (Thompson *et al.* 1998; Thompson 2002). In several different species, which occur in the different parts of the geographic range of the genus and in different taxonomic sections, mean female frequency is high (above 50%) and extremely variable among populations, as illustrated by *T. vulgaris* from southern France and *Thymus zygis* and *Thymus mastichina* from central and northern Spain (Fig. 5.7; Dommée *et al.* 1978; Manicacci *et al.* 1998).

In *T. vulgaris*, as in other gynodioecious species (Table 5.2), females produce on average 2–3 times more seeds than hermaphrodites (Assouad *et al.* 1978; Thompson and Tarayre 2000). Two lines of evidence suggest that resource compensation may be an important cause of these differences in maternal fertility of females and hermaphrodites of *T. vulgaris*. First, even when females and hermaphrodites only receive outcross pollen, females produce roughly twice as many seeds as hermaphrodites (Thompson and Tarayre 2000), a result which is thought to be due to sexual selection

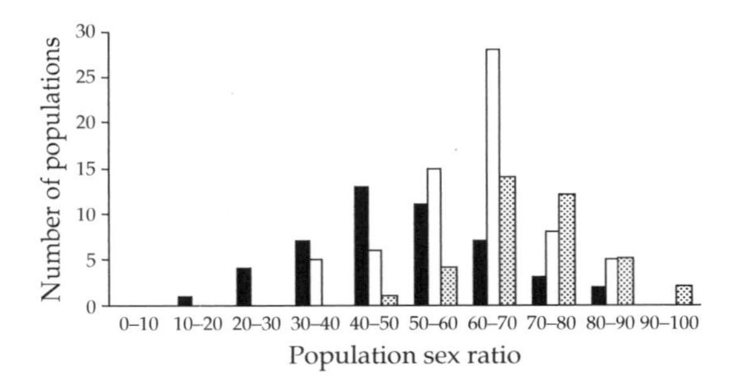

Figure 5.7 Sex ratio variation in three gynodioecious *Thymus* species in southern France and Spain (redrawn with permission from Manicacci *et al.* 1998). Closed bars: *T. zygis*, open bars: *T. vulgaris* and dotted bars: *T. mastichina*.

and subsequent specialization in male function by hermaphrodites (Couvet *et al.* 1985*b*). Second, pollen production and female fertility in hermaphrodites shows a negative correlation among families indicating a genetically based trade-off between male and female function (Atlan *et al.* 1992; Ehlers and Thompson 2004*a*).

Inbreeding depression may also contribute to the female fertility advantage. Hermaphrodites are self-compatible and despite protandry (anthers dehisce before stigmas are receptive in a given flower) may self-pollinate due to pollen movement between flowers on the same plant. Plants often bear many hundreds of open flowers at a given moment, and a short walk through the garrigues in early May is enough to convince anyone that the predominant pollinator (*A. mellifera*) visits many flowers on a given plant during each visit (Brabant *et al.* 1980). Butterflies also visit flowers and disperse pollen over longer distances. In fact, selfing rates on hermaphrodites vary among populations with the highest rates of selfing in populations with the highest female frequencies, below ~60% females, selfing rates are very low (Valdeyron *et al.* 1977; Thompson and Tarayre 2000). This positive correlation, traditionally interpreted as evidence that gynodioecy may be maintained due to its positive effect on outcrossing, may result from the reduced abundance and spatial isolation of hermaphrodites (among females) when female frequencies reach very high frequencies. Selfing is followed by significant inbreeding depression which markedly reduces the contribution of selfed offspring to subsequent generations (Assouad *et al.* 1978; Bonnemaison *et al.* 1979; Perrot *et al.* 1982; Thompson and Tarayre 2000; Thompson *et al.* 2004), hence an outcrossing advantage may contribute to the maintenance of high frequencies of females. The fact heterozygosity values of adult plants in a sample of 23 populations are unrelated to female frequencies concord with this idea (Tarayre and Thompson 1997). However, although outcrossing is no doubt important in the ecological adaptation of thyme to different environments around the Mediterranean Basin, and may contribute to the female fitness advantage in some populations, it may not be the principal cause of high female frequencies since selfing rates are low at female frequencies of up to 60%.

In colonizing populations thyme frequently occurs as dense patches (3–4 m in diameter) each composed of almost exclusively female plants (Dommée *et al.* 1983). As a result, female frequency is exceptionally high in many young populations which establish in early successional stages after the abandonment of vineyards or other cultivated crops or following a forest fire. Thyme usually establishes in old-field succession within about a decade following abandonment, perhaps earlier following a fire in areas where thyme was previously present. Its establishment usually follows an initial stage of colonization by annuals (see Chapter 4). After a period of population establishment (usually around 10–15 years) the frequency of females may decline (Belhassen *et al.* 1989).

These observations stimulated much research on how the spatial structure of sex-determining genes in the mosaic landscape occupied by this pioneer species affects plant performance, relative fertility, and sex ratio variation in thyme. In this species, sexual phenotype is governed by a complex interaction between nuclear and cytoplasmic (mitochondrial) genes (Belhassen *et al.* 1991), an interaction which probably involves several cytoplasmic sterility types and a range of restorer alleles. The interaction between cytoplasmic male sterility genes and nuclear restorer alleles may vary across the landscape if the different types of genes show spatial variation in frequency (Couvet *et al.* 1986, 1998). In the extreme case of spatial structure, a cytoplasmic male sterility type may occur in a population where the nuclear alleles that restore its male fertility are absent. This situation causes high female frequency because sex determination is cytoplasmic: all the offspring of a female will be female. Sex determination can be locally cytoplasmic if founder events during colonization of new sites cause the absence of restorer alleles for the limited number of cytoplasmic types that are present in founder populations. Subsequent immigration of nuclear restorer genes, via pollen or seeds, combined with a possible limitation on female seed set due to a paucity of pollen donors, will later cause a decline in female frequency.

The pattern of sex variation in thyme described above fits this hypothesis. Is there evidence of spatial variation in sex-determining genes and their interaction? A combination of data from molecular studies and experimental pollinations indicate that in thyme, the frequencies of cytoplasmic and nuclear restorer genes vary dramatically across the mosaic of thyme patches in the garrigue landscape of southern France. First, there is a sharp spatial differentiation in cytoplasmic genes among patches of females in young populations and among populations, which fits the idea that founder effects reduce the diversity of male-sterility genes in colonist populations (Manicacci *et al.* 1996; Tarayre *et al.* 1997). Second, experimental pollination of plants in an insect-free glasshouse by Couvet *et al.* (1985*a*) and Belhassen *et al.* (1991) have shown that male fertility restoration is greatest when females are pollinated with pollen from a hermaphrodite present in their original population compared to when the pollen source is a hermaphrodite from a different population. This result suggests that restorer gene frequency is variable among populations. In addition, Manicacci *et al.* (1997) reported marked variation in the frequency of hermaphrodites ('restoration rate') in the offspring of females transplanted into different populations. Two of the five females had high restoration rates in their original populations and one female was relatively well restored in her home population and, to a lesser extent, in the closest other population (~1 km distant), but not in other populations distant by more than 10 km. The frequency of cytoplasmic and restorer genes thus appears to show marked localized spatial variation among populations.

Thyme is a pioneer species of open habitats which is fairly rapidly excluded during secondary succession by over-shading and competition from grasses and small shrubs. In addition, thyme populations are often subject to disturbances which cause population extinction to be quite frequent. Human activities make up an important part of this extinction risk. In the early 1990s, with several colleagues, I marked thyme plants in nearly 40 populations for the different experimental studies then underway on thyme. I have followed these populations each year. In less than 15 years, five (16%) have gone extinct and three others have been dramatically reduced in size (numbers of plants and spatial extent). A total of 26% of populations have suffered important perturbations as a result of human activities. The landscape inhabited by thyme in the region where the above cited work has been carried out is a mosaic of clearings in low evergreen and deciduous oak woodland and discrete patches of garrigues separated by forest or cultivated fields and villages. The temporal population dynamics of the species in relation to human activities and the spatial habitat mosaic in which its populations occur may thus facilitate the development of biased sex ratios as the genetic combinations of cytoplasmic and nuclear genes are reshuffled during cycles of extinction and re-colonization. The precise combination of restorer genes and maternal cytoplasm may also affect a range of other traits that impact on the stability of gynodioecy in this and other species (Box 5.5).

The high rates of population turnover may mean that selection is unable to precisely adjust traits at the population level. In thyme, despite striking sex ratio variation among populations, female frequency is not related to increased gender allocation to male function in hermaphrodites (Manicacci *et al.* 1998; Thompson and Tarayre 2000), as theory predicts (Lloyd 1976). However, the predicted correlation between female frequency and hermaphrodite male function is observed among three species (Manicacci *et al.* 1998), indicative that high levels of genetic divergence are necessary to allow for the evolution of gender variation. An important point here is that nuclear–cytoplasmic interactions can cause a positive correlation between the frequency of hermaphrodites in a given family and the male function of hermaphrodites (Gigord *et al.* 1999), which will further limit the evolution of male function in hermaphrodites. This may have a bearing on why dioecy has not evolved from gynodioecy in *Thymus*, a genus in which there are many gynodioecious species, but no dioecious species. Finally, individual hermaphrodites vary their gender in response to seasonal changes in the availability of female flowers that results from gender specific flowering phenology (Ehlers and Thompson 2004*a*). Instead of showing genetic differentiation in gender among populations, hermaphrodites may phenotypically adjust male function in response to the number of open female flowers at a given time in a population, a form of adaptive plasticity in response to temporal sex ratio variation.

5.5 Pollination ecology and the evolution of style-length polymorphisms

5.5.1 Variation on a heterostylous theme

Heterostyly is a floral polymorphism in which the morphs differ in the sequence of heights at which stigmas and anthers are positioned within the flowers such that there is reciprocal positioning of male and female sex organs (Fig. 5.8). The genetic control of this polymorphism is relatively simple, in most species short-styled plants (S-morph) are heterozygous (Ss) and long-styled plants (L-morph) homozygous recessive (ss) for a series of closely linked genes (or super-gene). This polymorphism first fascinated Darwin (1877) who devoted the majority of his book on 'The different forms of flowers on plants of the same species' to the existence, function, and evolutionary significance of heterostyled flowers. Now known to occur in at least 28 angiosperm families (Barrett *et al.* 2000), heterostyly is one of the most visible and fascinating examples of convergent evolution in plants.

Heterostyly is primarily a polymorphism in style length and anther position. In addition, in many distylous species the floral polymorphism is linked to a di-allelic incompatibility system. This genetic linkage prevents self-fertilization and also cross-fertilization among plants of the same morph, hence, only crosses among plants of a different morph produce viable seeds. This is known as disassortative mating. In addition, pollen size and number, and stigma morphology often differ among morphs. The most common of such ancillary polymorphisms is the occurrence of variation in pollen size and number whereby the L-morph produces significantly more but smaller-sized pollen than the S-morph. These differences in pollen size have allowed many workers to count the relative numbers of large and small pollen grains on stigmas of the different morphs, work which has frequently shown that disassortative pollination frequently exceeds assortative pollination, strong evidence that the function

Box 5.5 The interaction between nuclear and cytoplasmic genes and its affects on performance, floral traits, and sex ratio variation (Figures redrawn from Thompson *et al.* 2002)

The combination of restorer genes and maternal cytoplasm can significantly affect the stability of gynodioecy. In both *Daphne laureola* in the Iberian peninsula (Alonso and Herrera 2001) and *Thymus vulgaris* in southern France (Thompson *et al.* 2004), variation in the performance of hermaphrodites' offspring has been interpreted to result, at least in part, from interactions between cytoplasmic sterility genes and nuclear restorer alleles. Such interactions can thus affect the relative performance of individuals within the sex phenotypes, and may cause quantitative variation in the gender of hermaphrodites (Thompson and Tarayre 2000).

In thyme, these interactions may also cause quantitative variation in sex expression. In addition to the qualitative differences between hermaphrodites (A) and females, the degree of stamen abortion varies in females (Thompson *et al.* 2002). Females with the least aborted anthers (type B and to a lesser extent type C) have flowers more similar in size to hermaphrodites than females (type D) with no trace of any anthers.

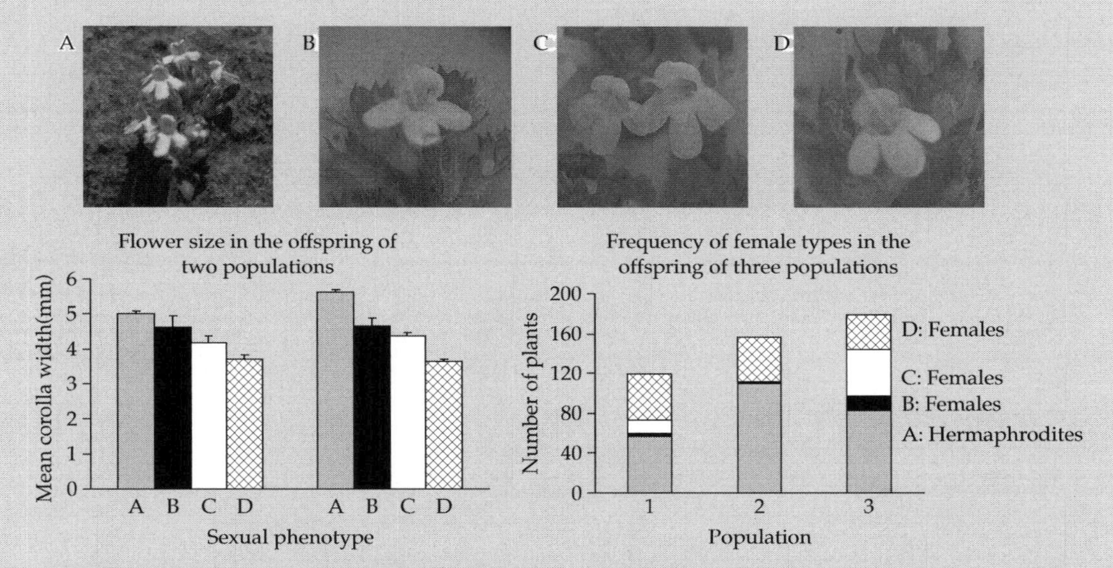

Although females tend to re-produce the same female type in their offspring, they most often produce all three types, as well as hermaphrodites. Type B females have a more hermaphrodite-biased offspring than type D females (type C females have intermediate values). These results suggest that nuclear-cytoplasmic interactions determine female phenotype and the capacity to produce hermaphrodite offspring, with a gradient in restoration of male fertility, such that type B females are close to the threshold where a plant would be hermaphrodite. The frequency of the female types particularly type B and C females varies among populations.

Interactions between nuclear and cytoplasmic genes may thus cause quantitative variation in floral phenotype and gender, and as a result of variation in space linked to the colonization dynamics of natural populations, create the conditions for wide-ranging sex ratio variation. Understanding the stability of gynodioecy may thus require a more precise understanding of the spatial dynamics of quantitative variation in sex expression within the two traditionally studied sex phenotypes.

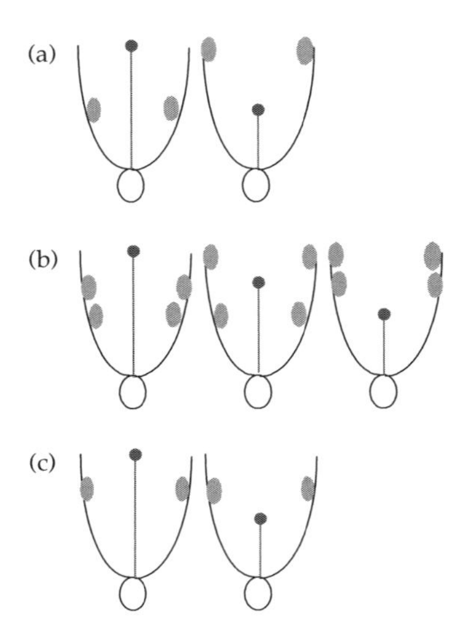

Figure 5.8 The three main forms of style-length polymorphism in flowering plants: (a) distyly and (b) tristyly—the two types of heterostyly, and (c) stigma-height dimorphism.

and maintenance of heterostyly are closely associated with its role in efficient pollen transfer and not just the avoidance of selfing (Box 5.6). Indeed, the reciprocal stigma and anther positions in the flowers of the different morphs allows for precise positioning of pollen on insect pollinators and thus its deposition on stigmas of the other morph(s). This precise plant–pollinator interaction may be facilitated by several other traits common in heterostyled flowers, such as tubular flowers with a nectar reward and stamens organized in one or two whorls (Lloyd and Webb 1992*a*).

Heterostyly may breakdown in association with the evolution of selfing (homostylous) variants. In the Mediterranean, variation and evolution of style-length polymorphisms has been observed in the genus *Linum* which includes several distylous species in the eastern Mediterranean (Dulberger 1973; Wolfe 2001), southern France (Thompson and Dommée 2000), and the Iberian peninsula (Nicholls 1985). In the western Mediterranean, *L. tenuifolium* has self-incompatible, diploid distylous populations at high elevation in the different Sierras of eastern Spain and self-compatible monomorphic populations in southern France and northern Italy (Nicholls 1985, 1986). Breakdown of the self-incompatibility system is not however associated with a typical homostylous plant, since the flowers of the self-compatible monomorphic taxon maintain the floral characteristics of a long-styled plant. The self-compatible form has a higher seed set than the self-incompatible races, illustrating that the extension of the range eastwards from the centre of diversity of this group in the Iberian peninsula may have been facilitated by the establishment of a self-compatible mating strategy.

How then does heterostyly evolve? Theory suggests two main evolutionary transitions for the appearance of heterostyly, of which the most recent is gaining increasing support (Box 5.6). This model, proposed by Lloyd and Webb (1992*a*,*b*), depends on the establishment of a transition polymorphism in which two mating types occur with different style lengths, but similar anther heights. This polymorphism is known as a stigma-height dimorphism. Although species with variation in style length, but lacking bimodal anther height, have been known to occur since Darwin (1877), in the early 1990s there had been no formal recognition that such variation represents a case of genetic polymorphism distinct from heterostyly. To my knowledge, only Heitz (1973) had previously referred to such a possible intermediate stage, whose morphology led him to give it the name of 'imperfect heterostyly' (p. 403). The intermediate stage (stigma-height dimorphism) is rare, probably because it rapidly evolves to distyly as a result of selection on anther height to favour proficient cross-pollination among morphs (Lloyd and Webb 1992*b*). These authors also point to the fact that many families with distylous species also contain species with flowers in which the stigma is positioned above the anthers, the possible ancestral type in their model.

The hunt was thus on for species which show qualitative style-length variation in the absence of anther-height variation. In the early 1990s, this search brought David Lloyd and Spencer Barrett, two prominent plant evolutionary biologists, to the Mediterranean, where they began to survey floral variation in *Narcissus* species across the Iberian peninsula, where previous work suggested the occurrence of heterostyly (Henriques 1887;

Box 5.6 The two principal theories for the evolution of distyly

(1) Charlesworth and Charlesworth (1979) proposed that, to avoid selfing, a di-allelic self-incompatibility system evolves in a monomorphic homostylous species.
The flower polymorphism only evolves after the incompatibility system in order to favour pollen transfer among two mating types.

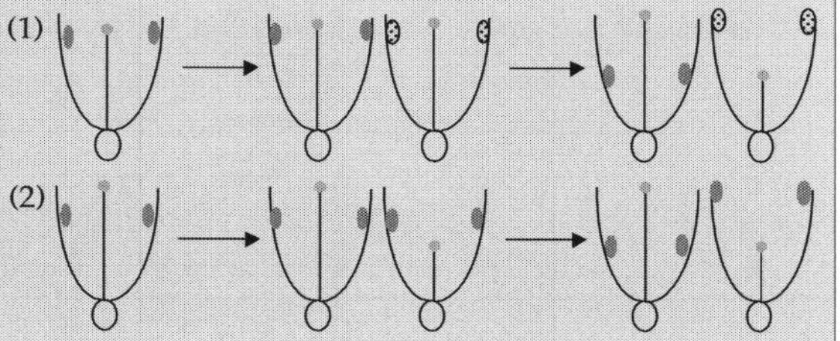

(1)
(2)

(2) More recently, Lloyd and Webb (1992*a*,*b*) proposed that the primary force acting on the evolution of distyly is selection for proficient cross-pollination, and not the avoidance of selfing. Here, the evolution of distyly is via a stigma-height dimorphism.

Although Darwin (1877) commented on how plants may have been 'rendered heterostylous to ensure cross-fertilization...' which... 'is highly important for the vigour and fertility of the offspring' (p. 258) he also recognized that 'there can hardly be a doubt that the relative length of... pistils and stamens... is an adaptation for the safe transportal by insects of the pollen from the one form to the other' (p. 253). Since then, the role of proficient pollen transfer (model 2) has gained strong support from the documentation of species with a stigma-height dimorphism (the intermediate stage) and studies which show that disassortative pollination exceeds assortative pollination in heterostylous species (Lloyd and Webb 1992*a*,*b*). Furthermore, as Barrett (2002*b*) points out, (a) self-incompatibility prevents selfing, whose avoidance cannot thus be the cause of the evolution of the floral polymorphism, and (b) the limitation of cross-fertilization to inter-morph matings may restrict opportunities for cross-pollination. This author argues that 'a more complete interpretation of the adaptive significance of heterostyly' (p. 276) is necessary, in particular the role of increased male fertility (proficient pollen transfer among morphs) and the avoidance of sexual interference between male and female sexual functions.

Once the floral polymorphism has evolved the incompatibility system may arise due to the adaptation of pollen of one morph to the stylar environment of the other morph (where it is most frequently transferred) and a loss of function in styles of the same morph (Darwin 1877; Lloyd and Webb 1992*a*,*b*). This final stage remains hypothetical.

Fernandes 1935; Lloyd *et al.* 1990). Detailed field investigations revealed that several species of *Narcissus* have populations with dimorphic style length (Barrett *et al.* 1996). Although the two morphs differ in style length and the sequence of heights at which stigmas and anthers are positioned in flowers, anther height shows only minor variation between the two morphs. Despite the resemblance to heterostyly, Barrett *et al.* (1996: 348) argued that such floral variation 'does not fully meet the criteria required to define that polymorphism'. Based on a detailed study of floral morphology, frequencies, and self-incompatibility status in *N. assoanus* and *N. dubius* in southern France, Baker *et al.* (2000*b*) also distinguished such variation from true distyly, and gave it the name of stigma-height dimorphism (Fig. 5.8; Box 5.7).

In addition to several species which have a stigma-height dimorphism (Dulberger 1964; Arroyo and Dafni 1995; Baker *et al.* 2000*b*;

Box 5.7 Floral biology of Narcissus assoanus

Narcissus assoanus occurs in stony garrigues and grassland communities from south-western Spain to south-east France. It is particularly abundant in upland pastures where it can form dense and extensive swards of yellow spreading as far as the eye can see. This species provides an intriguing and very practical study system to study the pollination ecology of a stigma-height dimorphism. Individual stems usually carry a single flower (or less-often two, or very rarely three, flowers). This makes it easy to do controlled pollinations and assess the total seed output of a plant. In southern France, this species is pollinated primarily by cleopatra butterflies (book cover) and hawkmoths and to a lesser extent by solitary bees.

The figure below shows stigma (open triangles) and anther (filled diamonds) positions (in millimetres above the tip of the ovary) for 45 plants from a natural population in southern France. In the L-morph, stigmas are positioned slightly above, or at the same position, as the anthers—note that each flower has two anther levels. In the S-morph, stigmas are positioned well below the anthers. The two floral morphs thus differ in their herkogamy, which is significantly greater in the S-morph (Baker *et al.* 2000*b*). The costs of self-pollination and patterns of pollen transfer may thus differ between the two morphs.

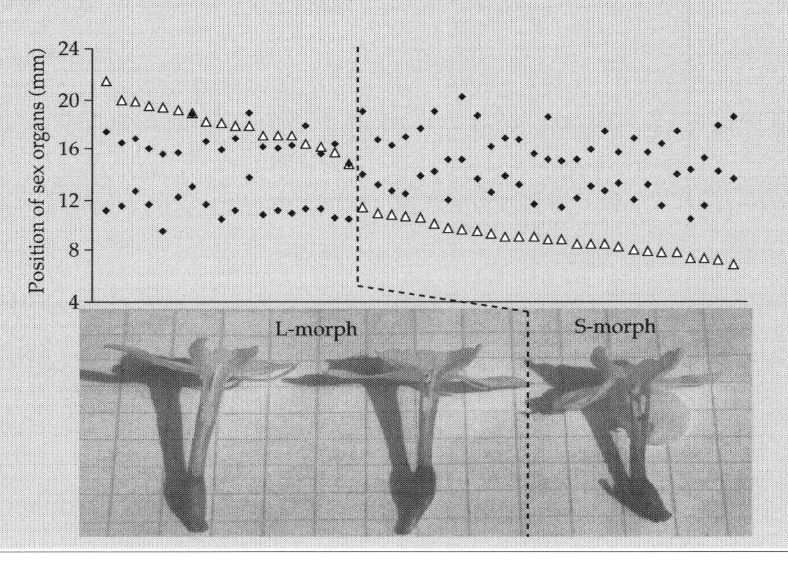

Arroyo *et al.* 2002), one *Narcissus* has been reported to be distylous (Arroyo and Barrett 2000) and one tristylous (Barrett *et al.* 1997). So, in *Narcissus* (Plate 4), three fundamentally different style-length polymorphisms occur. Furthermore, several species have the approach herkogamous morphology (Barrett *et al.* 1996), which phylogenetic evidence indicates to be the ancestral constitution for the evolution of a stigma-height dimorphism (Graham and Barrett 2004). The transition from a stigma-height dimorphism to heterostyly has occurred in the section Apodanthae which contains several species with a stigma-height dimorphism and one distylous species, *Narcissus albimarginatus*, which is endemic to the mountains of North Africa (Pérez *et al.* 2003; Graham and Barrett 2004). An interesting feature of the evolution of distyly and tristyly in *Narcissus* is the convergent evolution of pendulous flowers in distylous *N. albimarginatus* and tristylous *Narcissus triandrus* (Plate 4) which occur in separate sections of the genus and which have independently evolved heterostyly (Pérez *et al.* 2003; Graham and Barrett 2004). The significance of this change in floral form remains to be elucidated. However, the above work, in particular the phylogenetic reconstructions, provides us with strong evidence for the Lloyd and Webb (1992*a*,*b*) model, at least in *Narcissus*.

Narcissus do not stand alone, several other genera contain Mediterranean species which have a stigma-height dimorphism and distylous congeners. In *Anchusa*, floral measurements on *A. undulata* suggest the occurrence of a stigma-height dimorphism (Selvi 1998), while other congeners in the Mediterranean, for example, *A. hybrida* (Dulberger 1970), have floral variation which is intermediate between stigma-height dimorphism (little or no variation in anther height) and distyly (complete reciprocity). In the genus *Lithodora* (another Boraginaceae), I have observed that *L. fruticosa* in southern France has a stigma-height dimorphism, whereas on Rhodes, *L. hispidula* is distylous. In the genus *Linum*, some species may have a stigma-height dimorphism (Darwin 1877; Heitz 1973) while other Mediterranean *Linum* are distylous (Nicholls 1985; Thompson and Dommée 2000). As in *Narcissus*, both stigma-height dimorphism and heterostyly occur in all three genera, hence the potential evolutionary significance of stigma-height dimorphism as an intermediate stage in the evolution of distyly (Box 5.6).

5.5.2 The evolution of stigma-height dimorphism: self-interference and pollen transfer

Theoretical models predict that the adaptive significance of a stigma-height dimorphism lies in its role in proficient pollen transfer and limited self-interference (Box 5.6). As for heterostyly, to maintain a stigma-height dimorphism the frequency of disassortative mating should exceed that of assortative mating (Lloyd and Webb 1992*b*; Barrett *et al.* 1996; Baker *et al.* 2000*c*). For a stable stigma-height dimorphism to be maintained, the first step is the establishment of a mutant S-morph without it going to fixation. This requires that the relative fitness of the morphs be frequency-dependent so that one morph is fitter than the other when the former is rare, but not when common. Three important features of *Narcissus* species with a stigma-height dimorphism are important here.

1. It is far from being clear how the difference in stigma height, without a corresponding anther-height dimorphism and reciprocal positioning of stigmas and anthers, promotes disassortative pollination to levels which significantly exceed assortative pollination, and thus produce the 1 : 1 morph ratios observed in populations of *N. assoanus* (Baker *et al.* 2000*b*) and *Narcissus papyraceus* (Arroyo *et al.* 2002).
2. Most *Narcissus* species with a stigma-height dimorphism, and tristylous *N. triandrus* (Barrett *et al.* 1997), are at least partially self-incompatible. A critical feature of self-incompatibility in these species is that it is late acting, self-pollen tubes grow to the base of the style where a form of signal system between self-pollen tubes and the ovules renders ovules unavailable for fertilization (Sage *et al.* 1999). If self-pollen arrives on a stigma prior to outcross pollen it may thus render ovules unavailable for cross-fertilization. Such 'ovule discounting' could significantly reduce opportunities for outcrossing.
3. Plants of the same morph can cross and produce viable offspring, there is no genetic linkage between floral morphology and the incompatibility system (Baker *et al.* 2000*c*; Arroyo *et al.* 2002).

The conditions which allow a stigma-height dimorphism to be maintained in *Narcissus* will depend on rates of within- and among-morph pollen transfer and interference based costs which cause ovule discounting. For populations in pollen-limited situations, a dimorphism is maintained if inter-morph pollination exceeds intra-morph pollination as follows (Lloyd and Webb 1992*b*; Barrett *et al.* 1996):

$$\frac{(q_{\mathrm{ls}}(v_{\mathrm{s}}/v_{\mathrm{l}}) + q_{\mathrm{sl}})}{2} > q_{\mathrm{ll}}, \tag{1}$$

$$\frac{(q_{\mathrm{ls}} + q_{\mathrm{sl}}(v_{\mathrm{l}}/v_{\mathrm{s}}))}{2} > q_{\mathrm{ss}}, \tag{2}$$

where l and s are the two morphs, q_{ll} , q_{ss} , q_{ls}, q_{sl} are the probabilities of pollen transfer within and among morphs and v_l and v_s are the average proportion of ovules available after ovule discounting in the L- and S-morphs, respectively. To understand the evolution of a stigma-height dimorphism thus requires information on both ovule discounting and rates of pollen transfer.

Studies of several *Narcissus* species have shown that prior self-pollination can prevent plants from producing seeds when outcross pollination is slightly delayed (Sage *et al.* 1999; Arroyo *et al.* 2002; Cesaro *et al.* 2004). In *N. assoanus*, based on conditions (1) and (2), a reduction in the number of ovules available for outcrossing in the L-morph (relative to the S-morph) would enable the S-morph to invade and for dimorphism to be maintained (Box 5.8). Hence, ovule discounting may have played a key role in the evolution and maintenance of stigma-height dimorphism.

What then do we know about the relative rates of disassortative and assortative pollination in species with a stigma-height dimorphism? Unfortunately the absence of dimorphic pollen size in *Narcissus* prevents evaluation of rates of intra- and inter-morph pollen be simply counting stigmatic pollen loads, as done in many distylous species. The assessment of seed set in experimentally manipulated plots in a natural population where natural pollinators assure pollen dispersal has however provided some important insights here.

First, Thompson *et al.* (2003*a*) found that maternal fertility is significantly less in plots monomorphic for the S-morph relative to plots monomorphic for the L-morph and dimorphic plots. This result suggests that pollen flow among plants of the S-morph is significantly less than among plants of the L-morph (i.e. $q_{ss} < q_{ll}$). Second, Cesaro and Thompson (2004) found that overall rates of disassortative pollination exceed those for assortative pollination, primarily because of high rates of pollen transfer from the L-morph to S-morph. This result may be caused by subtle differences in floral morphology: whereas the uppermost anthers are similarly positioned in both morphs, in the L-morph the lowermost anthers are significantly lower in the flower than those of the S-morph (Baker *et al.* 2000*b*; Box 5.7). The positioning of anthers in the L-morph may favour its male function since it has upper anthers positioned close to the stigma height of the L-morph and lower anthers closer to the position of the stigma in the S-morph.

The evolution of a stigma-height dimorphism may thus be driven more by beneficial effects on male fitness and efficient cross-pollination associated with reduced male–female interference than the genetic consequences of inbreeding (Barrett 2002*b*). So one could ask: why is stigma-height dimorphism frequent in *Narcissus* but rare elsewhere? In their models, Lloyd and Webb (1992*a*,*b*) argue that a stigma-height dimorphism is a transient stage which rapidly evolves to distyly because of strong selection for proficient cross-pollination (among morphs). Hence its rarity. In *Narcissus*, all cross-pollinations are compatible, a feature which may weaken selection for the evolution of complete reciprocity, as suggested for *N. triandrus* (Barrett *et al.* 2004*b*). In *N. assoanus*, subtle variations in anther position may improve male function of the L-morph and maintain the maternal fertility of the S-morph, hence the strength of selection for complete reciprocity may also be weak, hence the stability of stigma-height dimorphism and the rarity of its transition to distyly in *Narcissus*.

5.5.3 Morph-ratio variation and frequency-dependent selection

In Mediterranean *Narcissus*, any variation in relative amounts of assortative and disassortative pollination may cause morph ratios to deviate from 1 : 1, particularly if the two morphs differ in seed set. Indeed, several *Narcissus* with a stigma-height dimorphism show striking patterns of morph-ratio variation (Box 5.9 and 5.10) with a suite of repeatable trends: (a) large numbers of populations with an L-biased morph ratio and occasional loss of the S-morph, (b) relative absence of populations with an S-biased morph ratio and no monomorphic S-morph populations, and (c) occasional 1 : 1 morph ratios, which are particularly apparent in *N. assoanus* and *N. papyraceus*. Likewise, in tristylous *N. triandrus* >90% of populations have a morph ratio biased towards the L-morph (Barrett *et al.* 2004*b*).

Box 5.8 Potential costs of self-interference in *Narcissus assoanus* (based on Cesaro *et al.* (2004), figure reproduced with permission)

In *N. assoanus*, simultaneous self- and cross-pollination (SP + CP) reduce seed set by 20% and self-pollination made 12 h or more prior to cross-pollination (SP + CP 12 h, 24 h, and 48 h) decreases seed set by 60%, to a level equivalent to that on selfing. Ovule discounting may thus occur.

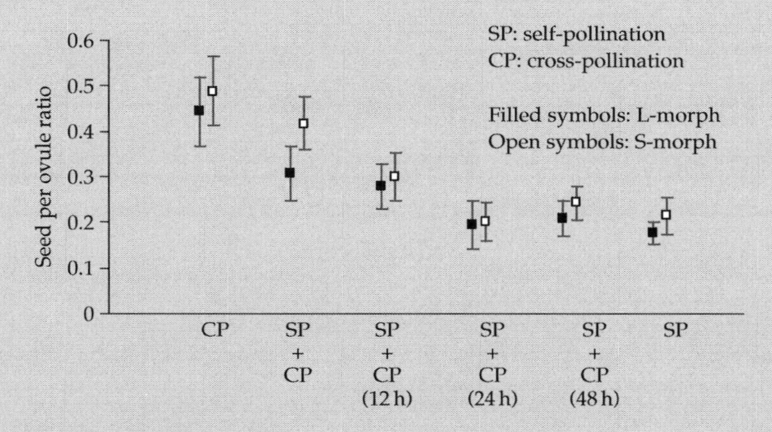

In *N. assoanus* there are morph-specific differences in how ovule discounting may occur.

First, the weak herkogamy of the L-morph causes greater levels of autonomous self-pollination in comparison with the S-morph. In addition, in the L-morph, ovule discounting can occur with simultaneous self-pollination (SP + CP) whereas in the S-morph, outcross pollen must be delayed by 12 h or more for ovule discounting to be significant. Despite this potential cost, there is no evidence that self-interference reduces maternal or paternal fitness in natural populations. The absence of a maternal fertility cost due to self-interference implies that the two style morphs avoid self-interference. The features of their floral biology indicate that the morphs differ in how they do so:

- The S-morph experiences almost no spontaneous deposition of self pollen, hence herkogamy is probably the principal mechanism that limits self-interference.
- The L-morph can incur autonomous self-pollination because of the close proximity of stigmas and anthers. However, unlike the S-morph this morph is protandrous, which may limit any costs of self-pollination.

These results provide an illustration of how traits may have evolved to reduce self-interference and thus support the idea that mating system evolution in plants is not just due to the avoidance of the genetic costs of selfing.

The existence of morph-ratio variation among populations provides an appropriate setting to evaluate the occurrence of frequency-dependent selection. Although such frequency-dependent selection is a key parameter in theoretical models, there is a relative lack of empirical demonstration of its role in the maintenance of plant sexual polymorphisms in the wild.

Evidence for the operation of frequency-dependent selection has been obtained in studies of *N. assoanus* and *N. triandrus*. In *N. assoanus*, Thompson *et al.* (2003*a*) found that in plots with morph frequencies biased in favour of the L-morph, seed set of the S-morph was significantly greater than that of the L-morph. This rare-morph fertility advantage is probably related to the proficient

Box 5.9 Morph-ratio variation in *Narcissus assoanus*

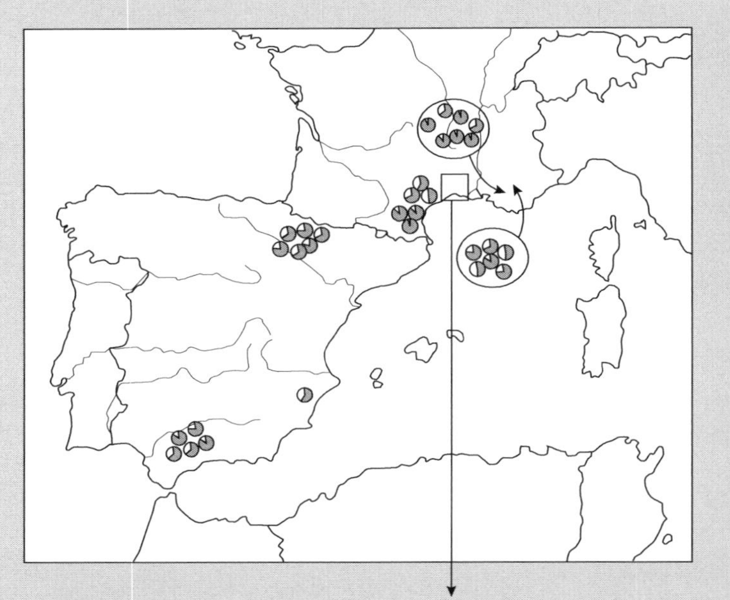

Examination of more than 100 populations of *N. assoanus* indicates that across the geographic range of this species most populations have a biased morph ratio due to a high proportion of the L-morph (filled portion of each circle) relative to the S-morph (open portion of each circle).

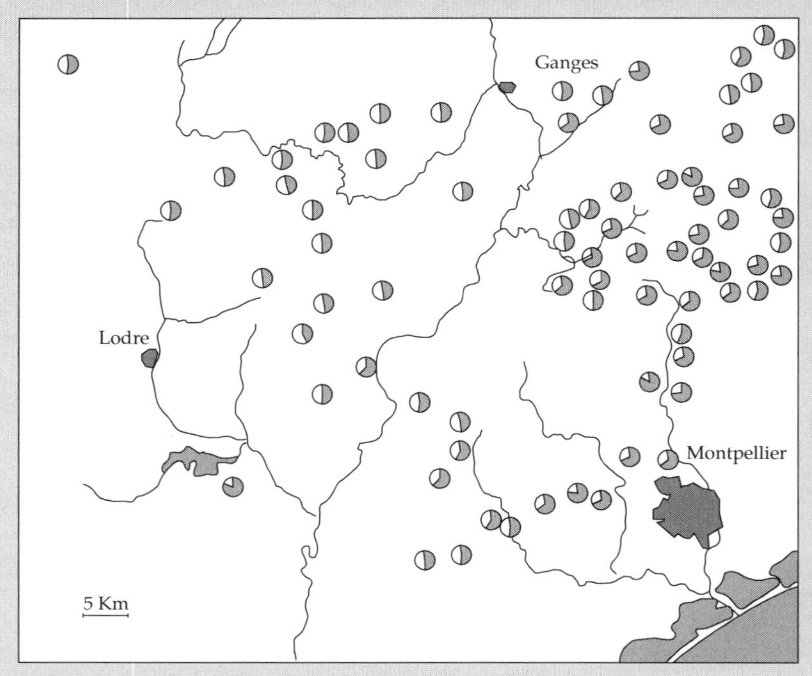

In southern France, populations contain either a 1 : 1 ratio of the two forms or have an L-biased morph ratio. In this region, morph-ratio variation shows a clear geographic pattern (see also Box 5.11). Whereas lowland garrigues (right and bottom of figure) landscape contains either L-biased populations or populations with a 1 : 1 morph ratio, all upland limestone plateau populations have a 1 : 1 morph ratio (upper left of figure).

Box 5.10 Morph-ratio variation in three closely related species of *Narcissus* section *tazetta*

In *N. tazetta* in Israel (Arroyo and Dafni 1995) and *N. papyraceus* (Arroyo *et al.* 2002) in southern Spain and North Africa, populations are either monomorphic for the L-morph or dimorphic (mostly L-biased). In both species, absence of the S-morph from some populations has been attributed to patterns of pollinator activity and reduced pollinator servicing of this morph in certain habitats. See Plate 4 for photos of these two species.

In *N. dubius* in southern France, only highly L-biased or monomorphic L-morph populations (mostly at the limits of this species distribution in the east) occur. Baker *et al.* (2000*b*,*c*) suggest that self-compatibility, in association with a capacity for autonomous self-pollination (i.e. ability to set seeds in the absence of pollinators) in the L-morph, may at least in part cause the low frequency and loss of the S-morph. Disassortative pollination may be difficult to achieve in this species, hence the lack of 1 : 1 morph ratios.

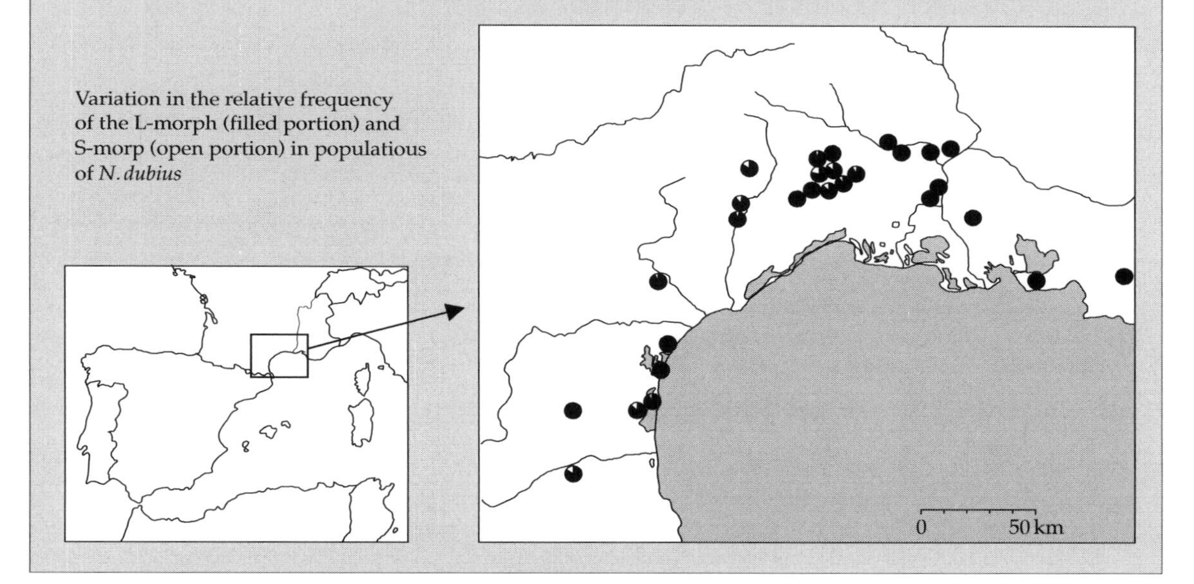

Variation in the relative frequency of the L-morph (filled portion) and S-morp (open portion) in populatious of *N. dubius*

transfer of pollen from the L-morph to the S-morph in this species (Cesaro and Thompson 2004). In *N. triandrus*, geographic variation in morph frequency is also closely associated with variation in floral morphology in a way which strongly suggest the operation of frequency-dependent effects (Barrett *et al.* 2004*b*).

Finally, the S-morph may occur at reduced frequency in different *Narcissus* species as a result of the stochastic loss of genes following population bottlenecks or founder events during colonization (Barrett 1993). In *Narcissus* the genetic control of stigma-height dimorphism probably involves a single Mendelian locus with the short-styled allele dominant to the long-styled allele, as has been demonstrated for *Narcissus tazetta* (Dulberger 1964), and as occurs in distylous species. Since the S-morph phenotype is governed by the expression of a dominant gene (S) it can be lost during colonization episodes if populations are founded by L-morph plants (ss). Dimorphism can only be subsequently expressed in such a population if the dominant

S allele originates by (very rare) local mutation or is introduced by pollen flow or seed migration. In contrast, populations founded by a combination of the S-morph (*Ss*) and the L-morph (*ss*), or only the S-morph, will segregate long-styled plants. Dulberger (1964) found no evidence for the occurrence of *SS* plants, hence the quasi-absence of populations with S-morph biased frequencies. The dominance relationships of alleles in association with episodes of colonization and population history could thus create the template for the occurrence of morph-ratio variation and the predominance of the L-morph in many populations. The finding of a positive relation between population size and morph ratio in both *N. assoanus* (Box 5.11) and *N. papyraceus* (Arroyo *et al.* 2002) indicate that colonization events or fragmentation of populations may contribute to the deviation of morph ratios from 1 : 1 as a result of stochastic loss of the dominant gene for short styles. Current work on *N. assoanus* indicates that spatial isolation and fragmentation may be more closely correlated with morph-ratio variation than population size. The spatial configuration of the mosaic of Mediterranean habitats could thus be of prime importance in the variability of the mating system.

5.6 Conclusions

The key message that I emphasize in this chapter concerns how the interplay between ecology and genetics on the spatial template of the Mediterranean mosaic landscape influences the ecological and evolutionary lability of reproductive traits in Mediterranean plants. I have discussed this theme in a wide variety of situations. First, in generalist pollination systems, or species which suffer intense herbivory during the reproductive process, a diversity of selection pressures may limit pollinator-mediated selection or cause disruptive selection on floral traits. Second, in gynodioecious species, the interplay between selection pressures (inbreeding depression and resource compensation), genetic features of the mating system (nuclear-cytoplasmic interactions), and the spatial dynamics of populations cause striking sex ratio variation. Third, in species with stylar polymorphisms, floral-trait and morph-ratio variation among populations are closely associated with patterns of pollen dispersal and transfer among plants and the spatial structure of populations in the landscape. These studies demonstrate the need for research which, at the interface of ecology and genetics, explicitly integrates the impact of spatial habitat configuration and the dynamics of natural populations in the Mediterranean landscape. A more refined consideration of how meta-population dynamics interact with spatial variation in ecological conditions and the successional development of Mediterranean plant communities would further our understanding of these issues.

Reproductive-trait variation is closely dependent on the strong abiotic and biotic selection pressures that characterize the Mediterranean region and the diversity and heterogeneity of plant–pollinator interactions. The spatial and temporal variation in pollinator abundance and the composition of pollinator assemblages described in various species may create marked spatio-temporal variation in pollen limitation and selection pressures on floral traits. The lack of close associations between insects and plant species with particular floral traits suggests weak or diversifying selection on floral traits by pollinators. One is tempted to predict that as pollinator diversity increases, the variance of traits associated with pollen removal and receipt may increase. Petanidou and Ellis (1993: 9) go as far as to suggest that the 'very diverse, and shifting pollinator fauna . . . (of some plants has) . . . brought the evolution of floral characters to a standstill'. The recent onset of the Mediterranean climate and its repeated oscillation towards cold dry conditions suggests, to quote J. Herrera (1988: 284), that 'the generalized nature of pollination systems may have been, and is today, another major factor contributing to the survival and invasive behaviour of many Mediterranean scrub species'. The potential for plant specialization can be further limited by herbivory which reduces the impact of any pollinator-mediated selection. This is not to say that specialization in plant–pollinator interactions does not occur, it does occur in several species. To appreciate this point, consider the following examples. First, figs and dwarf palms have maintained their tight species-specific mutualisms despite repeated

Box 5.11 Population characteristics and morph-ratio variation in *Narcissus assoanus* (graph redrawn from Baker *et al.* 2000*b*)

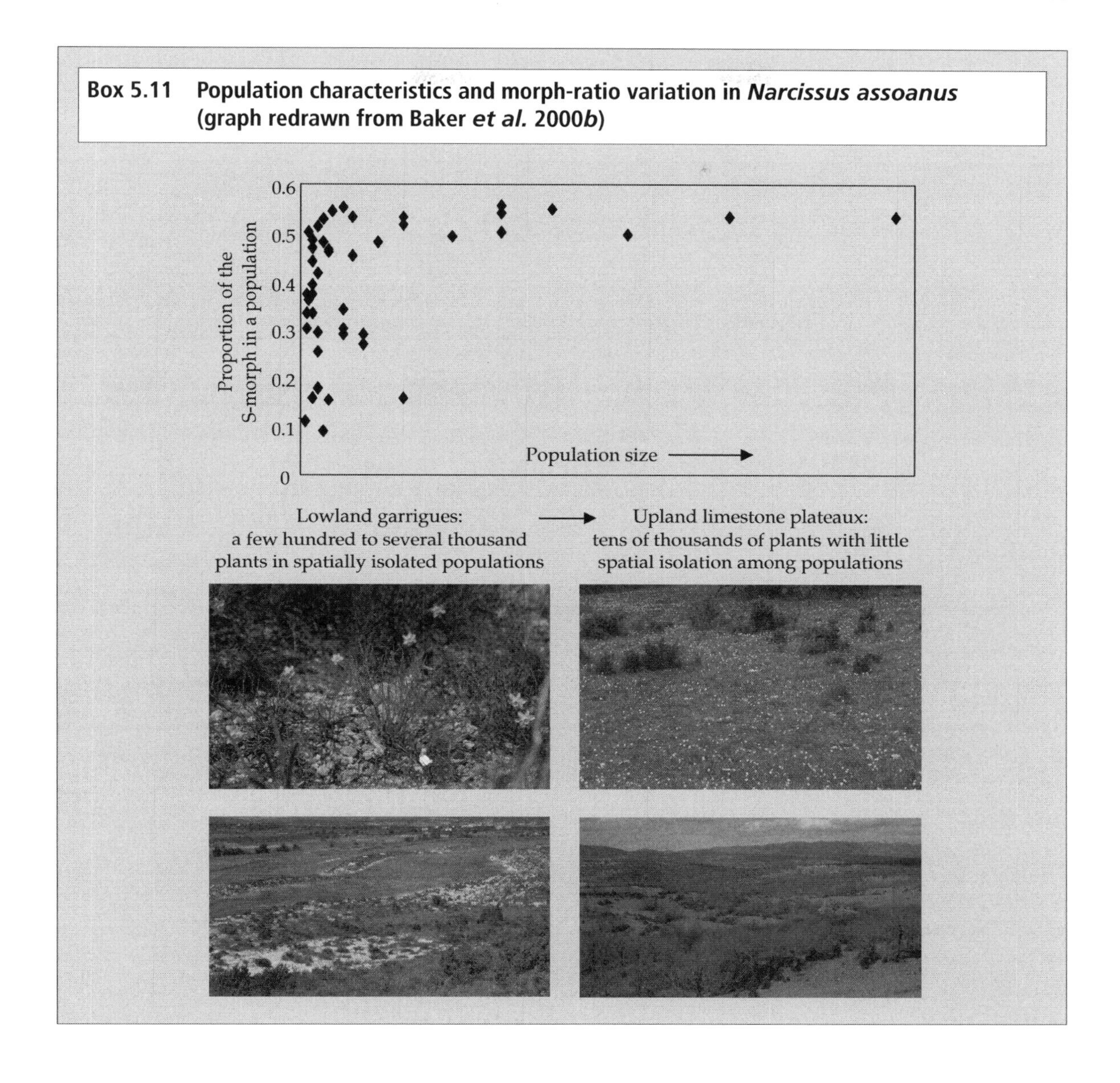

range contractions and expansions since the onset of a Mediteranean-climate regime. Second, the evolutionary dynamics of stylar polymorphisms illustrate how precise flower–pollinator interactions, in association with the spatial configuration of habitats in the heterogeneous Mediterranean mosaic landscape, influence variation and evolution of a fairly specialized floral morphology. Third, in Mediterranean orchids, the evolution of sibling species with different floral morphologies may be related to geographic variation in pollination strategies (Cozzolino *et al.* 2001), an idea that would be worthy of future investigation.

Finally, sclerophyllous vegetation in the Mediterranean flora (C.M. Herrera 1992*a*; Aronne and Wilcock 1994*b*), mostly the Pre-Pliocene group

(Quézel and Médail 2003), and other Mediterranean-type ecosystems (Arroyo and Uslar 1993) comprise many obligately outcrossing (dioecious) species. The avoidance of the genetic consequences of inbreeding and the genetic variability associated with outcrossing have no doubt been important for the evolutionary persistence of many of these species. However, this is just one side of the coin. The evolution of sexual systems is also strongly driven by selection for effective pollen dispersal and mating success by male function and by the costs of self-pollination in this respect. Several examples discussed in this chapter give support to the view that the traditional acceptance of inbreeding depression as the essential selection pressure acting on mating-system evolution should be replaced by a more complete and balanced view of the role of diverse ecological features of pollination (see Barrett 2002*a*). Placing floral trait and gender variation in the ecological context of (a) pollen dispersal and transfer among plants in a population and (b) the need to avoid male–female conflicts provides us with a more balanced view of the causes of the variability and evolution of plant reproduction. Precise and manipulative experimental studies of some of the study systems described in this chapter would be most worthwhile in this context.

CHAPTER 6

Ecology and evolution of domesticated and invasive species

... the introduction of plants and animals from one part of the world to another ... are potentially much more informative than most laboratory experimental work, since they have faced the introduced species, not with some simple defined change in selective conditions, but with a whole new ecological system in which the species has to find a place for itself.

C.H. Waddington (1965: 1)

6.1 Migration with man

What could be more dramatic for a plant species, other than its extinction, than a sudden change in its distribution as a result of long-distance dispersal across previous geographic barriers to migration? Such events have occurred, albeit at a very low rate, in nature, as the flora of oceanic islands attest. However, in the last 10,000 years in the Mediterranean, and in the last few hundred years on a global scale, plants have been moved around at an ever increasing rate. Indeed, in the last 500 years such episodes of long-distance dispersal have all of a sudden become far more frequent, in association with the increased movement of humans around the world. This movement of plants from one region to another or from one part of the world to another has provided evolutionary ecologists with test situations which can be used to examine the ecological and genetic factors which regulate populations, limit distributions, and determine how plants respond to a novel environment. Two main types of introduction provide insights into how species find their place in new systems, one concerns the cultivation and domestication of plants in different regions, the second involves the accidental or purposeful introduction of species to new regions, where they may subsequently become invasive.

Throughout this book I have emphasized the interplay between geology, climate, and human activities for plant evolution in the Mediterranean region. Human activities represent an integral ecological parameter and a significant evolutionary force in the Mediterranean Basin due to their long history and prevalence in the landscape. I have thus placed much emphasis on the role of human activities in the modification of environmental constraints and local selection pressures and in shaping the spatial configuration of populations in the landscape. In this chapter, I will further this discussion of the ecological and evolutionary role of human activities by recourse to the study of plant domestication and plant invasions, that is, situations where selection pressures are directly modified as a result of human activities and human-induced migration has dramatically altered patterns of colonization and gene flow. By moving plants around the world and placing them in completely new environments man has pushed species into new evolutionary trajectories.

Cultivation and domestication began early in the Mediterranean. The early evolution of crop plants in the hands of humans was an essential element in the development and organization of human societies in the Neolithic. Some of our most famous and cherished crop plants and fruit trees first became important in the eastern Mediterranean. The tale of their early cultivation, domestication, and migration around the Mediterranean, is a fascinating account of evolution in a new ecological setting. There have been three editions of the classic book on the domestication of plants on the eastern fringes of the Mediterranean (Zohary and Hopf 2000) and some excellent recent reviews of how molecular markers have enabled a more probing evaluation of where, when, and how domestication began in the Mediterranean (e.g. Salamini *et al.* 2002; Breton *et al.* 2004). In this chapter, I will discuss in detail the history of domestication and cultivation in a few selected examples of Mediterranean plants to illustrate the relative importance of new selection pressures and migration events for plant evolution under domestication. I will broaden this discussion by reference to the multitude of garden plants whose origin lies somewhere in the Mediterranean region. Its amazing that we still know so little about the biology, ecology, and genetic variation of the wild species of these groups in the Mediterranean landscape.

Many plants have been accidentally or purposefully introduced to new regions in association with human movements. Many introduced plants stay put or spread little from sites of introduction, be they in the wild or in gardens, while many go extinct in their new range. Others spread away from the site of introduction to become invasive species. This distinction between introduced (or exotic) species in a region and invasive species actively expanding their range is important.

Biological invasions have attracted the interest of ecologists and evolutionary biologists alike. Sparked by the work of Elton (1958) and the 1964 symposium on 'The genetics of colonizing species', which then became a classic book in evolutionary biology (Baker and Stebbins 1965), research on invasive species has developed and focused on three main issues: the potential effects of invasive species on biodiversity, whether it is possible to predict invasiveness of introduced species, and whether some ecological systems are more invasion prone than others. In addition, the introduction of a species to a new region, where previously it was excluded by geographic barriers to dispersal, usually involves a sharp and dramatic alteration in the selection pressures (e.g. novel abiotic conditions, new pollination environment, and absence of specialized herbivores). How plants respond to the new environment allows for the study of plant evolution and the factors which regulate persistence and distribution limits (Richardson and Bond 1991).

I believe that no balanced treatment of plant evolution in the Mediterranean could be made without reference to the issues of domestication and biological invasions. My discussion of domesticated and invasive species in this chapter is focused on the history and evolution of domesticated plants and the population ecology and evolution of invasive species in a Mediterranean environment. Throughout the chapter I will emphasize the role of genetic changes linked to processes such as hybridization in ecological adaptation and the evolution of both domesticated and invasive plants.

6.2 The evolutionary history of domesticated plants

6.2.1 Differentiation under domestication

Darwin (1859) first recognized how the process of domestication provides insightful illustrations of evolution in association with strong selection pressures. He thus began his treatise on the origin of species with a chapter concerned with 'variation under domestication', which allowed him to introduce his new theory with examples people were familiar with. Plant domestication involves conscious cultivation over several generations and the sampling of 'useful' or 'desired' traits, and thus just a small part of the variation present (a process akin to natural selection in the wild). As a result, domesticated populations have their genetic variability altered such that they deviate from wild congeners as the species becomes more and more useful to human consumers. In the Mediterranean, selection pressures associated with domestication have been acting on a small number of species for almost as long as more 'natural' selection pressures associated with the last continued spell of Mediterranean climate, that is, the few thousands of years since the return of a Mediterranean climate after the last bout of glaciation. Domestication has also involved dispersal, the evolutionary force which counters divergence due to selection and drift.

6.2.2 Early cultivation and the onset of domestication in the Fertile Crescent

The domestication of a number of important fruit and cereal crops began very early in the lands which border the eastern extremes of the Mediterranean Sea; far earlier than in the western Mediterranean, or elsewhere in temperate Europe. The identification of plant remains in archaeological sites provides some of the best evidence for where and when particular plants were first domesticated around the shores of the Mediterranean (Zohary and Spiegel-Roy 1975; Zohary and Hopf 2000). Most domesticated plants, especially those used as fruit and grain crops, differ dramatically from wild relatives in the size of their fruits and seeds. Hence their remains can be identified in well-dated sites, providing strong evidence of wild or domesticated fruit consumption and the cultivation of newly domesticated species. What appear to be wild remains in archaeological sites as early as ~4,000 BP suggest that fruits such as olives, grapes, and figs were probably consumed by humans prior to their cultivation and domestication (Zohary and Spiegel-Roy 1975). The same is also true for cereals such as wheat, barley, and rye (Zohary and Hopf 2000; Salamini *et al.* 2002; Pozzi *et al.* 2004). As these authors illustrate, initial cultivation followed soon after the harvesting of wild populations. It was only once initial cultivation became a regular feature that domestication began.

Like most of the topics I have discussed in this book, the evolutionary process of domestication showed strong spatial variation in its onset and intensity. Whereas in the Fertile Crescent the onset of domestication started early in the Neolithic, in the western Mediterranean the process began a few thousand years later (Diamond 1997). This author illustrates how this regional differentiation in the onset of domestication is due to geographic differences in five main factors that determined the potential for domestication to be successful.

1. A decline in available foods in the wild (wild mammals and wild fruits) would have discouraged hunter–gatherer type behaviour.
2. Climate change at the end of the Pleistocene would have allowed the spatial expansion of habitats appropriate for wild cereals (the ancestors of several modern-day crops such as wheat and barley) and perhaps made large harvests possible. As I outlined in Chapter 1, this is likely to have been earlier in the eastern Mediterranean than in the western parts of the region.
3. Cumulative developments of tools and technologies which represented 'the unconscious first steps of domestication' (Diamond 1997: 111) were locally present.
4. A rise in human population densities coupled with necessary sharp increases in rates of food production.
5. The presence of a number of potential crops.

I would argue that the most critical determining factor influencing historical patterns of domestication in the Mediterranean region concerns the plants themselves. When analysed together, botanical, genetic, and archaeological evidence suggest that agriculture originated in an area encompassed by the 'Fertile Crescent' region in the eastern part of the Mediterranean Basin between the Euphrates and the Tigris Rivers after 12,000 BP (Lev-Yadun *et al.* 2000; Salamini *et al.* 2002). This zone includes the eastern shores of the Mediterranean and the steppe zones of the Irano-Turanian floristic province. In addition, many crops are thought to have been initially domesticated in just a small core area, or 'cradle of agriculture' (Lev-Yadun *et al.* 2000), within the Fertile Crescent, near the upper reaches of the Tigris and Euphrates rivers. The wild progenitors of the Neolithic founder crops (einkorn wheat, emmer wheat, barley, lentil, pea, bitter vetch, and chickpea) as well as flax occur together only in this core area of the Fertile Crescent (Harlan and Zohary 1966; Lev-Yadun *et al.* 2000). Some of the important woody species were also to be found in more forested zones, for example, wild olives (Kislev *et al.* 1992). Although some of the early domesticated species occurred individually in the wild elsewhere in the Mediterranean, probably nowhere else was there a coexistence of several useful species, which became the 'founder' crops of early cultivation. For cultivation (deliberate planting and harvesting of either wild or domesticated forms) and domestication (the process of selecting key traits that ultimately transformed wild forms into cultivated crops) to develop, it would have been necessary to have a number of diverse staple crops and not just one or two species. The co-occurrence of a number of potential crops may thus have been the determining stimulus for early domestication in this region.

Wild species in the eastern Mediterranean were thus pre-domesticated over a large area in and around the Fertile Crescent. The archaeological record illustrates well how humans harvested the wild forms of cereals, from natural stands prior to cultivation and directed domestication (Zohary and Hopf 2000). Eight thousand years ago the choice of what to grow was not governed by molecular genetics or the availability of seeds distributed and controlled by multinational corporations. Early farmers had no model to guide them, and thus did not know for sure that they were heading towards improved and tastier varieties. They were probably aware however that trees with big fruits could produce progeny which also bore big fruits, which were perhaps easier to harvest and more vigorous. Initial selection thus involved easily observable traits such as size, fleshiness, oiliness, and of course taste (Diamond 1997). Automatically, early farmers began the domestication process for several cereals that remain integral elements of domesticated cereals in many parts of the world.

The diffusion of plants out of this area, westwards across the Mediterranean and into temperate Europe and Asia probably began ~8,000 BP (Salamini *et al.* 2002). From initial centres of domestication many species have migrated with humans to sites of secondary domestication, others have been subject to independent episodes of domestication in different regions, and as a result some have come back into contact with wild relatives, from which they have long been isolated and with which they have hybridized to produce introgressed and feral forms (see later). The movement of such species around the Mediterranean has been facilitated by the fact that several of the early domesticated species naturally occurred under the seasonally dry climate. In addition, some species occurred naturally in various parts of the Mediterranean region. For example, although olives and figs were first domesticated in the Fertile Crescent (Zohary and Spiegel-Roy 1975), they existed elsewhere in the Mediterranean, hence they could be easily transplanted. Furthermore, secondary domestication was a distinct possibility and hybridization among previously isolated stocks became possible. Analyses of the genetic structure of domesticated and naturalized species using the powerful modern tools of molecular biology have given us detailed insights into the history of such processes and have shed light on the intimate history of cultivated plants and human civilizations around the Mediterranean region.

Neolithic agriculture in the Fertile Crescent was based on three cereals (einkorn wheat, emmer wheat, and barley), four pulses (lentil, pea, chickpea, and bitter vetch), and flax. All of these occur wild in and around the central core area of the Fertile Crescent, several stretching eastwards into the steppes and semi-deserts of south-west Asia, although only wild flax has a circum-Mediterranean distribution (Harlan and Zohary 1966; Zohary and Hopf 2000). Most of these crops are self-compatible and also practice occasional outcrossing. The combination of reproductive isolation and enhanced character transmission associated with predominant self-pollination and genetic flexibility due to occasional outcrossing has no doubt enhanced their suitability as domestic crops.

Wild forms of wheat and barley have been detected in several eastern Mediterranean Pre-Neolithic (8,000–18,000 BP) and Neolithic archaeological sites, and ancient tool finds suggest that emmer wheat was harvested as far back as almost 10,000 BP (and other crops a little later) in this region (Lev-Yadun *et al.* 2000; Zohary and Hopf 2000). Pulses, because of their agronomic advantage (they do not require fixable nitrogen, and in fact add it to the soil) and their dietary importance, accompanied wheat and barley through the initial stages of cultivation and domestication in the central core area of the Fertile Crescent (Zohary and Hopf 2000).

Large-scale harvesting of wild crops in a sort of pre-domesticated state is thought to have been an important step towards domestication. The presence of wild species of einkorn and emmer wheat, barley and rye in the Fertile Crescent region and the presence of their seeds in the early archaeological sites represent strong evidence that these species were harvested in and around the Fertile Crescent region prior to the evolution of domesticated variants.

The domestication of cereal crops in the eastern Mediterranean has been monophyletic in many species (Salamini *et al.* 2002). For example, molecular phylogenetic evidence (based on allele frequencies at AFLP marker loci) indicates that cultivated einkorn wheat was domesticated from a lineage of wild einkorn wheat that currently exists only in a small area in the Karacadağ mountains of south-east Turkey (Heun *et al.* 1997). The genetic founder stocks that gave rise to domesticated emmer wheat (Özkan *et al.* 2002) and barley (Badr *et al.* 2000) have similarly been traced to small areas of the Fertile Crescent region, supporting the idea of a single core area of plant domestication (Lev-Yadun *et al.* 2000). The domestication of barley, slightly to the west of the wheat domestication area, may have depended on the importation of a procedure from the core region in south-east Turkey. Plant remains and seeds in archaeological excavations provide further evidence that early agriculture occurred in this small part of the Fertile Crescent (Zohary and Hopf 2000; Salamini *et al.* 2002). During these early stages of cultivation, and later in the western Mediterranean when the founder crops were introduced, human populations no doubt continued to act as hunter–gatherers and fishermen, in addition to their new role as farmers (Courtin 2000).

6.2.3 Evolution of agricultural crops: selection and dispersal

The new role of human populations as farmers was based on their use of two central evolutionary processes: selection of useful forms and their dispersal to new areas.

As soon as humans began re-planting their harvests, they began to act as a very strong selective agent on their new crop, since the cultivated field is a very different environment and the human selection pressures very different to those in the wild. Even in the early stages of cultivation, most of the pioneer cultivators would probably have carefully chosen the seeds they sowed, since human population persistence depended on the success of their crop. This initial domestication was probably quite rapid (Diamond 1997). Another essential point here is that selection during early domestication may have been very different to more contemporary selection in terms of its strength and intensity. Simply having high yield may have been less important than assuring that yields were regular among years, despite severe interannual variation in climate. In contrast to more recent phases of domestication, which have caused the progressive reduction in genetic variation in cultivated species (see below), the initial steps towards domestication probably had little effect on genetic diversity, since initial cultivation would have required enormous quantities, hence only at particular loci would variation have been reduced in the newly cultivated populations.

Domestication of wild cereals in the eastern Mediterranean (which have small naturally dispersed seeds) involved a common process of selection associated with distinct trait modifications (Harlan *et al.* 1973; Pozzi *et al.* 2004). Traits related to dispersal and seed characteristics were no doubt of primary importance. First, selection for increased seed recovery caused the evolution of non-shattering inflorescences and a more determinate growth (i.e. unbranched culms). Second, reduced selection for the maintenance of seed traits associated with (a) dispersal in space produced a reduction in size of bracts, bristles, glumes, and awns and (b) dispersal in time reduced germination inhibitors and produced crops with less dormancy. Selection then followed for increased seed production (i.e. increased seed set, enhanced fertility of otherwise sterile or male flowers, and/or increased inflorescence size or number, larger seeds with a higher carbohydrate : protein ratio, and a flowering phenology less affected by day length).

This domestication of wheat involved selection on (a) major genes that control seed and inflorescence traits and photoperiodic flowering and (b) genes with quantitative effects on traits such as height, tillering, and architecture (Box 6.1). Such genes often have a fairly simple mode of inheritance and pleiotropic effects on other traits, causing the

Box 6.1 Genetic control of key domestication traits (based on Salamini *et al.* 2002; Pozzi *et al.* 2004)

There are two main traits linked to how plants can be harvested that have been central to the domestication process. To be harvested, a plant must not have effective means for dispersal. Domestication has thus been targeted on variants that have lost their ability to scatter their seeds far and wide. The early domestication of peas, poppies, flax, and lentils in the eastern Mediterranean may all have involved initial selection of reduced dispersal traits and extended seed retention on maternal plants. In this way humans selected for mutant genes otherwise strongly selected against in wild populations. The genetic control of such traits has been extensively studied in wheat.

First, wheat can be classified according to how the ear shatters at maturity. Whereas domesticated wheat has a non-brittle rachis, that keeps the ear intact after maturation, wild forms usually have a brittle rachis that shatters at maturity into disarticulated spikelets, that can then disperse individually. This trait is controlled by two genes, whose dominant alleles combine to produce the brittle rachis in wild forms. In polyploids, the genetics of rachis fragility is more complicated, sometimes with a polygenic basis and a pleiotropic effect on other traits within the ear.

Second, all wild wheats and cultivated einkorn, emmer wheat, and Timopheev's wheat have a tenacious glume, a trait which characterizes 'hulled' wheat. In contrast, the hexaploid cultivated wheat and tetraploid hard wheat have a 'soft', glume which releases a naked seed at maturity. Having a soft glume is critical because it is this trait which allows the free release of the seeds and which thus characterizes the 'free-threshing' wheats at the tetraploid and hexaploid level. The harvesting of the so-called hulled wheats gives rise to spikelets and not grains, which then have to be freed by pounding or some other method. The development of free-threshing wheat varieties gave rise to a crop in which the grain is released on harvesting, a final and major stage in wheat domestication. This trait has a polygenic control involving several quantitative trait loci. One of these loci contains the famous *Q* factor which produces square wheat inflorescences of good threshability and seed characteristics that facilitate efficient harvesting.

The genetic control of other important domestication traits, for example, seed size, has been found to be under complex polygenic control, often involving several chromosomal regions and pleiotropic effects.

One of the recent and most dramatic genetic changes in wheat has been the creation of genetically improved varieties. One of the most striking examples concerns the dwarf cultivars developed during the 'Green Revolution' as a result of reduction in stalk height (by about 50%) and a doubling of yield relative to earlier and taller cultivars (Borlaug 1983). This increased yield was obtained as a result of reduced damage (taller varieties suffered greatly from flattening during inclement weather) and more efficient biomass conversion associated with high-dosage fertilization. The genes responsible code for proteins that regulate the synthesis and/or the pathway of growth hormone synthesis (Salamini 2003).

rapid evolution of suites of traits, the so-called 'domestication syndrome' (Box 6.2). Such selection has produced convergent evolution of similar traits in grasses such as wheat and barley and pulses (Box 6.3), indicative that the genes in question were homologous (Pozzi *et al.* 2004). The mapping and cloning of genes, standard tools in modern plant breeding programmes, now indicate that this domestication syndrome was originated by a few major pleiotropic genes with a sudden and dramatic effect on phenotype followed by the accumulation of numerous minor mutations. In addition, many traits behave as threshold traits, in which changes in gene control and regulation must pass a given threshold to alter the developmental programme (Pozzi *et al.* 2004). These authors also

Box 6.2 Evolution of traits in the domestication syndrome

The domestication of plants involves artificial selection on a variety of growth and yield traits. Several patterns of trait evolution in domesticated species have been documented, most noteworthy are those related to reduced dispersal and defence characters. Selected traits important for domestication are often controlled by genes with a pleiotropic effect that have a simultaneous effect on a number of traits: for example, softening of the glume, reduction in ear length, increased number of spikelets per ear, and a tougher rachis. It is the combination of these traits that renders the free-threshing wheats so useful in the contemporary farming environment.

Artificial selection, usually very strong, may cause slow-growing long-lived trees such as olives, to reduce their allocation of resources to the production of secondary compounds, if the latter are costly to produce (see Chapter 4). In a study of biomass and growth of wild and cultivated olives growing in an area of Italy with a high ungulate browsing pressure, Massei and Hartley (2000) reported that browsed olives produced high densities of shoots which resemble spines (due to the absence of leaves) and leaves of small size. These structural responses to browsing were even more marked than chemical changes, which were nevertheless significant and varied seasonally. In spring, during periods of active growth, cultivated olives had longer shoots and lower levels of leaf phenolics than oleasters. Hence the process of domestication may have created a strong selection pressure for high-yielding varieties with less defence capacity.

Sexual reproduction has also been a trait of primary interest in the cultivation of domesticated variants. For example, male sterility (present in many wild species—Chapter 5) has been of great use to plant breeders, since all the seed produced by female plants in such species is outcrossed, and thus likely to have good vigour. At the other end of the stick, hermaphrodism and self-compatibility are of utmost importance in other domesticated species in order to preserve useful combinations of traits. Cultivated vines, whose tiny flowers are both hermaphrodite and self-compatible and thus very different from their closest wild relatives (which are dioecious) are a case in point.

As Diamond (1997) argued, early farmers did not just exert strong selection pressures on perceptible traits that directly interested them, they also facilitated the evolution of a suite of different traits in domesticated variants as a result of indirect selection. Subtle changes in particular traits, as a result of strong selection, have thus had multiple consequences. Examination of the genetic control of the diverse traits involved in the domestication process of different cereals has revealed a remarkable similarity in their genetic basis (Box 6.1). This has no doubt favoured the convergent evolution of the domestication syndrome in species with a long history of genetic isolation.

discuss how mutant genes with major effects on traits linked to domestication may have been rapidly fixed very early in the domestication history of cereals in the Near East. The selective advantage to the rapid evolution of tightly linked genes may have been a critical element in the domestication process (Le Thierry d'Ennequin *et al.* 1999). Such genes will have had a strong selective advantage in a new environment where selection and migration were in the hands of humans.

Selection was employed by humans at different stages of the domestication process and in a variety of ways. After the initial reliance on wild harvests at the beginning of the Neolithic, humans began to cultivate wild species as their first seeds were collected and replanted in newly opened areas. Following this, domestication of particular types (e.g. non-shattering, regular, and/or high-yields) began gradually either as they were unavoidably chosen or consciously selected due to high seed yield and/or a more regular contribution to harvests. Once this domestication process was fully in operation, cultivation became more efficient. Although it is clear that cultivation of wild plants preceded

Box 6.3 Trait variation and selection during domestication in barley and pulses

The domestication of cultivated barley, *Hordeum vulgare* subsp. *vulgare* another founder crop in Neolithic agriculture, from wild progenitors in the eastern Mediterranean has involved the evolution of a similar array of traits as in wheat (Zohary and Hopf 2000; Salamini *et al.* 2002; Pozzi *et al.* 2004). Barley accompanied wheat during the selection and dispersal of new forms and the hundreds of modern varieties encompassing several thousand landraces. In barley, the evolution of a non-shattering rachis, a polystichous spike, increased seed weight, and non-tenacious glumes (naked seeds) have been the major phenotypic shifts in its domestication from wild *Hordeum vulgare* subsp. *spontaneum*. Again the genetics of these traits are fairly well known. The brittleness of the ears, for example, is governed by a mutation in one of two tightly linked genes, such that non-shattering phenotype (in all cultivated forms) is caused by recessive alleles at one or both of these loci. Wild barley occurs in several parts of the eastern Mediterranean, and eastwards into the drier steppes and semi-deserts, hence multiple domestication events may have occurred in different areas. Its tolerance of dry saline and resource-poor soils have made barley an important element in human consumption.

From the very early days of Neolithic agriculture, the domestication of wheat and barley went hand in hand with that of several legumes, the other elements of the founder crops in the Fertile Crescent. As in cereals, trait evolution in legumes has been characterized by the selection of a syndrome of domestication traits (Zohary and Hopf 2000), in particular the loss of dispersal ability, both in space and time. In cultivated legumes the characteristic bursting of the seed pod is almost absent, preventing seed dispersal and allowing seeds to be harvested easily. In peas and lentils the genetic basis of this loss of dispersal ability involves a single recessive gene, hence the evolution and fixation of this trait was rapidly accomplished. In addition, the dormancy which characterizes seeds of wild species has been lost. Whereas most wild eastern Mediterranean pulses have a fairly impermeable and thick seed coat which allows for a fairly prolonged dormancy, cultivated varieties have a more permeable seed that germinates rapidly on sowing. In the initial phases of cultivation of the wild progenitors of modern day cultivated varieties, there would have been strong selection on mutations facilitating rapid germination. There has also been strong positive selection on seed size, a dramatic reduction in the climbing habit, and diminished allocation of resources to toxic secondary compounds in cultivated legumes.

domestication, perhaps over a fairly long period, it is not completely certain whether initial domestication involved unconscious or conscious selection of plants. As discussed by Salamini *et al.* (2002) unconscious selection during the cultivation of wild progenitors could account for the polygenic inheritance of many of the traits in domesticated wheat. Finally, more recent selection of favourable alleles of a small number of 'domestication genes', particularly free-threshing varieties, and the march towards high-yielding varieties, have further advanced the domestication process.

Finally, domesticated plants provide an excellent illustration of a central principle of evolutionary ecology, directional selection reduces genetic variation. Apart from a few abrupt and brief events of accidental or purposeful hybridization, and the initial phases of domestication based on large numbers of different plant stocks, evolution under domestication has been associated with a common trend in cereal grasses: the progressive decline in genetic variation in the cultivated plants (Zohary and Hopf 2000; Pozzi *et al.* 2004). Reduced genetic diversity is of course problematic since it may compromise future programmes of development, particularly those whose aim is sustainability.

Vigorous forms of founder crops thus most likely first arose in a restricted core area of the Fertile Crescent and then migrated with human populations to displace wild progenitors and perhaps other (less vigorous) domesticated forms. Given that genetic evidence strongly suggests a monophyletic origin of founder crops (see above), it is likely that superior varieties of founder crops replaced locally cultivated wild crops as humans moved the former around the eastern Mediterranean. Widespread cultivation but monophyletic origins can thus be reconciled (Salamini *et al.* 2002). The introduction and domination of vigorous cereal varieties, particularly of barley and wheat, probably occurred repeatedly in different parts of the Mediterranean. Although several crops (wheat and barley and perhaps lentils and peas) may have simultaneously dispersed roughly to the western Mediterranean from the Near East, cereals tended to take over.

All the evidence for dispersal suggests that it occurred rapidly, starting in the early Neolithic. In the western part of the Fertile Crescent near the central core area of south-east Turkey, domesticated einkorn wheat remains are abundant in excavations that date to ~9,500 BP. This crop rapidly migrated to Cyprus, Greece, and the Balkans where its remains have been traced to ~8,000 BP. Emmer wheat may have arrived in the western Mediterranean a little earlier and was probably the principal cereal of the European Neolithic and Bronze Ages (Zohary and Hopf 2000). In southern France, there is evidence of cereal cultivation (large quantities of carbonized wheat and barley seeds were found along with diverse tools and grinding 'equipment') in the Cardial Period (~7,000 BP) and lentils and peas prior to this date (Courtin 2000). This is strong evidence that climatic conditions were favourable to the development of agriculture in this region in the early Neolithic and that long-distance dispersal from the eastern Mediterranean occurred in a roughly similar window of time for different crops. Rapid migration of early domesticated varieties across the Mediterranean no later than the Early Neoliothic clearly occurred. The rapidity of domesticated crops' dispersal across to the western Mediterranean indicates that it occurred over water. The large gaps in the spread of domesticated crops across the northern shores also suggest that long-distance dispersal under domestication occurred in boats and not by land. Human populations had long since been navigating around the shores of the Mediterranean, populating islands such as Corsica and Sardinia since at least 14,000 BP, bringing with them domesticated cereals in the early Neolithic (Courtin 2000) and then boat-loads of wine in Etruscan amphores and, no doubt, many useful seeds and plants.

There was also spatial heterogeneity in dispersal patterns of different varieties of wheat during the dawn ages of domestication. The wheat introduced to southern France was club wheat, a hexaploid variety of bread wheat (Courtin 2000), and not emmer wheat, which was dispersed early in the Neolithic through Greece and the Balkans. For some reason, free-threshing wheat varieties more rapidly established, in place of/or instead of, hulled varieties such as emmer wheat in the western Mediterranean compared to the Balkans (Zohary and Hopf 2000).

6.2.4 Hybridization and polyploidy in wheat domestication

In addition to selection and dispersal, two other important evolutionary process, polyploidy and hybridization, have also strongly influenced the evolution of cereals. Wheat species provide a particularly clear example of how these processes have stimulated genetic change and brought new material available for the domestication process to continue. In wild (Chapter 3) and cultivated (Table 6.1) species, ploidy level varies from diploid ($2n = 14$) through tetraploid ($4n = 28$) to the hexaploid ($6n = 42$) level.

The most ancient domestication occurred at the diploid level, where cultivated einkorn wheat arose from wild einkorn in the core area of the Fertile Crescent (Heun *et al.* 1997). Although einkorn was one of the earliest, perhaps the first (Salamini *et al.* 2002), domesticated wheats, it survives only as a relic crop, now only rarely planted in the Mediterranean Basin.

At the tetraploid level, cultivated emmer wheat was derived from wild emmer wheat. Phylogenetic analysis of the domesticated tetraploids, hulled emmer wheat and free-threshing hard wheat, shows

Table 6.1 A synopsis of wild and cultivated *Triticum* and *Aegilops* and their important domestication traits

Biological species	Common name	Ploidy level (genome)	Important domestication traits
T. monococcum subsp. *monococcum*	Wild einkorn	2*x* (AA)	Hulled wheat; brittle rachis
T. monococcum subsp. *boeoticum*	Cultivated einkorn wheat	2*x* (AA)	Hulled wheat; non-brittle rachis
T. uratu	Wild *T. uratu*	2*x* (AA)	Hulled wheat; brittle rachis
Aegilops tauschii	Wild *A. tauschii*	2*x* (DD)	Hulled wheat; brittle rachis
T. turgidum subsp. *dicoccoides*	Wild emmer wheat	4*x* (AABB)	Hulled wheat; brittle rachis
T. turgidum subsp. *dicoccum*	Cultivated emmer wheat	4*x* (AABB)	Hulled wheat; non-brittle rachis
T. turgidum subsp. *durum*	Hard wheat	4*x* (AABB)	Free-threshing; non-brittle rachis
T. timopheevii subsp. *araraticum*	Wild Timopheev's wheat	4*x* (AAGG)	Hulled wheat; brittle rachis
T. timopheevii subsp. *timopheevii*	Cultivated Timopheev's wheat	4*x* (AAGG)	Hulled wheat; non-brittle rachis
T. aestivum subsp. *spelta*	Cultivated spelt wheat	6*x* (AABBDD)	Hulled wheat; non-brittle rachis
T. aestivum subsp. *vulgare*	Cultivated bread wheat	6*x* (AABBDD)	Free-threshing; non-brittle rachis
T. compactum	Club wheat	6*x* (AABBDD)	Free-threshing; non-brittle rachis

Source: Zohary and Hopf (2000); Salamini *et al.* (2002).

that they are genetically very similar to one another and that they have a common origin from wild emmer wheat that now occurs in south-east Turkey (Özkan *et al.* 2002; Salamini *et al.* 2002). These analyses of AFLP allele frequencies also show that the wild populations of emmer wheat in south-east Turkey are genetically more similar to the domesticated polyploids than to other wild emmer populations. Other molecular analyses (Dvořák *et al.* 1993) have revealed that the donor of the *A* genome in cultivated emmer wheat was a wild *Triticum uratu*-like diploid, and not einkorn wheat. The occurrence of seeds in ancient excavation sites (Zohary and Hopf 2000) indicate that these tetraploid crops, including the free-threshing types, evolved very early in the domestication of wheat. Whereas emmer wheat was probably one of the most important crops in the Fertile Crescent from 10,000 BP to the Bronze Age, only one tetraploid wheat, free-threshing hard wheat, is widely cultivated today (Zohary and Hopf 2000; Salamini *et al.* 2002). Cultivated tetraploid Timopheev's wheat is endemic to a small region in Georgia and has been of little significance to the process of domestication and dispersal of wheat in the Mediterranean (Zohary and Hopf 2000).

One of the most intriguing stories in wheat domestication concerns the evolution of hexaploid free-threshing bread wheat, our principal contemporary crop. At some stage in the domestication process, cultivated tetraploids were dispersed into the geographic range of wild *Aegilops tauschii*, where hybridization between the crop and the wild species produced hexaploid wheat, from which the bread wheats *Triticum aestivum* subsp. *spelta* and *T. aestivum* subsp. *vulgare* were developed (Dvořák *et al.* 1998; Zohary and Hopf 2000). Such hybridization occurs in many crop-wild species groups (Box 6.4) and in this case involves allo-polyploidization, which may have occurred more than once. It is also possible that the free-threshing trait in hexaploids was directly inherited from free-threshing tetraploids.

6.2.5 The history of olive and chestnut domestication

The olive tree

The olive tree is perhaps the most emblematic of all Mediterranean plants. Wild species of *Olea* existed in the Mediterranean Basin in the Late Tertiary (Quézel 1978), long before the Mediterranean climate evolved. Olive trees, accompanied by grapevines, figs, and dates, were one of the initial fruit crops to be domesticated in the Fertile Crescent. Olive cultivation began at least 6,000 BP in the eastern Mediterranean (Zohary and Spiegel-Roy 1975; Galili *et al.* 1997) and almost as long ago

Box 6.4 Hybridization between cultivated plants and wild Mediterranean species

A key evolutionary process in the domestication of plants is hybridization between crops and wild relatives (Harlan *et al.* 1973; Ellstrand *et al.* 1999). In the Mediterranean, such hybridization has contributed to the evolution of new crops (e.g. hexaploid bread wheat) and the escape of genes from cultivated plants into wild feral forms, for example, in olives, the occurrence of a feral form of einkorn wheat in the Balkans (Salamini *et al.* 2002) and new allele combinations in *Quercus suber* (cork oak) when in contact with *Quercus ilex* (Toumi and Lumaret 1998). Hybridization among cultivated and wild taxa may not always produce feral forms since a long history of cultivation in the presence of wild species may produce hybrids of poor value for humans and thus their counter-selection (Zamir *et al.* 1984).

That hybridization between crop and wild species can commonly produce viable progeny illustrates that wild and domesticated gene pools are (a) not sufficiently different for their reproductive isolation and (b) not fully separated from one another by the process of speciation. As domestication proceeds, populations became isolated by cultivation and ecologically fine-tuned landraces and cultivars evolve. The evolution in isolation of such domesticated variants has occurred in the absence of selection for genetic barriers to crossing. In wheat, natural species already show distinct possibilities for hybridization (Chapter 3). It is thus not surprising to discover many examples of hybridization between domesticated and wild wheat in and around cultivated fields (Ellstrand *et al.* 1999).

An important question concerning the occurrence of crop to weed gene flow is the role played by such hybridization in the evolution of wild populations. First, in addition to the selective pressures imposed by domestication and those of the natural environment, genetic differentiation between co-occurring crops and their wild relatives will be greatly modified by crop to weed gene flow and variation. Second, weeds may become more aggressive and/or resistant to herbicides. Third, the genetic pollution of wild plants could go so far as to cause the extinction of a truly wild species.

In *Medicago sativa* in the western Mediterranean, many natural populations in Spain have maintained fairly original morphological features despite their parapatry with cultivated 'populations' (Jenczewski *et al.* 1998*b*). Crop and wild populations of this species are outcrossing and partially self-incompatible, have the same ploidy level and similar pollinators and flowering times. The maintenance of morphological integrity thus stimulated Jenczewski *et al.* (1999) to evaluate the relative importance of differences in selection pressures on crop and wild populations and levels of gene flow for the differentiation of wild and crop varieties. The comparison of patterns of population differentiation within and among wild and cultivated gene pools for genetic markers and quantitative traits revealed two clear results:

(1) the occurrence of intermediate populations with respect to both allozymes and quantitative traits, that is, crop to weed gene flow and the existence of hybrid populations;
(2) genetic differentiation of natural populations from cultivated landraces for quantitative traits but not allozymes in two populations, suggesting that natural selection prevents the establishment of 'cultivated traits' in populations that are otherwise genetically introgressed with cultivated marker genes.

Since reproductive isolation is not a feature of these populations, it would appear that contrasting and strong selection pressures are critical for genetic differentiation among crop and wild taxa and thus limit the introgression of crop traits into wild populations.

in the western Mediterranean (Terral and Arnold-Simard 1996; Terral 2000). Since the Bronze Age, the economic development of many areas has been associated not just with the development of cereals but also the cultivation of olives. Bread and olive oil have been staple diets in many rural areas of the Mediterranean, and still go hand in hand in many Mediterranean cafés and bars as a late morning snack. Initially based on fruit collection from wild oleasters in the eastern Mediterranean, which were present in this region ~19,000 BP (Kislev *et al.* 1992), domestication primarily involved a shift to reliance on the vegetative multiplication of selected high performing cultivars.

Mediterranean olives (*Olea europaea* subsp. *europaea*) are but one subspecies of a highly variable species complex containing several disjunct subspecies. Mediterranean olives comprise 'a complex of wild forms, weedy types, and cultivated varieties' (Zohary and Spiegel-Roy 1975: 321). Mediterranean olives include not only the cultivated olive (*O. europaea* subsp. *europaea* var. *europaea*) but also a wild form, the oleaster (*O. europaea* subsp. *europaea* var. *sylvestris*), which inhabits thermophilous vegetation around the Mediterranean Basin. The distinction between the two is often fraught with difficulty as a result of hybridization which has produced weedy or 'feral' forms that occur in many disturbed areas near cultivated groves or on previously cultivated sites. Cultivated olives are morphologically very difficult to distinguish from wild oleaster, a distinction which requires the analysis of multiple traits that contribute to the shape (but not the size) of the olive stone (Terral *et al.* 2004) and the presence of spiny juvenile branches.

Cultivated and wild olives differ markedly in their levels of genetic diversity and geographic differentiation. Most cultivated olives have genetic diversity values that are only a subset of those observed in wild olives, with feral forms having intermediate values (Lumaret and Ouazzani 2001; Breton *et al.* 2004). The genetic variability of oleasters is often superior to that of cultivated olives by virtue of the presence of novel nuclear genotypes (Ouazzani *et al.* 1993) or cytoplasmic haplotypes (Besnard *et al.* 2002) not present in cultivated forms and significantly reduced heterozygosity in cultivated gene pools (Lumaret *et al.* 2004). These results suggest that selection during domestication has dramatically reduced genetic variation in cultivated forms, and that inbreeding may have occurred during this process of domestication. However, some cultivated olives can show high levels of genetic diversity as a result of crosses among cultivars and hybridization with wild oleaster (Ouazzani *et al.* 1996). The study of variation in cytoplasmic genes within and among the different subspecies of *O. europaea* has revealed that Mediterranean olives contain haplotypes from two distinct maternal lineages; one unique to *O. europaea* subsp. *europaea*, the other containing both this subspecies and another subspecies in Africa. This result may be due to a hybrid origin of subsp. *europaea* (Besnard and Bervillé 2000) or secondary hybridization between the two subspecies. Western populations of oleaster have higher genetic diversity than eastern populations, more evidence of an origin associated with other subspecies in North Africa and subsequent spread of particular lineages to the east (Breton *et al.* 2004). Hybridization among subspecies and among cultivated and wild forms has been an integral component of the evolution and domestication of Mediterranean olives.

Oleasters show strong patterns of differentiation in genetic markers (Fig. 6.1; Besnard *et al.* 2002) and stone shape (Terral *et al.* 2004). The main pattern involves variation among regions, particularly between the eastern and western Mediterranean. As I discussed in Chapter 3 for other native species, such patterns indicate persistence and differentiation of wild forms in independent glacial refugia. For wild olives, three of these potential refugia, in north-west Africa, the western Mediterranean, and the eastern Mediterranean, are characterized by distinct molecular types for a range of genetic markers (Breton *et al.* 2004). It is also possible that oleasters survived the Pleistocene glaciations on Corsica, Sardinia, and/or Sicily.

Interestingly, the cytotype characteristic of eastern populations was also found, at lower frequency, in wild olives that occur close to olive groves in some western Mediterranean sites (Fig. 6.1; Besnard *et al.* 2002). In addition, there is linkage

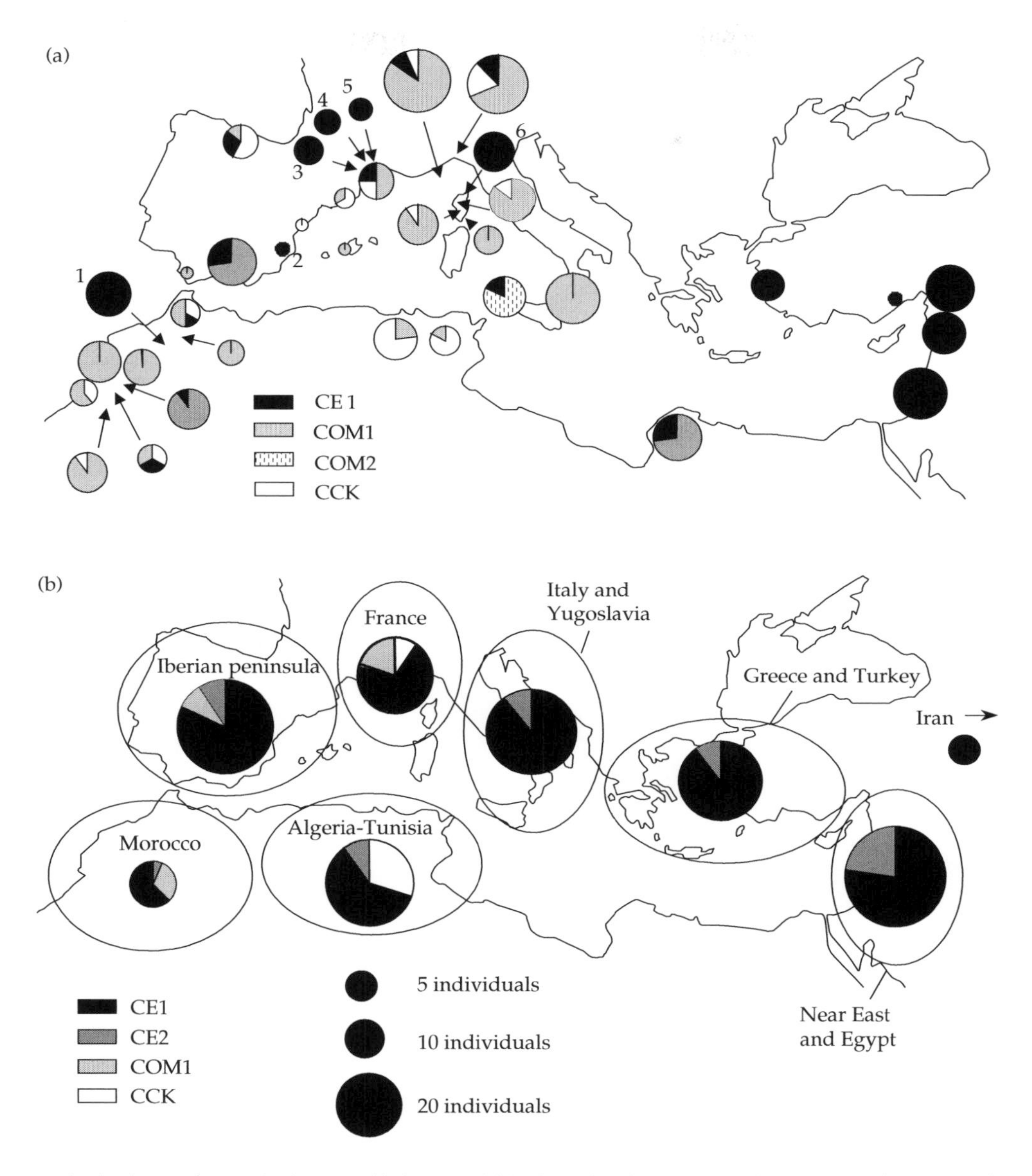

Figure 6.1 The distribution of cpDNA haplotypes in (a) oleaster and (b) cultivated Mediterranean olive (*O. europaea* subsp. *eoropaea*) (reproduced with permission from Besnard *et al.* 2002).

disequilibrium between this chlorotype and nuclear genes that are characteristic of oleasters in the eastern Mediterranean (Besnard and Bervillé 2000). These results suggest recent introduction of this cytotype from eastern locations via cultivated forms, and that oleasters near olive groves in many western Mediterranean sites represent recently naturalized feral populations. This, does not however mean that all oleasters are of hybrid origin or that they represent escaped cultivated forms. Several forests around the Mediterranean Basin, particularly in the western Mediterranean, also contain genuinely wild oleaster trees (Lumaret and Ouazzani 2001). These authors revealed that oleaster trees in some western Mediterranean sites have marker genes linked to genes that control the juvenile phase, suggesting that they would be unsuitable for domestication. The interpretation here is that such trees are genuinely wild oleasters, unpolluted by genes from cultivated varieties, as their geographic isolation from cultivated groves suggests. The suspicions of Zohary and Spiegel-Roy (1975) have thus been confirmed, there are genuinely wild oleasters in several regions, even in the western Mediterranean.

Most cultivated olives are of eastern origin, hence there is little east–west genetic differentiation among cultivated olives (Besnard *et al.* 2002). Some olive cultivars in the western Mediterranean, for example, on Corsica and in eastern Spain, arose however from direct domestication of local oleasters (Contento *et al.* 2002; Breton *et al.* 2004; Terral *et al.* 2004). Analysis of stone shape and size in archaeological remains and comparison of wild and cultivated varieties suggests that the domestication of olives in Spain was based on local wild populations and dates back to the Chalcolithic/Bronze Age, that is, prior to the introduction of cultivated plants from the Near East (Terral *et al.* 2004). However, despite multiple independent events of domestication based on local oleaster populations in different regions, that is, selection in different regions, a long history of migration with man and diversification of olive varieties which were then spread by the Phonecians, Greeks, Romans, and more contemporary humans populations, has blurred patterns of regional differentiation.

The European chestnut

The European chestnut, *Castanea sativa*, provides another illustrative example of evolution during domestication in the Mediterranean region. As in olives, the true distribution of wild chestnuts has been a topic of discussion among botanists and phytogeographers (Quézel and Médail 2003). Following westward migration of an Asian ancestral taxa probably during the Tertiary into northern Turkey, *C. sativa* is now widespread in many countries around the northern shores of the Mediterranean, as a result of human introductions, particularly during Roman times (Zohary and Hopf 2000). The status of populations in Greece and south-west Turkey remain debated, although it is not impossible that they are indigenous (Quézel and Médail 2003). However, further west, even the large expanses of chestnut forest in the Cévennes (southern France) or on Corsica are the result of initial plantations. The absence of chestnut pollen from the Sub-Boreal on Corsica strongly points to this species, an integral element of mid-elevation vegetation in large expanses on this island, being introduced to the island (Reille *et al.* 1999). The first pollen records of chestnut date to 7,000 BP in northern Spain and 5,500 BP in Italy followed by rapid expansion from about 3,000–2,000 BP (Huntley and Birks 1983). This expansion coincides with the Roman and post-Roman colonization of many Mediterranean parts of Europe. The present day distribution of chestnuts in Europe reached about 1,000 BP.

Chestnut forests occur over a wide range of climatic conditions extending from Mediterranean summer-dry zones into temperate Europe. In Turkey, where indigenous populations are thought to occur, chestnut occurs in a humid bioclimate near the Black Sea, in a summer-dry Mediterranean climate and in the north-west part of the country where the climate tends to a more continental regime. Across western Europe this species has been planted well outside of its natural ecological and geographic range. In addition, management has been variable in space and time, varying from neatly planted orchards (including some hybrid varieties) to what appear to be natural forests in the low mountains of many areas.

Zohary and Hopf (2000) point out that Turkish populations are distinct from the rest of the range in western Europe and that this region acted as a principal refuge during the Pleistocene glaciations. For these authors Turkish populations represented the source of western European chestnut and that even populations in Greece are not indigenous but were also introduced by humans. Hence migration of this species may have been associated with human movements long before the Romans began transporting chestnuts across western Europe. Studies of variation in a range of nuclear and chloroplast genetic markers in addition to morphometric analyses across the range of this species in Mediterranean Europe have allowed biologists to explore the origins of chestnut in the western Mediterranean in more depth. Several important points relating to the history of chestnut migration can be gleaned from these studies (Villani *et al.* 1991, 1992, 1994, 1999; Fineschi *et al.* 2000).

1. Analysis of cpDNA reveals low geographic structure in genetic variation, whereas nuclear markers show greater spatial structure, particularly in Turkey, where a prominent east–west pattern of differentiation has been revealed. As a result, this species provides an uncommon example of a pollen/seed flow ratio close to one. This result is no doubt the result of massive dispersal of seeds and vegetative fragments by humans and a long history of cultivation.
2. Turkish and Spanish populations have higher genetic diversity than French or Italian populations for nuclear markers, whereas populations in Italy contain a mixture of cpDNA haplotypes which are found either to the east or to the west, but are not common to the full distribution of the species. This probably results from human introductions during the Roman period.
3. For nuclear markers, Turkish populations are genetically distinct from populations across the rest of Europe. However, some cpDNA haplotypes which occur in the Iberian peninsula and Italy are absent from Turkey. This suggests that the Iberian peninsula may have served as a glacial refuge for this species. Chestnuts may thus have been present in the western Mediterranean longer than previously suggested.
4. The analysis of variation in fruit traits concords with the data on isozyme variation in Turkey, and suggests marked differentiation of eastern and western populations. In addition, populations in western Turkey are more similar to those in Italy than they are to populations in eastern Turkey.

To recapitulate, long-distance colonization associated with westward migration with humans from western Turkey, after establishment of populations in the latter region from eastern Turkey, best explain patterns of genetic variation in Mediterranean chestnut populations. Large-scale plantation in western Turkey probably hastened the genetic differentiation with populations in the centre and east of Turkey and served as a basis for rapid and fairly uniform spread of trees to the west.

In epilogue, in many areas the exploitation of chestnut forests has dramatically diminished in the last 100 years as a result of socioeconomic changes in rural systems, a trend greatly exacerbated by the debilitating effects of chestnut blight and ink root disease. This decline has stimulated some authors to question the sustainability of chestnut forests and to analyse their 'resilience' in the face of changing management and pathogen pressure (Leonardi *et al.* 2001).

6.2.6 Migration, differentiation, and hybridization in cork oak, *Quercus suber*

Cork oak is an evergreen species indigenous to the western part of the Mediterranean Basin, primarily the Iberian peninsula and North Africa, and also in a narrow belt of disjunct areas in coastal areas of southern France, western Italy, and Calabria and the islands of Corsica, Sardinia, and Sicily (Quézel and Médail 2003). Like olives, its distribution is limited to the low-elevation Mediterranean zones with mild winters. It is *the* species which allows a sustained economical production of cork, mostly from stands in the Iberian peninsula. The harvesting of cork oak begins when trees are about 25 years old (in traditional human populations this tree would have been one that was planted for peoples' children, unlike olives which were for grand children, . . .). Their cork can then be stripped once every ~10 years, and trees can live a few hundred years. The production of cork oak has been intimately associated with that of wine, nowadays good quality wine. However, cork oak stands are not only used for the production of cork as a stopper for good quality wine, they also allow for cropping, grazing, firewood, and other functions in the stand and the cork has a wide variety of uses due to its natural function as an insulating agent (Chapter 4).

The use of *Q. suber* by humans has given rise to two important processes discussed repeatedly in this section: a reduction in genetic variation in some stands and hybridization with congeners.

First, genetically distinct populations occur in different parts of the range of *Q. suber* (Toumi and Lumaret 1998). In the Iberian peninsula (and southern France), populations have relatively high diversity and low levels of differentiation among populations. Elsewhere, allelic richness is lower and there is greater differentiation among populations. This geographic structure may be the consequence of genetic bottlenecks during post-glacial re-colonization (see Chapter 3), which may have been further exacerbated by local selection by human populations. Alternatively, the occasional transport of acorns from a small number of trees with either a high yield of cork or which produce good quality cork in association with the migration of humans around the western Mediterranean could have produced this pattern. Similar patterns of reduced genetic variability, thought to be at least partly due to human-induced patterns of migration, have been detected in other oak species used by human populations, for example, *Quercus frainetto* which has edible acorns (Petit *et al.* 2002).

Second, *Q. suber* may hybridize with its congener *Quercus ilex*. In mixed stands of the two species there is good evidence from studies of allozyme diversity that introgression of *Q. ilex* genes into *Q. suber* (but not the reverse) has occurred (Elena Rossello *et al.* 1992; Toumi and Lumaret 1998). Such unidirectional introgression suggests that following an initial and perhaps rare hybridization event, there has been backcrossing primarily with parental *Q. suber*. This has profound ramifications for the genetic status of cork oak which grow in semi-natural conditions with or near *Q. ilex* populations or in mixed oak plantations.

Quercus suber is a naturally outcrossing tree, hence natural populations are likely to harbour important levels of genetic variability in traits related to oak production, particularly those in the Iberian peninsula where genetic marker diversity is highest. Studies of genetic differentiation among provenances for chemical and morphological traits could thus be favourably associated with management strategies for cork oak plantations.

6.2.7 Mediterranean geophytes: evolution under cultivation

Cyclamen, Narcissus, Crocus, Galanthus, and many other geophyte genera have diversified in the different parts of the Mediterranean region (and beyond). In these genera, a large number of cultivars now exist. What is fascinating here is that only a small selection of wild plants have been brought into cultivation and used to create the diversity of cultivated forms. In the genus *Cyclamen*, all of the diversity one can see in the florist's shop has arisen from wild *Cyclamen persicum* (Grey-Wilson 1997), whose natural distribution is limited to the eastern Mediterranean. In *Crocus*, which contains roughly four times the number of wild species of *Cyclamen*, only three species groups, all of which occur in the Mediterranean, are at the basis of the wide range of diversity one can see in cultivation (Matthew 1982). A similar tale can probably be told for other geophytic genera such as *Narcissus* from the Mediterranean and *Galanthus* from the Caucasus.

Evolution of these groups in cultivation has often seen the range of flower size and colour increase spectacularly by the establishment of pure lines of new flower colour variants (in which desired traits become fixed) with high vegetative vigour and then hybridization and chromosome change. This is particularly apparent in *Cyclamen*, where an additional and key trait which has been modified by selection is time to flowering. Perennial herbaceous geophytes have a notoriously long juvenile period and can often take a few years to flower. In controlled conditions wild *C. persicum* can flower within two growing seasons, if pampered. This period has been dramatically shortened in cultivation, and some of the contemporary cultivars can be raised to flowering in < 9 months.

The economic significance of bulb cultivars, produced from a few species in many Mediterranean geophytes, suggests to me that coordinated research involving several countries on the ecology, genetics, and conservation of wild geophytes across the Mediterranean Basin is now necessary. Such programmes exist or have existed in recent years for important tree species, including those discussed above. There has been a great deal of effort to reduce illegal trade of bulbs, and some of this has begun to pay off. In parallel, the horticultural industry, genetic resource bodies, international conservation groups and grant awarding governmental institutions would do well to jointly stimulate and co-ordinate research on (a) the conservation of the wild species thus far used to produce cultivated varieties, and (b) patterns of variation and the evolutionary processes which have produced such variation, in the wild.

6.2.8 Farming for fragrance: cultivation of aromatic plants

No overview of evolutionary process involved in the domestication of Mediterranean plants would be complete with reference to the aromatic plants of the Lamiaceae which perfume the garrigues, maquis, phrygana, or tomillares landscape of different regions in the Mediterranean Basin. Several aromatic plants, in particular lavender and thyme, are known to have been used by the Greeks and Romans, for similar reasons as those today: folk medicine, baths and toiletry, perfumes, etc. During medieval plagues they found a use in fumigation. For most species, cultivation is a secondary enterprize, by farmers with other primary crops. There are nevertheless a diversity of species which have begun to develop an important economic interest as their uses have diversified: honey, essential oils, and frozen foods have all developed alongside the important use of aromatic plants in the perfume industry. The most famous is of course lavender.

During the nineteenth century the scattered plantations of lavender in the high plains of Provence in southern France became more organized as the local perfume industry in Grasse developed and flourished. The domestication of lavender was based initially on cultivated forms of wild *Lavandula angustifolia*, which occurs naturally above ~500 m, from the Iberian peninsula to Greece. The ability to vegetatively propagate plants of this species produced several commercial clones of interest for their floral bouquet (vivid blue flowers which remain attached to the inflorescence when they dry) and essential oil. This procedure has of course been associated with low levels of genetic diversity across the landscape, hence the severe problems of die-back experienced in many areas. As more recognition of the importance of hybridization and the maintenance of genetic diversity in these species infiltrates the commercial development process, this problem may diminish. In the lowland plains, cultivation is now based on clones of a sexually sterile hybrid, the lavandin: the result of a cross between *L. angustifolia* and *Lavandula latifolia* (which is also indigenous to the western Mediterranean but generally at lower elevation). This vegetatively vigorous sterile hybrid produces a high yield of essential oil. It does however have a fairly narrow genetic resource base. A geographic study of potential sites of naturally occurring hybridization and the cultivation of a range of spontaneous sterile hybrids could overcome this deficiency. The differences in distribution and performance of distinct chemical variants within species of Lamiaceae discussed in Chapter 4 are of glaring significance here.

6.3 Invasive species in a Mediterranean environment

In his classic textbook Elton (1958: 51) first focused attention on the simple fact that 'one of the primary reasons for the spread and establishment of species has been quite simply the movement around the world by man of plants, especially those intentionally brought for crops or garden ornament or forestry . . .'. Many plants introduced in association with human activities to a new region or continent, that is, across a natural barrier to dispersal, have then spread from their sites of introduction to spread across large areas, sometimes in fairly natural habitat, most often in areas disturbed by human activities. These are what we call invasive species. The massive interchange of cultivated and weedy species among the five different Mediterranean-climate regions of the world means that several of these regions have been overrun by introduced species from the Mediterranean Basin. Indeed, some very serious invasive species have made the leap, in the hands of humans, from one Mediterranean-climate region to another, where they pose a serious threat to natural biodiversity (Box 6.5).

Box 6.5 Effects of invasions on biodiversity

The long-distance dispersal of plants over barriers to their natural dispersal in association with human activities has given rise to a major biological problem due to the spread of many species away from the site of introduction to invade new areas (Heywood 1989; Vitousek 1990). The problem of biological invasions is complex, it occurs at different levels, from the population to the ecosystem, and on a range of spatial scales. Although patterns consistent with niche shifts or competitive exclusion in native species as a result of invasion by an exotic plant have not been reported (Crawley 1987), there are several important biodiversity consequences of plant invasions. For example, the arrival of a dominant invasive species can cause a decline in local species diversity and because invasive species usually cover large areas there is a serious risk of homogenization of floras (reduced β-diversity) as communities contain a smaller number of more widespread species. Significant effects of plant invasions on local biodiversity may also occur as a result of alteration of the trophic structure in an ecosystem.

Some examples of the biodiversity consequences of plant invasions in the Mediterranean include the following.

1. The invasion of coastal vegetation by clonal plants such as *Carpobrotus edulis* and other succulents such as *Agave americana* and *Opuntia ficus-indica* has had negative effects on natural vegetation in northern Spain and on the western Mediterranean islands (Casasayas Fornell 1990; Vilà and Muñoz 1999; Suehs *et al.* 2001). In California invasive *Carpobrotus* have fewer native plant species associated with them than native *Carpobrotus* (Albert *et al.* 1997). Mats of invasive *Carpobrotus* in some coastal sites in California can be up to 55 cm deep, and thus have no seedling recruitment or other species within the clump (D'Antonio 1993).
2. Pine plantations in southern Spain are associated with low plant species diversity and a reduced presence of heathland endemic species (Andrés and Ojeda 2002).
3. Domesticated crops and introduced invasive species may also alter the ecology and evolutionary dynamics of native plants. For example, fleshy-fruited domesticated crops in the Mediterranean region (such as olives discussed above, but also those in suburban zones such as *Pyracantha* and *Cotoneaster*) and may alter the behaviour of native bird dispersers (Debussche and Isenmann 1990), with potential effects on the dispersal of native shrubs.
4. The introduction of species can cause hybridization with native populations, with clear evolutionary consequences and conservation implications (Vilà *et al.* 2000).

Plant evolution may also be affected by the invasion of other groups such as insect herbivores and pollinators. For example, the introduction of *Bombus terrestris* in sites in the eastern Mediterranean may introduce constraints on the pollination of plant species, since this species of bumblebee may often behave as a nectar robber which reduces the reward available for native pollinators (Dafni and Shmida 1996).

In order to predict future invasions and control them we need to understand why particular species in certain regions become invaders. Three questions are at the heart of this issue:

- Many introduced species do not become invasive, hence one can ask: what are the biological traits which promote the invasive capacity of an introduced species?
- Like any colonization event (Chapter 4), invasion is a context-dependent process. What then are the properties of the indigenous community/ecosystem which determine whether an ecological system will be invasion prone or resistant to invasion?
- Does evolutionary change promote invasion?

In this section I discuss the problem of biological invasions in the Mediterranean region and the invasiveness of Mediterranean species in other parts of the world.

6.3.1 Can we predict invasiveness based on biological traits?

Several authors have discussed the biological attributes which characterize invasive species.

1. In ruderal species in highly disturbed and agricultural settings, the most weedy species have a similar array of characteristics. First, annual weeds are predominantly selfers, allowing for rapid local differentiation and adaptation. Second, the combination of continuous and high seed production, rapid germination, and seedling growth in favourable environments and a high tolerance of environmental variation are the traits which make for an 'ideal weed' (Baker 1965). These traits illustrate the potential role of propagule pressure in promoting invasion success (Lonsdale 1999). Lepart and Debussche (1991) illustrate examples of the importance of propagule pressure in the Mediterranean region.

2. In perennial herbaceous plants, vegetative reproduction and plasticity are important traits favouring invasion of new sites (Baker 1965; Thompson 1991). Vegetative reproduction can be particularly important where it permits the invader to either stabilize the local soil or to pre-empt sites and prevent the establishment of competitors (Crawley 1987). The strategy of invasion by *Ailanthus altissima* in the Mediterranean illustrates this point (Lepart and Debussche 1991).

3. In wind-dispersed trees, for example, *Pinus* species, a short juvenile period and a short interval between large seed crops favour rapid population growth and local dispersal and, along with small seeds, are associated with invasiveness (Rejmánek and Richardson 1996). This has been illustrated for pines in the Mediterranean region (Barbero *et al.* 1990). Such traits also favour re-colonization by indigenous pine trees (Chapter 4), illustrating the fact that invasive plants have traits which facilitate colonization in a parallel fashion to those of pioneer species in early succession.

4. The genetic attributes of particular species and traits linked to the mating system, which ultimately determines the organization of genetic variation within and among populations, may also be important in the invasion process (Barrett and Richardson 1986; Gray 1986).

5. In pine trees, Rejmánek and Richardson (1996) have shown that invasiveness has a phylogenetic component, with invasive species clearly concentrated in the subgenus *Pinus* and non-invasive species in the subgenus *Strobus*. In some areas of the Mediterranean, for example, the Balearic islands (Vilà and Muñoz 1999), several families (Solanaceae, Amaranthaceae, Iridaceae, and Euphorbiaceae) are overrepresented in exotic species, indicative of a phylogenetic component to invasion success.

There are thus many characteristics which contribute to the capacity for a plant to become invasive, hence, a range of solutions are open to take up the challenge of a new environment. There is no single best strategy, although a combination of effective dispersal, high phenotypic plasticity, and high interspecific competitive ability is likely to be of prime importance. So, it is rather difficult to provide a general statement concerning traits which favour invasion, and it is not surprising that closely related species may differ dramatically in their invasive potential, due to subtle differences in traits and combinations of traits in relation to the environment of the introduced range. A precise illustration of this can be seen in Box 6.6.

Predicting invasion based on traits is also difficult due to the important effect of chance events, for example, in a study of exotic species on five large islands in the Mediterranean, Lloret *et al.* (2003) found that most exotic species only occur on a single island. Nevertheless these authors also reported that dispersal attributes appear as important correlates of invasion success and that factors determining the rate of invasion at the regional level, may also be important at smaller spatial scales, as correlated patterns of local and regional abundance of invasive species would suggest.

6.3.2 What makes an ecosystem invasion prone?

Invasion success is context dependent. Indeed, the history of invasion of plants in many regions

Box 6.6 Biological differences between closely related species can interact with a range of environmental factors to influence invasion success

The colonization of a Mediterranean old-field by *Conyza* subsp. in southern France just two years after the removal of vine plants

Thébaud *et al.* (1996) studied two invasive *Conyza* species, *Conyza canadensis* and *C. sumatrensis* in the French Mediterranean region. The two species both have a very similar set of the 'ideal weed' traits identified by Baker (1965). They differ nevertheless in patterns of habitat occupation: whereas *C. canadensis* is limited to recently abandoned old-fields and disturbed sites, *C. sumatrensis* colonizes both early and mid-successional habitats and is more abundant than its congener in sites where they co-occur. To assess the causes of these differences in invasion success, the above authors transplanted seedlings of the two species into three contrasting Mediterranean old-fields which differed in age since abandonment: 6 months, 4 years, and 17 years, respectively. In each type of old-field, seedling performance and reproductive success of adult plants were assessed in relation to competition with neighbouring plants, nutrient and water stress, and herbivory. This experiment clearly showed that *C. sumatrensis* has unarguably a superior competitive ability and a higher capacity for water and resource uptake than *C. canadensis*. Competition, in relation to spatio-temporal heterogeneity of nutrient and water stress, may be of paramount importance for persistence in Mediterranean old-fields and thus the differential distribution of the two species.

has shown that local 'environmental circumstances extrinsic to the invader are often the most important factor in determining its success' (Newsome and Noble 1986: 15). This idea underlies much research into the question of what makes an ecological system invasion prone. A brief survey of this literature reveals that several features of local communities and the abiotic environment may be associated with resistance to invasion, disturbance being of primary importance (Box 6.7). An important point here is the complexity of the interaction between demographic traits such as competitive

Box 6.7 Features of local communities and the abiotic environment which may be associated with resistance to invasion

Many factors influence whether an ecological system will suffer invasions.

Disturbance. Invasive species are often most abundant in disturbed areas where the dominant competitors have been removed, overall cover reduced, and potential pests eliminated (Crawley 1987; Rejmánek 1989). Competitive exclusion by native species may be of major importance to the resistance to invasion (Keane and Crawley 2002).

Species richness. Ever since Elton's (1958) classic, there has been much interest in the idea that diverse communities may be more resistant to invasions than species-poor communities. In a review of this issue, Levine and D'Antonio (1999) show that although theoretical models consistently concord with this prediction, the empirical data are variable. Studies which have experimentally manipulated diversity have not furnished a general relationship. In fact some of the most diverse habitats may be particularly prone to invasion (Levine and D'Antonio 1999; Lonsdale 1999), perhaps due to spatial covariation with other ecological parameters which promote invasion or their better resistance to disturbance, which may be more directly linked to invasion success.

Type of plant formation and species identity. Several authors have examined whether different types of formation, or the presence of different combinations of groups of species, differ in their resistance to invasion. These studies show that the spatial heterogeneity of the biotic environment may limit invasion success in some areas.

Reduced pest pressure. In their native habitats, many species are subject to high levels of herbivory, parasitism, and flower and seed predation. Introduction to new areas is often accompanied by a loss of these regulatory agents, which may allow plants to become better competitors or produce more seed and thus persist locally and spread more rapidly.

Spatial habitat heterogeneity. This is a key issue, which I discuss in Box 6.9.

ability which may facilitate site pre-emption and propagule pressure and the nature of the communities into which species invade.

Ecosystems in the Old World, are thought to be more 'resistant' to invasions than those of the southern hemisphere (Di Castri 1990). However, this is only true when variation due to different native species diversity levels are accounted for (Lonsdale 1999). Both in Mediterranean Europe and North-Africa the overall number of exotic non-native plants that have established populations is very low compared to other Mediterranean-climate regions (Di Castri 1990; Le Floc'h *et al.* 1990; Quezel *et al.* 1990; Blondel and Aronson 1999). Although plants from the Mediterranean Basin have successfully colonized other Mediterranean-type ecosystems following their introduction in association with human movements, the reverse direction of invasion is less frequent.

The asymmetry of invasions among the Mediterranean-climate regions is no doubt caused by several factors. First, human patterns of colonization and the direction of trade have been asymmetric, occurring primarily towards the other Mediterranean-climate regions. Although this may have contributed to the asymmetry it does not fully explain why introductions to the Mediterranean region have produced relatively fewer invasions. Introductions have indeed occurred. One precise example concerns the establishment of an exotic flora, the so-called 'Flora Juvenalis' near Montpellier in southern France, where many species were introduced with the flourishing wool trade at the Port Juvenal site south of Montpellier in the eighteenth and early nineteenth centuries (Le Floc'h 1991). As this author relates, the number of known exotic species in this area reached ~450 by the middle of the nineteenth century. This number declined dramatically to ~10 species at the start of the twentieth century in association with a regression of the wool trade in the region. Only six species were reported in the middle of the twentieth century, all of which were confined to a single ruderal community. Only one species appears to have spread further.

A second important contributing factor to the asymmetry of invasions concerns the long history of human activities in the Mediterranean Basin, which has created evolutionary pressures that have no doubt selected for weedy plants capable of exploiting highly disturbed and cultivated areas. Grasses and composites, two families with large numbers of invasive species in the different Mediterranean-climate regions (Groves 1991), have been taken from regions with a long history of selection pressures imposed by agricultural activities to others where this selection regime is much more recent. Such species would thus have been pre-adapted for human disturbed habitats in the climates of the other Mediterranean regions of the world (Di Castri 1991; Groves 1991). The export of such species into newly disturbed areas, beginning in the sixteenth century, would thus have provided the impetus for the rapid invasion of what were already weedy species in the western Mediterranean at that time. The shorter history of agriculture and disturbance in the other Mediterranean-climate zones has probably caused the low number of possible invasions into the Mediterranean Basin.

The long history of disturbance and agriculture in the Mediterranean flora will thus have been important elements in the relatively low numbers of plant invaders in Mediterranean ecosystems (Di Castri 1990). In addition, Mediterranean landscapes have seen many episodes of plant migration and colonization of species with very different biogeographical affinities. It is as if the flora were full of ancient invaders that prevent contemporary introduced species from becoming highly invasive. In addition, human activities have long been present in the Mediterranean region, allowing for long-standing adaptation to human disturbance in the native flora and thus a better resistance to invasive species. Lets not forget, in addition to these factors, the fact that the climate of the Mediterranean Basin is not completely identical to that of the other Mediterranean-climate regions, lacking any oceanic buffering of extreme climatic events and experiencing cold temperatures in winter (Di Castri 1991). This climatic difference could also affect the asymmetry of invasions.

The species which have become invasive following introduction to the Mediterranean region have diverse origins (Table 6.2). The majority are not, as one may have thought, from other Mediterranean-climate zones. The most important sources are the North American temperate climate zones (Quézel *et al.* 1990) and the tropics (Dafni and Heller 1990). In a study of the exotic plant species on the islands of Majorca, Sardinia, Corsica, Malta, and Crete, Lloret *et al.* (2003) found that exotic species of non-European origin were more abundant than those of European origin, both in terms of local abundance on a given island and regional

Table 6.2 The diverse origins of invasive species in the Mediterranean Basin

Source region	Examples of invasive species
South Africa (mostly the Cape region)	*Oxalis pes-caprae, Carpobrotus edulis, C. acinaciformis, Senecio inaequidens*
Temperate flora of the Old World	*Ailanthus altissima, Buddleja davidii, Artemisia* spp. *Impatiens* spp.
Temperate climatic zones of North America	*Amaranthus retroflexus, Opuntia ficus-indica, Solidago* spp., *Xanthium* spp., *Robinia pseudoacacia, Phytolacca americana, Baccharis halmifolia, Amorpha fruticosa, Conyza canadensis*
Central and South America (temperate zone and neotropics)	*Solanum eleaginifolium, Oxalis latifolia, Amaranthus deflexus, Cortaderia selloana, Conyza bonariensis, Paspalum dilatatum*
Subtropical and tropical regions of Africa and south-east Asia	*Pennisetum villosum, Sorghum* spp.

abundance (proportion of islands where species are present).

In the Mediterranean, as in many other regions, the majority of invasive species become established in unstable and highly disturbed anthropogeneic habitats with a low average level of natural plant cover (Casayas Fornel 1990; Quézel *et al.* 1990). The removal of dominant species (which may be strong competitors in the regeneration niche) as a result of disturbance in such habitats may be a critical component of this invasiveness, although reduced competition is not the only cause since disturbance also removes natural enemies and can directly alter abiotic resource levels in the soil, particularly when associated with fires (Richardson and Bond 1991). Natural vegetation in the Mediterranean has thus suffered less from invasive species than other regions, albeit with 'some remarkable exceptions' (Quézel *et al.* 1990: 51), some of which are illustrated in Box 6.8.

6.3.3 Population ecology and evolutionary change in invasive species

The above discussion indicates that the interplay of intrinsic and extrinsic factors will be critical in determining invasion success. An important point here is that invasion success will be dependent on the local environmental conditions. Resource status, community structure, and the abundance of herbivores, parasites, and pollinators may all co-vary among local sites, creating a mosaic environment for the invasive species. Invasion success will depend on the capacity of the invader to respond to the new challenges imposed by this heterogeneous environment (Box 6.9). At the heart of this interplay, and the theme for this section, is the potential for evolutionary change in the introduced range, where new selection pressures are encountered and population characteristics (primarily size) are dramatically altered. In fact plant invasions come with a number of advantages for the study of evolutionary ecology: (a) there is usually some knowledge of the history of the colonization process, in terms of dates of introduction and observations of spread, (b) although frequently flawed by lack of replication they provide an experiment in natural conditions, (c) compared with the study of rare species, which also illustrate issues pertaining to persistence and distribution limits (Chapter 2), there is ample material for experimental work, and one does not work with the worry of exacerbating rarity by manipulating one's study material.

Evolutionary change in invasive populations can occur in relation to the primary evolutionary forces of genetic drift in founder populations, opportunities for hybridization, polyploidy, and novel selection pressures in the introduced range (primarily due to altered pest and predator regulation, differences in pollinators, and changes in abiotic resource status). Much of the response to spatial and temporal habitat heterogeneity boils down to either evolving adaptations to the new challenge or buffering environmental extremes and stress by phenotypic plasticity. Hence, the genetic and ecological processes which promote or limit biological invasions are those associated with natural colonization and evolution (Chapters 3 and 4). Whether local adaptation will contribute to invasion success will depend on the environmental variation faced by the invader, the amount and spatial organization of genetic variation in the introduced populations and the genetic basis of fitness trait variation in relation to variation in other traits (Barrett and Richardson 1986). My focus here is on examples from the Mediterranean region which provide insights into the nature of evolutionary change in species in their introduced range and whether such changes contribute to invasive success.

A primary form of evolutionary change concerns morphological adjustments associated with a shift in strategy from reduced defence to improved competitive ability. Based on a survey of plant species introduced to California from Europe, Crawley (1987) first remarked on a trend for plant species to be larger in their introduced range than in their home range more often than they are smaller in the introduced range. Since then other authors have developed the idea that due to reduced resource allocation to defensive compounds associated with reduced pest pressure outside of the native range, plants may re-allocate saved resources to competitive ability, and thus be larger in the introduced range (Blossey and Nötzold 1995). Although there is some evidence

Box 6.8 Spectacular examples of rapid colonization and invasion of semi-natural habitats. Three main types of habitat tend to be colonized by invasive species

(a) Woodland vegetation along watercourses. Here one can observe a fairly diverse array of exotic species: for example, *Impatiens, Solidago, Lonicera*, and *Oenothera* (Quézel *et al.* 1990). The flowering ash *Fraxinus ornus* (bearing white flowers in photo) has rapidly spread down the Hérault and Vis river system in southern France after its introduction to a single site (Thébaud and Debussche 1991).

(b) Upland grasslands in southern France. For example, *Pinus nigra* is spreading rapidly from planted areas (background) across open grassland on the open limestone plateau areas of southern France (Lepart *et al.* 2001). (Photo kindly supplied by J. Lepart and P. Marty).

(c) Coastal vegetation. Invasion by clonal plants such as *Carpobrotus acinaciformis, C. edulis* and hybrids between them, as well as other succulents, for example, *Agave americana* and *Opuntia ficus-indica*, are a feature of many coastal areas (Casasayas Fornell 1990; Suehs *et al.* 1991). Photo (kindly supplied by C. Suehs) shows mats of *C. edulis* on the Cape Medes on the island of Pourquerolles (southern France).

Box 6.9 Invasion success can depend on the capacity of the invader to respond to a variable environment

Several examples illustrate the role of spatial and temporal environmental heterogeneity in Mediterranean settings for the limitation of invasive species.

First, the invasion of different coastal habitats in California by *Carpobrotus edulis* studied by Carla d'Antonio and her colleagues clearly shows the context-dependent nature of invasion success. Introduced from South Africa in the early twentieth century, *C. edulis* has become a serious threat to the Mediterranean coastal vegetation of California. In a study of the factors limiting germination, establishment and growth of this species in grassland, backdune, and coastal scrub communities, D'Antonio (1993) reported that invasion is restricted by aridity stress, herbivory, and lack of seedling recruitment in dense vegetation, but that once established this species can out-compete native grasses. As a result of reduced herbivory, dune communities were less invaded than scrub communities. Disturbance and fire also facilitate the process of invasion by invasive *Carpobrotus* (D'Antonio *et al.* 1993). In this example, invasion success is highly dependent on the combination of local abiotic and biotic conditions and is thus 'a context-specific process' (D'Antonio 1993: 92). The identity of different groups of species may also influence invasion success in Mediterranean communities (Prieur-Richard *et al.* 2000). To predict invasion potential thus requires some knowledge of the local plant and animal communities and the interaction between the invader and this environmental heterogeneity, which may limit invasion into natural communities, and create divergent selection pressures on the invader.

Second, in the Mediterranean, spatio-temporal heterogeneity in plant recruitment is likely to complicate our ability to assess invasive potential. As shown in the resprouting tussock grass *Ampelodesmos mauritanica*, which has recently colonized coastal shrubland areas in many areas of the western Mediterranean, there can be strong spatial and temporal variation in reproductive output, which makes it difficult to assess long-term dynamics (Vilà and Lloret 2000). However in this system, sporadic episodes of rainfall or resource availability after fire, typical components of the Mediterranean environment, may dramatically increase the seed crop of this species and thus its rates of recruitment. Drought stress may nevertheless be a major limitation on the expansion of this species.

Finally, Mésleard *et al.* (1993) reported a reversal of competitive effects of the introduced grass *Paspalum paspalodes* on a native *Aelurops* in the Camargue wetlands of Mediterranean France. Whereas at low salinity the introduced species may eliminate the native species, the lack of high salt tolerance of the introduced species caused a reversal of the competitive balance. This low tolerance of high salinity means that in the absence of human-induced flooding, the introduced species loses its invasive capacity.

for changes in size and allocation, the generality of this pattern has not been established. A comparative study of plant size for species in their introduced range compared to their native range by Thébaud and Simberloff (2001) has shown that in general there is no trend for increased size in the introduced range, either for plants introduced to North America from Europe or vice versa, although several species do show the predicted increase in size. Likewise, although comparative evaluation has shown that invasive species tend to have stronger competitive effects on native species than vice versa, it is far from being clear whether the former generally have a better competitive ability (Vilà and Weiner 2004). Indeed, in natural habitats, one would predict that native species have superior competitive ability and thus resist invasion, unless the invasive species have a very strong advantage in terms of 'enemy release' (Keane and Crawley 2002). This resistance will of course be totally lost in disturbed habitats where the invasive species may be better adapted to local conditions and be free of competitors and pests. Indeed the invasion of coastal wetlands in Mediterranean France by *Paspalum paspalodes* depends on human-induced flooding which gives this species a competitive advantage over native species, which are better adapted to high salinity in the absence of flooding (Mésleard *et al.* 1993). This illustrates that differences in competitive ability among introduced and native species can be context dependent (Box 6.9).

In Mediterranean plants the few studies that have been done on this issue provide some encouraging results for the hypothesis of evolutionary change in the introduced range. For example, in a study of introduced *Trifolium subterraneum* in Australia, Olivieri *et al.* (1991) reported that two extreme phenotypes may have evolved. One phenotype has few unbranched and long stolons and many flowers, while the other has many stolons, short internodes, and few flowers. Plants in the Mediterranean region (sampled on Corsica) have an intermediate phenotype. The two phenotypes in the introduced range may thus represent the result of selection for the colonization of open sites or persistence and high competitive ability, respectively. Reduced resistance in the introduced range to pathogens only present in the home range has also been documented in Mediterranean *Carduus* introduced to Australia (Olivieri 1984). However, in the exotic flora of some of the big Mediterranean islands, Lloret *et al.* (2003) reported that taller plants do not have a greater regional abundance than smaller plants. More precise experimental comparisons of competitive ability will thus be necessary (Vilà and Weiner 2004).

Hybridization is a key process in plant evolution (Chapter 3) which may commonly occur among cultivated and native species (Box 6.5) and may produce successful invaders due to hybrid vigour resulting from enhanced vegetative growth and escape from herbivores (Thompson 1991; Abbott 1992; Vilà and D'Antonio 1998*c*; Ellstrand and Schierenbeck 2000). Hybridization is often not a one-off event, it continues if hybrid forms are viable, and back crossing and introgression subsequent to the initial hybridization event may be critical for future adaptive evolution (Rieseberg and Wendel 1993). Hybridization may occur at different moments in the history of an invasion. First it may occur among two native species in their native range to produce plants which become successful invaders elsewhere. Second it may occur in the introduced range, either among two introduced species or among an introduced and native congener, to produce a new taxon with invasive potential. Several examples either in the Mediterranean region or involving Mediterranean plants which have been moved out of their native range illustrate these scenarios.

Senecio inaequidens. This species provides an illustration of hybridization prior to introduction and invasion by hybrids in a new region. In Chapter 3, I discussed work done by R.J. Abbott and H.P. Comes which has shown how hybridization in the Mediterranean has produced hybrid *Senecio* unable to complete speciation and establish stable populations due to a lack of reproductive isolation with the parents. Once introduced to a new region the hybrids have however become successful colonists. An example of invasion in the Mediterranean region by a hybrid Senecio, in allopatry from its progenitors, which is of current concern is *S. inaequidens*. This species is native to South-Africa. Following its introduction in sheep's wool in the late nineteenth century/early twentieth century into south-west France and other localities in Europe, this species has, in the last 30 years, begun colonizing many sites away from the site of introduction, including the Mediterranean part of France (Guillerm *et al.* 1990). All sampled populations in Europe, including Mediterranean France, appear to contain only polyploid individuals, whereas both diploids and polyploids occur in South Africa where a mixed cytotype zone has also been discovered (Lafuma *et al.* 2003). The genome size of the polyploids suggest that they arose by hybridization among previously isolated diploids in this contact zone. These authors thus propose that introductions have involved vigorous allo-polyploid material which in reproductive isolation has been able to spread rapidly. The hybrid origin of the polyploid may have contributed to its remarkable vigour in a new and variable environment, including Mediterranean France, where the species is actively expanding its range.

Onopordum. *Onopordum* thistles introduced to Australia from the Mediterranean region illustrate how hybridization has probably occurred after the transport of a native species out of the Mediterranean with a second introduced species. In Australia, two introduced species, *Onopordum acanthium* (Scotch thistle) and *Onopordum illyricum* (Illyrian thistle), form important invasive weeds (Briese *et al.* 1990). These two species occur in separate sections of the genus and in different clades (O'Hanlon 2001), but have the same chromosome number of $2n = 34$. *O. illyricum* occurs in Mediterranean garrigue-type habitat around the northern shores of the Mediterranean region from Portugal to southern Bulgaria, while *O. acanthium* has a more temperate distribution northwards into Atlantic and continental Europe. The two species are not commonly found together although where their natural distributions do show overlap in Mediterranean Europe, such as on the Causses du Larzac limestone plateau in southern France, the presence of morphologically intermediate plants present suggests that the two species form a narrow hybrid zone (O'Hanlon 2001). In the introduced range, the distributions of the two species overlap more clearly and many sites contain plants with genetic profiles (based on nuclear AFLP analyses) that span the range of intermediate variation possible between the two species (O'Hanlon *et al.* 1999). Hybridization is the most likely cause of this variation in the introduced range. It has also been found that most infestations in the introduced range were comprised of hybrid genotypes, which have established by secondary dispersal from neighbouring sites rather than primary contact between the parental species (O'Hanlon 2001). So although some hybrids may have been introduced to Australia, hybridization and introgression following introduction of the parents has probably been more important. As for cases of natural hybridization in *Senecio* (Chapter 3) genetic stabilization of hybrids may be favoured in the new conditions of the introduced range (novel ecological conditions and relaxed pest and parasite pressure). This example illustrates the importance of isolation for hybrid success. The contrasting fates of hybrid genotypes where the two parental species overlap naturally in Mediterranean Europe and in the introduced range provide strong support for the idea that hybrid establishment and speciation may be ecologically driven.

Rhododendron ponticum. This is a serious invasive species in the British Isles where it illustrates a Mediterranean plant introduced to a new region where it has probably hybridized with a second introduced species. *R. ponticum* has a highly disjunct natural distribution, with populations in the south-eastern Iberian peninsula, around the southern and eastern fringes of the Black Sea and in the Lebanon. The source of invasive material was no doubt the western Mediterranean populations, ~90% of accessions from different parts of the introduced range in the British Isles have a cpDNA haplotype that occurs in native material from Spain, and 10% in Portugal (Milne and Abbott 2000). In addition, these authors found evidence for hybridization with another introduced species from North America, whose germplasm represents ~5% that of naturalized *R. ponticum*. Since it has a better cold tolerance, the high frequency of its germplasm in invasive *R. ponticum* in Scotland suggest that hybridization has been associated with the transfer of adaptive genes in the new range. The capacity of *Rhododendron* species for natural hybridization (Chapter 3) supports such a claim.

Carpobrotus. Species of *Carpobrotus* provide a classic example of the role of hybrid vigour in promoting a biological invasion in Mediterranean ecosystems. A well-documented example concerns the spread of mat-forming perennial *Carpobrotus* in coastal habitats in California and the Mediterranean Basin. In California, hybridization between exotic *C. edulis* and native *C. chilensis* has produced a range of introgressed and highly vigorous morphotypes that are abundant in many coastal areas (Albert *et al.* 1997; Gallagher *et al.* 1997). These hybrid forms are more vigorous (in terms of vegetative growth) than the two parents and show enhanced survival in the presence of herbivory in three different coastal environments compared to the native species but only in one of the three habitats when compared to their exotic parental species (Vilà and D'Antonio 1998*a*).

Exotic *C. edulis* and the hybrid derivatives also have enhanced fruit attractivity for native frugivores and higher seed survival after gut passage than the native species (Vilà and D'Antonio 1998*b*), as well as higher levels of allozyme variation (Gallagher *et al.* 1997). Several vegetative (vigour and resistance to herbivory) and reproductive characters, plus higher levels of genetic variability, may thus facilitate the spread of the invasive species and their aggressive hybrids, with the hybrids potentially more invasive than the introduced species in some habitats.

Hybridization involving exotic *Carpobrotus* has also been detected in recent work in the Mediterranean Basin where *C. edulis* and *C. acinaciformis* have colonized back-dunes, maritime cliff-tops, coastal scrublands and grasslands, mostly on acid substrates in the western Mediterranean (Suehs *et al.* 2001). This time the hybridization has occurred among two exotic species, no native *Carpobrotus* being present. These species were introduced to the Marseille Botanical Garden in the nineteenth century and have also been planted for dune stabilization. The two taxa became naturalized in the twentieth century and following a latent phase (or lag-time) which characterizes many invasions, the two species have become invasive, particularly on small islands in the western Mediterranean, where they threaten native vegetation (Vilà and Muñoz 1999; Suehs *et al.* 2001). In a recent study of genetic variability of exotic *Carpobrotus* on the island of Bagaud in the Hyères archipelago off the coast of southern France, Suehs *et al.* (2004*a*) demonstrated the presence of intermediate individuals resulting from the introgression of genes from *C. edulis* into *C. acinaciformis*. The population described as *C. acinaciformis* contains a large number of introgressed plants which outnumber the parental genotypes and form a population of recent introgressants which, given the allopatry of *C. edulis* may become a single population of stabilized introgressants over time. Both species have high levels of genetic diversity compared to other clonal plants, despite the fact that they may have incurred genetic bottlenecks during their introduction. As predicted for introgressant populations (Rieseberg and Wendel 1993), hybridization may have enhanced genetic diversity, with potential effects on vigour and adaptive potential in the introduced range.

The two introduced *Carpobrotus* species differ markedly in their reproductive biology (Suehs *et al.* 2004*b*), in a way which illustrates once more how subtle differences in biological traits among closely related introduced species may alter the process of invasion (see also Box 6.6). *C. edulis* has a greater capacity for uniparental reproduction by selfing and agamospermy than *C. acinaciformis* which only produces a significant number of seeds when it is pollinated by *C. edulis* or after open pollination in a natural population (Fig. 6.2). Since the study population used for *C. acinaciformis* is the site containing many introgressed individuals, open pollination is likely to involve some form of hybridization due to the presence of *C. edulis* genes in the majority of the plants at this site (Suehs *et al.* 2004*a*). These results strongly suggest that sexual reproduction by *C. acinaciformis* requires genes from *C. edulis* to be

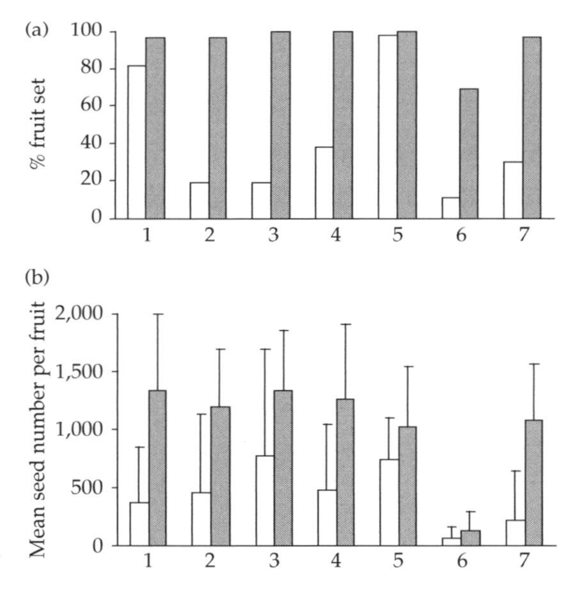

Figure 6.2 The reproductive performance of *C. acinaciformis* (open bars) and *C. edulis* (closed bars) in terms of (a) percentage of fruit set and (b) mean seed number per fruit in seven different treatments: (1) open pollination, (2) spontaneous selfing, (3) hand selfing, (4) hand outcrossing, (5) hybridization, (6) agamospermy, and (7) spontaneous selfing with insecticide. Reproduced with permission from Suehs *et al.* (2004*b*).

successful, a feature that will facilitate introgression where the species occur together. The authors conclude that both *C. edulis* and *C. acinaciformis* should be considered as harmful invasive plants in the Mediterranean Basin, the former because of the flexibility of its mating system and high seed production, and the latter because of its strong clonality, hybrid vigour, and potential for continued introgression of *C. edulis* genes. In a comparison of populations on mainland France and the island of Bagaud in the Hyères archipelago, Suehs *et al.* (2004*c*) found no differences in capacity for uniparental reproduction in *C. edulis* that is indicative of adaptive evolution in the two different contexts. The lack of differences is probably explained by the long generation time of this clonal plant and its recent introduction (~150 years ago).

The differences in reproductive biology of these two invasive *Carpobrotus* in the western Mediterranean point to a need for independent control strategies, and the elimination of sites where the two species occur together is a clear priority, especially since the population containing a hybrid swarm is currently expanding (Suehs *et al.* 2004*a*). An important additional point here is that interactions with mammalian seed dispersers which are attracted by *Carpobrotus* fruits may enhance seedling recruitment, hence a direct trophic interaction may also facilitate invasion success (Bourgeois *et al.*, IMEP-Marseille, unpublished manuscript).

Arrhenatherum. Hybridization may be compromised due to differences in the ploidy level of introduced species (often polyploids) and wild relatives, which may be diploid progenitors of cultivated varieties. As I discussed in Chapter 3, triploid inviability in polyploid groups can cause selection for reproductive isolation among local diploids and polyploids. In *Arrhenatherum elatius* the spread of tetraploids as a pasture fodder plant, and its capacity to colonize open, disturbed areas has brought this cytotype into secondary contact with natural populations of the Mediterranean patro-endemic diploid in the southern Pyrenees (Box 6.10). Triploids are extremely rare in such sites (<1% of plants in the zone) and did not flower in cultivation (C. Petit *et al.* 1997), suggesting strong selection for reproductive isolation among diploids and polyploids prior to fertilization (i.e. pre-zygotic isolation). The two cytotypes show almost complete divergence in flowering time, both in parapatry or allopatry. So although flowering time differences may serve to isolate the two taxa in the contact zone, the differences have not evolved due to selection for reproductive isolation, they existed prior to contact (probably because the polyploid occurs at lower altitudes than the diploid and thus flowers earlier). In contrast, both cytotypes showed higher selfing rates in the contact zone. This result points to selection for selfing in the contact zone, in both cytotypes. In allopatric populations, the diploids showed no evidence for selfing, whereas tetraploids apparently practice significant rates of selfing in the absence of reproductive contact with diploids. It is thus possible that the high selfing rate of diploids in the contact zone has evolved after introgression of selfing genes from the polyploid.

6.3.4 Human assisted range extensions in the Mediterranean

A very particular case of invasions concerns the human-induced spread of native Mediterranean species outside of their natural range but within the Mediterranean region. This is particularly apparent in a number of species that have become naturalized (and which in some cases are actively spreading) following their introduction to new sites. This is particularly evident for species introduced to the western Mediterranean from the eastern part of the region, but less common in the opposite direction. Some examples, in addition to the cultivated crops discussed in the early part of this chapter, include *Phlomis fruticosa*, *Ornithogallum nutans*, *Platanus orientalis*, *Fraxinus ornus*, *Dittrichia viscosa*, and *Cercis siliquastrum*, plus a bunch of weedy species that usually grow in and around cultivated fields. The geographic asymmetry of this invasion is no doubt primarily the result of associations with the migration of agricultural activities and crops towards the west although I would suggest that environmental variation across the Mediterranean may also contribute.

The flowering ash, *F. ornus*, illustrates well this process of invasion just outside of the native range

Box 6.10 Reproductive isolation in a diploid–polyploid contact zone in the southern Pyrenees (from C. Petit *et al.* 1997)

In *Arrhenatherum elatius*, diploid subsp. *sardoum* occurs in pine forests and open scree slopes in mountain areas of Corsica and northern Spain. This diploid is a disjunct patro-endemic which is closely related to the widespread tetraploid subsp. *elatius*. The latter occurs naturally in open areas and has been spread by man as a fodder crop. It has since colonized road-sides and disturbed areas around fields in the western Mediterranean and northwards across temperate Europe. The two cytotypes occur in secondary contact in northern Spain, for example, on El Turbon mountain (a).

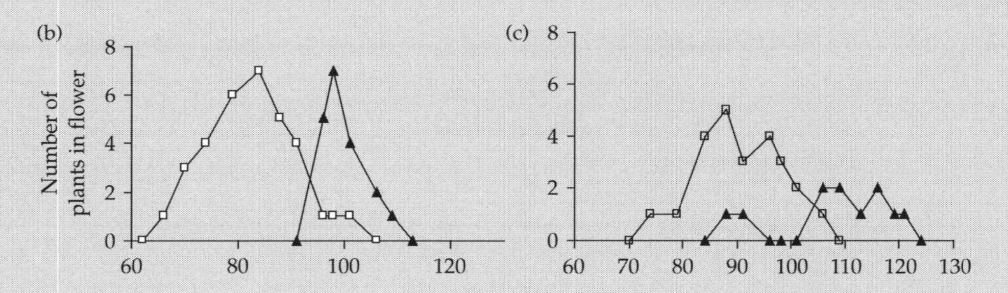

Flowering time of tetraploids (open squares) and diploids (closed triangles), from the contact zone (b) and from single cytotype allopatric sites (c) when cultivated in a common garden.

(d) Selfing rates (s) are higher in parapatric populations than in allopatric populations of both cytotypes

Diploids	Tetraploids
Parapatric: $s = 0.36$	Parapatric: $s = 0.46$
Allopatric: $s = 0.07$	Allopatric: $s = 0.30$

but within the Mediterranean region (Box 6.8). This species is native in many parts of the eastern Mediterranean, Italy, south-east Spain, and on three Tyrhhenian islands (Corsica, Sardinia, and Sicily). In southern France it does not naturally occur west of the maritime Alps. In the 1920s a small number of plants, perhaps from the same tree in the Montpellier Botanical gardens, were introduced to a small arboretum in the Vis valley (~40 km north of Montpellier). A study of the extant distribution of *F. ornus* in the Vis and its confluent Hérault river system in southern France, has shown that this species has spread rapidly since its localized introduction (Thébaud and Debussche 1991). These authors found that this species has spread ~60 km downstream in ~65 years. Immediately downstream (<1 km) from the site of introduction plants are very dense, with more than 750 trees in the first kilometre from its site of planting. However, there has been very little colonization upstream of the site of introduction or upstream of the site of confluence of the two rivers. Dispersal by water of what one would have predicted to be wind-dispersed samaras would appear to have greatly facilitated the rapid spread of this species. In addition, the capacity of plants for vegetative regeneration after disturbance may have facilitated establishment after autumnal flooding (which has probably greatly facilitated long-distance dispersal down the river system). The process of colonization has not been slowed by the fact that this species contains non-seed producing male plants (Dommée *et al.* 1999). The pattern observed in this species has general implications for our understanding of the rapidity of species range expansions following post-glacial climatic warming—perhaps other species also dispersed along river systems in this manner.

Dittrichia viscosa is a circum-Mediterranean shrubby Asteraceae which has expanded its range in association with human activities, particularly in the ruderal parts of recently urbanized areas where it thrives despite harsh water and mineral stress (Wacquant 1990). In this species, populations on limestone have a better tolerance of iron deficiency than those on more acid soils, which appears to be linked to the reducing capacity of the roots. However, despite the occurrence of this habitat-linked variation, most plants possess a wide ecological amplitude when grown on different substrates. These results indicate that plasticity may be more important than ecological specialization (in addition to its high fecundity and germination capacity) for the colonization of anthropogenic habitats outside of its natural range.

6.3.5 The pertinence of recent invasions

To finish this chapter, and without wishing to labour a seemingly obvious point, I will emphasize how invasive species can improve our understanding of plant evolution in the Mediterranean. As Barrett and Richardson (1986: 29) pointed out 'one of the major difficulties in the study of the biology of invasions is the lack of information on the early stages of colonization'. During these early stages, the process of selection will be acting on available genetic variation, hence the study of species actively colonizing a region provide appropriate systems for the study of the factors which limit species distributions and those which promote evolution in invasive populations. Persual of the recent compilation of introduced species in France (Muller 2001) illustrates that several species introduced to Europe now have populations in both non-Mediterranean temperate regions and geographically close sites in the Mediterranean region. In the Mediterranean the invasion process is recent or just beginning. Three examples illustrate this rather interesting situation:

(1) *Senecio inaequidens* (discussed earlier) has been colonizing the Mediterranean region for less than 50 years;
(2) *Ambrosia artemisifolia*, introduced to France from North America in the second half of the nineteenth century and now a major pest in central France has begun colonizing the Mediterranean area from the north (A. Martin, CEFE-CNRS, Montpellier, unpublished data);
(3) *Cortaderia selloana* and *Baccharis halmifolia* are currently spreading across the Camargue wetlands (A. Chapentier, Tour de Valat Biological Station, unpublished data) and also exist in south-west France under an Atlantic climatic regime.

Such examples provide for the study of evolution in action, as plants colonize new regions and thus provide a means to assess the role of adaptation and plasticity in the differentiation of populations in ecologically contrasting situations. Populations of exotic species in different parts of their introduced range provide ideal model systems to test the evolutionary response of plants to environmental variation, and in particular the extreme seasonality and other environmental conditions which characterize the Mediterranean. The observation that several invasive species do not have a phenological strategy (e.g. germination in spring and flowering during summer) typical of Mediterranean plants is particularly intriguing here.

6.4 Conclusions

My objective in this chapter has been to outline how species introductions and domestication provide us with some fascinating models for the study of plant evolution in the Mediterranean.

European civilizations first developed in the Mediterranean. The domestication of grasses and fruit trees in the eastern Mediterranean was a revolution for humankind and the development of western civilizations. Sometime after the end of the last Pleistocene glaciation, as the climate warmed and open grasslands with wild cereals replaced *Artemesia* steppe, human groups, then living as hunter–gatherers in the eastern Mediterranean, gradually adopted a more sedentary lifestyle as they began to harvest and then cultivate wild plants. This occurred primarily, at least initially, in one tiny area within the Fertile Crescent of the eastern Mediterranean. The 'Neolithic Revolution' had begun, as had a new evolutionary trajectory in many grasses, legumes, and fruit trees, as well as in many ruderal weedy species.

The study of the evolutionary history of domesticated plants is a particularly pertinent example of how research benefits from pluri-disciplinarity. By the combined application of technical skills, knowledge, and research methods in genetics, archaeology, and phylogeography it has been possible to obtain information on the 'where, when, and why' of early domestication and the evolutionary processes that have fashioned contemporary crops and fruit trees. Studies of genetic variation in crop–wild species complexes illustrate the complex biogeographic history of domestication in some emblematic Mediterranean cereal and fruit crops. For example, genetic differentiation due to regional differences among oleaster populations and local selection of cultivars has been countered by the diffusion of cultivars in association with human movements down the ages. Cultivation changed the local selection pressures, and long-distance migration introduced variants to a new selection regime in a novel environment. Just as geological history and climate have fashioned the environment for natural selection to operate in wild populations, cultivation and migration with man have drastically altered the selection pressures and evolutionary potential of plants under domestication.

The Mediterranean flora has been the source of a whole range of plants that have accompanied the development of human civilizations. In addition to the founder crops of agriculture, which were all originally domesticated in the Near East prior to their dispersal westwards, the Mediterranean is also home to numerous other wild progenitors of fibre and oil-producing seed plants, such as sesame (*Sesamum indicum*), poppy (*Papaver somniferum*), and false flax (*Camelina sativa*) (Zohary and Hopf 2000). One can also add to the cereals and the fruit crops discussed in this chapter an immense diversity of edible plants whose wild progenitors were first sampled in the Mediterranean, figs and dates being obvious examples (Zohary and Spiegel-Roy 1975). The domestication of a variety of aromatic plants such as the lavenders on the high plains of Provence and the essential ingredients of '*herbes de Provence*' (wild thyme, rosemary, etc.) in lowland areas, has also occurred. Once again this has often involved a progressive reduction of the genetic resource of wild species to a few important cultivars with occasional hybridization events producing new cultivated forms. If we extend the list to plants with other uses a whole book could be written. An important message here is that many cultivated herbaceous plants, in particular geophytes such as *Cyclamen*, *Iris*, *Crocus*, *Galanthus*, and *Narcissus*, have been key elements in horticultural development, and the variety of cultivated forms well known to botanists and gardeners come from just a small number of wild species. Indeed, the biology, ecology, and genetics of natural populations of the wild plants have, until recently, received surprisingly little attention. Tree species, such as the oaks, olives, and chestnuts discussed in this chapter, represent a major component of the European forest resource. Their economic and cultural importance has allowed for the development and funding of several large-scale research projects involving several countries in order to further knowledge of the genetic structure of wild and cultivated stands and the ecology of regeneration in such stands. Such research is invaluable for the sustainable use and the development of new varieties to further evolution under domestication. Similar programmes on some of the Mediterranean geophytes which delight people in their gardens all over Europe would be more than worthwhile.

Domesticated species represent an important cultural component of many Mediterranean landscapes. The famous historian F. Braudel (1986: 9, my translation) commented that

> ... the Mediterranean is an ancient crossroads. For thousands of years, humans, domestic animals, cars, diverse merchandise, ships, ideas and religions have converged on the region to enrich its history. Even plants have come flowing into the region. You think they are Mediterranean? Well, with the exception of ... [some native species] ... established very early in the region a large majority originated far from the Mediterranean Sea.

However, as I have discussed here, with only ~250 naturalized species (i.e. 1% of the total number of native species present) the Mediterranean flora has suffered less than many regions from plant invasions, while providing a source of potential invaders for other Mediterranean parts of the world. This asymmetry is not just a result of differential movements of humans and the direction of trade among these regions. By far the greater part of alien plants live in habitats where the natural vegetation has been drastically modified by human activities. These habitats are also very simplified where the complexity of natural conditions has been reduced to a small number of severe limitations on plant populations. The long history of human activities in the Mediterranean Basin may have selected genetic variation in many species pre-adapting them to the new disturbances in other regions.

The ecological factors which determine why some species are invasive in a particular region are complex, making it difficult to make predictions concerning invasions on the basis of a small number of simple generalizations. When it comes to prediction, the ecological sophistication that would allow predictions about the possible spread of introduced species remains to be achieved, primarily as a result of the complex interactions between plant traits and effects of spatial and temporal habitat heterogeneity. What is clear is that introductions continue at unprecedented rates, faster than the development of any legislation and control measures. Invasions are currently proceeding into the Mediterranean region, hence spectacular invasions are more and more likely to occur and threaten native communities and ecosystems. If ever European legislation finally comes around to the problem of invasive species and the necessary controls on their spread, I would plead that the evolutionary process of hybridization and its biological consequences be given explicit treatment, since hybridization is a key element of many invasions, with clear implications for the conservation of natural systems (Vilà *et al.* 2000). Finally, the current occurrence of species whose introduced range includes both the Mediterranean and non-Mediterranean portion of some European countries provides an ideal situation to further our understanding of plant evolution in the Mediterranean.

Conclusions: Endemism, adaptation, and conservation

> Endemic species are of great interest to the student of plant life for they help to elucidate both the past history of the flora and its continuing development. Endemics may also be distinctive and of considerable beauty, and enhanced by their uniqueness they are often sought after by naturalists, who in consequence have a special responsibility to ensure that they continue to survive.
>
> O. Polunin (1980: 23–26)

To understand plant evolution requires research at the interface of ecology and genetics. This interactive science requires a sound knowledge of regional history and how this history has modulated ecological and evolutionary processes. I have written a book which is centred on how the process of plant evolution is modulated by regional history and spatial habitat variation. To do so, I have discussed plant evolution in relation to the history of the Mediterranean region and contemporary changes in environmental conditions linked to the superposition of human activities on a highly heterogeneous ecological landscape. Past and present are inexorably linked in any such study of evolution. One of my primary objectives has been to bring together two main poles of research in evolutionary ecology:

- the evolution of diversity, and in particular endemism
- population ecology and adaptation in relation to habitat heterogeneity.

Plant diversity in the Mediterranean is based on the coexistence, within a single flora, of plants of diverse biogeographical origins (Chapter 1) and a large number of narrow endemic species (Chapter 2). The contemporary Mediterranean flora has developed from the persistence of lineages derived from the Tertiary (subtropical) woody vegetation, the differentiation of Mediterranean elements and colonization by elements of the Irano-Turanian, Pontic, and Arcto-Tertiary floras. It has developed and differentiated in response to ancient isolation by geographical and geological barriers and more recent distribution changes as a result of climatic oscillations (Chapters 2 and 3). The message from these chapters is that geology and climate can be seen to form the basis for our understanding of diversity in Mediterranean-climate regions (Cowling *et al.* 1996) because they have provided the template for the operation of diverse evolutionary processes, that is, drift, selection, gene flow, hybridization, and chromosomal evolution. As molecular dating techniques develop, our understanding of this evolution will improve.

The Mediterranean mosaic is an ideal place to study population differentiation and adaptation. As I illustrate in Chapters 4 and 5, the highly localized spatial heterogeneity of ecological conditions and the mosaic configuration of habitats in the landscape greatly impact on the generation of patterns of population variation. The spatial heterogeneity of dispersal patterns across different habitats, the role of abiotic and biotic factors on regeneration and population persistence and spatial variability of reproductive traits in relation to pollinators and herbivores all contribute to spatial pattern, with potentially long-lasting effects on plant population dynamics. Such factors are essential elements of our understanding of plant evolution in the Mediterranean mosaic.

It remains for me to conclude. To do so, I have decided to recapitulate the key points and themes which emerge from my discussion of plant evolution in a particular context: the future conservation of plant biodiversity in the Mediterranean region. In many regions of the Mediterranean Basin, high rates of endemism go hand in hand with high rates of extinction risk, at least in terms of the percentage of endangered species (Médail and Quézel 1997; Ramade 1997). One could say that this is nothing new: extinctions have occurred massively and repeatedly in the history of the Mediterranean region (Chapter 1). These extinctions occurred however on a timescale which allowed for immigration and the evolution of new species. What is new is that contemporary risks may cause future extinction rates to exceed rates of immigration and the evolution of new species. In addition, on this timescale invasive species may cause a dramatic homogenization of the differentiation diversity that characterizes the flora of the Mediterranean Basin (Chapter 6). The threats are thus very real, even though the recent rates of extinction in the flora may be low (Greuter 1994).

The subject of Mediterranean plant conservation has been the focus of whole books (Gomez-Campo 1985) and journal special issues (see the 1995 volume of *Ecologia Mediterranea* dedicated to the understanding and conservation of Mediterranean island floras). There are also many standard research papers on the conservation of Mediterranean plant species. I do not intend to provide a comprehensive review of this literature. My aim here is to set the themes of this book, focused until now on plant evolutionary ecology, in the context of conservation biology.

While I fully realize the need for conservation biology to incorporate the ecosystem processes on which the maintenance of biodiversity depends, I will focus here on the conservation of endemic plant species. Although many endemic species are unlikely to be those whose removal and loss has the greatest impact on ecosystem function, they do provide an objective means of assessing the conservation value of particular habitats and thus the establishment of habitat directives. By focusing on their populations, both in terms of spatial variation and local dynamics, threats can be identified, not just for a particular species but whole communities. It is individual species that provide us with molecules (Chapter 4) and crops (Chapter 6), and the Mediterranean has been an amazing source of such plants. I believe that scientists in the field of evolutionary ecology do indeed have, as the citation which heads this chapter suggests, a critical responsibility, if not to go out and manage rare plant populations, at least to draw attention to key issues for future conservation plans. The key issue which I develop in these final pages is that the conservation of species should only be envisaged by addressing the conservation of ecological processes and evolutionary potential. In this context I argue that conservation plans directed at endemic species are an essential step in the conservation of whole ecological systems.

First, conservation plans require a strong scientific appreciation of the ecological processes which impact on population persistence and the dynamics of populations in the landscape. In Chapter 2, I discussed evidence for the idea that endemism is associated with ecological differentiation. In Chapter 3, I furthered this argument, by illustrating cases where ecological differentiation may be associated with species divergence. The fact that endemic species occur in habitats which differ from those of more widespread species, that is, primarily in open rocky habitats with fewer coexisting species or in crevices and cliffs, is of critical importance to their conservation. Although there is still no direct evidence that such differentiation is associated with specialization, any future losses of natural habitat are likely to be a primary detrimental factor for their persistence. Conservation of the habitat and the ecological factors influencing population persistence is thus a crucial issue for the development of well-adapted conservation guidelines for narrow endemic species. However, although the conservation issue for a number of endemic species is not that populations are declining, but more that they do not occur elsewhere, legal protection of the habitat is only a first step since it alone is unlikely to be a sufficient guarantee of success, since in some cases demographic factors clearly threaten population viability.

Since many endemic species occur in rocky habitats and/or in and around cliffs (Chapter 2), I use an example from cliff habitats to illustrate the importance of a process oriented basis to conservation ecology. I discussed in Chapter 4 the important role played by facilitation in the natural process of succession in Mediterranean plant communities following the abandonment of human activities. Facilitation may also be important in limestone cliffs where endemism is rife. For example, Escudero (1996) suggests that the establishment of nano-phanerophytes in cliffs in the Iberian peninsula depends on a process of facilitation whereby colonist chamaephytes mediate soil accumulation in crevices and hollows. The nano-phanerophytes can grow to much greater size and their root development probably contributes to a shattering of the rock and erosion, allowing for re-colonization by the pioneer chamaephytes. The cliff faces which are home to so many endemic species in the Mediterranean thus represent a complex mosaic of discrete micro-communities which succeed one another in a colonization, succession, extinction cycle. In a similar fashion, the microhabitat dependent spatial variation in reproductive success of narrow endemic *Erodium* (Albert *et al.* 2001) discussed in Chapter 4 clearly illustrates that maintaining the species in a localized mosaic of habitat heterogeneity is a critical aspect of endemic species conservation. This is salient to the idea brought up in Chapters 2 and 4 that the stable habitats where populations of many endemic species have persisted may represent groups of local subpopulations that differ markedly in their dynamics. A sort of highly localized metapopulation, and one in which microenvironmental variation among patches is a key element in their dynamics.

Conservation programmes for natural populations of endemic plants must thus consider the spatial scale of habitat heterogeneity. To conserve populations, one must quantify the ecological interactions which currently regulate population viability and species distribution. In this respect, spatial variability of demographic parameters associated with dispersal, regeneration, and reproduction are key parameters to incorporate in the study and conservation of endemic species and their habitats (e.g. Hampe and Arroyo 2002; Mejías *et al.* 2002; Traveset *et al.* 2003; Fréville *et al.* 2004), as are the high levels of floral herbivory and seed predation detected in several Mediterranean endemic species (Chapter 5). Plant species interact with, and indeed their survival depends on, other trophic levels, such as soil microorganisms and insect pollinators. The devastation of insect populations in association with the use of pesticides and herbicides and other human activities may thus have dramatic consequences for the long-term persistence of many Mediterranean plant species. Experimental tests of the effects of different management regimes, themselves developed on the basis of demographic data concerning the factors which limit populations, will be of critical importance in this respect (Heywood and Iriondo 2003). This issue is also pertinent to the restoration of highly degraded ecosystems in many areas of the Mediterranean, particularly in North Africa. Such ecosystem restoration requires understanding of which plant traits permit re-colonization and the role of community-level interactions in the development and restoration of degraded habitats.

Many endemic plants have persisted through the ages. The Tertiary endemics discussed in Chapter 2 are a prime example. Most of the literature on these ancient endemic species classes them as 'Tertiary relicts'. However, they also illustrate the need for an approach based on the conservation of evolutionary potential since some of their isolated populations may be of great value for future evolution. Two examples discussed in Chapter 3 illustrate this point. First, relict populations of *Frangula alnus* in southern Spain and northern Morocco, which have not contributed to post-glacial re-colonization of non-Mediterranean temperate Europe, contain a unique store of genetic variation, not present elsewhere in the range of this species (Hampe *et al.* 2003). Second, peripheral and isolated populations of the argan tree, *Argania spinosa*, endemic to North-West Africa, are primary sources of rare alleles in this species (El Mousadik and Petit 1996*a*,*b*). The importance of such a unique genetic composition of peripheral and isolated populations in conservation biology should thus be recognized (Petit *et al.* 1998). In this context there is a need to document the spatial organization of genetic diversity, both in terms of genetic marker diversity and adaptive trait variation (Chapters 3 and 4). When combined with demographic data and experimental studies of the ecological interactions which limit population viability (Schemske *et al.* 1994; Oostermeijer *et al.* 2003), such information opens the way for reintroduction and reinforcement actions to be planned and put into action within the spatial context of socioeconomic factors which will ultimately determine their success.

The other side of the coin features endemic species with a more recent origin. Climate oscillations in the last 2 million years have been critical here, causing range contractions, which have allowed for divergence in isolation, and range expansions, which have brought populations back into contact and thus facilitated the origin of novel hybrid variants (Chapter 3). Such hybridization plays a major role as a genetic stimulus for ecological adaptation and evolution and for the occurrence of chromosomal evolution which may further determine differentiation and divergence. Areas of contact are thus prime sites for conservation policy based on evolutionary potential. In other situations they may require control if protected species suffer genetic 'pollution' as a result of hybridization with widespread species and/or invasive species. I would suggest that there is a distinct need for sites of potential and known hybridization in the Mediterranean to be synthesized in a database which identifies three main classes: (a) sites of risk for endemic species, (b) sites of risk for the evolution of invasive species, and (c) potential sites of future evolution in the native flora.

A final point in the elaboration of a conservation policy based on evolutionary potential concerns the evolutionary significance of peripheral or marginal populations, which may also have a unique genetic make-up and adaptive potential relative to populations in more central parts of the range. Marginal microhabitats may not only be of value for the persistence of Mediterranean endemic species, but also the divergence of new variants. In many cases such isolates may contain the endemic species of the future. It would thus be wise to integrate some recognition of this idea into conservation policy, despite the difficulty one may have in convincing decision-makers of the value of isolated and small populations of otherwise widespread species. It is interesting to note here that although narrow endemic species in the western Mediterranean have shown a relative stability in terms of population persistence, rare species whose affinities lie with the Euro-Siberian floristic province, and which are at their range limits in the Mediterranean region, have shown a regression in population numbers in southern France in the last 100 years (Lavergne *et al.* 2004*c*). Marginal populations may thus be those which incur a high risk of extinction, hence their significance in any conservation plan.

Some strong arguments can thus be developed to argument a conservation policy centred on endemic species in the Mediterranean Basin. First, and by definition, endemic plants do not occur elsewhere, hence we do indeed have a great responsibility for their conservation. Second, narrow endemism is a key feature of the high species richness in the Mediterranean Basin. Third, the Mediterranean Basin is home to many aromatic plants (Chapter 4) used in folk medicine (Vokou *et al.* 1993; San Miguel 2003) and the wild relatives of some of the most important domesticated cereals, legumes, and fruit trees (Chapter 6). The spatial and genetic variability of essential oils and other traits of value to humans also provide a basis for conservation plans. The natural history-based and resource-based approaches to conservation are thus both relevant to plant biodiversity in the Mediterranean.

A major conservation issue at the present time is the regional disparity of human impacts on vegetation within the Mediterranean Basin, particularly from north to south across the sea where current patterns of human impact, land-use, and socioeconomic change are very different. North of the Mediterranean Sea, forests are spreading by a process of natural re-colonization in the back country of several countries which is the result of socioeconomic changes in land use and spatial changes in the demography of human populations. To the south of the Mediterranean Sea, many different types of forest vegetation (firs, pines, oaks, thermo-Mediterranean woodland with endemic olives and pistacias, etc.) are in some way threatened by over use and destruction (Quézel and Médail 2003), posing serious problems for their sustainable use and conservation. This north–south disparity in land use and human activities has been highlighted by studies demonstrating striking regional differences on either side of the Straits of Gibraltar in both (a) the population dynamics of diverse endangered species, for example, the insectivorous *Drosophyllum lusitanicum* (Garrido *et al.* 2003) and trees such as *Juniperus thurifera* (Gauquelin *et al.* 1999) and (b) the biodiversity status of heathlands and oak woodlands (Ojeda *et al.* 1996*a*; Marañón *et al.* 1999). No rational plan of conservation can be established without an adequate integration of the geographic variation of human activities, their impact and the socioeconomic causes of such variation.

In addition to regional disparities, the impact of human activities on Mediterranean ecosystems can vary in time and changes in human perception of the landscape may dramatically alter its conservation value. A striking illustration of this point has been documented for an ecosystem which I have repeatedly mentioned in this book, namely the '*Grands Causses*' limestone plateaux which fringe the Mediterranean region in southern France. Until the middle to late twentieth century, these open grasslands were considered as degraded landscapes (as a result of grazing and cultivation) that merited large-scale reforestation in order to improve ecosystem function (Table C.1; Lepart *et al.* 2000). The abandonment of much extensive grazing and cultivation, from the late nineteenth century onwards, has allowed the massive spread of woody species and a natural process of reforestation has occurred in many areas (Chapter 4). At the same time, the perception of the landscape changed, with the development of a strong link between the patrimonial value of what has become a cultural landscape and traditional farming activities. The perception of the forest as a reference ecosystem, that should be restored in the landscape, has thus, in the space of <30 years, been progressively overtaken by a conception based on a cultural value of open grassland in the landscape. The latter view has gained importance with the realization that the return of the forest may in fact cause a decline in biodiversity (Chapter 4). The conservation objectives and perspectives for this region have thus been dramatically altered.

Table C.1 The changing perception of the '*Grands Causses*' limestone grasslands in southern France which have accompanied the abandonment of traditional farming activities and the natural progression of woodlands in the twentieth century

	Late nineteenth century to ~1970	1970–2000
1. Reference ecosystem	Natural forest	Open grasslands
2. Environmental arguments	Improved ecosystem function (e.g. reduced soil erosion)	Increased species richness and landscape diversity
3. Cultural landscape	Little or no appreciation of the cultural value of open grasslands	Development of a link between the open landscape and traditional farming activities
4. Perception of the landscape	Promote the return of the forest	Maintain a mosaic of grassland and open shrubs

Source: Lepart *et al.* (2000).

Having argued for a unity in the flora, I would also argue for a unity in our approach to its conservation. The difficulty which arises here is that ~20 different countries have part of their territory within the Mediterranean zone. Many endemic and endangered species are not limited to one country, they often cross administrative borders, and some form part of the disjunct endemics which I introduced in Chapter 2. The marked differences in the socioeconomic status and impact of human activities in different regions will greatly complicate the development of a rational and united approach to plant conservation. Several countries which have part of their territory in the Mediterranean region now have 'red lists' of endangered species. Where species have biogeographic limits which do not follow administrative limits it will be important to adopt a unified approach to their conservation. This approach could base itself on objectives developed as a part of the European Plant Conservation Strategy, such as (a) understand, document, and conserve diversity, (b) provide guidelines for sustainable use and control measures concerning plant introductions and potentially invasive species, (c) implement education and promote awareness of

Mediterranean plant diversity (see Heywood and Iriondo 2003).

Conservation issues must be addressed by reference to the biogeographic history, spatial ecology, and population dynamics of the taxa and should, where possible, avoid decisions that arise due to administrative partitioning of a given range. Also at stake is the need to involve local people. A good example here concerns the significant decline in illegal collecting and trading of Mediterranean geophytes in the eastern Mediterranean due to the implementation of cultivation programmes and trade by local populations (see Quézel and Médail, 2003). Conservation biology is a multidisciplinary science, in the Mediterranean it should also be international and unified, a procedure which could not advance without close coordination between policy and the local people in regions where such policies are to be implemented. A first step here should be the implementation of coordinated international research on the population ecology, diversity, and evolution of different groups of plants. I have provided some examples of what is currently known on the ecology, diversity, and evolution of Mediterranean geophytes, aromatic plants, palaeo-endemic, and neo-endemic species and invasive plants. Much of the work has been the fruit of individual groups of researchers or small-scale collaboration. What is now needed, in a similar vein to what has been done on tree species such as oaks, chestnuts, and olives, is coordinated international research on such groups of plants involving research teams from different countries and disciplines in association with natural history and conservation societies, amateur botanists, and decision makers.

A final word. If we are to understand what is likely to happen to the ecological balance of plant diversity in the Mediterranean flora we need to examine the past, present, and future modifications to ecological factors. The barriers to spread imposed by tectonic activity and climate change have been critical to the richness of endemic species in the Mediterranean Basin. As Elton (1958) commented (in the context of invasive plants), if plants and animals had developed in a world where dispersal limitation due to geographic barriers did not prevent the spread of organisms, the potential for future change under the impact of human activities would be far less, due to widespread nature of most species in such a world. Understanding future evolution in the flora of the Mediterranean thus requires rigorous appreciation of the past constraints that have limited the spread of species and given us what many marvel at today: high rates of narrow endemism in many parts of the Mediterranean Basin. Improving our understanding of the ecological, genetic, and historical causes of the origin and maintenance, as well as the loss, of endemic plant species will help formulate better science-based biodiversity policies. My argument is that the conservation of endemic species should be maintained as a priority since it will complement the effectiveness of measures taken at the scale of the habitat and the ecosystem. This priority will require the development of well-adapted measures and incentives as well as the implementation of legal restrictions. All of this will require sound scientific knowledge of spatial variation in the ecology and genetics of the plants and the historical factors which have shaped their present-day distribution. The conservation of plant biodiversity must thus be spatially oriented and not just concerned with populations in protected areas. In writing this book I hope that I have in some way made a useful contribution to this issue.

Species list

List of plant taxa mentioned in the text

Nomenclature follows Flora Europaea (author abbreviations were obtained from the website). Exceptions are detailed at the end of the list.

Adenostoma fasciculatum J.R. Forst & G. Forst (Compositae) 147
Aegilops cylindrica Host (Gramineae) 216
Aegilops tauschii = *Aegilops cylindrica* Host (Gramineae) 216
Aesculus hippocastaneum L. (Hippocastanaceae) 31
Agave americana L. (Agavaceae) 224, 230
Ailanthus altissima (Mill.) Swingle (Simaroubaceae) 225, 228, 230
Allium grosii Font Quer (Liliaceae) 52
Alnus cordata (Loisel.) Loisel. (Betulaceae) 53
Alnus viridis (Chaix) DC. subsp. *viridis* 77
Alnus viridis (Chaix) DC. subsp. *suaveolens* (Req.) P.W. Ball 77
Amaranthus deflexus L. (Amaranthaceae) 228
Amaranthus retroflexus L. 228
Ambrosia artemisifolia L. (Compositae) 237
Amorpha fruticosa L. (Leguminosae) 228
Ampelodesmos mauritanica (Poir.) T. Durand & Schinz (Gramineae) 231
Anchusa crispa Viv. (Boraginaceae) 183
Anchusa hybrida = *Anchusa undulata* L. subsp. *hybrida* (Ten.) Cout. 199
Anchusa undulata L. 199
Anthoxanthum odoratum L. (Gramineae)
Anemone coronaria L. (Ranunculaceae) 120
Anemone palmata L. 101
Anthyllis hermanniae L. (Leguminosae) 24
Anthyllis montana L. 76
Anthyllis vulneraria L. 145
Antirrhinum lopesianum Rothm. (Scrophulariaceae) 88
Antirrhinum microphyllum Rothm. 70, 88
Antirrhinum mollissimum = *Antirrhinum hispanicum* Chav. subsp. *mollissimum* (Rothm.) D.A. Webb 70, 88
Antirrhinum valentinum Font Quer 70
Aquilegia viscosa Gouan (Ranunculaceae) 60, 179, Plate 3
Aquilegia vulgaris L. 60, 179, Plate 3
Arbutus unedo L. (Ericaceae) 35, 115
Arenaria balearica L. (Caryophyllaceae) 53, 77
Argania spinosa (L.) Skeels (Sapotaceae) 30, 70, 81, 244
Arrhenatherum elatius (L.) P. Beauv. ex J. Presl & C. Presl (Gramineae) 140, 235–236
Arrhenatherum elatius (L.) P. Beauv. ex J. Presl & C. Presl subsp. *elatius* 140, 235–236
Arrhenatherum elatius (L.) P. Beauv. ex J. Presl & C. Presl subsp. *sardoum* (Em.Schmid) Gamisans 40, 235–236
Armeria filicaulis (Boiss.) Boiss. (Plumbaginaceae) 94
Armeria splendens (Lag. & Rodr.) Webb 94
Armeria villosa Girard 94
Arum italicum Mill. (Araceae) 177
Arum pictum L.f. 77
Asphodelus albus = *Asphodelus albus* Mill. (Liliaceae) 171
Astragalus hamosus L. (Leguminosae) 148

Baccharis halmifolia L. (Compositae) 228, 237
Berberis aetnensis C. Presl (Berberidaceae) 53
Brachypodium distachyon (L.) P. Beauv. (Gramineae) 116
Brachypodium phoenicoides (L.) Roem. & Schult. 148
Brassica insularis Moris (Cruciferae) 70, 79
Bromus erectus Huds. (Gramineae) 116, 164
Bromus fasciculatus C. Presl 116
Bromus hordeaceus L. 70
Bromus intermedius Guss. 70
Bromus lanceolatus Roth 70

Bromus squarrosus L. 70
Buddleja davidii Franch. (Buddlejaceae) 228
Buxus sempervirens L. (Buxaceae) 131, 133, 136

Camelina sativa (L.) Crantz (Cruciferae) 238
Campanula calaminthifolia Lam. (Campanulaceae) 48
Campanula dichotoma L. 92
Campanula saxatilis L. subsp. *cytherea* Rech. f. & Phitos 77
Cardamine chelidonia L. (Cruciferae) 53
Cardamine pratensis L. 100, 103
Carduus cephalanthus Vis. (Compositae) 53
Carex glauca = *Carex flacca* Schreb. subsp. *flacca* (Cyperaceae) 60
Carex olbiensis Jord. 60
Carpobrotus acinaciformis = *Carpobrotus edulis* (L.) N.E. Br. var. *rubescens* Druce (Aizoaceae) 228, 230, 234, 235
Carpobrotus chilensis (Molina) N.E. Br. 233
Carpobrotus edulis (L.) N.E. Br. 224, 228, 231, 233, 234, 235
Castanea sativa Mill. (Fagaceae) 27, 220
Cedrus atlantica (Endl.) Carrière (Pinaceae) 27
Cedrus libani A. Rich. 27
Centaurea corymbosa Pourr. (Compositae) 59, 60
Centaurea maculosa Lam. subsp. *maculosa* 60, 70, 80
Centaurea maculosa Lam. subsp. *albida* (Lecoq & Lamotte) Dostál 80
Cephalaria squamiflora (Sieber) Greuter (Dipsacaceae)
Cephalaria squamiflora (Sieber) Greuter subsp. *squamiflora* 42
Cephalaria squamiflora (Sieber) Greuter subsp. *balearica* (Willk.) Greuter 42
Cerastium soleirolii Ser. ex Duby (Compositae) 41, 51
Ceratonia siliqua L. (Leguminosae) 30, 115
Cercis siliquastrum L. (Leguminosae) 31, 235
Chamaerops humilis L. (Palmae) 149, 169
Chrysanthemum tomentosum = *Leucanthemopsis alpina* (L.) Heywood subsp. *tomentosa* (Loisel.) Heywood (Compositae) 40
Cistus albidus L. (Cistaceae) 148
Cistus ladanifer L. var. *albiflorus** 153
Cistus ladanifer L. var. *maculata** 153
Cistus libanotis L. 171
Cistus pouzolzii = *Cistus varius* Pourr. 60
Cistus monspeliensis L. 60
Cneorum tricoccon L. (Cneoraceae) 123
Conyza bonariensis (L.) Cronquist (Compositae) 228
Conyza canadensis (L.) Cronquist 226, 228
Conyza sumatrensis (Retz) E.H. Walker 226
Coridothymus capitatus = *Thymus capitatus* (L.) Hoffmanns. & Link (Labiatae) 148, 153
Cornus sanguinea L. (Cornaceae) 184
Cortaderia selloana (Schult. & Schult. f.) Asch. & Graebn. (Gramineae) 228, 237
Crepis sancta (L.) Babc. (Compositae) 123, 139, 182
Crocus olbanus Siehe (Iridaceae) 122
Cupressus sempervirens L. (Cupressaceae) 142
Cyclamen africanum Boiss. & Reut. (Primulaceae) 122, 124
Cyclamen balearicum Willk. 41, 42, 51, 60, 68, 70, 75, 76, 77, 79, 80, 81, 82, 83, 84, 85, 86, 87, 107, 124, 184, Plate 1
Cyclamen cilicium Boiss. & Heldr. 124
Cyclamen coum Mill. 122, 124
Cyclamen creticum (Dörfler) Hildebr. = *Cyclamen repandum* Sibth. & Sm. subsp. *creticum* (Dörfler) Debussche & Thompson *comb. nov.* 42, 69, 70, 77, 79, 80, 81, 82, 83, 84, 124, Plate 1
Cyclamen cyprium Kotschy 124
Cyclamen graecum Link 124
Cyclamen hederifolium Aiton 70, 122, 124, Plate 1
Cyclamen intaminatum (Meikle) Grey-Wilson 124
Cyclamen libanoticum Hildebrand 124
Cyclamen mirabile Hildebrand 124
Cyclamen persicum Mill. 122, 124, 222
Cyclamen pseudibericum Hildebrand 124
Cyclamen purpurascens Mill. 122
Cyclamen repandum Sibth. & Sm. 42, 60, 68, 69, 70, 77–79, 80, 81, 82, 83, 84, 85, 86, 87, 107, 122, 124, Plate 1
Cyclamen repandum Sibth. & Sm. subsp *creticum* (Dörffer) Debussche & Thompson *comb. nov.* 42, 69, 70, 77, 79–84, 122, Plate 1
Cyclamen repandum Sibth. & Sm. subsp. *repandum* 42, 82, 83, 84, 85, 86, 87, Plate 1
Cyclamen repandum Sibth. & Sm. subsp. *peloponnesiacum* Grey-Wilson 42, 82, 83, Plate 1
Cyclamen repandum Sibth. & Sm. subsp. *rhodense* (Meikle) Grey-Wilson 42, 82, 83
Cyclamen rohlfsianum Aschers. 82, 122, 124
Cyclamen somalense Thulin & Warfa 82

Cyclamen trochopterantum Schwarz 124
Cymbalaria aequitriloba (Viv.) A. Chev. (Scrophulariaceae) 52
Cymbalaria hepaticaefolia (Poir.) Wettst. 41
Cytisus aeolicus Guss. ex Lindl. (Leguminosae) 70

Dactylis glomerata L. (Gramineae) 41, 96, 98, 100, 115, 149
Dactylis glomerata L. subsp. *ibizensis* Stebbins & Zohary 96
Dactylis glomerata L. subsp. *juncinella* (Bory) Stebbins & Zohary 96
Dactylis glomerata L. subsp. *reichenbachii* Hausm. ex Dalla Torre & Sarnth.) Stebbins & Zohary 96
Daphne laureola L. (Thymelaeaceae) 191, 195
Daphne rodriguezii Texidor 52
Delphinium pictum Willd. (Ranunculaceae) 52, 53, 104
Delphinium requienii DC. 104
Dianthus fruticosus L. (Caryophyllaceae) 48
Dittrichia viscosa (L.) Greuter (Compositae) 235, 237
Dorycnium suffruticosum = *Dorycnium pentaphyllum* Scop. subsp. *pentaphyllum* (Leguminosae) 148
Drosophyllum lusitanicum (L.) Link (Droseraceae) 244

Ecballium elaterium (L.) A. Rich. (Cucurbitaceae) 70, 186, 188
Echium vulgare L. (Boraginaceae) 191
Erica arborea L. (Ericaceae) 28, 35
Erodium corsicum Léman (Geraniaceae) 77
Erodium reichardii (Murray) DC. 77
Erucaria hispanica (L.) Druce (Cruciferae) 116, 148
Eryngium amorginum Rech. f. (Umbelliferae) 48
Erysimum candicum Snogerup subsp. *candicum* (Cruciferae) 88, 89
Erysimum candicum Snogerup subsp. *carpathum* Snogerup 89
Erysimum corinthium (Boiss.) Wettst. 87, 88
Erysimum grandiflorum Desf. 103
Erysimum mediohispanicum = *Erysimum nevadense* Reut. subsp. *mediohispanicum* (Polatschek) P.W. Ball 178, Plate 3
Erysimum naxense Snogerup 87, 89
Erysimum rhodium Snogerup 87, 89
Erysimum senoneri (Heldr. & Sart.) Wettst. subsp. *senoneri* 87, 89
Erysimum senoneri (Heldr. & Sart.) Wettst. subsp. *amorginum* Snogerup 87, 89
Erysimum senoneri (Heldr. & Sart.) Wettst. subsp. *icarium* Snogerup 89
Euphorbia dendroides L. (Euphorbiaceae) 35

Fagus moesiaca (K.Malý) (Fagaceae) 71
Fagus sylvatica L. 70, 74
Ferula communis L. (Umbelliferae)
Ficus carica L. (Moraceae) 168, 169
Forsythia europaea Degen & Bald. (Oleaceae) 31
Frangula alnus Mill. (Rhamnaceae) 55, 56, 74, 109, 123, 178, 242
Fraxinus angustifolia Vahl (Oleaceae) 133
Fraxinus ornus L. 138, 189, 230, 235, 237

Galium corsicum Spreng. (Rubiaceae) 51
Genista corsica (Loisel.) DC. (Leguminosae) 41
Genista pilosa L. 60
Genista villarsii = *Genista pulchella* Vis. 60
Geranium purpureum Vill. (Geraniaceae) 136
Globularia cambessedesii Willk. (Globulariaceae) 52

Haberlea rhodopensis Friv. (Gesneriaceae) 55
Hedera helix L. (Araliaceae) 112, 140
Helianthemum squamatum (L.) Pers. (Cistaceae) 109
Helichrysum amorginum Boiss. & Orph. (Compositae) 48
Helleborus foetidus L. (Ranunculaceae) 177–178, 183
Helleborus lividus Aiton subsp. *lividus* 77
Helleborus lividus Aiton subsp. *corsicus* (Willd.) Tutin 77
Herniaria latifolia Lapeyr. subsp. *latifolia* (Caryophyllaceae) 77
Herniaria latifolia Lapeyr. subsp. *litardierei* Gamisans 77
Hirschfeldia incana (L.) Lagr.-Foss. (Cruciferae) 191
Hordeum vulgare L. subsp. *vulgare* (Gramineae) 214
Hordeum vulgare L. subsp. *spontaneum* Thell. = *Hordeum spontareum* K. Koch
Hordeum spontaneum K. Koch 116, 119, 214
Hormathophylla pyrenaica (Lapeyr.) Cullen & T.R. Dudley (Cruciferae) 60
Hormathophylla spinosa (L.) P. Küpfer 60, 171, 179, 181, Plate 3

Hyacinthus orientalis L. (Liliaceae) 122
Hymenocarpus circinatus (L.) Savi (Leguminosae) 148

Ilex aquifolium L. (Aquifoliaceae) 112
Inula candida (L.) Cass. (Asteraceae) 48
Iris spuria L. (Iridaceae) 60
Iris xiphium L. 60

Jankaea heldreichii (Boiss.) Boiss. (Gesneriaceae) 55–56
Jasminum fruticans L. (Oleaceae) 126, 128, 171–176, 185, Plate 3
Juniperus communis L. (Cupressaceae) 126, 127, 136, 137
Juniperus thurifera L. 244

Knautia dipsacifolia Kreutzer (Dipsacaceae) 96

Lactuca viminea (L.) J. Presl & C. Presl subsp. *chondrilliflora* (Boreau) Bonnier (Compositae) 182–183
Lactuca viminea (L.) J. Presl & C. Presl subsp. *ramosissima* (All.) Bonnier 182–183
Lathyrus cirrhosus Ser. (Leguminosae) 60
Lathyrus latifolius L. 60
Laurus azorica (Seub.) Franco (Lauraceae) 55
Laurus nobilis L. 115
Lavandula angustifolia Mill. (Labiatae) 152, 223, Plate 3
Lavandula latifolia Medik. 104, 109, 116, 118, 125, 130, 148, 152, 172–174, 177, 223
Lavandula stoechas L. 152
Lilium carniolicum Bernh. ex W.D.J. Koch (Liliaceae) 105, 106
Lilium martagon L. 60
Lilium pomponium L. 105
Lilium pyrenaicum Gouan 60, 105, 106
Linum arboreum L. (Linaceae) 48
Linum tenuifolium L. 196
Lithodora fruticosa (L.) Griseb. (Boraginaceae) 199
Lithodora hispidula (Sibth. & Sm.) Griseb. 199
Liquidambar orientalis Mill. (Hamamelidaceae) 31, 55, 56
Lobularia maritima (L.) Desv. (Cruciferae) 125, 171
Lonicera pyrenaica L. (Caprifoliaceae) 24, 58, 60
Lonicera xylosteum L. 60
Lysimachia ephemerum L. (Primulaceae) 60
Lysimachia vulgaris L. 60

Majorana syriaca L. (Labiatae) 149
Medicago citrina (Font Quer) Greuter (Leguminosae) 183
Medicago minima (L.) Bartal. 148
Medicago sativa L. 70, 219
Medicago truncatula Gaertner 70, 119
Melica amethystina = *Melica bauhinii* All. (Gramineae) 60
Melica ciliata L. 60
Mentha piperata L. (Labiatae) 150
Mentha spicata L. 153, 156
Mercurialis annua L. (Euphorbiaceae) 101, 102, 187, 188
Mercurialis corsica Coss. 40
Muscari comosum (L.) Mill. (Liliaceae) 104

Nananthea perpusilla (Loisel.) DC. (Compositae) 40
Narcissus albimarginatus D. & U. Muller-Doblies (Amaryllidaceae) 198, Plate 4
Narcissus assoanus Dufour 172, 177, 178, 197–205, Plate 4
Narcissus dubius Gouan 60, 184, 197, 203, Plate 4
Narcissus longispathus Pugsley 71, 177
Narcissus papyraceus Ker Gawl. 199, 200, 203, 204, Plate 4
Narcissus tazetta L. 60, 203, Plate 4
Narcissus triandrus L. 115, 199, 200, 201, 203, Plate 4
Naufraga balearica Constance & Cannon (Umbelliferae) 40, 53
Nerium oleander L. (Apocynaceae) 109, 171
Nigella arvensis L. subsp. *aristata* (Sibth. & Sm.) Nyman (Ranunculaceae) 88, 90, 91
Nigella arvensis L. subsp. *brevifolia* Strid 88, 90
Nigella arvensis L. subsp. *glauca* (Boiss.) N.Terracc. 90, 91
Nigella carpatha Strid 90
Nigella degenii Vierh.* 90, 91
Nigella doerfleri Vierh. 90, 91
Nigella icarica Strid 90

Odontites jaubertiana (Boreau) D.Dietr. ex Walp. (Scrophulariaceae) 60
Odontites lutea (L.) Clairv. 60
Olea europaea L. subsp. *europaea* (Oleaceae) 30, 73, 109, 115, 127, 130, 218, 219
Onopordum acanthium L. (Compositae) 233
Onopordum illyricum L. 233
Opuntia ficus-indica (L.) Mill (Cactaceae) 224, 228, 230

Ophrys fusca Link (Orchidaceae) 97, 149
Ophrys lutea (Gouan) Cav. 97, 149
Ophrys sphegodes Mill. 150
Orchis caspia Trautv. (Orchidaceae) 171
Orchis insularis = *Dactylorhiza sambucina* (L.) Soó subsp. *insularis* (Sommier) Soó (Orchidaceae) 77
Origanum onites L. (Labiatae) 157
Origanum vulgare L. 153, 157
Ornithogalum umbellatum L. (Liliaceae) 98–99, 102
Ornithogalum nutans L. 235
Osyris alba L. (Santalaceae) 171
Osyris quadripartita Salzm. ex Decne. 127, 128, 130
Oxalis latifolia Kunth (Oxalidaceae) 228
Oxalis pes-caprae L. 228

Paeonia arietina = *Paeonia mascula* (L.) Mill. subsp. *arietina* (G. Anderson) Cullen & Heywood (Paeoniaceae) 95
Paeonia broteroi Boiss. & Reut. 95, 171, 178, Plate 3
Paeonia cambessedesii (Willk.) Willk. 95
Paeonia clusii Stern 95
Paeonia coriacea Boiss. 95
Paeonia mascula (L.) Mill. 95, 138
Paeonia mascula (L.) Mill. subsp. *arietina* (G. Anderson) Cullen & Heywood 95
Paeonia officinalis L. 95, 138
Paeonia peregrina Mill. 95
Paeonia rhodia = *Paeonia clusii* Stern subsp. *rhodia* (Stearn) Tzanoud. 95
Papaver somniferum L. (Papaveraceae) 238
Paspalum dilatatum Poir. (Gramineae) 228
Paspalum paspalodes (Michx.) Scribn. 231, 232
Pastinaca latifolia (Duby) DC. (Umbelliferae) 77
Pastinaca lucida L. 77
Pennisetum villosum R.Br. ex Fresen. (Gramineae) 228
Petrocoptis grandiflora Rothm. (Caryophyllaceae) 172
Phillyrea angustifolia L. (Oleaceae) 189
Phillyrea latifolia L. 126, 127, 133, 134, 137
Phlomis fruticosa L. (Labiatae) 235
Phoenix theophrasti Greuter (Palmae) 19, 55
Phytolacca americana L. (Phytolaccaceae) 228
Phyteuma charmelii Vill. (Campanulaceae) 60
Phyteuma orbiculare L. 60
Picris hieracioides L. (Compositae) 136
Pinguicula corsica Bernard & Gren. (Lentibulariaceae) 40
Pinguicula vallisneriifolia Webb 140
Pinus brutia Ten. (Pinaceae) 30, 54, 131, 141, 142
Pinus halepensis Mill. 27, 30, 54, 114, 132, 133, 135, 137, 138, 141, 142, 153, 159
Pinus maritima Mill. 79
Pinus nigra J.F. Arnold 30, 42, 137, 138, 230
Pinus nigra J.F. Arnold subsp. *nigra* 42
Pinus nigra J.F. Arnold subsp. *dalmatica (Vis.) Franco* 42
Pinus nigra J.F. Arnold subsp. *laricio* (Poir.) Maire 25, 42, 53
Pinus nigra J.F. Arnold subsp. *mauretanica** 42
Pinus nigra J.F. Arnold subsp. *pallasiana* (Lamb.) Holmboe 42
Pinus nigra J.F. Arnold subsp. *salzmanii* (Dunal) Franco 42
Pinus pinaster Aiton 27, 74, 119, 138
Pinus sylvestris L. 134, 135, 137
Pistacia lentiscus L. (Anacardiaceae) 30, 115, 127, 134, 136, 153
Pistacia terebinthus L. 24, 35, 134
Plantago media L. (Plantaginaceae) 100
Plantago psyllium = *Plantago afra* L. 148
Plantago subulata L. 41
Platanus orientalis L. (Platanaceae) 235
Poa bulbosa L. (Gramineae) 120
Polygonum maritimum L. (Polygonaceae) 60
Polygonum romanum Jacq. 60
Prunus mahaleb L. (Rosaceae) 126, 130

Quercus alnifolia Poech (Fagaceae) 68, 71, 88
Quercus calliprinos = *Quercus coccifera* L. 30, 88
Quercus canariensis Willd. 27, 62
Quercus coccifera L. 35, 54, 55, 68, 70, 114, 134, 141
Quercus frainetto Ten. 222
Quercus ilex L. 24, 25, 27, 30, 35, 54, 70, 73, 114, 115, 121, 137, 141, 146, 153, 219, 224
Quercus ithaburensis Decaisne 123, 141
Quercus lusitanica Lam. 141
Quercus pubescens Willd. 114, 115, 131, 136, 137, 138
Quercus rotundifolia = *Quercus ilex* L. subsp. *rotundifolia* (Lam.) Tab. Morais 73
Quercus suber L. 25, 27, 62, 70, 73, 121, 141, 217, 221, 222, 223, 224
Quercus toza = *Quercus pyrenaica* Willd. 27

Ramonda myconi (L.) Rchb. (Gesneriaceae) 55
Ramonda nathaliae Pančić & Petrović 55
Ramonda serbica Pančić & Petrović 55
Reseda jacquinii Rchb. (Resedaceae) 60

Reseda phyteuma L. 60
Rhamnus alaternus L. (Rhamnaceae) 24, 127
Rhamnus alpinus L. 24
Rhamnus ludovici-salvatoris Chodat 52, 109, 136, 177
Rhamnus lycioides L. 134
Rhododendron ponticum L. (Ericaceae) 55, 73, 233
Robertia taraxacoides (Loisel.) DC. (Compositae) 51
Robinia pseudoacacacia L. (Leguminosae) 228
Romulea bulbocodium (L.) Sebast. & Mauri (Iridaceae) 191
Rosmarinus officinalis L. (Labiatae) 146, 148, 153, 158
Rumex acetosella L. (Polygonaceae) 100
Ruscus aculeatus L. (Liliaceae) 123
Ruta angustifolia Pers. (Rutaceae) 164
Ruta chalepensis L. 164
Ruta corsica DC. 40, 153
Ruta graveolens L. 150, 164
Ruta montana (L.) L. 164

Sambucus nigra L. (Caprifoliaceae) 112
Satureja cordata = *Micromeria filiformis* (Aiton) Benth. subsp. *filiformis* (Labiatae) 79
Satureja horvatii Cespare Silic 115
Satureja thymbra L. 148, 171
Saxifraga cochlearis Rchb. (Saxifragaceae) 73, 88
Saxifraga crustata Vest 73
Saxifraga moncayensis D.A. Webb 24, 58
Saxifraga paniculata Mill. 73, 88
Scabiosa albocincta Greuter (Dipsacaceae) 42
Scabiosa cretica L. 42, 104
Scabiosa hymetia Boiss. & Spruner 42
Scabiosa minoana (P.H. Davis) Greuter 42
Scabiosa variifolia Boiss. 42
Scilla autumnalis L. (Liliaceae) 104
Scirpus maritimus L. (Cyperaceae) 185
Scrophularia heterophylla Willd. (Scrophulariaceae) 48
Senecio aegyptius L. (Compositae) 72
Senecio aethnensis Jan ex DC. 72, 93
Senecio chrysanthemifolius = *Senecio siculus* All. 72, 93
Senecio flavus (Decne.) Sch. Bip. 93
Senecio flavus (Decne.) Sch. Bip. subsp. *flavus* 72, 93
Senecio flavus (Decne.) Sch. Bip. subsp. *breviflorus* Kadereit = *Senecio mohavensis* subsp. *breviflorus* (Kadereit) M. Coleman *comb. nov.* 72, 93
Senecio gallicus Chaix 70, 72, 74, 75, 88
Senecio glaucus L. 70, 72, 93, 97, 116
Senecio inaequidens DC. 228, 232, 237
Senecio leucanthemifolius Poir. 72
Senecio lividus L. 72
Senecio malacitanus = *Senecio linifolius* L. 72
Senecio mohavensis A. Gray subsp. *mohavensis* 93
Senecio mohavensis subsp. *breviflorus* (Kadereit) M. Coleman *comb. nov.* 93
Senecio nebrodensis L. 72
Senecio petraeus Boiss. & Reut. 88
Senecio rupestris Waldst. & Kit. 70, 74
Senecio sylvaticus L. 72
Senecio squalidus L. 72, 93
Senecio vernalis Waldst. & Kit. 70, 72, 97, 116
Senecio viscosus L. 72
Senecio vulgaris L. 72, 123
Sesamum indicum L. (Pedaliaceae) 238
Sesleria insularis Sommier subsp. *cordata* (Gramineae) 77
Sesleria insularis subsp. *insularis**
Silene ammophila Boiss. & Heldr. subsp. *carpathae* Chowdhuri (Caryophyllaceae) 77
Silene cambessedesii Boiss. & Reut. 52
Silene diclinis (Lag.) M. Laínz 71, 79
Silene saxifraga L. 24, 58
Silene vulgaris (Moench) Garcke 191
Solanum eleaginifolium Cav. (Solanaceae) 228
Soleirolia soleirolii (Req.) Dandy (Urticaceae) 40, 53, 77
Stachys corsica Pers. (Labiatae) 51
Staehelina dubia L. (Compositae) 148
Stipa tenacissima L. (Gramineae) 136
Stipa tortilis = *Stipa capensis Thumb.* 142
Styrax officinalis L. (Styracaceae) 31

Tetraclinis articulata (Vahl) Mast. (Cupressaceae) 30, 55, 142
Teucrium marum L. (Labiatae) 53
Teucrium polium L. 148
Thalictrum minus L. (Ranunculaceae) 60
Thalictrum tuberosum L. 60
Thlaspi caerulescens J. Presl & C. Presl 144, 145, 146
Thymelaea hirsuta (L.) Endl. (Thymelaeaceae) 190
Thymelaea passerina (L.) Coss. & Germ. 189
Thymelaea velutina = *Thymelaea myrtifolia* (Poir.) D.A. Webb 52
Thymus albicans = *Thymus tomentosus* Willd. (Labiatae) 154

Thymus antoninae Rouy & Coincy 154
Thymus baeticus Boiss. ex Lacaita 154
Thymus bracteatus Lange ex Cutanda 154
Thymus caespititius Brot. 154
Thymus camphoratus Hoffmanns. & Link 154
Thymus capitatus (L.) Hoffmanns. & Link 148, 149
Thymus carnosus Boiss. 154
Thymus funkii = *Thymus longiflorus* Boiss. 154
Thymus herba-barona Loisel 154
Thymus hirtus Willd. 154
Thymus hyemalis Lange 154
Thymus leptophyllus Lange 154
Thymus longiflorus Boiss. 154
Thymus loscosii Willk. 71, 154
Thymus mastichina L. 154, 192
Thymus membranaceus Boiss. 154
Thymus moroderi = *Thymus longiflorus* Boiss. 154
Thymus nitens Lamotte 60, 154
Thymus orospedanus H. del Villar 154
Thymus piperella L. 154
Thymus pulegioides L. 60, 154
Thymus serpylloides Bory 154
Thymus villosus L. 154
Thymus vulgaris L. 70, 109, 121, 148, 149, 152, 153, 154, 155–156, 158–164, 191–195, Plate 2
Thymus wilkomii Ronniger 154
Thymus zygis L. 154, 158, 192
Thymus zygis subsp. *gracilis** 154, 158
Thymus zygis subsp. *sylvestris** 154, 158
Thymus zygis subsp. *zygis** 154, 158
Trifolium subterraneum L. (Leguminosae) 232
Triticum aestivum L. (Gramineae) 216
Triticum aestivum L. subsp. *spelta* 216
Triticum aestivum L. subsp. *vulgare* Host. 216
Triticum compactum Host 216
Triticum dicoccoides = *Triticum turgidum* L. *subsp. dicoccoides* Aschers 71, 119
Triticum monococcum L. subsp. *monococcum* 216
Triticum monococcum L. subsp. *boeoticum* Boiss. 216
Triticum timopheevi Zhuk. 216
Triticum timopheevii Zhuk. subsp. *timopheevii* 216
Triticum timopheevii Zhuk. subsp. *araraticum* Jakubz 216
Triticum turgidum L. subsp. *dicoccum* Schübl. 216
Triticum turgidum L. subsp. *dicoccoides* Aschers 216
Triticum turgidum L. subsp. *durum* Desf 216
Triticum uratu Tuman 216
Tulipa oculus-solis St.Amans (Liliaceae) 101

Urginea maritima (L.) Baker (Liliaceae) 172

Viburnum tinus L. (Caprifoliaceae) 112, 140
Viola cazorlensis Gand. (Violaceae) 177, Plate 3
Viola corsica Nyman 41
Viola stolonifera = *Viola odorata* L. 52

Withania somnifera (L.) Dunal (Solanaceae) 153

Zelkova abelicea (Lam.) Boiss. (Ulmaceae) 19, 31, 55
Zelkova sicula Di Pasquale, Garfi, & Quézel 31, 55, 56

Exceptions to nomenclature used by Flora Europaea and species not present in Flora Europaea were named as follows

- The International Plant Names Index was used for exotic species not listed in Flora Europaea or species not present in the European part of the Mediterranean.
- Grey-Wilson (1997) was used for *Cyclamen* not in Flora Europaea, except for *C. creticum* which is treated as in Debussche & Thompson (2002).
- The Real Jardin Botánico (Madrid) website was used (www.programanthos.com/anthos.asp) for *Hormathophylla*, *Medicago citrina*, and *Narcissus assoanus*, Di Pasquale *et al.* (1992) for *Zelkova sicula*, Matthew (1982) for *Crocus olbanus*, Quézel & Médail for Mediterranean *Quercus* not in Flora Europaea, Salamini *et al.* (2002) for *Triticum*, Coleman *et al.* (2001) for *Senecio mohavensis*.

* Sub-specific taxa discussed in the text not yet recognized in Flora Europaea.

Bibliography

Abbas, H., M. Barbero, and R. Loisel (1984). 'Réflexions sur le dynamisme actuel de la régénération du Pin d'Alep en région méditerranéenne dans les Pinèdes incendiées de Provence calcaire'. *Ecologia Méditerranea* **10**: 85–104.

Abbott, R.J. (1992). 'Plant invasions, interspecific hybridization and the evolution of new plant taxa'. *Trends in Ecology and Evolution* **7**: 401–405.

Abbott, R.J., J.K. James, J.A. Irwin, and H.P. Comes (2000). 'Hybrid origin of the Oxford ragwort *Senecio squalidus* L'. *Watsonia* **23**: 123–138.

Abbott, R.J., J.K. James, D.G. Forbes, and H.P. Comes (2002). 'Hybrid origin of the Oxford ragwort *Senecio squalidus* L: morphological and allozyme differences between *S. squalidus* and *S. rupestris* Waldst. anf Kit'. *Watsonia* **24**: 17–29.

Abraham, A., I. Kirson, E. Glotter, and D. Lavie (1968). 'A chemotaxonomic study of *Withania somnifera* (L.) Dun'. *Phytochemistry* **7**: 957–962.

Achérar, M., J. Lepart, and M. Debussche (1984). 'La colonisation des friches par le pin d'Alep (*Pinus halepensis* Mill.) en Languedoc méditerranéen'. *Oecologia Plantarum* **19**: 179–189.

Achérar, M. and S. Rambal (1992). 'Comparative water relations of four Mediterranean oak species'. *Vegetatio* **99–100**: 177–184.

Ackerman, J.D., E.J. Meléndes-Ackerman, and J. Salguero-Faria (1997). 'Variation in pollinator abundance and selection on fragrance phenotypes in an epiphytic orchid'. *American Journal of Botany* **84**: 1383–1390.

Affre, L. and J.D. Thompson (1997*a*). 'Variation in the population genetic structure of two *Cyclamen* species on the island of Corsica'. *Heredity* **78**: 205–214.

Affre, L. and J.D. Thompson (1997*b*). 'Population genetic structure and levels of inbreeding depression in the Mediterranean island endemic *Cyclamen creticum*'. *Biological Journal of the Linnaean Society* **60**: 527–549.

Affre, L. and J.D. Thompson (1998). 'Floral trait variation in four *Cyclamen* (Primulaceae) species'. *Plant Systematics and Evolution* **212**: 279–293.

Affre, L. and J.D. Thompson (1999). 'Variation in levels of self-fertility, inbreeding depression and levels of inbreeding in four *Cyclamen* species'. *Journal of Evolutionary Biology* **12**: 113–122.

Affre, L., J.D. Thompson, and M. Debussche (1995). 'The reproductive biology of the Mediterranean endemic *Cyclamen balearicum*.' *Botanical Journal of the Linnaean Society* **118**: 309–330.

Affre, L., J.D. Thompson, and M. Debussche (1997). 'Genetic structure of continental and island populations of the Mediterranean endemic *Cyclamen balearicum* (Primulaceae)'. *American Journal of Botany* **84**: 437–451.

Agence Méditerranéenne de l'Environnement (2000). 'La forêt en Languedoc-Roussillon, hier, aujourd'hui, demain'. *La Lettre de l'Environnement* **37**: 4–5.

Ainouche, M., M.-T. Misset, and A. Huon (1995). 'Genetic diversity in Mediterranean diploid and tetraploid *Bromus* L. (section *Bromus* Sm.) populations'. *Genome* **38**: 879–888.

Akman, Y., P. Quézel, O. Ketenoglu, and L. Kurt (1993). 'Analyse syntaxonomique des forêts de *Liquidambar orientalis* en Turquie'. *Ecologia Méditerranéa* **24**: 63–71.

Albert, M.E., C.M. D'Antonio, and K.A. Schierenbeck (1997). 'Hybridization and introgression in *Carpobrotus* spp. (Aizoaceae) in California: I. Morphological evidence'. *American Journal of Botany* **84**: 896–904.

Albert, M.J., A. Escudero, and J.M. Iriondo (2001). 'Female reproductive success of narrow endemic *Erodium paularense* in contrasting microhabitats'. *Ecology* **82**: 1734–1747.

Alcántara, J.M., P.J. Rey, F. Valera, J.E. Gutiérrez, and A.M. Sánchez-Lafuente (1997). 'Temporal pattern of seed dispersal of Wild Olive (*Olea europaea* var *sylvestris*): its effect on intraspecific competition'. *Lagascalia* **19**: 583–590.

Alcántara, J.M., P.J. Rey, F. Valera, and A.M. Sánchez-Lafuente (2000). 'Factors shaping the seed fall pattern of a bird dispersed plant'. *Ecology* **81**: 1937–1950.

Alexander, J.C.M. (1979). 'Mediterranean species of *Senecio* sections *Senecio* and *Delphinifolius*'. *Notes of the Royal Botanic Gardens Edingburgh* **37**: 387–428.

Aliotta, G., G. Cafiero, V. De Feo, and R. Sacchi (1994). 'Potential allelochemicals from *Ruta graveolens* L. and their action on radish seeds'. *Journal of Chemical Ecology* **20**: 2761–2755.

Alomar, G., M. Mus, and J.A. Rossello (1997). *Flora endèmica de les balears*. Palma, Consell Insuar de Mallorca.

Alonso, C. and C.M. Herrera (1996). 'Variation in herbivory within and among plants of *Daphne laureola* (Thymeleaceae): a correlation with plant size and architecture'. *Journal of Ecology* **84**: 495–502.

Alonso, C. and C.M. Herrera (2001). 'Neither vegetative nor reproductive advantages account for high frequency of male-steriles in southern Spanish gynodioecious *Daphne laureola* (Thymelaeaceae)'. *American Journal of Botany* **88**: 1016–1024.

Alvarez, W. (1976). 'A former continuation of the Alps'. *Geological Society of America Bulletin* **87**: 891–896.

Alvarez, W., T. Coccoza, and T.C. Wezel (1974). 'Fragmentation of the alpine orogenic belt by microplate dispersal'. *Nature* **248**: 309–314.

Amiot, J., Y. Salmon, C. Colin, and J.D. Thompson, 'Differential resistance to freezing in a chemically polymorphic plant *Thymus vulgaris*'. Unpublished ms.

Anderberg, A.A. (1993). 'Phylogeny and subgeneric classification of *Cyclamen* L. (Primulaceae)'. *Kew Bulletin* **49**: 455–467.

Anderberg, A.A., I. Trift, and M. Källersjö (2000). 'Phylogeny of *Cyclamen* L. (Primulaceae): evidence from morphology and sequence data from the internal transcribed spacers of nuclear ribosomal DNA'. *Plant Systematics and Evolution* **220**: 147–160.

Anderson, E. and G.L. Stebbins (1954). 'Hybridization as an evolutionary stimulus'. *Evolution* **8**: 378–388.

Andersson, S. (1997). 'Genetic constraints on phenotypic evolution in *Nigella* (Ranunculaceae)'. *Biological Journal of the Linnaean Society* **62**: 519–532.

Andre, K. (1992). I. La dissémination des graines par les oiseaux en région mediterranéenne: II. Le role des invertébrés herbivores dans le maintien de la richesse spécifique des friches. Maitrise de Biologie des Organismes et des Populations.

Andrés, C. and F. Ojeda (2002). 'Effects of afforestation with pines on woody plant diversity of Mediterranean heathlands in southern Spain'. *Biodiversity and Conservation* **11**: 1511–1520.

Andrieu, E. (2002). Conséquences de la dynamique forestière sur la persistance, la reproduction et l'installation d'une espèce végétale polycarpique à longue durée de vie. Le cas de la Pivoine officinale (*Paeonia officinalis* L.). Mémoire D.E.A. Biologie de l'Evolution et Ecologie. Université Montpellier II Sciences et Techniques du Languedoc—Ecole Nationale Agronomique, Montpellier.

Anstett, M.-C. (1999). 'An experimental study of the interaction between the dwarf palm (*Chamerops humilis*) and its floral visitor *Derelomus chaeropsis* throughout the life cycle of the weevil'. *Acta Oecologia* **20**: 551–558.

Anstett, M.-C., M. Gibernau, and M. Hossaert-McKey (1998). 'Partial avoidance of female inflorescences of a dioecious fig by their mutualistic pollinating wasps'. *Proceedings of the Royal Society of London* **265**: 45–50.

Anstett, M.-C., M. Hossaert-McKey, and F. Kjellberg (1997). 'Figs and fig pollinators: evolutionary conflicts in a coevolved mutualism'. *Trends in Ecology and Evolution* **12**: 94–99.

Antonovics, J. (1968). 'Evolution of closely adjacent plant populations. VI. Manifold effects of geneflow'. *Heredity* **23**: 507–524.

Antonovics, J. (1976). 'The nature of limits to natural selection'. *Annals of the Missouri Botanical Garden* **63**: 224–247.

Araña, V. and R. Vegas (1974). 'Plate tectonics and volcanism in the Gibraltar Arc'. *Tectonophysics* **24**: 197–212.

Arianoutsou, M. (2001). 'Landscape changes in Mediterranean ecosystems of Greece: implications for fire and biodiversity issues'. *Journal of Mediterranean Ecology* **2**: 165–178.

Arianoutsou, M. and C. Radea (2000). 'Litter production and decomposition in *Pinus halepensis* forests'. In *Ecology, biogeography, and management of* Pinus halepensis *and* P. brutia *forest ecosystems in the Mediterranean Basin*, G. Ne'eman and L. Trabaud (eds.). Leiden, Backhuys Publishers: 183–190.

Armbruster, W.S. (1996). 'Evolution of floral morphology and function: an integrative approach to adaptation, constraint, and compromise in *Dalechampia* (Euphorbiaceae)'. In *Floral biology: studies on floral evolution in animal-pollinated plants*, D.G. Lloyd and S.C.H. Barrett (eds.). New York, Chapman & Hall: 241–272.

Armbruster, W.S., J.J. Howard, T.P. Clausen, E.M. Debevec, J.C. Loquvam, M. Matsuki, B. Cerendelo, and F. Andel (1997). 'Do biochemical exaptations link evolution of plant defense and pollination systems? Historical hypotheses and experimental tests with *Dalechampia* vines'. *American Naturalist* **149**: 461–484.

Arnold, M.L. (1997). *Natural hybridization and evolution*. Oxford, Oxford University Press.

Aronne, G. and C.C. Wilcock (1994*a*). 'First evidence of myrmechory in fleshy-fruited shrubs of the Mediterranean region'. *New Phytologist* **127**: 781–788.

Aronne, G. and C.C. Wilcock (1994*b*). 'Reproductive characteristics and breeding system of shrubs of the Mediterranean region'. *Functional Ecology* **8**: 69–76.

Aronne, G., C.C. Wilcock, and P. Pizzolongo (1993). 'Pollination biology and sexual differentiation of *Osyris alba* (Santalaceae) in the Mediterranean region'. *Plant Systematics and Evolution* **188**: 1–16.

Aronson, J., J. Kigel, and A. Shmida (1993). 'Reproductive allocation strategies in desert and Mediterranean populations of annual plants grown with and without water stress'. *Oecologia* **93**: 336–342.

Arroyo, J. (1990*a*). 'Geographic variation in flowering phenology in twenty-six common shrubs in SW Spain'. *Flora* **184**: 43–49.

Arroyo, J. (1990*b*). 'Spatial variation of flowering phenology in the Mediterranean shrublands of southern Spain'. *Israel Journal of Botany* **39**: 249–262.

Arroyo, J. (1997). 'Plant diversity in the region of the strait of Gibraltar: a multilevel approach'. *Lagascalia* **19**: 393–404.

Arroyo, J. and S.C.H. Barrett (2000). 'Discovery of distyly in *Narcissus* (Amaryllidaceae)'. *American Journal of Botany* **87**: 748–751.

Arroyo, J., S.C.H. Barrett, R. Hidalgo, and W.W. Cole (2002). 'Evolutionary maintenance of stigma-height dimorphism in *Narcissus papyraceus* (Amaryllidaceae)'. *American Journal of Botany* **89**: 1242–1249.

Arroyo, J. and A. Dafni (1995). 'Variations in habitat, season, flower traits and pollinators in dimorphic *Narcissus tazetta* L. (Amaryllidaceae) in Israel'. *New Phytologist* **129**: 135–145.

Arroyo, J. and T. Marañón (1990). 'Community ecology and distributional spectra of Mediterranean shrublands and heathlands in Southern Spain'. *Journal of Biogeography* **17**: 163–176.

Arroyo, M.T.K. and P. Uslar (1993). 'Breeding systems in a temperate Mediterranean-type climate montane scerophylous forest in central Chile'. *Botanical Journal of the Linnaean Society* **111**: 83–102.

Assouad, M.W., B. Dommée, and G. Valdeyron (1978). 'Reproductive capacities in the sexual forms of the gynodioecious species *Thymus vulgaris* L'. *Biological Journal of the Linnaean Society* **77**: 29–39.

Atlan, A., P.-H. Gouyon, T. Fournial, D. Pomente, and D. Couvet (1992). 'Sex allocation in an hermaphroditic plant: the case of gynodioecy in *Thymus vulgaris* L'. *Journal of Evolutionary Biology* **5**: 189–203.

Audus, L.J. and A.H. Cheetham (1940). 'Investigations on the significance of ethereal oils in regulating leaf temperatures and transpiration rates'. *Annals of Botany* **4**: 465–483.

Augspurger, C. (1979). 'Irregular rain cues and the germination and seedling survival of a Panamanian shrub (*Hybanthus prunifolius*)'. *Oecologia* **44**: 53–59.

Axelrod, D.I. (1958). 'Evolution of the Madro-Tertiary Geoflora'. *Botanical Review* **24**: 433–509.

Axelrod, D.I. (1967). 'Drought, diatrophism, and quantum evolution'. *Evolution* **21**: 201–209.

Axelrod, D.I. (1972). 'Edaphic aridity as a factor in angiosperm evolution'. *American Naturalist* **106**: 311–320.

Axelrod, D.I. (1975). 'Evolution and biogeography of the Madrean-Tethyan sclerophyll vegetation'. *Annals of the Missouri Botanical Garden* **62**: 284–334.

Ayasse, M., F. Schiestel, H.F. Paulus, C. Lofstedt, B. Hansson, F. Ibarra, and W. Francke (2000). 'Evolution of reproductive strategies in the sexually deceptive orchid species *Ophrys sphegodes*: how does flower-specific variation of odor signals influence reproductive success?'. *Evolution* **54**: 1998–2006.

Azizian, D. and D.F.F.L.S. Cutler (1982). 'Anatomical, cytological and phytochemical studies on *Phlomis* L. and *Eremostachys* Bunge (Labiatae)'. *Botanical Journal of the Linnaean Society* **85**: 249–281.

Bachiri Taoufiq, N. (2000). Les environnements marins et continentaux du corridor rifain au Miocène supérieur d'après la palynologie. Thése de Doctorat d' Etat. Casablanca, Université de Casablanca: 206.

Badr, A., K. Müller, R. Schäfer-Pregl, H. El Rabey, S. Effgen, H.H. Ibrahim, C. Pozzi, W. Rohde, and F. Salamini (2000). 'On the origin and domestication history of barley'. *Molecular Biology and Evolution* **17**: 499–510.

Baker, A.M., S.C.H. Barrett, and J.D. Thompson (2000*a*). 'Variation of pollen limitation in the early-flowering Mediterranean geophyte *Narcissus assoanus* (Amaryllidaceae)'. *Oecologia* **124**: 529–535.

Baker, A.M., J.D. Thompson, and S.C.H. Barrett (2000*b*). 'Evolution and maintenance of stigma-height dimorphism in *Narcissus*: I. Floral variation and style-morph ratios'. *Heredity* **84**: 502–513.

Baker, A.M., J.D. Thompson, and S.C.H. Barrett (2000*c*). 'Evolution and maintenance of stigma-height dimorphism in *Narcissus*: II. Fitness comparisons between style morphs'. *Heredity* **84**: 514–524.

Baker, H.G. (1961). 'The adaptation of flowering plants to nocturnal and crepuscular pollinators'. *Quarterly Review of Biology* **36**: 64–73.

Baker, H.G. (1965). 'Characteristics and modes of origin of weeds'. In *The genetics of colonizing species*, H.G. Baker and G.L. Stebbins (eds.). New York, Academic Press: 147–168.

Baker, H.G. (1966). 'The evolution, functioning and breakdown of heteromorphic incompatibility systems. 1. The Plumbaginaceae.' *Evolution* **20**: 349–368.

Baker, H.G. and G.L. Stebbins (eds.) (1965). *The genetics of colonizing species*. New York, Academic Press.

Banthorpe, D.V., B.V. Charlwood, and M.J.O. Francis (1972). 'The biosynthesis of monoterpenes'. *Chemical Review* **72**: 115–155.

Barbero, M. (1967). 'L'endémisme dans las Alpes Maritimes et ligures'. *Bulletin de la Société Botanique de France* **114**: 179–199.

Barbero, M., A. Benabid, C. Peyre, and P. Quézel (1981). 'Sur la présence au Maroc de *Laurus azorica* (seub.) Franco'. *Annal. Jard. Bot. Madrid* **37**: 467–472.

Barbero, M., G. Bonin, R. Loisel, F. Miglioretti, and P. Quézel (1987*a*). 'Impact of forest fires on structure and architecture of Mediterranean ecosystems'. *Ecologia Méditerranea* **13**: 39–50.

Barbero, M., G. Bonin, R. Loisel, F. Miglioretti, and P. Quézel (1987*b*). 'Incidence of exogenous factors on the regeneration of *Pinus halepensis* after fires'. *Ecologia Méditerranea* **13**: 51–56.

Barbero, M., G. Bonin, R. Loisel, and P. Quézel (1990). 'Changes and disturbances of forest ecosystems caused by human activities in the western part of the Mediterranean Basin'. *Vegetatio* **87**: 151–173.

Barbero, M., J. Guidicelli, R. Loisel, P. Quezel and E. Terzian (1982). 'Etude des biocénoses des mares et ruisseaux temporaires à éphémérophytes dominants en région méditerranéenne française'. *Bulletin d'Ecologie* **13**: 387–400.

Barbero, M., R. Loisel, and P. Quézel (1992). 'Biogeography, ecology and history of Mediterranean *Quercus ilex* ecosystems'. *Vegetatio* **99–100**: 19–34.

Barbero, M., R. Loisel, P. Quézel, D.M. Richardson, and F. Romane (1998). 'Pines of the Mediterranean Basin'. In *Ecology and biogeography of pinus*, D.M. Richardson (ed.). Cambridge, Cambridge University Press.

Barbero, M. and P. Quézel (1975). 'Végétation culminale du mont Ventoux; sa signification dans une interprétation phytogéographique des Préalpes083 mériodionales'. *Ecologia Méditerranéa* **1**: 3–33.

Barbero, M. and P. Quézel (1980). 'La végétation forestière de Crète'. *Ecologia Mediterranea* **5**: 175–210.

Barbero, M. and P. Quézel (1990). 'La déprise rurale et ses effets sur les superficies forestières dans la région Provence-Alpes-Côte-d'Azur'. *Bulletin de la Société Linnéenne de Provence* **41**: 77–88.

Barrett, S.C.H. (1993). 'The evolutionary biology of tristyly'. *Oxford Surveys in Evolutionary Biology* **9**: 283–326.

Barrett, S.C.H. (2002*a*). 'The evolution of plant sexual diversity'. *Nature Reviews Genetics* **3**: 237–284.

Barrett, S.C.H. (2002*b*). 'Sexual interference of the floral kind'. *Heredity* **88**: 154–159.

Barrett, S.C.H., A.L. Case, and G.B. Peters (1999). 'Gender modification and resource allocation in subdioecous *Wurmbea dioica* (Liliaceae)'. *Journal of Ecology* **87**: 123–137.

Barrett, S.C.H., W.W. Cole, J. Arroyo, M.B. Cruzan, and D.G. Lloyd (1997). 'Sexual polymorphisms in *Narcissus triandrus* (Amaryllidaceae): is this species tristylous?'. *Heredity* **78**: 135–145.

Barrett, S.C.H., W.W. Cole, and C.M. Herrera (2004*a*). 'Mating patterns and genetic diversity in the wild daffodil *Narcissus longispathus* (Amaryllidaceae)'. *Heredity* **92**: 459–465.

Barrett, S.C.H., L.D. Harder, and W.W. Cole (2004*b*). 'Correlated evolution of floral morphology and mating-type frequencies in a sexually polymorphic plant'. *Evolution* **58**: 964–975.

Barrett, S.C.H., L.K. Jesson, and A.M. Baker (2000). 'The evolution and function of stylar polymorphisms in flowering plants'. *Annals of Botany* **85** (Supplement A): 253–265.

Barrett, S.C.H., D.G. Lloyd, and J. Arroyo (1996). 'Stylar polymorphisms and the evolution of heterostyly in *Narcissus* (Amaryllidaceae)'. In *Floral biology: studies on floral evolution in animal-pollinated plants*, D.G. Lloyd and S.C.H. Barrett (eds.). New York, Chapman & Hall: 339–376.

Barrett, S.C.H. and B.J. Richardson (1986). 'Genetic attributes of invading species'. In *Ecology of biological invasions: an Australian perspective*, R.H. Groves and J.J. Burdon (eds.). Canberra, Australian Academy of Science: 21–33.

Bartholomew, B.L., L.C. Eaton, and P.H. Raven (1973). '*Clarkia rubundica*: a model of plant evolution in semiarid regions'. *Evolution* **27**: 505–517.

Basker, D. and E. Putievsky (1978). 'Seasonal variation in the yields of herb and essential oil in some Labiatae species'. *Journal of Horticultural Science* **53**: 179–183.

Baskin, Y. (1994). 'California's ephemeral vernal pools may be a good model for speciation'. *BioScience* **44**: 384–388.

Bazile-Robert, E., J.P. Suc, and J.L. Vernet (1980). *Les flores méditerranéennes et l'histoire climatique depuis le Pliocène*. Fondation Louis Emberger. La mise en place, l'évolution et la caractérisation de la flore et de la végétation circum-méditerranéennes, Montpellier 9–10 April, Naturalia Monspeliensa.

Beard, J.S., A.R. Chapman, and P. Gioia (2000). 'Species richness and endemism in the Western Australian flora'. *Journal of Biogeography* **27**: 1257–1268.

Beker, R., A. Dafni, D. Eisikowitch and U. Ravid (1989). 'Volatiles of two chemotypes of *Majorana syriaca* L. (Labiatae) as olfactory cues for the honeybee'. *Oecologia* **79**: 446–451.

Belhassen, E., B. Dommee, A. Atlan, P.-H. Gouyon, D. Pomente, M.W. Assouad, and D. Couvet (1991). 'Complex determination of male sterility in *Thymus vulgaris* L.: genetic and molecular analysis'. *Theoretical and Applied Genetics* **82**: 137–143.

Belhassen, E., L. Trabaud, D. Couvet, and P.-H. Gouyon (1989). 'An example of nonequilibrium processes: gynodioecy of *Thymus vulgaris* L. in burned habitats'. *Evolution* **43**: 662–667.

Bell, T.L. and F. Ojeda (1999). 'Underground starch storage in *Erica* species of the Cape Floristic Region—differences between seeders and respoutersare the resprouters'. *New Phytologist* **144**: 143–152.

Ben-Hod, G., J. Kigel, and B. Steinitz (1988). 'Dormancy and flowering in *Anemone coronaria* L. as affected by photoperiod and temperature'. *Annals of Botany* **61**: 623–633.

Bentzer, B. and M. Ellmer (1975). 'A case of stable chromosomal polymorphism in *Leopoldia comosa* (Liliaceae)'. *Hereditas* **81**: 127–132.

Berenbaum, M.R., J.K. Nitao, and A.R. Zangerl (1991). 'Adaptive significance of furanocoumarin diversity in *Pastinaca sativa* (Apiaceae)'. *Journal of Chemical Ecology* **17**: 207–215.

Bergstrom, G. (1978). 'Role of volatile chemicals in Ophrys-pollinator interactions'. In *Biochemical aspects of plant and animal coevolution*, J.B. Harborne (ed.). London, Academic Press: 207–231.

Bertini, A. (1994). 'Messinian–Zanclean vegetation and climate in North-Central Italy'. *Historical Biology* **9**: 3–10.

Bertness, M.D. and R. Callaway (1994). 'Positive interactions in communities'. *Trends in Ecology and Evolution* **9**: 191–193.

Bertoldi, R., D. Rio, and R. Thunell (1989). 'Pliocene–Pleistocene vegetational and climatic evolution of the South-Central Mediterranean'. *Palaeogeography, Palaeoclimatology, and Palaeoecology* **72**: 263–265.

Bertrand C., N. Fabre, C. Moulis, and J.-M. Bessière (2003). Composition of the essential oils of *Ruta corsica* DC. *Journal of Essentail Oil Research* **15**: 88–89.

Besnard, G. and A. Bervillé (2000). 'Multiple origins for Mediterranean olive (*Olea europaea* L. subsp. *europaea*) based upon mitochondrial DNA polymorphisms'. *Comptes Rendus de l'Académie des Sciences séries III* **323**: 173–181.

Besnard, G., B. Khadari, P. Baradat, and A. Bervillé (2002). '*Olea europaea* (Oleaceae) phylogeography based on chloroplast DNA polymorphism'. *Theoretical and Applied Genetics* **104**: 1353–1361.

Bessais, E. and J. Cravatte (1988). 'Les écosystèmes végétaus pliocènes de Catalognes mériodionale: variations latitudinales dans le domaine nord-ouest méditerranéen'. *Geobios* **21**: 49–63.

Bessedik, M. (1985). Reconstitution des environnements miocènes des régions nord-ouest méditerranéennes à partir de la palynologie. Montpellier, USTL: 162 pp.

Bessedik, M., P. Guinet, and J.-P. Suc (1984). 'Données paleofloristiques en Méditerranée nord-occidentale depuis l'Aquitanien'. *Révue de Paléobiologie* **Volume spécial**: 25–31.

Beug, H.J. (1967). 'On the forest history of the Dalmatian coast'. *Review of Palaeobotany and Palynology* **2**: 271–279.

Beug, H.J. (1975). 'Changes of climate and vegetation belts in the mountains of Mediterranean Europe during the Holocene'. *Biuletyn Geol.* **19**: 101–110.

Biju-Duval, B., J. Dercourt, and X. Le Pichon (1976). 'La genèse de la Méditerranée'. *La Recherche* **7**: 811–822.

Billès, G., J. Cortez, and P. Lossaint (1975). 'Etude comparative de l'activité biologique des sols sous peuplements arbustifs et herbacés de la garrigue méditerranéenne: I. Minéralisation du carbone et de l'azote'. *Révue d'Ecologie et biologie du Sol* **12**: 115–139.

Billès, G., P. Lossaint, and J. Cortez (1971). 'L'activité biologique des sols dans les écosystèmes méditerranéens: II.—Minéralisation de l'azote'. *Révue d'Ecologie et biologie du Sol* **8**: 533–552.

Blanchard, J.W. (1990). *Narcissus . A guide to wild daffodils*. Woking, Alpine Garden Society.

Bliss, S.A. and P.H. Zedler (1998). 'The germination process in vernal pools: sensitivity to environmental conditions and effects on community structure'. *Oecologia* **113**: 67–73.

Bliss, S.A., P.H. Zedler, J.E. Keeley, and K.A.M.T. (1998). 'A floristic survey of the temporary wetlands in the Mediterranean-climate region of Chile'. In *Wetlands for the future*, A.J. McComb and J.A. Davis (eds.): 219–228.

Blondel, J. and J. Aronson (1999). *Biology and wildlife of the Mediterranean region*. Oxford, Oxford University Press.

Blossey, B. and R. Nötzold (1995). 'Evolution of increased competitive ability in invasive nonindigenous plants: a hypothesis'. *Journal of Ecology* **83**: 887–889.

Boaz, M., U. Plitmann, and C.C. Heyn (1994). 'Reproductive effort in desert versus Mediterranean crucifers: the allogamous *Erucaria rostrata* and *E. hispanica* and the autogamous *Erophila minima*'. *Oecologia* **100**: 286–292.

Bocquet, G., B. Widler, and H. Kiefer (1978). 'The Messinian model—a new outlook for the floristics and systematics of the Mediterranean area'. *Candollea* **33**: 269–287.

Boettcher, S.E. and P.J. Kalisz (1990). 'Single-tree influence on soil properties in the mountains of eastern Kentucky'. *Ecology* **71**: 1365–1372.

Bond, W.J. (1983). 'On alpha diversity and the richness of the cape flora: a study in southern Cape fynbos'. In *Mediterranean-type ecosystems: the role of nutrients*, F.J. Kruger, D.T. Mitchell, and J.U.M. Jarvis (eds.). Berlin, Springer-Verlag: 337–356.

Bonis, A., J. Lepart, and P. Grillas (1995). 'Seed bank dynamics and coexistence of annual macrophytes in a temporary and variable habitat'. *Oikos* **74**: 81–92.

Bonnemaison, F., B. Dommée, and P. Jacquard (1979). 'Etude expérimentale de la concurrence entre formes sexuelles chez le thym, *Thymus vulgaris* L'. *Oecologia Plantarum* **14**: 85–101.

Bonnin, I., T. Huguet, M. Gherardi, J.-M. Prosperi, and I. Olivieri (1996). 'High level of polymorphism and spatial structure in a selfing plant species, *Medicago truncatula* (Leguminosae), shown using RAPD markers'. *American Journal of Botany* **83**: 843–855.

Borlaug, N.E. (1983). 'Contributions of conventional plant breeding to food production'. *Science* **219**: 689–693.

Borrill, M. and R. Lindner (1971). 'Diploid–tetraploid sympatry in *Dactylis* (Gramineae)'. *New Phytologist* **70**: 1111–1124.

Bosch, J. (1992). 'Floral biology and pollinators of three co-occurring *Cistus species* (Cistaceae)'. *Botanical Journal of the Linnaean Society* **109**: 39–55.

Bosch, J., J. Retena, and X. Cerdá (1997). 'Flowering phenology, floral traits and pollinator composition in a herbaceous Mediterranean plant community'. *Oecologia* **109**: 583–591.

Bottner, P. (1982). 'Evolution des sols et conditions bioclimatiques méditerranéennes'. *Ecologia Méditerranéa* **8**: 115–134.

Bou Dagher-Kharrat, M., G. Grenier, M. Bariteau, S. Brown, S. Siljak-Yakovlev, and A. Savouré (2001). 'Karyotype analysis reveals interspecific differentiation in the genus *Cedrus* despite genome size and base composition constancy'. *Theoretical and Applied Genetics* **103**: 846–854.

Bousquet, J.-C. (1997). *Géologie du Languedoc Roussillon*. Montpellier, Les Presses du Languedoc.

Box, E.O. (1997). 'Plant life forms and Mediterranean environments'. *Annali di Botanica* **XLV**: 7–42.

Brabant, P., P.H. Gouyon, G. Lefort, G. Valdeyron, and P. Vernet (1980). 'Pollination studies in *Thymus vulgaris* L. (Labiatae)'. *Oecologia Plantarum* **15**: 37–44.

Braudel, F. (1986). *La Méditerranée: les hommes et l'héritage*. Paris: Flammarion.

Brenac, P. (1984). 'Végétation et climat dans la Campanie du Sud (Italie) au pliocène final d'après l'analyse pollinique des dépôts de Camerota'. *Ecologia Mediterranea* **10**: 207–216.

Bretagnolle, F. and J.D. Thompson (1995). 'Gametes with the somatic chromosome number: mechanisms of their formation and role in the evolution of autopolyploid plants'. *New Phytologist* **129**: 1–22.

Bretagnolle, F. and J.D. Thompson (1996). 'An experimental study of ecological differences in winter growth between sympatric diploid and autotetraploid *Dactylis glomerata*'. *Journal of Ecology* **84**: 343–351.

Bretagnolle, F. and J.D. Thompson (2001). 'Phenotypic plasticity in sympatric diploid and autotetraploid *Dactylis glomerata* L'. *International Journal of Plant Science* **162**: 309–316.

Breton, C., G. Besnard, and A. Bervillé (2004). 'Using multiple molecular markers to understand olive phylogeography'. In *Documenting domestication: new genetic and archaeological paradigms*,, M.A. Zeder, D. Decker-Walters, D. Dradlay, and B. Smith (eds.). Washington DC, Smithsonian Press: in press.

Briese, D.T. (2000). 'Impact of the *Onopordum capitulum* weevil *Larinus latus* on seed production by its host-plant'. *Journal of Applied Ecology* **37**: 238–246.

Briese, D.T., D. Lane, B.H. Hyde-Wyatt, J. Crocker, and R.G. Diver (1990). 'Distribution of thistles of the genus *Onopordum* thistles in Australia'. *Plant Protection Quaterly* **5**: 23–27.

Burban, C., R.J. Petit, E. Carcreff, and H. Jactel (1999). 'Rangewide variation of the maritime pine bast scale *Matsucoccus feytaudi* Duc. (Homoptera: Matsucoccidae) in relation to the genetic structure of its host'. *Molecular Ecology* **8**: 1593–1602.

Cain, M.L., B.G. Milligan, and A.E. Strand (2000). 'Long-distance seed dispersal in plant populations'. *American Journal of Botany* **87**: 1217–1227.

Calsbeek, R., J.N. Thompson, and J.E. Richardson (2003). 'Patterns of molecular evolution and diversification in a biodiversity hotspot: the California floristic province'. *Molecular Ecology* **12**: 1021–1029.

Cardona, M.A. and J. Contandriopoulos (1979). 'Endemism and evolution in the islands of the Western Mediterranean'. In *Plants and islands*, D. Bramwell (ed.). London, Academic Press Inc.: 133–169.

Carlquist, S. (1974). *Island Biology*. New York, Columbia University Press.

Carmo, M.M., S. Frazao, and F. Venancio (1989). 'The chemical composition of Portuguese *Origanum vulgare* oils'. *Journal of Essential Oil Research* **2**: 69–71.

Carrión, J. (2001). 'Dialectec with climatic interpretations of Late-Quaternary vegetation history in Mediterranean Spain'. *Journal of Mediterranean Ecology* **2**: 145–156.

Casal, M. (1987). 'Post-fire dynamics of shrubland dominated by papilionaceae plants'. *Ecologia Mediterranea* **13**: 87–98.

Casasayas Fornell, T. (1990). 'Widespread adventive plants in Catalonia'. In *Biological invasions in Europe and the Mediterranean Basin*, F. Di Castri, A.J. Hansen, and M. Debussche (eds.). Dordrecht, Kluwer Academic Press: 85–104.

Castola, V., A. Bighelli, and J. Casanova (2000). 'Intraspecific chemical variability of the essential oil of *Pistacia lentiscus* L. from Corsica'. *Biochemical Systematics and Ecology* **28**: 79–88.

Cesaro, A.C. and J.D. Thompson (2004). Darwin's cross-promotion hypothesis and the evolution of stylar polymorphism. *Ecology Letters* **7**: 1209–1215.

Cesaro, A.C., S.C.H. Barrett, S. Maurice, B. Vaissière, and J.D. Thompson (2004). 'An experimental evaluation of self-interference in *Narcissus assoanus*: functional and evolutionary implications'. *Journal of Evolutionary Biology* **17**: 1367–1376.

Charlesworth, D. (1984). 'Androdioecy and the evolution of dioecy'. *Biological Journal of the Linneaen Society* **23**: 333–348.

Charlesworth, D. (1999). 'Theories of the evolution of dioecy'. In *Gender and sexual dimorphism in flowering plants*, M.A. Geber, T.E. Dawson, and L.F. Delph (eds.). Berlin, Springer-Verlag: 33–60.

Charlesworth, D. and B. Charlesworth (1979). 'A model for the evolution of distyly'. *American Naturalist* **114**: 467–498.

Charnov, E.L. (1982). *'The theory of sex allocation'*. Princeton, NJ, Princeton University Press.

Charpentier, A., P. Grillas, and J.D. Thompson (2000). 'The effects of population size limitation on fecundity in mosaic populations of the clonal macrophyte *Scirpus martitimus*'. *American Journal of Botany* **87**: 502–507.

Cheptou, P.-O., A. Berger, A. Blanchard, C. Collin, and J. Escarré (2000*a*). 'The effects of drought stress on inbreeding depression in four populations of the Mediterranean outcrossing plant *Crepis sancta* (Asteraceae)'. *Heredity* **85**: 294–302.

Cheptou, P.-O., E. Imbert, J. Lepart, and J. Escarré (2000*b*). 'Effects of competition on lifetime estimates of inbreeding depression in the outcrossing plant *Crepis sancta* (Asteraceae)'. *Journal of Evolutionary Biology* **13**: 522–531.

Cheptou, P.-O., J. Lepart, and J. Escarré (2002). 'Mating system variation along a successional gradient in the allogamous and colonizing plant *Crepis sancta* (Asteraceae)'. *Journal of Evolutionary Biology* **15**: 753–762.

Cherchi, A. and L. Montadert (1982). 'Oligo-Miocene rift of Sardinia and the early history of the Western Mediterranean Basin'. *Nature* **298**: 736–739.

Chesson, P.L. and R.R. Warner (1981). 'Environmental variability promotes species coexistence in lottery competitive models'. *American Naturalist* **117**: 923–943.

Cheylan, J.P. (1990). Evolution des profils communaux de 1936 à 1990. *In Atals Permanent du Languedoc-Roussillon*. Montpellier, GIP RECLUS.

Cisky, O. and G. Seufert (1999). 'Terpenoid emissions of Mediterranean oaks and their relation to taxonomy'. *Ecological Applications* **9**: 1138–1146.

Cita, M.B. (1982). 'The Messinian salinity crisis: a review'. In *Alpine-Mediterranean Geodynamics*, K.H. Berckhemer and K.J. Hsü (eds.). Washington DC, American Geophysical Union: 113–140.

Cody, M.L. and H.A. Mooney (1978). 'Convergence versus nonconvergence in Mediterranean-climate ecosystems'. *Annual Reviews of Ecology and Systematics* **9**: 265–321.

Cohen, C.R. (1980). 'Plate tectonic model for the Oligo-Miocene evolution of the western Mediterranean'. *Tectonophysics* **68**: 283–311.

Cohen, D. and A. Shmida (1993). 'The evolution of floral display and reward'. *Evolutionary Biology* **68**: 81–120.

Colas, B., I. Olivieri, and M. Riba (1997). '*Centaurea corymbosa*, a cliff dwelling species tottering on the brink of extinction: a demographic and genetic study'. *Proceedings of the National Academy of Science USA* **94**: 3471–3476.

Coleman, M., D.G. Forbes, and R.J. Abbott (2001). 'A new subspecies of *Senecio mohavensis* (Compositae) reveals old–new world species disjunction'. *Edinburgh Journal of Botany* **58**: 389–403.

Coleman, M., A. Liston, J.W. Kadereit, and R.J. Abbott (2003). 'Repeat intercontinental dispersal and Pleistocene speciation in disjunct Mediterranean and desert *Senecio* (Asteraceae)'. *American Journal of Botany* **90**: 1446–1454.

Coley, P.D., J.P. Bryant, and F.S.I. Chapin (1985). 'Resource availability and plant herbivore defense'. *Science* **230**: 895–899.

Comborieu-Nebout, N. (1993). 'Vegetation response to upper pliocene glacial/intergalacial cyclicity in the central Mediterranean'. *Quaternary Research* **40**: 228–236.

Comes, H.P. and R.J. Abbott (1998). 'The relative importance of historical events and gene flow on the population structure of a Mediterranean ragwort, *Senecio gallicus*'. *Evolution* **52**: 355–367.

Comes, H.P. and R.J. Abbott (1999*a*). 'Reticulate evolution in the Mediterranean species complex of *Senecio* sect. *Senecio*: uniting phylogenetic and population level approaches. In *Molecular systematics and plant evolution*, P.M. Hollingsworth, R.M. Bateman, and R.J. Gornall (eds.). London, Taylor & Francis: 171–198.

Comes, H.P. and R.J. Abbott (1999*b*). 'Population genetic structure and gene flow across arid versus mesic environments: a comparative study of two parapatric *Senecio* species from the near east'. *Evolution* **53**: 36–54.

Comes, H.P. and R.J. Abbott (2000). 'Random amplified polymorphic DNA (RAPD) and quantitative trait analyses across a major phylogeographical break in the Mediterranean ragwort *Senecio gallicus* Vill. (Asteraceae)'. *Molecular Ecology* **9**: 61–76.

Comes, H.P. and R.J. Abbott (2001). 'Molecular phylogeography, reticulation, and lineage sorting in Mediterranean *Senecio* sect. *Senecio* (Asteraceae)'. *Evolution* **55**: 1943–1962.

Connell, J.H. and R.O. Slatyer (1977). 'Mechanisms of succession in natural communities and their role in community stability and organization'. *American Naturalist* **111**: 1119–1144.

Contandriopoulos, J. (1962). Recherches sur la flore endémique de la Corse et sur ses origines, Faculté des Sciences de Montpellier.

Contandriopoulos, J. (1984). 'Differentiation and evolution of the genus *Campanula* in the Mediterranean region'. In *Plant biosystematics*, W.F. Grant (ed.). Toronto, Academic Press: 141–158.

Contandriopoulos, J. (1990). *Spécificitè de l'endémisme corse*. Biogeographical aspects of insularity, Rome 18–22 May 1987, Accademia Nazionale dei Lincei.

Contandriopoulos, J. and M.A. Cardona (1984). 'Caractère original de la flore endémique des Baléares'. *Botanica Helvetica* **94**: 101–132.

Conte, L., A. Troia, and G. Cristofolini (1998). 'Genetic diversity in *Cytisus aeolicus* Guss. (Leguminosae), a rare endemite of the Italian flora'. *Plant Biosystems* **132**: 239–249.

Contento, A., M. Ceccarelli, M. Gelati, F. Maggini, L. Baldoni, and P. Cionini (2002). 'Diversity of *Olea* genotypes and the origin of cultivated olive'. *Theoretical and Applied Genetics* **104**: 1229–1238.

Conti, E., D.E. Soltis, T.M. Hardig, and J. Schneider (1999). 'Phylogenetic relationships of the silver saxifrages (*Saxifraga*, Sect. *Ligulatae* Haworth): implications for the evolution of substrate specificity, life histories and biogeography'. *Molecular Phylogenetics and Evolution* **13**: 536–555.

Costich, D.E. (1995). 'Gender specialization across a climatic gradient: experimental comparison of monoecious and dioecious *Ecballium*'. *Ecology* **76**: 1036–1050.

Costich, D.E. and F. Galán (1988). 'The ecology of monoecious and dioecious subspecies of *Ecballium elaterium* (L.) Richard (Cucurbitaceae): I. Geographic distribution and its relationship to climatic conditions in Spain'. *Lagascalia* **15**: 697–710.

Costich, D.E. and T.R. Meagher (1992). 'Genetic variation in *Ecballium elaterium* (Cucurbitaceae): breeding system and geographic distribution'. *Journal of Evolutionary Biology* **5**: 589–601.

Courtin, J. (2000). *Les premiers paysans du Midi*. Paris, La maison des roches.

Couvet, D. (1982). Contribution à l'étude des polymorphismes chémotypique et sexuels. Montpellier, DEA USTL: 51.

Couvet, D., F. Bonnemaison, and P.H. Gouyon (1986). 'The maintenance of females among hermaphrodites: the importance of nuclear–cytoplasmic interactions'. *Heredity* **57**: 325–330.

Couvet, D., P.-H. Gouyon, F. Kjellberg, and G. Valdeyron. (1985). 'La différenciation nucleo-cytoplasmique entre populations: une cause de l'existence de males-steriles dans les populations naturelles de thym'. *Comptes Rendus de l'Académie des Sciences Paris* **300**: 665–668.

Couvet, D., J.P. Henry, and P.H. Gouyon (1985). 'Sexual selection in hermaphroditic plants: the case of gynodioecy'. *The American Naturalist* **126**(2): 294–299.

Couvet, D., O. Ronce, and C. Gliddon (1998). 'The maintenance of nucleocytoplasmic polymorphism in a metapopulation: the case of gynodioecy'. *American Naturalist* **152**: 59–70.

Cowan, I.R. and G.D. Farquhar (1977). 'Stomatal behaviour and environment'. *Advances in Botanical Research* **4**: 117–128.

Cowling, R. and D. Richardson (1995). *Fynbos. South Africa's unique floral kingdom*. Vlaeberg, Fernwood Press.

Cowling, R.M. and B.M. Cambell (1980). 'Convergence in vegetation structure in the Mediterranean communities of California, Chile and South Africa'. *Vegetation* **43**: 191–197.

Cowling, R.M. and P.M. Holmes (1992). 'Endemism and speciation in a lowland flora from the Cape Floristic Region'. *Biological Journal of the Linnaean Society* **47**: 367–383.

Cowling, R.M. and E.T.F. Witowski (1994). 'Convergence and non-convergence of plant traits in climatically matched sites in Mediterranean Australia and South Africa'. *Australian Journal of Ecology* **19**: 220–232.

Cowling, R.M., E.T.F. Witkowski, A.V. Milewski, and K.R. Newbey (1994). 'Taxonomic, edaphic and biological aspects of narrow endemism on matched sites in Mediterranean South Africa and Australia'. *Journal of Biogeography* **21**: 651–664.

Cowling, R.M., P.W. Rundel, B.B. Lamont, M.K. Arroyo, and M. Arianoutsou (1996). 'Plant diversity in Mediterranean-climate regions'. *Trends in Ecology and Evolution* **11**: 362–366.

Cozzolino, S., S. Aceto, P. Caputo, A. Widmer, and A. Dafni (2001). Speciation processes in eastern Mediterranean *Orchis* s.l. species: molecular evidence and the role of pollination biology. *Israel Journal of Plant Science* **49**: 91–103.

Cozzolino, S., S. D'Emerico, and A. Widmer (2004). 'Evidence for reproductive isolate selection in Mediterranean orchids: karyotype differences compensate for the lack of pollinator specificity'. *Proceedings of the Royal Society of London B* (suppl.) *Biology Letters* **271**: 259–262.

Crawley, M.J. (1987). 'What makes a community invasible'. In *Colonisation, succession and stability*, A.J. Gray, M.J. Crawley, and P.J. Edwards (eds.). Oxford, Blackwell Scientific Publications: 429–453.

Crisp, M.D., S. Laffan, H.P. Linder, and A. Monro (2001). 'Endemism in the Australian flora'. *Journal of Biogeography* **28**: 183–198.

Croteau, R. (1987). 'Biosynthesis and catabolism of monotepenoids'. *Chemical Review* **87**: 929–954.

D'Antonio, C.M. (1993). 'Mechanisms controlling invasion of coastal plant communities by the alien succulent *Caprobotus edulis*'. *Ecology* **74**: 83–95.

D'Antonio, C.M., D.C. Odion, and C.M. Tyler (1993). 'Invasion of maritime chaparral by the introduced succulent *Carpobrotus edulis*'. *Oecologia* **95**: 14–21.

D'Emerico, S., S. Cozzolino, G. Pellegrino, D. Pignone, and A. Scrugli (2002). 'Karyotype structure, supernumerary chromosomes and heterochromatin distribution suggest a pathway of karyotype evolution in *Dactylorhiza* (Orchidaceae)'. *Botanical Journal of the Linnaean Society* **138**: 85–91.

Dafni, A. (1983). 'Pollination of *Orchis caspia*—a nectarless plant which deceives the pollinators of nectariferous species from other plant families'. *Journal of Ecology* **71**: 467–474.

Dafni, A. (1996). 'Autumnal and winter pollination adaptations under Mediterranean conditions'. *Bocconea* **5**: 171–181.

Dafni, A. and P. Bernhardt (1990). 'Pollination of terrestrial orchids of southern Australia and the Mediterranean region. Systematic, ecological and evolutionary implications'. *Evolutionary Biology* **24**: 193–252.

Dafni, A. and R. Dukas (1986). 'Insect and wind pollination in *Urginea maritima* (Liliaceae)'. *Plant Systematics and Evolution* **137**: 137.

Dafni, A. and D. Heller (1990). Invasions of adventive plants in Israel. *Biological invasions in Europe and the Mediterranean Basin*, F. Di Castri, A.J. Hansen, and M. Bebussche (eds.). Dordrecht, Kluwer Academic Press: 135–160.

Dafni, A. and Y. Ivri (1981*a*). 'Floral mimicry between *Orchis israelitica* Baumann and Dafni (Orchidaceae) and *Bellevalia flexuosa* Boiss. (Liliaceae)'. *Oecologia* **49**: 229–232.

Dafni, A. and Y. Ivri (1981*b*). 'The flower biology of *Cephalanthera longifolia* (Orchidaceae)—pollen limitation and facultative floral mimicry'. *Plant Systematics and Evolution* **137**: 229–240.

Dafni, A. and E. Werker (1982). 'Pollination ecology of *Sternbergia clusiana* (Ker-Gawler) Spreng. (Amaryllidaceae)'. *New Phytologist* **91**: 571–577.

Dafni, A. and C. O'Toole (1994). 'Pollination syndromes in the Mediterranean: generalizations and peculiarities'. In *Plant–animal interactions in Mediterranean-type ecosystems*, M. Arianoutsou and R.H. Groves (eds.). Dordrecht, Kluwer Academic Publishers: 125–135.

Dafni, A. and A. Shmida (1996). 'The possible ecological implications of the invasion of *Bombus terrestris* (L.) (Apidae) at Mt Carmel, Israel. In: *The conservation of bees* (eds. Matheson A., Buchmann S.L., O'Toole C., Westrich P., Williams I.H.), pp. 183–200. The Linnaean Society, London.

Dafni, A., D. Cohen, and F. Noy-Meir (1981). 'Life-cycle variation in geophytes'. *Annals of the Missouri Botanical Garden* **68**: 652–660.

Dafni, A., A. Schmida, and M. Avishai (1981). 'Leafless-autumnal flowering geophytes in the Mediterranean region—phytogeographical, ecological and evolutionary aspects'. *Plant Systematics and Evolution* **137**: 181–193.

Daget, P. (1977*a*). 'Le bioclimat méditerranéen: caractères généraux, modes de caractérisation'. *Vegetatio* **34**: 1–20.

Daget, P. (1977*b*). 'Le bioclimat méditerranéen: analyse des formes climatiques par le système d'Emberger'. *Vegetatio* **34**: 87–103.

Dahl, A.E., A.-B. Wassgren, and G. Bergstrom (1990). 'Floral scents in *Hypecoum* sect. *Hypecoum* (Papaveraceae): chemical composition and relevance to taxonomy and mating system'. *Biochemical Systematics and Ecology* **18**: 157–168.

Dahlgren, G. and L. Svensson (1994). 'Variation in leaves and petals of *Ranunculus* subgenus *Batrachium* on the

Aegean Islands, analysed by multivariate analysis'. *Botanical Journal of the Linnean Society* **14**: 253–270.

Dallman, P.R. (1998). *Plant life in the world's Mediterranean climates*. Oxford, Oxford University Press.

Damesin, C. and S. Rambal (1995). 'Field study of leaf photosynthetic performance by a Mediterranean deciduous oak tree (*Quercus pubescens*) during a severe summer drought'. *New Phytologist* **131**: 159–167.

Damesin, C., S. Rambal, and R. Joffre (1998). 'Co-occurrence of trees with different leaf habit: a functional approach on Mediterranean oaks'. *Acta Oecologica* **19**: 195–204.

Darwin, C. (1859). *The origin of species*. London, John Murray.

Darwin, C.R. (1877). *The different forms of flowers on plants of the same species*. London, John Murray.

Davis, P.H. (1951). 'Cliff vegetation in the Eastern Mediterranean'. *Journal of Ecology* **39**: 63–93.

de Bolòs, O. and R. Molinier (1960). 'Vue d'ensemble de la végétation des îles Baléares'. *Végétatio* **17**: 251–270.

de Saporta, G. (1863). 'Etudes sur la végétation du sud-est de la Fran,ce à l'époque tertiaire'. *Annals de Science Naturelles* **4**: 5–125.

Debain, S. (2003). *L'expansion de* Pinus sylvestris *et de* Pinus nigra *sur le Causse Méjean : param'etres démographiques et interactions biotiques* PhD thesis, Ecole Nationale Supérieure Agronomique de Montpellier.

Debain, S., T. Curt, and J. Lepart (2003). 'Seed mass, seed dispersal capacity and seedling performance in a *Pinus sylvestris* population'. *Ecoscience* **10**: 168–175.

Debain, S., T. Curt, J. Lepart, and B. Prevosto (2003). 'Variation of reproduction and seed dispersal in an invasive species: the case of *Pinus sylvestris* L. expansion in the Causse Méjean, southern France'. *Journal of Vegetation Science* **14**: 509–516.

Debussche, M. (1985). 'Rôle des oiseaux disséminateurs dans la germination des graines de plantes à fruits charnus en région méditerranéenne'. *Acta Oecologica Oecologia Plantarum* **6**: 365–374.

Debussche, M. (1988). 'La diversité morphologique des fruits charnus en Languedoc méditerranéen: relations avec les caractéristiques biologiques et la distribution des plantes, et avec les disséminateurs'. *Acta Oecologica/-Oecologia Generalis* **9**: 37–52.

Debussche, M. and P. Isenmann (1989). 'Fleshy fruit characters and the choices of bird and mammal seed dispersers in a Mediterranean region'. *Oikos* **56**: 327–338.

Debussche, M. and P. Isenmann (1990). 'Introduced and cultivated fleshy-fruited plants: consequences of a mutualistic Mediterranean plant-bird system'. In *Biological invasions in Europe and the Mediterranean Basin*, F. Di Castri, A.J. Hansen, and M. Bebussche (eds.). Dordrecht, Kluwer Academic Press: 399–416.

Debussche, M. and P. Isenmann (1992). A Mediterranean bird disperser assemblage: composition and phenology in relation to fruit availability. *Revue d'Ecologie (Terre et Vie)* **47**: 411–432.

Debussche, M. and P. Isenmann (1994). Bird-dispersed seed rain and seedling establishment in patchy Mediterranean vegetation. *Oikos* **69**: 414–426.

Debussche, M. and J. Lepart (1992). Establishment of woody plants in Mediterranean old fields: opportunity in space and time. *Landscape Ecology* **6**: 133–145.

Debussche, M. and J.D. Thompson (2000). Les populations à fleurs blanches de *Cyclamen repandum* Sibth. & Sm. en Corse. *Bulletin de la Société Botanique du Centre-Ouest* **31**: 1–8.

Debussche, M. and J.D. Thompson (2002). 'Morphological differentiation among closely related species with disjunct distributions: a case study of Mediterranean *Cyclamen* L. subgenus *Psilanthum* (Primulaceae)'. *Botanical Journal of the Linnaean Society* **132**: 133–144.

Debussche, M. and J.D. Thompson (2003). 'Habitat differentiation between two closely related Mediterranean plant species, the endemic *Cyclamen balearicum* and the widespread *C. repandum*'. *Acta Oecologia* **24**: 35–45.

Debussche, M., J. Lepart, and J. Molina (1985). La dissémination des plantes à fruits charnus par les oiseaux: rôle de la structure de la végétation et impact sur la succession en région méditerranéenne. *Acta Oecologica/Oecologia Generalis* **6**: 65–80.

Debussche, M., J. Cortez, and I. Rimbault (1987). 'Variation in fleshy fruit composition in the Mediterranean region: the importance of ripening season, life-form, fruit type and geographical distribution'. *Oikos* **49**: 244–252.

Debussche, M., J. Escarré, J. Lepart, C. Houssard, and S. Lavorel (1996). 'Changes in Mediterranean plant succession: old fields revisited'. *Journal of Vegetation Science* **7**: 519–526.

Debussche, M., J. Lepart, and A. Dervieux (1999). 'Mediterranean landscape changes: evidence from old postcards'. *Global Ecology and Biogeography* **8**: 3–15.

Debussche, M., G. Debussche, and J. Lepart (2001). 'Changes in the vegetation of *Quercus pubescens* woodland after cessation of coppicing and grazing'. *Journal of Vegetation Science* **12**: 81–92.

Debussche, M., E. Garnier, and J.D. Thompson (2004). 'Exploring the causes of variation in phenology and morphology in Mediterranean geophytes: a genus wide study of *Cyclamen* L'. *Botanical Journal of the Linnaean Society* 134: 145: 469–484.

Delph, L.F. and P. Mutikainen (2003). 'Testing why the sex of the maternal parent affects seedling survival in a gynodioecious species'. *Evolution* **57**: 231–239.

D'Émerico, S., S. Cozzolino, G. Pellegrino, D. Pignone, and A. Scrugli. 2002 'Karyotype structure, supernumerary chromosomes and heterochromatin distribution suggest a pathway of karyotype evolution in *Dactylohiza* (Orchidaceae)'. *Botanical Journal of the Linnaean Society* **138**: 85–91.

Demesure, B., B. Comps, and R. Petit (1996). 'Chloroplast DNA phylogeography of the common beech (*Fagus sylvatica* L.) in Europe'. *Evolution* **50**: 2515–2520.

Demiriz, H. and T. Baytop (1985). 'The Anatolian peninsula'. *In Plant conservation in the Mediterranean area*, C. Gomez-Campo, (ed.). Dordrecht, Dr W. Junk Publishers: 113–121.

den Nijs, J.C.M. (1983). 'Biosystematic studies of the *Rumex acetosella* complex (Polygonaceae) VI. South-eastern Europe, including a phylogenetic survey'. *Bot. Jahrb. Syst.* **104**: 33–90.

den Nijs, J.C.M., K. Sorgdrager, and J. Stoop (1985). 'Biosystematic studies of the *Rumex acetosella* complex. IX. Cytogeography of the complex in the Iberian peninsula and taxonomic discussion'. *Botanica Helvetica* **95**: 141–157.

Dewey, J.F., M.L. Helman, E. Turco, D.H.W. Hutton, and S.D. Knott (1989). Kinematics of the western Mediterranean, In *Alpine tectonics*, M.P. Coward, D. Dietrich, and R.G. Park (eds.). London, Geological Society Special Publication. Vol. **45**: 265–283.

Di Castri, F. (1973). 'Climatographical comparisons between Chile and the western coast of North America'. In *Mediterranean type ecosystems. Origin and structure*, F. Di Castri and H.A. Mooney (eds.). Heidelberg; Berlin, Springer-Verlag: 21–36.

Di Castri, F. (1990). 'On invading species and invaded ecosystems; the interplay of historical chance and biological necessity'. In *Biological invasions in Europe and the Mediterranean Basin*, F. Di Castri, A.J. Hansen, and M. Debussche (eds.). Dordrecht, Netherlands, Kluwer Academic Publishers: 3–16.

Di Castri, F. (1991). 'An ecological overview of the five regions of the world with a Mediterranean climate'. In *Biogeography of Mediterranean invasions*, R.H. Groves and F. Di Castri (eds.). Cambridge, Cambridge University Press: 3–16.

Di Pasquale, G., G. Garfì, and P. Quézel (1992). 'Sur la présence d'un *Zelkova* nouveau en Sicile sud-orientale (Ulmaceae)'. *Biocosme mésogéen* **8–9**: 401–409.

Diamond, J. (1997). *Guns, germs and steel. A short history of everybody for the last 13,000 years*. London, Chatto & Windus.

Díaz Lifante, Z. (1996). 'A karyological study of *Asphodelus* L. (Asphodeleaceae) from the western Mediterranean'. *Botanical Journal of the Linnaeaen Society* **121**: 285–344.

Dobson, H.E.M., J. Arroyo, G. Bergstrom, and I. Groth (1997). 'Interspecific variation in floral fragrances within the genus *Narcissus* (Amaryllidaceae)'. *Biochemical Systematics and Ecology* **25**: 685–706.

Dommée, B., M.W. Assouad, and G. Valdeyron (1978). 'Natural selection and gynodioecy in *Thymus vulgaris* L'. *Botanical Journal of the Linnaean Society* **77**: 17–28.

Dommée, B., J.L. Guillerm, and G. Valdeyron (1983) Régime de reproduction et hétérozygotie des populations de *Thymus vulgaris L.*, dans une succession postculturale. *Comptes Rendus de l'Académie des Sciences de Paris* **296**: 111–114.

Dommée, B., N. Denelle, and J.-L. Bompar (1988). *Le genre* Thymelaea *Endl. et plus spécialement l'espèce* Thymelaea hirsuta *(L.) Endl.: sexualité et distribution géographique*. Symposium International de Botanique, Lleida: 1–6.

Dommée, B., J.-L. Bompar, and N. Denelle (1990). 'Sexual tetramorphism in *Thymelaea hirsuta* (Thymeleaceae): evidence of the pathway from heterodichogamy to dioecy at the infraspecific level'. *American Journal of Botany* **77**: 1449–1462.

Dommée, B., A. Biascamanno, N. Denelle, J.-L.Bompar, and J.D. Thompson (1995). 'Sexual tetramorphism in *Thymelaea hirsuta* (Thymeleaceae): morph ratios in open pollinated progeny'. *American Journal of Botany* **82**: 734–740.

Dommée, B., A. Geslot, J.D. Thompson, M. Reille, and N. Denelle (1999). 'Androdioecy in the entomophilous tree *Fraxinus ornus* (Oleaceae)'. *New Phytologist* **143**: 419–426.

Dubar, M., J.P. Ivaldi, and M. Thinon (1995). 'Feux de forêt méditerranéens: une histoire de pins'. *La Recherche* **273**: 188–189.

Dubois, S., P.-O. Cheptou, C. Petit, P. Meerts, M. Poncelet, X. Vekemans, C. Lefébvre, and J. Escarré (2003). 'Genetic structure and mating systems of metallicolous and nonmetallicolous populations of *Thlaspi caerulescens* in south France, Belgium and Luxembourg'. *New Phytologist* **157**: 633–641.

Dufaÿ, M. and M.-C. Anstett (2004). 'Cheating is not always punished: killer female plants and pollination by deceit in the dwarf palm *Chamaerops humilis*'. *Journal of Evolutionary Biology* in press.

Dufaÿ, M., M. Hossaert-McKey, and M.-C. Anstett (2003). 'When leaves act like pollinators: how dwarf palms attract their pollinators?'. *Ecology Letters* **6**: 28–34.

Dufaÿ, M., M. Hossaert-McKey, and M.-C. Anstett (2004). 'Temporal and sexual variation of leaf-produced pollinator attracting odours in the dwarf palm *Chamaerops humilis*'. *Oecologia* in press.

Dufour-Dror, J.-M. (2002). 'A quantitative classification of Mediterranean mosaic-like landscapes'. *Journal of Mediterranean Ecology* **3**: 3–12.

Duggen, S., K. Hoernie, P. van den Bogard, L. Rüpke, and J.P. Morgan (2003). 'Deep roots of the Messinian salinity crisis'. *Nature* **422**: 602–606.

Duhme, F. and T.M. Hinckley (1992). 'Daily and seasonal variation in water relations of macchia shrubs and trees in France (Montpellier) and Turkey (Antalya)'. *Vegetatio* **100**: 185–198.

Dulberger, R. (1964). 'Flower dimorphism and self-incompatibility in *Narcissus tazetta* L'. *Evolution* **18**: 361–363.

Dulberger, R. (1970). 'Floral dimorphism in *Anchusa hybrida* Ten'. *Israel Journal of Botany* **19**: 37–41.

Dulberger, R. (1973). 'Distyly in *Linum pubescens* and *Linum mucronatum*'. *Botanical Journal of the Linneaen Society* **66**: 117–126.

Dulberger, R. and A. Horovitz (1984). 'Gender polymorphism in flowers of *Silene vulgaris* (Moench) Garcke (Caryophyllaceae)'. *Botanical Journal of the Linnaean Society* **89**: 101–117.

Durand, B. (1963). 'Le complexe *Mercurialis annua* L. s.l.: une étude biosystématique'. *Annales des Sciences Naturelles, Botanique, Paris* **12**: 579–736.

Dvořák, J., P. Di Terlizzi, H.B. Zhang, and P. Resta (1993). 'The evolution of polyploid wheats: identification of the A genome donor species'. *Genome* **36**: 21–31.

Dvořák, J., M.C. Luo, Z.L. Yang, and H.B. Zhang (1998). The structure of the *Aegilops tauschii* genepool and the evolution of hexaploid wheat. *Theoretical and Applied Genetics* **97**: 657–670.

Ehlers, B. and J.D. Thompson (2004*a*). 'Temporal variation in sex allocation in hermaphrodites of gynodioecious *Thymus vulgaris* L'. *Journal of Ecology* **92**: 15–23.

Ehlers, B., and J.D. Thompson (2004*b*). Do co-occuring plant species adapt to one another? The response of *Bromus erectus* to the presence of different *Thymus vulgaris* chemotypes. *Oecologia* **141**: 511–518.

Ehrendorfer, F. (1980). 'Polyploidy and distribution'. In *Polyploidy: biological relevance*, W.H. Lewis (ed.). New York & London, Plenum Press: 45–60.

Ehrlich, P.R. and P.H. Raven (1969). 'Differentiation of populations'. *Science* **165**: 1228–1232.

El-Keblawly, A., J. Lovett Doust, L. Lovett Doust, and K.H. Shaltout (1995). 'Labile sex expression and dynamics of gender in *Thymelaea hirsuta*'. *Ecoscience* **2**: 55–66.

El Mousadik, A. and R.J. Petit (1996*a*). 'Chloroplast DNA phylogeography of the argan tree of Morocco'. *Molecular Ecology* **5**: 547–555.

El Mousadik, A. and R.J. Petit (1996*b*). 'High level of genetic differentiation for allelic richness among populations of the argan tree [*Argania spinosa* (L.) Skeels] endemic to Morocco'. *Theoretical and Applied Genetics* **92**: 832–839.

Elena-Rossello, J.A., R. Lumaret, E. Cabrera, and H. Michaud (1992). 'Evidence for hybridization between sympatric holm-oak and cork-oak in Spain based on diagnostic enzyme markers'. *Vegetatio* **99–100**: 111–118.

Ellstrand, N.C. and K.A. Schierenbeck (2000). 'Hybridization as a stimulus for the evolution of invasiveness in plants'. In *Variation and evolution in plants and microorganisms. Toward a new synthesis 50 years after Stebbins*, F.J. Ayala, W.M. Fitch, and M.T. Clegg (eds.). Washington DC, National Academy Press: 289–309.

Ellstrand, N.C., H.C. Prentice, and J.F. Hancock (1999). 'Gene flow and introgression from domesticated plants into their wild relatives'. *Annual Review of Ecology and Systematics* **30**: 539–563.

Elton, C.S. (1958). *The ecology of invasions by animals and plants*. London, Methuen.

Emberger, L. (1930*a*). 'La végétation de la région méditerranéenne, esai d'une classification des groupements végétaux'. *Revue Générale de Botanique* **42**: 641–662.

Emberger, L. (1930*b*). 'La végétation de la région méditerranéenne, esai d'une classification des groupements végétaux'. *Revue Générale de Botanique* **42**: 705–721.

Emberger, L. (1930*c*). 'Sur une formule climatique applicable en géographie botanique'. *Comptes Renus de l'Academie des Sciences Paris* **191**: 389–390.

Eron, Z. (1987). 'Ecological factors restricting the regeneration of *Pinus brutia* in Turkey'. *Ecologia Méditerranea* **13**: 57–67.

Escarré, J., C. Houssard, and M. Debussche (1983). 'Evolution de la végétation et du sol après abandon cultural en région méditerranéenne: étude de la succession dans les Garrigues Montpelliérais (France)'. *Acta Oecologica: Oecologia Plantarum* **4**: 221–239.

Escarré, J., J. Lepart, F.X. Sans, J.J. Santuc, and V. Gorse (1999). 'Effects of herbivory on the growth and reproduction of *Picris hieracioides* in the Mediterranean region'. *Journal of Vegetation Science* **10**: 101–110.

Escarré, J., C. Lefèbvre, W. Gruber, M. Leblanc, J. Lepart, Y. Riviére, and B. Delay (2000). 'Zinc and cadmium hyperaccumulation by *Thlaspi caerulescens* J. and C. Presl from metalliferous and non-metalliferous sites in the Mediterranean area: implications for phytoremediation'. *New Phytologist* **145**: 429–437.

Escudero, A. (1985). Efectos de arboles aislados sobre las propriedades quimicos del suelo. *Revue d'Ecologie et de Biologie du Sol* **22**: 149–159.

Escudero, A. (1996). 'Community patterns on exposed cliffs in a Mediterranean calcareous mountain'. *Vegetatio* **125**: 99–110.

Escudero, A., R.C. Somolinos, J.M. Olano, and A. Rubio (1999). 'Factors controlling the establishment of *Helianthemum squamatum*, an endemic gypsophite of semi-arid Spain'. *Journal of Ecology* **87**: 290–302.

Fauquette, S., P. Quézel, J. Guiot, and J.P. Suc (1998). 'Signification bio-climatique de taxons-guides du Pliocène méditerranéen'. *Geobios* **31**: 151–169.

Favarger, C. and J. Contandriopoulos (1961). 'Essai sur l'endémisme'. *Bulletin de la Socité Botanique Suisse* **77**: 383–408.

Favarger, C. and S. Siljak-Yakovlev (1986). *A propos de la classification des taxons endémiques basée sur la cytotaxonomie et la cytogénétique*. Colloque Internationale de Botanique Pyrenéenne.

Fennane, M. and M. Ibn Tattou (1998). 'Catalogue des plantes vasculaires, rares, menacées ou endémiques du Maroc'. *Bocconea* **8**: 1–243.

Ferguson, D. and T. Sang (2001). 'Speciation through homoploid hybridization between allotetraploids in peonies (*Paeonia*)'. *Proceedings of the National Academy of Sciences (USA)* **98**: 3915–3919.

Fernandes, A. (1935). 'Remarque sur l'hétérostylie de *Narcissus triandrus* et de *N. reflexus*'. *Bol. Soc. Brot. Sér. 2* **10**: 5–15.

Figueroa, M.E. and A.J. Davy (1991). 'Response of Mediterranean grassland species to changing rainfall'. *Journal of Ecology* **79**: 925–941.

Fineschi, S., D. Taurchini, F. Villani, and G.G. Vendramin (2000). 'Chloroplast DNA polymorphism reveals little geographic structure in *Castanea sativa* Mill. (Fagaceae) throughout southern European countries'. *Molecular Ecology* **9**: 1495–1503.

Fox, M.D. (1995). 'Impact of interannual and seasonal rainfall variability on plant community composition across a rainfall gradient'. In *Time scales of biological responses to water constraints*, J. Roy, J. Aronson, and F. Di Castri (eds.). Amsterdam, SPB Academic Publishing: 56–70.

Fraenkel, G.S. (1959). 'The raison d'être of secondary plant substances'. *Science* **129**: 1460–1470.

Frérot, H., C. Petit, C. Lefébvre, W. Gruber, C. Collin, and J. Escarré (2003). 'Zinc and cadmium accumulation in controlled crosses between metallicolous and non-metallicolous populations of *Thlaspi caerulescens* (Brassicaceae)'. *New Phytologist* **157**: 643–648.

Fréville, H., B. Colas, J. Ronfort, M. Riba, and I. Olivieri (1998). Predicting endemism from population structure of a widespread species: case study in *Centaurea maculosa* Lam. (Asteraceae)'. *Conservation Biology* **12**: 1–10.

Fréville, H., B. Colas, M. Riba, H. Caswell, A. Mignot, E. Imbert, and I. Olivieri (2004). 'Spatial and temporal demographic variability in the endemic plant *Centaurea corymbosa* (Asteraceae)'. *Ecology* **85**: 694–703.

Fritz, R.S. (1990). 'Effects of genetic and environmental variation on resistance of willow to sawflies'. *Oecologia* **82**: 325–332.

Fuertes Aguilar, J., J.A. Rosselló, and G. Nieto Feliner (1999). 'Molecular evidence for the compilospecies model of reticulate evolution in *Armeria*'. *Systematic Biology* **48**: 735–754.

Galili, E., D.J. Stanley, J. Sharvit, and M. Weinstein-Evron (1997). 'Evidence for earliest olive oil production in submerged settlements of the coast, Israel'. *Journal of Archaeological Science* **24**: 1141–1150.

Gallagher, K.G., K.A. Schierenbeck, and C.M. D'Antonio (1997). 'Hybridization and introgression in *Carpobrotus* spp. (Aizoaceae) in California: II. Isozyme evidence'. *American Journal of Botany* **84**: 905–911.

Gamisans, J. (1991). *La végétation de la Corse*. Aix-en-Provence. Edisud.

Gamisans, J. and D. Jeanmonod (1995). 'La flore de Corse: Bilan des connaissances, intérêt patrimonial et état de conservation'. *Ecologia Mediterranea* **21**: 135–148.

Garcia-Ramos, G. and M. Kirkpatrick (1997). 'Genetic models of adaptation and gene flow in peripheral populations'. *Evolution* **51**: 21–28.

García, M.B., R. Zamora, J.A. Hódar, J.M. Gómez, and J. Castro (2000). 'Yew (*Taxus baccata* L.) regeneration is facilitated by fleshy-fruited shrubs in Mediterranean environments'. *Biological Conservation* **95**: 31–38.

García, D., R. Zamora, J.M. Gómez, and J. Hódar (2001). 'Frugivory at *Juniperus communis* depends more on population characteristics than on individual attributes'. *Journal of Ecology* **89**: 639–647.

Garnier, E., J. Cortez, G. Billès, M.-L. Navas, C. Roumet, M. Debussche, G. Laurent, A. Blanchard, D. Aubry, A. Bellmann, C. Neill, and J.-P. Toussaint (2004). 'Plant functional markers capture ecosystem properties during secondary succession'. *Ecology* **85**: 2630–2637.

Garrido, J.L., P. Rey, X. Cerda, and C.M. Herrera (2002). 'Geographical variation in diaspore traits of an ant-dispersed plant (*Helleborus foetidus*): are ant community composition and diaspore traits correlated?'. *Journal of Ecology* **90**: 449–455.

Garrido, B., A. Hampe, T. Marañon, and J. Arroyo (2003). 'Regional differences in land use affect population performance of the threatened insectivorous plant *Drosophyllum lusitanicum* (Droseraceae)'. *Diversity and Distribution* **9**: 335–350.

Gauquelin, T., V. Bertaidière, t. Mo, N., W. Badri, and J.F. Asmodé (1999). 'Endangered stands of thuriferous stands juniper in the western Mediterranean Basin: ecological status, conservation and management'. *Biodiversity and Conservation* **8**: 1479–1498.

Gentry, A.H. (1986). 'Endemism in tropical versus temperate plant communities'. In *Conservation Biology. The science of scarcity and diversity*, M.E. Soulé (ed.). Sunderland, MA, Sinauer Associates Inc.: 153–181.

Geslot, A. (1983). 'Contribution à la connaissance biosystématique des campanules de la sous-section Heterophylla (Wit.) Fed.: I. - étude de la biologie de la reproduction des campanules pyrénéennes à feuilles hétéromorphes'. *Rev. Gén. Bot.* **90**: 185–220.

Gibernau, M., H.R. Busher, J.E. Frey, and M. Hossaert-Mckey (1997). 'Volatile extracts from extracts of syconia (figs) of *Ficus carica* (Moraceae)'. *Phytochemistry* **46**: 241–244.

Gielly, L., M. Debussche, and J.D. Thompson (2001). 'Geographic isolation and evolution of endemic *Cyclamen* in the Mediterranean: insights from chloroplast *trn*L (UAA) intron sequence variation'. *Plant Systematics and Evolution* **230**: 75–88.

Gigord, L., C. Lavigne, J.A. Shykoff, and A. Atlan (1999). 'Evidence for effects of restorer genes on male and female reproductive functions of hermaphrodites in the gynodioecious species *Thymus vulgaris* L. *Journal of Evolutionary Biology* **12**: 596–604.

Gitzendanner, M.A. and P.S. Soltis (2000). 'Patterns of genetic variation in rare and widespread plant congeners'. *American Journal of Botany* **87**: 783–792.

Givnish, T.J. (1980). 'Ecological constraints on the evolution of breeding systems in seed plants: dioecy and dispersal in gymnosperms'. *Evolution* **34**: 959–972.

Glyphis, J.P. and G.M. Puttick (1989). 'Phenolics, nutrition, and insect herbivory in some garrigues and macquis plant species'. *Oecologia* **7**: 259–263.

Goldblatt, P. (1978). 'An analysis of the flora of southern Africa. Its characteristics, relationships and origins'. *Annals of the Missouri Botanical Garden* **65**: 369–436.

Golenberg, E.M. and E. Nevo (1987). 'Multilocus differentiation and population structure in a selfer, wild emmer wheat, *Triticum dicoccoides*'. *Heredity* **58**: 451–456.

Gollan, T., N.C. Turner, and E.-D. Schulze (1985). 'The responses of stomata and leaf gas exchange to vapour pressure deficits and soil water content: III. In the sclerophyllous species *Nerium oleander*'. *Oecologia* **65**: 356–362.

Gómez-Campo, C. (ed.) (1985). *Plant conservation in the Mediterranean area*. Dordrecht, Dr W. Junk Publishers.

Gómez-Campo, C. and J. Malato-Beliz (1985). 'The Iberian peninsula'. In *Plant conservation in the Mediterranean area*, C. Gomez-Campo (ed.). Dordrecht, Dr W. Junk Publishers: 47–70.

Gómez, J.M. (1993). 'Phenotypic selection on flowering synchrony in a high mountain plant, *Hormathophylla spinosa* (Cruciferae)'. *Journal of Ecology* **81**: 605–613.

Gómez, J.M. (2000). 'Effectiveness of ants as pollinators of *Lobularia maritima*: effects on main sequential fitness components of the host plant'. *Oecologia* **122**: 90–97.

Gómez, J.M. (2003). 'Herbivory reduces the strength of pollinator-mediated selection in the Mediterranean herb *Erysimum mediohispanicum*: consequences for plant specialization'. *American Naturalist* **162**: 242–256.

Gómez, J.M. and R. Zamora (1992). 'Pollination by ants: consequences of the quantitative effects on a mutualistic system'. *Oecologia* **91**: 410–418.

Gómez, J.M. and R. Zamora (1999). 'Generalization versus specialization in the pollination system of *Hormatophyllum spinosa* (Cruciferae)'. *Ecology* **80**: 796–805.

Gómez, J.M. and R. Zamoro (1996). 'Wind pollination in high mountain populations of *Hormatophyllum spinosa* (Cruciferae)'. *American Journal of Botany* **83**: 580–585.

Gómez, J.M. and R. Zamora (2000). 'Spatial variation in the selective scenarios of *Hormatophylla spinosa* (Cruciferae)'. *American Naturalist* **155**: 657–668.

Gómez, J.M., R. Zamoro, J.A. Hódar, and D. Garciá (1996). 'Experimental study of pollination by ants in Mediterranean high mountain and arid habitats'. *Oecologia* **105**: 236–242.

Gömöry, D., L. Paule, R. Brus, P. Zhelev, Z. Tomovic, and J. Gracans (1999). 'Genetic differentiation and phylogeny of beech of the Balkan peninsula'. *Journal of Evolutionary Biology* **12**: 746–754.

González-Martínez, S.C., R. Alía, and L. Gil (2002). 'Population genetic structure in a Mediterranean pine (*Pinus pinaster* Ait.): a comparison of allozyme markers and quantitative traits'. *Heredity* **89**: (199–206.

Gouyon, P.H., P. Fort, and G. Caraux (1983). 'Selection of seedlings of *Thymus vulgaris* by grazing slugs'. *Journal of Ecology* **71**: 299–306.

Gouyon, P.H., P. Vernet, J.L. Gaillerm and G. Valdeyron (1986). 'Polymorphisms and environment: the adaptive value of the oil polymorphisms in *Thymus vulgaris* L'. *Heredity* **57**: 59–66.

Graham, S.W. and S.C.H. Barrett (2004). 'Phylogenetic reconstruction of the evolution of stylar polymorphisms in *Narcissus* (Amaryllidaceae)'. *American Journal of Botany* **91**: 1007–1021.

Granger, R. and J. Passet (1973). '*Thymus vulgaris* L. spontané de France: races chimiques et chemotaxonomie'. *Phytochemistry* **12**: 1683–1691.

Granger, R., J. Passet, and G. Teulade-Arbousset (1973). 'Plantes medicinales à essence et chimiotaxonomie'. *Riv. Italiana* **55**: 353–356.

Granger, R., J. Passet, G. Teulade-Arbousset, and P. Auriol (1973). 'Types chimiques de *Thymus nitens* endémique Cévénol'. *Plantes médicinales et phytothérapie* **7**: 225–233.

Grant, V. and K.A. Grant (1965). *Flower pollination of the phlox family*. New York, Columbia University Press.

Gratani, L., P. Marzi, and M.F. Crescente (1992). 'Morphological adaptations of *Quercus ilex* leaves in the Castelporziano forest'. *Vegetatio* **100**: 155–161.

Gravano, E., R. Desotgiu, C. Tani, F. Busotti, and P. Grossoni (2000). 'Structural adaptations in leaves of two Mediterranean evergreen shrubs under different climatic conditions'. *Journal of Mediterranean Ecology* **1**: 165–170.

Gray, A.G. (1986). 'Do invading species have definable genetic characteristics?' *Philosophical Transactions of the Royal Society (London)* **314**: 655–674.

Gray, A.J. (1993). 'The vascular plant pioneers of primary successions: persistence and phenotypic plasticity'. In *Primary succession on land*, J. Miles and D.W.H. Walton (eds.). Oxford, Blackwell Scientific. Vol. 12: 179–191.

Greuter, W. (1972). 'The relict element of the flora of Crete and its evolutionary significance'. In *Taxonomy, phytogeography and evolution*, D.H. Valentine (ed.). London, Academic Press: 161–177.

Greuter, W. (1979). 'The origins and evolution of island floras as exemplified by the Aegean archipelago'. In *Plants and islands*, D. Bramwell (ed.). London, Academic Press Inc.: 87–106.

Greuter, W. (1991). 'Botanical diversity, endemism, rarity, and extinction in the Mediterranean area: an analysis based on the published volumes of Med-Checklist'. *Bot. Chron.* **10**: 63–79.

Greuter, W. (1994). 'Extinction in Mediterranean areas'. *Philosophical Transactions of the Royal Society Series B* **344**: 41–46.

Grey-Wilson, C. (1997). *Cyclamen. A guide for gardeners, horticulturists, and botanists*. London, B.T. Batsford Ltd.

Griffin, J.R. (1973). Xylem sap tension in three woodland oaks of central California. *Ecology* **54**: 152–159.

Grillas, P. and J. Roché (1997). La végétation des marais temporaires. Arles, Tour de Valat Biological Station.

Grillas, P., C. van Wijck, and A. Bonis (1991). 'Life history traits: a possible cause for the higher frequency of occurrence of *Zannichellia pedunculata* than of *Zannichellia obtusifollia* in temporary marshes'. *Aquatic Botany* **42**: 1–13.

Grison-Pigé, L., J.-M. Bessière, T.C.J. Turlings, F. Kjellberg, J. Roy, and M. Hossaert-Mckey (2001). 'Limited intersex mimicry of floral odour in *Ficus carica*'. *Functional Ecology* **15**: 551–558.

Grove, A.T. and O. Rackham (2001). *The nature of Mediterranean Europe. An ecological history*. New Haven & London, Yale University Press.

Groves, R.H. (1991). 'The biogeography of Mediterranean plant invasions'. In *Biogeography of Mediterranean invasions*, R.H. Groves and F. Di Castri (eds.). Cambridge, Cambridge University press: 427–452.

Grubb, P.J. (1977). 'The maintenance of species richness in plant communities: the importance of the regeneration niche'. *Biological Reviews* **52**: 107–145.

Guenther, E. (1948–52). *The essential oils*. New York, Van Nostrand.

Guillerm, J.L., E. le Floc'h, J. Maillet, and C. Boulet (1990). 'The invading weeds within the western Mediterranean Basin'. In *Biological invasions in Europe and the Mediterranean Basin*, F. Di Castri, A.J. Hansen, and M. Bebussche (eds.). Dordrecht, Kluwer Academic Press: 61–84.

Guitián, J. and J.M. Sanchez (1992). 'Seed dispersal spectra of plant communities in the Iberain peninsula'. *Vegetatio* **98**: 157–164.

Guitián, J., P. Guitián, and M. Medrano (1998). 'Floral biology of the distylous Mediterranean shrub *Jasminum fruticans* (Oleaceae)'. *Nordic Journal of Botany* **18**: 195–201.

Gutiérrez Larena, B., J. Fuertes Aguilar, and G. Nieto Feliner (2002). 'Glacial induced altitudinal migrations in *Armeria* (Plumbaginaceae) inferred from patterns of chloroplast DNA haplotype sharing'. *Molecular Ecology* **11**: 1965–1974.

Hampe, A. (2003). 'Large scale geographical trends in fruit traits of vertebrate-dispersed temperate plants'. *Journal of Biogeography* **30**: 487–496.

Hampe, A. (2004). 'How to be a relict in the Mediterranean: the ecology of reproduction and regeneration of *Frangula alnus* subsp. *baetica*'. PhD thesis. Universidad de Sevilla: Sevilla.

Hampe, A. and J. Arroyo (2002). 'Recruitment and regeneration in populations of an endangered South Iberian tertiary relict tree'. *Biological Conservation* **107**: 263–271.

Hampe, A. and F. Bairlein (2000). 'Modified dispersal-related traits in disjunct populations of bird-dispersed *Frangula alnus* (Rhamnaceae): a result of its Quaternary distribution shifts'. *Ecography* **23**: 603–613.

Hampe, A., J. Arroyo, P. Jordano, and R.J. Petit (2003). 'Rangewide phylogeography of a bird-dispersed Eurasian shrub: contrasting Mediterranean and temperate glacial refugia'. *Molecular Ecology* **12**: 3415–3426.

Harlan, J.M., J.M.J. deWet, and E.G. Price (1973). 'Comparative evolution of cereals'. *Evolution* **27**: 311–325.

Harlan, J.R. and D. Zohary (1966). 'Distribution of wild wheats and barley'. *Science* **153**: 1074–1080.

Harrison, A.T., E. Small, and H.A. Mooney (1971). 'Drought relationships and distribution of two Mediterranean-climate California plant communities'. *Ecology* **52**: 869–875.

Harvey, P.H. and M.D. Pagel (1991). *The comparative method in evolutionary biology*. Oxford, Oxford University Press.

Hättenschwiler, S., F. Miglietta, A. Raschi, and C. Körner (1997). 'Morphological adjustments to elevated CO_2 in mature *Quercus ilex* trees growing around natural CO_2 springs'. *Acta Oecologica* **18**: 361–365.

Hegazy, A.K. and N.M. Eesa (1991). 'On the ecology, insect seed predation, and conservation of a rare and endemic plant species: *Ebenus armitagei*'. *Conservation Biology* **5**: 317–324.

Heitz, B. (1973). 'Hétérostylie et spéciation dans le groupe *Linum perenne*'. *Annales des Sciences Naturelles, Botanique, Paris* **14**: 385–406.

Hengeveld, R. (1989). *Dynamics of biological invasions*. London, Chapman & Hall.

Henriques, J.A. (1887). Observaçoes sobre algumas especies de *Narcissus*, entrecados em Portugal. *Bol. Soc. Brot.* **5**: 168–174.

Hermenger, R.P., W.A. Watts, J.R.M. Allen, B. Huntley, and S.C. Fritz (1996). 'Vegetation history and climate of the last 15,000 years at Laghi di Monticchio, southern Italy'. *Quaternary Science Reviews* **15**: 113–132.

Herms, D.A. and W.J. Mattson (1992). 'The dilemma of plants: to grow or defend'. *Quarterly Review of Biology* **67**: 283–335.

Herrera, C.M. (1982*a*). 'Breeding systems and dispersal-related maternal reproductive effort of southern Spanish bird-dispersed plants'. *Evolution* **36**: 1299–1314.

Herrera, C.M. (1982*b*). 'Seasonal variation in quality of fruits and diffuse coevolution between plants and avian dispersers'. *Ecology* **63**: 773–775.

Herrera, C.M. (1984*a*). 'Adaptation to frugivory of Mediterranean avian seed dispersers'. *Ecology* **65**(2): 609–617.

Herrera, C.M. (1984*b*). 'A study of avian frugivores, bird-dispersed plants, and their interaction in Mediterranean scrublands'. *Ecological Monographs* **54**(1): 1–23.

Herrera, C.M. (1984*c*). 'The annual cycle of *Osyris quadripartita*, a hemiparasitic dioecious shrub of Mediterranean shrublands'. *Journal of Ecology* **72**: 1065–1078.

Herrera, C.M. (1987). 'Components of pollinator "quality": comparative analysis of a diverse insect assemblage'. *Oikos* **50**: 70–90.

Herrera, C.M. (1988*a*). 'The fruiting ecology of *Osyris quadripartita*: individual variation and evolutionary potential'. *Ecology* **69**: 233–249.

Herrera, C.M. (1988*b*). 'Plant size, spacing patterns, and host–plant selection in *Osyris quadripartita*, a hemiparasitic dioecious shrub'. *Journal of Ecology* **76**: 995–1006.

Herrera, C.M. (1988*c*). 'Variation in mutualisms: the spatio-temporal mosaic of a pollinator assemblage'. *Biological Journal of the Linnean Society* **35**: 95–125.

Herrera, C.M. (1989). 'Pollinator abundance, morphology, and flower visitation rate: analysis of the'quantity' component in a plant-pollinator system'. *Oecologia* **80**: 241–248.

Herrera, C.M. (1990*a*). 'Daily patterns of pollinator activity, differential pollinator effectiveness, and floral resource availability in a summer-flowering Mediterranean shrub'. *Oikos* **58**: 277–288.

Herrera, C.M. (1990*b*). 'The adaptedness of the floral phenotype in a relict endemic, hawkmoth pollinated violet. 1. Reproductive correlates of floral variation'. *Biological Journal of the Linnaean Society* **40**: 263–274.

Herrera, C.M. (1990*c*). 'The adaptedness of the floral phenotype in a relict endemic, hawkmoth pollinated violet. 2. Patterns of variation among disjunct populations'. *Biological Journal of the Linnaean Society* **40**: 275–291.

Herrera, C.M. (1991). 'Dissecting factors responsible for individual variation in plant fecundity'. *Ecology* **72**: 1436–1448.

Herrera, C.M. (1992*a*). 'Historical effects and sorting processes as explanations for contemporary ecological patterns; character syndromes in Mediterranean woody plants'. *American Naturalist* **140**: 421–446.

Herrera, C.M. (1992*b*). 'Interspecific variation in fruit shape: allometry, phylogeny, and adaptation to dispersal agents'. *Ecology* **73**: 1832–1841.

Herrera, C.M. (1992*c*). 'Individual flowering time and maternal fecundity in a summer flowering Mediterranean shrub: making the right prediction for the wrong reason'. *Acta Oecologia* **13**: 13–24.

Herrera, C.M. (1993). 'Selection on floral morphology and environmental determinants of fecundity in a hawk moth-pollinated violet'. *Ecological Monographs* **63**: 251–275.

Herrera, C.M. (1995*a*). 'Microclimate and individual variation in pollinators: flowering plants are more than their flowers'. *Ecology* **76**: 1516–1524.

Herrera, C.M. (1995*b*). 'Floral biology, microclimate, and pollination by ectothermic bees in an early-blooming herb'. *Ecology* **76**: 218–228.

Herrera, C.M. (1995*c*). 'Plant-vertebrate seed dispersal systems in the Mediterranean: ecological, evolutionary and historical determinants'. *Annual Review of Ecology and Systematics* **26**: 705–727.

Herrera, C.M. (1996). 'Floral traits and plant adaptation to insect pollinators: a devils advocate approach'. In *Floral Biology. Studies on floral evolution in animal-pollinated plants*, D.G. Lloyd and S.C.H. Barett (eds.). New York, Chapman & Hall: 65–87.

Herrera, C.M. (1998). 'Long-term dynamics of Mediterranean frugivorous birds and fleshy fruits: a 12 year study'. *Ecological Monographs* **68**: 511–538.

Herrera, C.M. (2000*a*). 'Flower-to-seedling consequences of different pollination regimes in an insect-pollinated shrub'. *Ecology* **81**: 15–29.

Herrera, C.M. (2000*b*). 'Individual differences in progeny viability in *Lavendula latifolia*: a long term field study'. *Ecology* **81**: 3036–3047.

Herrera, C.M. (2000*c*). 'Measuring the effects of pollinators and herbivores: evidence for non-additivity in a perennial herb'. *Ecology* **81**: 2170–2176.

Herrera, C.M. (2001). 'Deconstructing a floral phenotype: do pollinators select for corolla integration in *Lavendula latifolia*?'. *Journal of Evolutionary Biology* **14**: 574–584.

Herrera, C.M. (2002). 'Topsoil properties and seedling recruitment in *Lavendula latifolia*: stage-dependence and spatial decoupling of influential parameters'. *Oikos* **97**: 260–270.

Herrera, C.M. and P. Jordano (1981). '*Prunus mahaleb* and birds: the high efficiency seed dispersal system of a temperate fruiting tree'. *Ecological Monographs* **51**: 203–218.

Herrera, J., P. Jordano, L. Lopez-Soria, and J.A. Amat (1994). Recruitment of a mast-fruiting, bird-dispersed tree: bridging frugivore activity and seedling establishment. *Ecological Monographs* **64**: 315–344.

Herrera, C.M., A. Sanchez-Lafuente, M. Medrano, J. Guitian, X. Cerda, and P. Rey (2001). 'Geographical variation in autonomous self-pollination levels unrelated to pollinator service in *Helleborus foetidus* (Ranunculaceae)'. *American Journal of Botany* **88**: 1025–1032.

Herrera, C.M., M. Medrano, P.J. Rey, A.M. Sánchez-Lafuente, M.B. García, J. Guitián, and M.A.G. (2002). 'Interaction of pollinators and herbivores on plant fitness suggests a pathway for correlated evolution of mutualism- and antagonism-related traits'. *Proceedings of the National Academy of Science (USA)* **99**: 16823–16828.

Herrera, J. (1985). 'Nectar secretion patterns in southern Spanish Mediterranean scrublands'. *Israel Journal of Botany* **34**: 47–58.

Herrera, J. (1986). 'Flowering and fruiting phenology in the coastal shrublands of Doñana, south Spain'. *Vegetatio* **68**: 91–98.

Herrera, J. (1987). 'Flower and fruit biology in southern Spanish Mediterranean shrublands'. *Annals of the Missouri Botanical Garden* **74**: 69–78.

Herrera, J. (1988). 'Pollination relationships in southern Spanish Mediterranean shrublands'. *Journal of Ecology* **76**: 274–287.

Herrera, J. (1991*a*). 'The reproductive biology of a riparian Mediterranean shrub, *Nerium oleander* L. (Apocynaceae)'. *Botanical Journal of the Linnaean Society* **106**: 147–172.

Herrera, J. (1991*b*). 'Herbivory, seed dispersal, and the distribution of a ruderal plant living in a natural habitat'. *Oikos* **62**: 209–215.

Herrera, J. (1991*c*). 'Allocation of reproductive resources within and among inflorescences of *Lavendula stoechas* (Lamiaceae)'. *American Journal of Botany* **78**: 789–794.

Heun, M., R. Schäfer-Pregl, D. Klawan, R. Castagna, M. Accerbi, B. Borghi, and F. Salamini (1997). 'Site of einkorn wheat domestication identified by DNA fingerprinting'. *Science* **278**: 1312–1314.

Hewitt, G.M. (1996). 'Some genetic consequences of ice ages, and their role in divergence and speciation'. *Biological Journal of the Linnaean Society* **58**: 247–276.

Heywood, V.H. (1989). 'Patterns, extents and modes of invasion by terrestrial plants'. In *Ecology of biological invasions: a global perspective*, J.A. Drake, F. Di Castri, R.H. Groves, F.J. Kruger, H.A. Mooney, M.Rejmanek, and M.H. Williamson, (eds.). New York, Wiley & Sons: 31–60.

Heywood, V.H. (1995). 'The Mediterranean flora in the context of world biodiversity'. *Ecologia Méditerranea* **21**: 11–18.

Heywood, V.H. and J.M. Iriondo (2003). 'Plant conservation: old problems, new perspectives'. *Biological Conservation* **113**: 321–335.

Hildebrand, F. (1898). *The genus* Cyclamen *L. A. systematic and biological monograph. English translation 1999*. The Cyclamen Society, London.

Hodgson, J.G. (1986). 'Commonness and rarity in plants with special reference to the sheffield flora: IV. A European context with particular reference to endemism'. *Biological Conservation* **36**: 297–314.

Holland, R.F. and S.K. Jain (1981). 'Insular biogeography of vernal pools in the Central Valley of California'. *American Naturalist* **117**: 24–37.

Horner, J.D. and W.G. Abrahamson (1992). 'Influence of plant genotype and environment on oviposition preference and offspring survival in a gallmaking herbivore'. *Oecologia* **90**: 323–332.

Horovitz, A. (1979). *The Quaternary of Israel*. London, Academic Press.

Horovitz, A. and J. Galil (1972). 'Gynodioecism in East Mediterranean *Hirschfeldia incana* (Cruciferae)'. *Botanical Gazette* **133**: 127–131.

Horovitz, A., J. Galil, J. Nevet, and J. Iofe (1972). 'Bulb habit and reproduction in different ploidy forms of *Tulipa oculus-solis* in Israel'. *Israel Journal of Botany* **21**: 185–196.

Houston, J.M. (1964). *The western Mediterranean world. An introduction to its regional landscapes*. New York, Frederik A. Praeger.

Hsü, K.J. (1971). 'Origin of the Alps and Western Mediterranean'. *Nature* **233**: 44–48.

Hsü, K.J. (1972). 'When the Mediterranean dried up'. *Scientific American* **227**: 27–36.

Hsü, K.J. (1973). *The Mediterranean was a desert*. Princeton, Princeton University Press.

Hsü, K.J., W.B.F. Ryan, and M.B. Cita (1973). 'Late Miocene dessication of the Mediterranean'. *Nature* **242**: 240–244.

Hubbell, S.P. (2001). *The unified neutral theory of biodiversity and biogeography*. Princeton, Princeton University Press.

Hulme, P.E. (1997). 'Post-dispersal seed predation and the establishment of vertebrate dispersed plants in Mediterranean scrublands'. *Oecologia* **111**: 91–98.

Huntley, B. and H.J.B. Birks (1983). *An atlas of past and present pollen maps for Europe: 0–13000 years ago*. Cambridge, Cambridge University Press.

Hurtrez-Boussès, S. (1996). 'Genetic differentiation among natural populations of the rare Corsican endemic *Brassica insularis* Moris: implications for conservation guidelines'. *Biological Conservation* **76**: 25–30.

Husband, B.C. and D.W. Schemske (1996). 'Evolution of the magnitude and timing of inbreeding depression in plants'. *Evolution* **50**: 54–70.

Hyder, P.W., E.L. Fredrickson, R.E. Estell, and M.E. Lucero (2002). 'Transport of phenolic compounds from leaf surface of creosotebush and tarbush to soil surface by precipitation'. *Journal of Chemical Ecology* **28**: 2475–2482.

Imbert, E., J. Escarré, and J. Lepart (1997). 'Seed heteromorphism in *Crepis sancta* (Asteraceae): performance of two morphs in contrasting environments'. *Oikos* **79**: 325–332.

Imbert, E., J. Escarré, and J. Lepart (1999*a*). 'Differentiation among populations for life history, morphology, head traits, and achene morph proportions in the heterocarpic species *Crepis sancta* (L.) Bornm. (Asteraceae)'. *International Journal of Plant Science* **160**: 543–552.

Imbert, E., J. Escarré, and J. Lepart (1999*b*). 'Local adaptation and non-genetic maternal effects among three populations of *Crepis sancta* (Asteraceae)'. *Ecoscience* **6**: 223–229.

Inderjit and K.M.M. Dakshini (1995). 'On laboratory bioassays in allelopathy'. *Botanical Review* **61**: 28–44.

Irwin, D.E. (2002). Phylogeographic breaks without geographic barriers to gene flow. *Evolution* **56**: 2383—2394.

Izhaki, I. and U.N. Safriel (1990). 'The effect of some Mediterranean scrubland frugivores upon germination patterns'. *Journal of Ecology* **78**: 56–65.

Izhaki, I., P.B. Walton, and U.N. Safriel (1991). 'Seed shadows generated by frugivorous birds in an eastern Mediterranean scrub'. *Journal of Ecology* **79**: 575–590.

Jackson, L.E. and J. Roy (1986). 'Growth patterns of Mediterranean annual and perennial grasses under simulated rainfall regimes of southern France and California'. *Acta Oecologia–Oecologia Plantarum* **7**: 191–212.

Janzen, D.H. (1979). 'How to be a fig'. *Annual Reviews of Ecology and Systematics* **10**: 13–51.

Jenczewski, E., M. Angevain, A. Charrier, G. Génier, J. Ronfort, and J.-M. Prosperi (1998*a*. 'Contrasting patterns of genetic diversity in neutral markers and agromorphological traits in wild and cultivated populations of *Medicago sativa* L. from Spain'. *Genetics Selection Evolution* **30**: S103–S119.

Jenczewski, E., J.-M. Prosperi, and J. Ronfort (1998*b*. 'Evidence for gene flow between wild and cultivated populations of *Medicago sativa* (Leguminosae) based on allozyme markers and quantitative traits'. *American Journal of Botany* **86**: 677–687.

Jenczewski, E., J.-M. Prosperi, and J. Ronfort (1999). 'Differentiation between natural and cultivated populations of *Medicago sativa* (Leguminosae) from Spain: analysis with random amplified polymorphic DNA (RAPD) markers and comparison to allozymes'. *Molecular Ecology* **8**: 1317–1330.

Jessen, K., S.T. Anderson, and A. Farrington (1959). 'The interglacial flora'. *Proceedings of the Royal Irish Academy* **60B**: 1–77.

Joffre, R. and S. Rambal (1993). 'How tree cover influences the water balance of Mediterranean rangelands'. *Ecology* **74**: 570–582.

Joffre, R. and S. Rambal (2002). 'Mediterranean Ecosystems'. In *Encyclopedia of Life Sciences*, Macmillan Publishers Ltd, Nature Publishing Group: 1–7.

Joffre, R., J. Vacher, C. de los Llanos, and G. Long (1988). 'The dehesa: an agrosilvopastoral system of the Mediterranean region with special reference to the Sierra Morena area of Spain'. *Agroforestry Systems* **6**: 71–96.

Joffre, R., S. Rambal, and C. Damesin (1999). 'Functional attributes in Mediterranean-type ecosystems'. In *Handbook of functional plant ecology*, F.I. Pugnaire and F. Valladares (eds.). New York, Marcel Dekker, Inc: 347–380.

Joffre, R., S. Rambal, J.P. Ratte (1999). 'The dehesa system of southern Spain and Portugal as a natural ecosystem mimic'. *Agroforestry Systems* **45**: 57–79.

Johnson, S.D. (1992). 'Climatic and phylogenetic determinants of flowering seasonality in the Cape flora'. *Journal of Ecology* **81**: 567–572.

Jones, K.N. (1970). 'Chromosome changes in plant evolution'. *Taxon* **19**: 172–179.

Jones, R.N. (1995). 'B chromosomes in plants'. *New Phytologist* **131**: 411–434.

Jordano, P. (1987). 'Avian fruit removal: effects of fruit variation, crop size, and insect damage'. *Ecology* **68**: 1711–1723.

Jordano, P. (1989). 'Pre-dispersal biology of *Pistacia lentiscus* (Anacardiaceae): cumulative effects on seed removal by birds'. *Oikos* **55**: 375–386.

Jordano, P. (1995). 'Frugivore mediated selection on fruit and seed size: birds and St Lucieff's cherry, *Prunus mahaleb*'. *Ecology* **76**: 2627–2639.

Jordano, P. and C. Herrera (1995). 'Shuffling the offspring: uncoupling and spatial discordance of multiple stages in vertebrate seed dispersal'. *Ecoscience* **2**: 230–237.

Jordano, P. and E. Schupp (2000). 'Seed disperser effectiveness: the quantity component and patterns of seed rain for *Prunus mahaleb*'. *Ecological Monographs* **70**: 591–615.

Kaiser, K. (1969). 'The climate of Europe during the quaternary ice age'. In *Quaternary geology and climate*, H.E. Wright (ed.). Washington DC, National Academy of Sciences: 10–37.

Kaminsky, R. (1981). 'The microbial origin of the allelopathic potential of *Adenostoma fasiculatum* H. & A'. *Ecological Monographs* **51**: 365–382.

Karamanoli, K., D. Vokou, U. Menkissoglu, and H.-I. Constantinidou (2000). 'Bacterial colonization of phyllosphere of Mediterranean aromatic plants'. *Journal of Chemical Ecology* **26**: 2035–2048.

Katz, D.A., B. Sneh, and J. Friedman (1987). 'The allelopathic potential of *Coridothymus capitatus* L. (Labiatae). Preliminary studies on the roles of the shrub in the inhibition of the annuals germination and/or to promote allelopathically active actinomycetes'. *Plant and Soil* **98**: 53–66.

Keane, R.M. and M.J. Crawley (2002). 'Exotic plant invasions and the enemy release hympothesis'. *Trends in Ecology and Evolution* **17**: 164–169.

Keeley, J.E. (1991). 'Seed germination and life history syndromes in the California chaparral'. *Botanical Review* **57**: 81–116.

Keeley, J.E., B.A. Morton, A. Pedrosa, and P. Trotter (1985). 'Role of allelopathy, heat and charred wood in the germination of chaparral herbs and suffrutescents'. *Journal of Ecology* **73**: 445–458.

Kislev, M.E., D. Nadel, and I. Carmi (1992). 'Epipalaeolithic (19,000 BP) cereal and fruit diet at Ohalo II, Sea of Galilee, Israel'. *Revue of Palaeobotany and Palynology* **73**: 161–166.

Kjellberg, F., P.-H. Gouyon, M. Ibrahim, M. Raymond, and G. Valdeyron (1987). 'The stability of the symbiosis between dioecious figs and their pollinators: a study of *Ficus carica* L. and *Blastophaga psenes* L'. *Evolution* **41**: 693–704.

Klein, W.M. (1970). 'The evolution of three diploid species of *Oenothera* subgenus *Anogra* (Oenotheraceae)'. *Evolution* **24**: 578–597.

Knight, R. (1973). 'The climatic adaptation of populations of cocksfoot (*Dactylis glomerata* L.) from southern France'. *Journal of Applied Ecology* **10**: 1–12.

Koechlin, B., S. Rambal, and M. Debussche (1986). 'Rôle des arbres pionniers sur la teneur en eau du sil en surface de friches de la région méditerranéenne'. *Acta Oecologica–Oecologia Plantarum* **7**: 177–190.

Kokkini, S. and D. Vokou (1989). '*Mentha spicata* chemotypes growing wild in Greece'. *Economic Botany* **43**: 192–202.

Kokkini, S., D. Vokou, and R. Karousou (1989). 'Essential oil yeild of Lamiaceae plants in Greece'. *Biosciences, Proceedings 11th Int. Congress Essential Oils, Fragrances and flavours*, Oxford & IBH.

Kokkini, S., D. Vokou, and R. Karousou (1991). 'Morphological and chemical variation of *Origanum vulgare* L. in Greece'. *Botanica Chronica* **10**: 337–346.

Krijgsman, W., F.J. Hilgen, I. Raffi, F.J. Sierro, and D.S. Wilson (1999). 'Chronology, causes and progression of the Messinian salinity crisis'. *Nature* **400**: 652–655.

Kropf, M., J.W. Kadereit, and H.P. Comes (2002). Late Quaternary distributional stasis in the submediterranean mountain plant *Anthyllis montana* L. (Fabeaceae) inferred from ITS sequences and amplified fragment length polymorphism markers. *Molecular Ecology* **11**: 447–463.

Kruckeberg, A.R. and D. Rabinowitz (1985). 'Biological aspects of endemism in higher plants'. *Annual Review of Ecology and Systematics* **16**: 447–479.

Krüsi, B.O. and M. Debussche (1988). 'The fate of flowers and fruits of *Cornus sanguinea* L. in three contrasting Mediterranean habitats'. *Oecologia* **74**: 592–599.

Kugler, H. (1977). 'On the pollination of Mediterranean springtime flowers'. *Flore* **66**: 43–64.

Kullenberg, B. (1976). 'The pollination of *Ophrys* orchids'. *Botanika Notisker* **129**: 11–19.

Kummerow, J. (1973). 'Comparative anatomy of sclerophylls of Mediterranean climatic areas'. In *Mediterranean type ecosystems. Origin and structure*, F. Di Castri and H.A. Mooney (eds.). Heidelberg; Berlin, Springer-Verlag: 157–167.

Kummerow, Z., M. Kummerow, and L. Trabaud (1990). 'Root biomass, root distribution and the fine root growth dynamics of *Quercus coccifera* L. in the garrigue of southern France'. *Vegetatio* **87**: 37–44.

Kunholtz-Lordat, G. (1958). 'L'écran vert'. *Mémoires du Museum National d'Histoire Naturelle série B Botanique* **9**: 1–276.

Kunin, W.E. and K.J. Gaston (1993). 'The biology of rarity: patterns, causes and consequences'. *Trends in Ecology and Evolution* **8**: 298–301.

Kunin, W.E. and A. Schmida (1997). 'Plant reproductive traits as a function of local, regional, and global abundance'. *Conservation Biology* **11**: 183–192.

Küpfer, P. (1981). 'Les processus de différenciation des taxons orophiles en Méditerranée occidentale'. *Annales Jard. Bot. Madrid* **37**: 321–337.

Kushnir, U., J. Galil, and M.W. Feldman (1977). 'Cytology and distribution of *Ornithogallum* in Israel: I. Section *Heliocharmos* Bak'. *Israel Journal of Botany* **26**: 63–82.

Kutiel, P., H. Lavee, and M. Shoshany (1995). 'Influence of a climatic gradient upon vegetation dynamics along a Mediterranean-arid transect'. *Journal of Biogeography* **22**: 1065–1071.

Kyparissis, A., Y. Petropoulou, and Y. Manetas (1995). 'Summer survival of leaves in a soft-leaved shrub (*Phlomis fruticosa* L., Labiatae) under Mediterranean field conditions: avoidance of photoinhibitory damage through decreased chlorophyll contents'. *Journal of Experimental Botany* **46**: 1825–1831.

Lafuma, L., K. Balkwill, E. Imbert, R. Verlaque, and S. Maurice (2003). 'Ploidy level and origin of the European invasive weed *Senecio inaequidens* (Asteraceae)'. *Plant Systematics and Evolution* **243**: 59–72.

Lavergne S., M. Debussche, and J.D. Thompson (2004*a*). Limitations on reproductive success in endemic *Aquilegia viscosa* (Ranunculaceae) relative to its widespread congener *Aquilegia vulgaris*: the interplay of herbivory and pollination. *Oecologia* in press.

Lavergne, S., J.D. Thompson, E. Garnier, and M. Debussche (2004*b*). The biology and ecology of endemic and widespread plants: A comparative study of trait variation in 20 congeneric pairs. *Oikos* **107**: 505–518.

Lavergne S., W. Thuiller, J. Molina, and M. Debussche (2004*c*). Environmental and human factors influence rare plant local occurrence, extinction and persistence: a 115 year study in the Mediterranean region. *Journal of Biogeography* **31**: in press.

Le Floc'h, E. (1991). 'Invasive plants of the Mediterranean Basin'. In *Biogeography of Mediterranean invasions*, R.H. Groves and F. Di Castri (eds.). Cambridge, Cambridge University Press: 67–80.

Le Floc'h, E., H.N. Le Houerou, and J. Mathez (1990). 'History and patterns of plant invasion in Northern Africa'. In *Biological invasions in Europe and the Mediterranean Basin*, F. Di Castri, A.J. Hansen, and M. Bebussche (eds.). Dordrecht, Kluwer Academic Press: 105–133.

Le Houerou, H.N. (1987). 'Vegetation wildfires in the Mediterranean Basin: evolution and trends'. *Ecologia Méditerranea* **13**: 13–24.

Leonardi, S., M. Rapp, and F. Romane (2001). 'The sustainability of chestnut forest in the Mediterranean region'. *Ecologia Méditerranéa* **26**: 1–179.

Lepart, J. and M. Debussche (1991). 'Invasion processes as related to succession and disturbance'. In *Biogeography of Mediterranean invasions*, R.H. Groves and F. Di Castri (eds.). Cambridge, Cambridge University Press: 159–177.

Lepart, J. and M. Debussche (1992). 'Human impact on landscape patterning: Mediterranean examples'. In *Landscape boundaries*, A.J. Hansen and F. Di Castri (eds.). Berlin, Springer-Verlag.

Lepart, J. and B. Dommée (1992). 'Is *Phillyrea angustifolia* L. (Oleaceae) an androdioecious species?' *Botanical Journal of the Linnaean Society* **108**: 375–387.

Lepart, J. and J. Escarré (1983). 'La succession végétale, mécanismes et modèles: analyse bibliographique'. *Bulletin d'Ecologie* **14**: 133–178.

Lepart, J., A. Dervieux, and M. Debussche (1996). Photographie diachronique et changement des paysages, un siecle de dynamique naturelle de la forêt à Saint-Bauzille-de-Putois. *Forêt Méditerranéenne* **17**: 63–80.

Lepart, J., A. Martin, P. Marty, and S. Debain (2001). 'La progression des pins sur les causses: un phénomène difficilement contrôlable?' *Forêt Méditerranéenne* **22**: 23–28.

Lepart, J., P. Marty, and O. Rousset (2000). 'Les conceptions normatives du paysage. Le cas des Grands Causses'. *Nature Sciences et Sociétés* **8**: 16–25.

Lerdau, M., M. Litvak, and R. Monson (1994). 'Plant chemical defense: monoterpenes and the growth-differentiation balance hypothesis'. *Trends in Ecology and Evolution* **9**: 58–61.

Le Thierry d'Ennequin, M.L.T., B. Toupance, T. Robert, B. Godelle, and P.-H. Gouyon (1999). 'Plant domestication: a model for studying the selection of linkage'. *Journal of Evolutionary Biology* **112**: 1138–1147.

Lev-Yadun, S., A. Gopher, and S. Abbo (2000). 'The cradle of agriculture'. *Science* **288**: 1602–1603.

Levin, D.A. (1975). 'Minority cytotype exclusion in local plant populations'. *Taxon* **24**: 35–43.

Levin, D.A. (1983). 'Polyploidy and novelty in flowering plants'. *The American Naturalist* **122**: 1–25.

Levin, D.A. (1993). 'Local speciation in plants: the rule not the exception'. *Systematic Botany* **18**: 197–208.

Levin, D.A. (2000). *The origin, expansion, and demise of plant species*. Oxford, Oxford University Press.

Levine, J.M. and C.M. D'Antonio (1999). 'Elton revisited: a review of evidence linking diversity and invasibility'. *Oikos* **87**: 15–26.

Lewis, H. (1962). 'Catastrophic selection as a factor in evolution'. *Evolution* **16**: 257–271.

Lewis, H. (1966). 'Speciation in flowering plants'. *Science* **152**: 167–172.

Lewontin, R.C. and R.C. Birch (1966). 'Hybridization as a source of variation for adaptation to new environments'. *Evolution* **20**: 315–336.

Lihová, J., A. Tribisch, and K. Marhold (2003). 'The *Cardamine pratensis* (Brassicaceae) group in the Iberian Peninsula: taxonomy, polyploidy and distribution'. *Taxon* **52**: 783–802.

Linder, H.P., M.E. Meadows, and R.M. Cowling (1992). 'History of the Cape flora'. In *The ecology of fynbos. Nutrients, fire and diversity*, R.M. Cowling (ed.). Cape Town, Oxford University Press: 113–134.

Linhart, Y.B. (1988). 'Intra-population differentiation in annual plants: III. The contrasting effects of intra- and interspecific competition'. *Evolution* **42**: 1047–1064.

Linhart, Y.B. and M.C. Grant (1996). 'Evolutionary significance of local genetic differentiation in plants'. *Annual Review of Ecology and Systematics* **27**: 237–277.

Linhart, Y.B. and J.D. Thompson (1995). 'Terpene-based selective herbivory by *Helix aspersa* (Mollusca) on *Thymus vulgaris* (Labiatae)'. *Oecologia* **102**: 126–132.

Linhart, Y.B. and J.D. Thompson (1999). 'Thyme is of the essence: biochemical variability and multi-species deterence'. *Evolutionary Ecology Research* **1**: 151–171.

Lloret, F. and G. Marí (2001). 'A comparison of the medieval and the current fire regimes in managed pine forests in Catalonia (NE Spain)'. *Forest Ecology and Management* **141**: 155–163.

Lloret, F., M. Verdú, N. Flores-Hernández, and A. Valiente-Baunet (1999). 'Fire and resprouting in Mediterranean ecosystems: insights from an external biogeographical region, the Mexican shrubland'. *American Journal of Botany* **86**: 1655–1661.

Lloret, F., F. Médail, G. Brundu, and P.E. Hulme (2003). 'Local and regional abundance of exotic plant species on Mediterranean islands: are species traits important'. *Global Ecology and Biogeography* **12**.

Lloyd, D.G. (1976). 'The transmission of genes via pollen and ovules in gynodioecious angiosperms'. *Theoretical Population Biology* **9**: 299–316.

Lloyd, D.G. and K.S. Bawa (1984). 'Modification of the gender of seed plants in varying conditions'. *Evolutionary Biology* **17**: 255–338.

Lloyd, D.G. and C.J. Webb (1992*a*). 'The evolution of heterostyly'. In *Evolution and function of heterostyly*, S.C.H. Barrett (ed.). Berlin, Springer-Verlag. Vol. 15: 151–178.

Lloyd, D.G. and C.J. Webb (1992*b*). The selection of heterostyly'. In *Evolution and function of heterostyly*, S.C.H. Barrett (ed.). Berlin, Springer-Verlag. Vol. 15: 179–207.

Lloyd, D.G., C.J. Webb, and R. Dulberger (1990). 'Heterostyly in species of *Narcissus* (Amaryllidaceae) and *Hugonia* (Linaceae) and other disputed cases'. *Plant Systematics and Evolution* **172**: 215–227.

Lo Gullo, M.A. and S. Salleo (1988). 'Different strategies of drought resistance in three Mediterranean sclerophyllous trees growing in the same environmental conditions'. *New Phytologist* **108**: 267–276.

Lobo, J.M., I. Castro, and J.C. Moreno (2001). 'Spatial and environmental determinants of vascular plant species richness distribution in the Iberian Peninsula and Balearic Islands'. *Biological Journal of the Linnaean Society* **73**: 233–253.

Lonsdale, W.M. (1999). 'Global patterns of plant invasions and the concept of invasibility'. *Ecology* **80**: 1522–1536.

López-Pujol, J., M. Bosch, J. Simon, and C. Blanché. (2004). 'Allozyme diversity in the tetraploid endemic *Thymus loscosii* (Lamiaceae)'. *Annals of Botany* **93**: 323–332.

Loreau, M. and C. de Mazancourt (1999). 'Should plants in resource-poor environments invest more in antiherbivore defence'. *Oikos* **87**: 195–200.

Loreto, F., P. Ciccioli, E. Brancaleoni, R. Valentini, M. De Lillis, O. Csiky, and G. Seufert (1998). 'A hypothesis on the evolution of isoprenoid emission by oaks based on the correlation between emission type and *Quercus* taxonomy'. *Oecologia* **115**: 302–305.

Lossaint, P. (1973). 'Soil-vegetation relationships in Mediterranean ecosystems of southern France. In *Mediterranean type ecosystems. Origin and structure*, F. di Castri and H.A. Mooney (eds.). Heidelberg; Berlin, Springer-Verlag: 199–210.

Loveless, A.R. (1961). 'A nutritional interpretation of scerophylly based on differences in the chemical composition of sclerphyllous and mesophytic leaves'. *Annals of Botany* **25**: 168–184.

Loveless, A.R. (1962). 'Further evidence to support a nutritional interpretation of scerophylly'. *Annals of Botany* **26**: 551–561.

Lumaret, R. (1988). 'Cytology, genetics, and evolution in the genus *Dactylis*'. *CRC Critical Reviews in Plant Sciences* **7**: 55–91.

Lumaret, R. and E. Barrientos (1990). 'Phylogenetic relationships and gene flow between sympatric diploid and tetraploid plants of *Dactylis glomerata* (Gramineae)'. *Plant Systematics and Evolution* **169**: 81–96.

Lumaret, R. and N. Ouazzani (2001). 'Ancient wild olives in Mediterranean forests'. *Nature* **413**: 700.

Lumaret, R., J.L. Guillerm, J. Delay, A. Ait Lhaj Loufti, J. Izco, and M. Jay (1987). 'Polyploidy and habitat differentiation in *Dactylis glomerata* L. from Galicia (Spain)'. *Oecologia* **73**: 436–446.

Lumaret, R., J.L. Guillerm, J. Maillet, and R. Verlaque (1997). 'Plant species diversity and polyploidy in islands of natural vegetation isolated in extensive cultivated lands'. *Biodiversity and Conservation* **6**: 591–613.

Lumaret, R., N. Ouazzani, H. Michaud, G. Vivier, M.-F. Deguilloux, and F. Di Giusto (2004). 'Allozyme variation of oleaster populations (wild olive tree) (*Olea europaea* L.) in the Mediterranean Basin'. *Heredity* **92**: 343–351.

Luque, T. (1995). 'Karyology of *Nonea medicus* (Boraginaceae) in Spain; relationships between genera of Boraginaceae Barbier & Mathez (Anchusae DC)'. *Botanical Journal of the Linnaean Society* **117**: 321–331.

Luque, T. and B. Valdés (1984). 'Karyological studies on Spanish Boraginaceae: *Lithospermum* L. *sensu lato*'. *Botanical Journal of the Linnaean Society* **88**: 335–350.

Mabry, T.J. and D.R.J. Difeo (1973). 'The role of the secondary plant chemistry in the evolution of the Mediterranean scrub vegetation'. In *Mediterranean type ecosystems. Origin and structure*, F. di Castri and H.A. Mooney (eds.). Heidelberg; Berlin, Springer-Verlag: 121–155.

MacArthur, R.H. and E.O. Wilson (1967). *The theory of island biogeography*. Princeton, Princeton University Press.

Macnair, M.R. (1987). 'Heavy metal tolerance in plants: a model evolutionary system'. *Trends in Ecology and Evolution* **2**: 354–359.

Macnair, M.R. (1993). 'The genetics of metal tolerance in vascular plants'. *New Phytologist* **124**: 541–559.

Maestre, F.T., S. Bautista, and J. Cortina (2003). 'Positive, negative, and net effects in grass-shrub interactions in Mediterranean semiarid grasslands'. *Ecology* **(2003)**: 3186–3197.

Mai, D.H. (1989). 'Development and regional differentiation of the European vegetation during the tertiary'. *Plant Systematics and Evolution* **162**: 79–91.

Major, J. (1988). 'Endemism: a botanical perspective'. In *Analytical biogeography. A integrative approach to the study of animal and plant distributions*, A.A. Myers and P.S. Giller (eds.). London, Chapman & Hall: 117–146.

Maldonado, A. (1985). 'Evolution of the Mediterranean Basin and a detailed reconstruction of the Cenozoic paleoceanography'. In *Western Mediterranean*, R. Margalef (ed.). Oxford, Pergamon Press: 17–59.

Manicacci, D., A. Atlan, and D. Couvet (1997). 'Spatial structure of nuclear factors involved in sex determination in the gynodioecious *Thymus vulgaris* L'. *Journal of Evolutionary Biology* **10**: 889–907.

Manicacci, D., A. Atlan, J.A. Elena-Rosselló, and D. Couvet (1998). 'Gynodioecy and reproductive trait variation in three *Thymus* species (Lamiaceae)'. *International Journal of Plant Science* **159**: 948–957.

Manicacci, D., D. Couvet, E. Belhassen, P.H. Gouyon, and A. Atlan (1996). 'Founder effects and sex ratio in the gynodioecious *Thymus vulgaris* L'. *Molecular Ecology* **5**: 63–72.

Marañon, M. (1986). 'Plant species richness and canopy effect in the Savanna-like 'dehesa' of S.W. Spain'. *Ecologia Méditerranea* **12**: 131–141.

Marañón, T., R. Ajbilou, F. Ojeda, and J. Arroyo (1999). 'Biodiversity of woody species in oak woodlands of southern Spain and northern Morocco'. *Forest Ecology and Management* **115**: 147–156.

Marchi, A., G. Appendino, I. Pirisi, M. Ballero, and M.C. Loi (2003). 'Genetic differentiation of two distinct chemotypes of *Ferula communis* (Apiaceae) in Sardinia (Italy)'. *Biochemical Systematics and Ecology* **31**: 1397–1408.

Margaris, N.S. (1976). 'Structure and dynamics in a phyrgana (east Mediterranean) ecosystem'. *Journal of Biogeography* **3**: 249–259.

Margaris, N.S. and D. Vokou (1982). 'Structural and physiological features of woody plants in phryganic ecosystems related to adaptive mechanisms'. *Ecologia Mediterranea* **T.VIII**: 449–459.

Martinez-Pallé, E. and G. Aronne (2000). 'Pollinator failure in Mediterranean *Ruscus aculeatus* L'. *Botanical Journal of the Linnaean Society* **134**: 443–452.

Marty, P., E. Pélaquier, B. Jaudon, and J. Lepart (2003). 'Spontaneous reforestation in a peri-Mediterranean landscape: history of agricultural systems and dynamics of woody species'. In *Environmental dynamics and history in Mediterranean regions*, E. Fouache (ed.). Paris, Elsevier: 179–186.

Mascle, J. and J.-P. Rehault (1991). 'Le destin de la Méditerranée'. *La Recherche* **22**: 188–196.

Massei, G. and S.E. Hartley (2000). 'Disarmed by domestication? Induced responses to browsing in wild and cultivated olive'. *Oecologia* **122**: 225–231.

Massei, G., S.E. Hartley, and P. Bacon (2000). 'Chemical and morphological variation of Mediterranean woody evergreen species: do plants respond to ungulate browsing?'. *Journal of Vegetation Science* **11**: 1–8.

Mateu-Andrés, I. (1999). 'Allozymic variation and divergence in three species of *Antirrhinum* L. (Scrophulariaceae—Antirrhineae)'. *Botanical Journal of the Linnaen Society* **131**: 187–199.

Mateu-Andrés, I. and J.G. Segarra-Moragues (2000). 'Population subdivision and genetic diversity in two narrow endemics of *Antirrhinum* L'. *Molecular Ecology* **9**: 2081–2087.

Matthew, B. (1982). *The Crocus*. The Royal Botanic Gardens, Kew and Timber Press.

Mayr, E. (1982). *The growth of biological thought*. Cambridge MA, The Belknap Press of Harvard University Press.

Mazurek, H. and F. Romane (1986). 'Dynamics of a young *Pinus pinaster* vegetation in a Mediterranean area: diversity and niche strategy'. *Vegetatio* **66**: 27–40.

McDonald, D.J. and R.M. Cowling (1995). 'Towards a profile of an endemic mountain fynbos flora: implications for conservation'. *Biological Conservation* **72**: 1–12.

McDonald, D.J., J.M. Juritz, and W.J. Knottenbelt (1995). 'Modelling the biological aspects of local endemism in South African fynbos'. *Plant Systematics and Evolution* (**195**: 137–147.

McNeilly, T. and J. Antonovics (1968). 'Evolution in closely adjacent plant populations. IV. Barriers to gene flow'. *Heredity* **23**: 205–1218.

McPherson, J.K. and C.H. Muller (1969). 'Allelopathic effects of *Adenostoma fasiculatum*, 'chamise', in the California chaparral'. *Ecological Monographs* **39**: 177–198.

Médail, F., H. Michaud, J. Molina, G. Paradis, and R. Loisel (1998). 'Conservation de la flore et de la végétation des marais temporaires dulçaquicoles et oligotrophes de France méditerranéenne'. *Ecologia Mediterranea* **24**: 119–134.

Médail, F. and P. Quézel (1997). 'Hot-spots analysis for conservation of plant biodiversity in the Mediterranean Basin'. *Annals of the Missouri Botanical Garden* **84**: 112–127.

Médail, F. and P. Quézel (1999). 'The phytogeographical significance of S.W. Morocco compared to the Canary Islands'. *Plant Ecology* **140**: 221–244.

Médail, F. and R. Verlaque (1997). 'Ecological characteristics and rarity of endemic plants from southeast France and Corsica: implications for biodiversity conservation'. *Biological Conservation* **80**: 269–281.

Médail, F. and E. Vidal (1998). 'Organisation de la richesse et de la composition floristiques d'îles de la Méditerranée occidentale (sud-est de la France)'. *Canadian Journal of Botany* **76**: 321–331.

Médail, F., S. Ziman, M. Boscaiu, J. Riera, M. Lambrou, E. Vela, B. Dutton, and F. Ehrendorfer (2002). 'Comparative analysis of biological and ecological differentiation of *Anemone palmata* L. (Ranunculaceae) in the western Mediterranean (France and Spain): an assessment of rarity and population persistence'. *Botanical Journal of the Linnaean Society* **140**: 95–114.

Medus, J. and A. Pons (1980). *Les prédecesseurs des végétaux Méditerranéens actuels jusqu'au debut du Miocène*. Fondation Louis Emberger. La mise en place, l'évolution et la caractérisation de la flore et de la végétation circumméditerranéennes, Montpellier 9–10 April, Naturalia Monspeliensa.

Mejías, J.A. (1993). 'Cytotaxonomic studies in the Iberian taxa of the genus *Lactuca* (Compositae)'. *Botanica Helvetica* **103**: 113–130.

Mejías, J.A. (1994). 'Self-fertility and associated flower head traits in the Iberian taxa of *Lactuca* and related genera (Asteraceae: Lactuceae)'. *Plant Systematics and Evolution* **191**: 147–160.

Mejías, J.A., J. Arroyo, and F. Ojeda (2002). 'Reproductive ecology of *Rhododendron ponticum* (Ericaceae) in relict Mediterranean populations'. *Botanical Journal of the Linnaean Society* **140**: 297–311.

Méndez, M. (1998). 'Modification of the phenotypic and functional gender in the monoecious *Arum italicum* (Araceae)'. *American Journal of Botany* **85**: 225–234.

Méndez, M. (2001). 'Sexual mass allocation in species with inflorescences as pollination units: a comparison between *Arum italicum* and *Arisaema* (Araceae)'. *American Journal of Botany* **88**: 1781–1785.

Méndez, M. and A. Diaz (2001). 'Flowering dynamics in *Arum italicum* (Araceae): relative role of inflorescence traits, flowering synchrony, and pollination context on fruit initiation'. *American Journal of Botany* **88**: 1774–1780.

Méndez, M. and A. Traveset (2003). 'Sexual allocation in single-flowered hermaphroditic individuals in relation to plant and flower size'. *Oecologia* **137**: 69–75.

Merino, J., C. Field, and H.A. Mooney (1982). 'Construction and maintenance costs of Mediterranean-climate evergreen and deciduous leaves: I. Growth and CO_2 exchange analysis'. *Oecologia* **53**: 208–213.

Mésleard, F., L. Tan Ham, V. Boy, C. van Wijck, and P. Grillas (1993). 'Competition between an introduced and an indigenous species: the case of *Paspalum paspalodes* (Michx) Schribner and *Aeluropus littoralis* (Gouan) in the Camargue (southern France)'. *Oecologia* **94**: 204–209.

Michaud, H., L. Toumi, R. Lumaret, T.X. Li, F. Romane, and F. Di Guisto (1995). 'Effect of geographical discontinuity on genetic variation in *Quercus ilex* L. (holm oak). Evidence from enzyme polymorphism'. *Heredity* **74**: 590–606.

Michaux, J., J.-P. Suc, and J.-L. Vernet (1979). 'Climatic inference from the history of the Taxodiaceae during the pliocene and the early pleistoccene in western Europe'. *Review of Palaeobotany and Palynology* **27**: 185–191.

Midgley, G.F., L. Hannah, R. Roberts, D.J. MacDonald, and J. Allsopp (2001). 'Have Pleistocene climatic cycles influenced species richness patterns in the greater Cape Mediterranean Region'. *Journal of Mediterranean Ecology* **2**: 137–144.

Milesi, S., B. Massot, E. Gontier, F. Bourgaud, and A. Guckert (2001). '*Ruta graveolens* L.: a promising species for the production of furanocoumarins'. *Plant Science* **161**: 189–199.

Milewski, A.V. (1983). 'A comparison of ecosystems in Mediterranean Australia and South Africa: nutrient poor sites at the Barrens and the Caledon coast'. *Annual Review of Ecology and Systematics* **14**: 57–76.

Milne, R.I. and R.J. Abbott (2000). 'Origin and evolution of invasive naturalized material of *Rhododendron ponticum* L. in the British Isles'. *Molecular Ecology* **9**: 541–556.

Milne, R.I., R.J. Abbott, K. Wolff, and D.F. Chamberlain (1999). 'Hybridization among sympatric species of *Rhododendron* (Ericaceae) in Turkey: morphological and molecular evidence'. *American Journal of Botany* **86**: 1776–1785.

Mitrakos, K. (1982). 'Winter low temperatures in Mediterranean type ecosystems'. *Ecologia Méditerranéa* **8**: 95–102.

Moody, M.E. and R.N. Mack (1988). 'Controlling the spread of plant invasions: the importance of nascent foci'. *Journal of Applied Ecology* **25**: 1009–1021.

Mooney, H.A. and E.L. Dunn (1970*a*). 'Photosynthetic systems of Mediterranean-climate shrubs and trees of California and Chile'. *American Naturalist* **104**: 447–453.

Mooney, H.A. and E.L. Dunn (1970*b*). 'Convergent evolution of Mediterranean-climate evergreen sclerophyll shrubs'. *Evolution* **24**: 292–303.

Morales, R. (2002). 'The history, botany and taxonomy of the genus *Thymus*'. In *Thyme: the genus Thymus*, E. Stahl-Biskup and F. Sáez (eds.). London, Taylor & Francis: 1–43.

Moreno, J.M. and W.C. Oechel, (eds.) (1994). *The role of fire in Mediterranean-type ecosystems*. New York, Springer-Verlag.

Moreno, J.M., A. Vazquez, and R. Velez (1998). 'Recent history of forest fires in Spain'. In *Large forest fires*, J.M. Moreno (ed.). Leiden, Backhuys: 159–186.

Moret, J. (1991). 'Les stratégies de reproduction du complexe polyploïde *Ornithogallum umbellatum* (Hyacinthaceae) en France'. *Bullétin du Muséum National d'Histoire Naturelle Paris 4e séries* **13**: 25–46.

Moret, J. and Y. Faverereau (1991). 'Etude de la balance reproduction sexuée/multiplication végétative dans un complexe polyploïde du genre *Ornithogallum* (Liliaceae) en méditerranée occidentale'. *Bullétin de la Société Botanique de France* **138**: 201–214.

Moret, J., Y. Faverereau, and R. Grenflot (1991). 'A biometric study of the *Ornithogallum umbellatum* (Hyacinthaceae) complex in France'. *Plant Systematics and Evolution* **175**: 73–86.

Moret, J. and N. Galland (1992). 'Phenetic, biogeographical, and evolutionary study of *Ornithogalum* subg. *Heliocharmos* (*Hyacinthaceae*) in the western Mediterranean basin'. *Plant Systematics and Evolution* **181**: 179–202.

Moret, J., A. Bari, A. leThomas, and P. Goldblatt (1992). 'Gynodioecy, herkogamy and sex-ratio in *Romulea bulbocodium* var. *dioica* (Iridaceae)'. *Evolutionary Trends in Plants* **6**: 99–109.

Morjan, C.L. and L.H. Rieseberg (2004). 'How species evolve collectively: implications of gene flow and selection for the spread of advantageous alleles'. *Molecular Ecology* **13**: 1341–1356.

Mourer-Chauviré, C. (1989). 'A peafowl from the Pliocene of Perpignan, France'. *Palaeontology* **32**: 439–446.

Muller, S. (2001). Les invasions biologiques causées par les plantes exotiques sur le territoire français métropolitain. Etat des connaissances et propositions d'actions. Paris, Minitère de l'Aménagement du Territoire et de l'Environnement. Direction de la Nature et des Paysages.

Murray, M.J. and D.E. Lincoln (1970). 'The genetic basis of acyclic oil constituents in *Mentha citrata* Ehrh'. *Genetics* **65**: 457–471.

Myers, N., R.A. Mittermeier, C.G. Mittermeier, D.F.G.A.B., and J. Kent (2000). 'Biodiversity hotspots for conservation priorities'. *Nature* **403**: 853–858.

Nakagawa, T., G. Garfì, M. Reille, and R. Verlaque (1998). 'Pollen morphology of *Zelkova sicula* (Ulmaceae), a recently discovered relic species of the European tertiary flora: description, chromosomal relevance, and palaeobotanical significance'. *Review of Palaeobotany Palynolology* **100**: 27–37.

Nathan, R., U.N. Safriel, I. Noy-Meir, and G. Schiller (1999). 'Seed release without fire in *Pinus halepensis*, a Mediterranean serotinous wind-dispersed tree'. *Journal of Ecology* **87**: 659–669.

Nathan, R., U.N. Safriel, I. Noy-Meir, and G. Schiller (2000). 'Spatiotemporal variation in seed dispersal and recruitment near and far from *Pinus halepensis* trees'. *Ecology* **81**: 2156–2169.

Navarro, L., J. Guitian, and P. Guitian (1993). 'Reproductive biology of *Petrocoptis grandiflora* Rothm. (Caryophyllaceae), a species endemic to Northwest Iberian Peninsula'. *Flora* **188**: 253–261.

Naveh, Z. (1967). 'Mediterranean ecosytems and vegetation types in California and Israel'. *Ecology* **48**: 445–459.

Naveh, Z. (1975). 'The evolutionary significance of fire in the Mediterranean region'. *Vegetatio* **29**: (199–208.

Ne'eman, G. (1993). 'Variation in leaf phenology and habit in *Quercus ithaburensis*, a Mediterranean deciduous tree'. *Journal of Ecology* **81**: 627–634.

Ne'eman, G. (2003). 'To be or not to be—the effect of nature conservation management on flowering of *Paeonia mascula* (L.) Miller in Israel'. *Biological Conservation* **109**: 103–109.

Nevo, E., D. Zohary, A.H.D. Brown, and M. Haber (1979). 'Genetic diversity and environmental associations of wild barley, *Hordeum spontaneum*, in Israel'. *Evolution* **33**: 815–833.

Nevo, E., E.M. Golenberg, A. Beiles, A.H.D. Brown, and D. Zohary (1982). 'Genetic diversity and environmental associations of wild wheat, *Triticum dicoccoides*, in Israel'. *Theoretical and Applied Genetics* **62**: 241–254.

Nevo, E., A. Beiles, N. Storch, H. Doll, and B. Anderson (1983). 'Microgeographic edaphic differentiation in hordein polymorphisms of wild barley'. *Theoretical and Applied Genetics* **64**: 123–132.

Nevo, E., A. Beiles, Y. Gutterman, N. Storch, and D. Kaplan (1984*a*). 'Genetic resources of wild cereals in Israel and vicinity: I. Phenotypic variation within and between populations of wild wheat *Triticum dicoccoides*'. *Euphytica* **33**: 717–735.

Nevo, E., A. Beiles, Y. Gutterman, N. Storch, and D. Kaplan (1984*b*). 'Genetic resources of wild cereals in Israel and vicinity: II. Phenotypic variation within and between populations of wild barley *Hordeum spontaneum*'. *Euphytica* **33**: 737–756.

Nevo, E., A. Beiles, and T. Krugman (1988*a*). 'Natural selection of allozyme polymorphisms: a microgeographical differentiation by edaphic, topographical, and temporal factors in wild emmer wheat (*Triticum dicoccoides*)'. *Theoretical and Applied Genetics* **76**: 737–752.

Nevo, E., A. Beiles, and T. Krugman (1988*b*). 'Natural selection of allozyme polymorphisms: a microgeographic climatic differentiation in wild emmer wheat (*Triticum dicoccoides*)'. *Theoretical and Applied Genetics* **75**: 529–538.

Newsome, A.E. and I.R. Noble (1986). 'Ecological and physiological characters of invading species'. In *Ecology of biological invasions: an Australian perspective*, R.H. Groves and J.J. Burdon (eds.). Canberra, Australian Academy of Science: 1–20.

Nicholls, M.S. (1985). 'The evolutionary breakdown of distyly in *Linum tenuifolium* (Linaceae)'. *Plant Systematics and Evolution* **150**: 291–301.

Nicholls, M.S. (1986). 'Variation and evolution in *Linum tenuifolium* (Linaceae)'. *Plant Systematics and Evolution* **153**: 243–258.

Niklewski, J. and W. Van Zeist (1970). 'A late quaternary pollen diagram from northwestern Syria'. *Acta Botanica Neerlandica* **19**: 737–754.

Noy-Meir, I., M. Gutman, and Y. Kaplan (1989). 'Responses of Mediterranean grassland plants to grazing and protection'. *Journal of Ecology* **77**: 290–310.

Nyman, Y. (1991). 'Crossing experiments within the *Campanula dichotoma* group (Campanulaceae)'. *Plant Systematics and Evolution* **177**: 185–192.

Obeso, J.R. (1992). 'Pollination ecology and seed set in *Asphodelus albus* (Liliaceae) in northern Spain'. *Flora* **187**: 219–226.

Ofir, M. and J. Kigel (1998). 'Abscisic acid involvement in the induction of summer dormancy in *Poa bulbosa*, a grass geophyte'. *Physiologia Plantarum* **102**: 163–170.

Ofir, M. and J. Kigel (1999). 'Photothermal control of the imposition of summer dormancy in *Poa bulbosa*, a perennial grass geophyte'. *Physiologia Plantarum* **105**: 633–640.

O'Hanlon, P. (2001). *Genetic and ecological investigations in the thistle genus* Onopordum *with special emphases on hybridization and evolution*. PhD thesis Canberra, Australian National University.

O'Hanlon, P.C., R. Peakall, and D.T. Briese (1999). 'Amplified fragment length polymorphism (AFLP) reveals introgression in weedy *Onopordum* thistles: hybridization and invasion'. *Molecular Ecology* **8**: 1239–1246.

Ojeda, F. (1998). 'Biogeography of seeder and resprouter *Erica* species in the Cape Floristic Region—where are the resprouters'. *Biological Journal of the Linnaean Society* **63**: 331–347.

Ojeda, F., J. Arroyo, and T. Marañón (1995). 'Biodiversity components and conservation of Mediterranean heathlands in southern Spain'. *Biological Conservation* **72**: 61–67.

Ojeda, F., J. Arroyo, and T. Marañón (1996*a*). 'Patterns of ecological, chronological and taxonomic diversity at both sides of the Strait of Gibraltar'. *Journal of Vegetation Science* **7**: 63–72.

Ojeda, F., T. Marañón, and J. Arroyo (1996*b*). 'Postfire regeneration of a Mediterranean Heathland in southern Spain'. *International Journal of Wildland Fire* **6**: 191–198.

Ojeda, F., J. Arroyo, and T. Marañón (1998). 'The phytogeography of European and Mediterranean heath species (Ericoideae), (Ericaceae): a quantitative analysis'. *Journal of Biogeography* **25**: 165–178.

Ojeda, F., T. Marañón, and J. Arroyo (2000*a*). 'Plant diversity patterns in the Aljibe Mountains (S. Spain): a comprehensive account'. *Biodiversity and Conservation* **9**: 1323–1343.

Ojeda, F., J. Arroyo, T. Marañón (2000*b*). 'Ecological distribution of four co-occuring Mediterranean heath species'. *Ecography* **23**: 148–159.

Ojeda, F., M.T. Simmons, J. Arroyo, T. Marañón, and R.M. Cowling (2001). 'Biodiversity in South African fynbos and Mediterranean heathland'. *Journal of Vegetation Science* **12**: 867–874.

Olesen, J.M. and P. Jordano (2002). 'Geographic patterns in plant-pollinator mutualistic networks'. *Ecology* **83**: 2416–2424.

Oliva, A., E. Lahoz, R. Contillo, and G. Aliotta (1999). 'Fungistatic activity of *Ruta graveolens* extract and its allelochemicals'. *Journal of Chemical Ecology* **25**: 519–526.

Olivieri, I. (1984). 'Effect of *Puccinia cardui-pycnocephali* on slender thistles (*Carduus pycnocephalus* and *C. tenuiflorus*)'. *Weed Science* **32**: 508–510.

Olivieri, I., P.H. Gouyon, and J.M. Prosperi (1991). 'Life-cycles of some Mediterranean invasive plants'. In *Biogeography of Mediterranean invasions*, R.H. Groves and F. Di Castri (eds.). Cambridge, Cambridge University Press: 131–143.

Ollerton, J. and A.J. Lack (1992). 'Flowering phenology: an example of relaxation of natural selection?' *Trends in Ecology and Evolution* **7**: 274–276.

Oostermeijer, J.G.B., S.H. Luitjen, and H.C.M. Den Nijs (2003). 'Integrating demographic and genetic approaches in plant conservation'. *Biological Conservation* **113**: 389–398.

Ornduff, R. (1969). 'Ecology, morphology, and systematics of *Jepsonia* (Saxifragaceae)'. *Brittonia* **21**: 286–298.

Ouazzani, N., R. Lumaret, P. Villemur, and F. di Guisto (1993). 'Leaf allozyme variation in cultivated and wild olive trees (*Olea europaea* L.)'. *Journal of Heredity* **84**: 34–42.

Ouazzani, N., R. Lumaret, and P. Villemur (1996). 'Genetic variation in the olive tree (*Olea europaea* L.) cultivated in Morocco'. *Euphytica* **91**: 9–20.

Ozenda, P. (1975). 'Sur les étages de végétation dans les montagnes du Bassin Méditerranéen'. *Documents de Cartographie Ecologique* **16**: 1–32.

Özkan, H., A. Brandolini, R. Schäfer-Pregl, and F. Salamini (2002). 'AFLP analysis of a collection of tetraploid wheats indicates the origin of emmer hard wheat domestication in Southeast Turkey'. *Molecular Biology and Evolution* **19**: 1797–1801.

Palamarev, E. (1989). 'Paleobotanical evidences of the Tertiary history and origin of the Mediterranean sclerophyll dendroflora'. *Plant Systematics and Evolution* **162**: 93–107.

Pannell, J. (1997*a*). 'Widespread functional androdioecy in *Mercurialis annua* L. (Euphorbiaceae)'. *Biological Journal of the Linnaean Society* **61**: 95–116.

Pannell, J. (1997*b*). 'Variation in sex ratios and sex allocation in androdioecious *Mercurialis annua*'. *Journal of Ecology* **85**: 57–69.

Pannell, J. (1997*c*). 'Mixed genetic and environmental sex determination in an androdioecious population of *Mercurialis annua*'. *Heredity* **78**: 50–56.

Pannell, J. and F. Ojeda (2000). 'Patterns of flowering and sex ratio variation in the Mediterranean shrub *Phillyrea angustifolia* (Oleaceae): implications for the maintenance of males with hermaphrodites'. *Ecology Letters* **3**: 495–502.

Parker, J.S., R. Lozano, and M.R. Rejon (1991). 'Chromosomal structure of populations of *Scilla autumnalis* in the Iberian peninsula'. *Heredity* **67**: 287–297.

Parsons, D.J. (1976). 'Vegetation structure in the Mediterranean scrub communities of California and Chile'. *Journal of Ecology* **64**: 435–447.

Paskoff, R.P. (1973). 'Geomorphological processes and characteristic land-forms in the Mediterranean regions of the world'. In *Mediterranean type ecosystems. Origin and structure*, F. Di Castri and H.A. Mooney (eds.). Heidelberg; Berlin, Springer-Verlag: 53–60.

Paulus, H.F. and C. Gack (1990). 'Pollinators as prepollinating isolation factors: evolution and speciation in *Ophrys* (Orchidaceae)'. *Israel Journal of Botany* **39**: 43–79.

Pausas, J.G. (1997). 'Resprouting of *Quercus suber* in NE Spain after fire'. *Journal of Vegetation Science* **8**: 703–706.

Pausas, J.G. (2001). 'Resprouting vs seeding—a Mediterranean perspective'. *Oikos* **94**: 193–194.

Pausas, J.G., R.A. Bradstock, D.A. Keith, J.E. Keeley, and G.G.C.o.T.E.F. Network (2004). 'Plant functional traits in relation to fire in crown-fire ecosystems'. *Ecology* **85**: 1085–1100.

Pausas, J.G. and M. Verdú (2004). 'Plant persistence traits in fire-prone ecosystems of the Mediterranean Basin'. *Oikos* in press.

Peñalba, M.C. (1994). 'The history of the Holocene vegetation in northern Spain from pollen analysis'. *Journal of Ecology* **82**: 815–832.

Perevolotsky, A. and N.G. Seligman (1998). 'Role of grazing in Mediterranean rangeland ecosystems'. *Bioscience* **48**: 1007–1017.

Pérez, R., P. Vargas, and J. Arroyo (2003). 'Convergent evolution of flower polymorphism in *Narcissus* (Amaryllidaceae)'. *New Phytologist* **161**: 235–252.

Pérez-Bañon, C., A. Juan, T. Petanidou, M.A. Marcos-García and M.B. Crespo (2003). 'The reproductive ecology of *Medicago citrina* (Font Quer) Greuter (Leguminosae): a bee-pollinated plant in Mediterranean islands where bees are absent'. *Plant Systematics and Evolution* **241**: 29–46.

Pérez Chiscano, J.L. (1985). Distribución geográfica de *Ecballium elaterium* (L.) Richard (Cucurbiataceae) en la Peninsula Iberica e Islas Baleares. *Studia Botanica* **4**: 57–77.

Perrot, V., D. Dommée, and P. Jacquard (1982). 'Etude expérimentale de la concurrence entre individus issus d'autofécondation et d'allofécondation chez le thym (*Thymus vulgaris* L.)'. *Acta Oecologica—Oecologia Plantarum* **3**: 171–184.

Petanidou, T. and W.N. Ellis (1993). 'Pollinating fauna of a phryganic ecosystem: composition and diversity'. *Biodiversity Letters* **1**: 9–22.

Petanidou, T. and E. Smets (1996). 'Does temperature stress induce nectar secretion in Mediterranean plants'. *New Phytologist* **133**: 513–518.

Petanidou, T. and D. Vokou (1990). 'Pollination and pollen energetics in Mediterranean ecosystems'. *American Journal of Botany* **77**: 986–992.

Petanidou, T. and D. Vokou (1993). 'Pollination ecology of Labiatae in a phryganic (east Mediterranean) ecosystem'. *American Journal of Botany* **80**: 892–899.

Petanidou, T., W.N. Ellis, N.S. Margaris, and D. Vokou (1995). 'Constraints on flowering phenology in a phryganic (east Mediterranean shrub) community'. *American Journal of Botany* **82**: 607–620.

Petanidou, T., A.J. van Laere, and E. Smets (1996). 'Change in floral nectar components from fresh to senescent flowers of *Capparis spinosa* (Capparidaceae), a nocturnally flowering Mediterranean shrub'. *plant systematics and Evolution* **199**: 79–92.

Petanidou, T., V. Goethals, and E. Smets (2000). 'Nectary structure of Labiatae in relation to their nectar secretion and characteristics in a Mediterranean shrub community—does flowering time matter?'. *Plant Systematics and Evolution* **225**: 103–118.

Petit, C. and J.D. Thompson (1997). 'Variation in the phenotypic response to light availability between diploid and tetraploid populations of the perennial grass *Arrhenatherum elatius* from open and woodland sites'. *Journal of Ecology* **85**: 657–667.

Petit, C. and J.D. Thompson (1998). 'Phenotypic selection and population differentiation in relation to habitat heterogeneity in *Arrhenatherum elatius*'. *Journal of Ecology* **86**: 829–840.

Petit, C. and J.D. Thompson (1999). 'Species richness and ecological range in relation to ploidy level in the flora of the Pyrenees'. *Evolutionary Ecology* **13**: 45–66.

Petit, C., P. Lesbros, X. Ge, and J.D. Thompson (1997). 'Variation in flowering phenology and outcrossing rate across a contact zone between diploid and tetraploid *Arrhenatherum elatius*'. *Heredity* **79**: 31–40.

Petit, C., H. Fréville, A. Mignot, B. Colas, M. Riba, E. Imbert, S. Hurtrez-Boussés, M. Virevaire, and I. Olivieri. (2001). 'Gene flow and local adaptation in two endemic plant species'. *Biological Conservation* **100**: 21–34.

Petit, R.J., N. Bahrman, and P. Baradat (1995). 'Comparison of genetic differentiation in maritime pine (*Pinus pinaster* Ait) estimated using isozymes, total proteins and terpenic loci'. *Heredity* **75**: 382–389.

Petit, R.J., E. Pineau, B. Demesure, R. Bacilieri, A. Ducouso, and A. Kremer (1997). 'Chloroplast DNA footprints of postglacial recolonization by oaks'. *Proceedings of the National Academy of Sciences of the USA* **94**: 9996–10001.

Petit, R.J., A. El Mousadik, and O. Pons (1998). 'Identifying populations for conservation on the basis of genetic markers'. *Conservation Biology* **12**: 844–855.

Petit, R.J., S. Brewer, S. Bordács, and a. others. (2002). 'Identification of refugia and post glacial colonisation routes of European white oaks based on chloroplast DNA and fossil pollen evidence'. *Forest Ecology and Management* **156**: 49–74.

Picó, F.X. (2000). 'Temporal variation in the female components of reproductive success over the extended flowering season of a Mediterranean perennial herb'. *Oikos* **89**: 485–492.

Picó, F.X. and J. Retana (2001). 'The flowering pattern of the perennial herb *Lobularia maritima*: an unusual case in the Mediterranean Basin'. *Acta Oecologica* **22**: 209–217.

Pignatti, S. (1978). 'Evolutionary trends in Mediterranean flora and vegetation'. *Vegetatio* **37**: 175–185.

Pigott, C.D. and S. Pigott (1993). 'Water as a determinant of the distribution of trees at the boundary of the Mediterranean zone'. *Journal of Ecology* **81**: 557–566.

Piñol, J., J. Terredas, and F. Lloret (1998). 'Climate warming, wildfire hazard, and wildfire occurrence in coastal eastern Spain'. *Climatic Change* **38**: 345–357.

Plitmann, U. (1981). 'Evolutionary history of old world lupines'. *Taxon* **30**: 430–437.

Polunin, O. (1980). *Flowers of Greece and the Balkans: a field guide*. Oxford, Oxford University Press.

Polunin, O. and B.E. Smythies (1973). *Flowers of south-west Europe: a field guide*. London, Oxford University Press.

Pomente, D. (1987). *Etude expérimentale génétique, écologique et écophysiologique du polymorphisme végétal: chémotypes et formes sexuelles du thym*. PhD thesis Montpellier, USTL.

Pons, A. (1981). 'The history of the Mediterranean shrublands'. In *Macquis and chaparral*, F. Di Castri, D.W. Goodall, and R.L. Sprecht (eds.). Amsterdam, Elsevier: 131–138.

Pons, A. (1984). 'Les changements de la végétation de la région Méditerrané durant la Pliocène et le Quartenaire en relation avec l'histoire du climat et l'action de l'homme'. *Webbia* **38**: 427–439.

Pons, A. and P. Quézel (1985). 'The history of the flora and vegetation and past and present human disturbance in the Mediterranean region'. In *Plant conservation in the Mediterranean area*, C. Gomez-Campo (ed.). Dordrecht, Dr W. Junk Publishers: 25–43.

Pons, A. and M. Reille (1988). 'The Holocene and Upper Pleistocene pollen record from Padul (Granada, Spain)'. *Palaeogeography, Palaeoclimatology, Palaeoecology* **66**: 243–263.

Pons, A. and J.P. Suc (1980). *Les témoignages de structures de végétation Méditerranéennes dans le passé antérieur à l'action de l'homme*. Fondation Louis Emberger. La mise en place, l'évolution et la caractérisation de la flore et de la végétation circum-méditerranéennes, Montpellier 9–10 April, Naturalia Monspeliensa.

Pons, A. and M. Thinon (1987). 'The role of fire from palaeoecological data'. *Ecologia Méditerranea* **13**: 3–11.

Pons, A., J.-P. Suc, M. Reille, and N. Combourieu-Nebout (1995). 'The history of dryness in regions with a Mediterranean climate'. In *Time scales of biological responses to water constraints*, J. Roy, J. Aronson and F. di Castri (eds.). Amsterdam, SPB Academic Publishing: 169–188.

Potts, S.G., A. Dafni, and G. Ne'eman (2001). 'Pollination of a core flowering shrub species in Mediterranean phrygana: variation in pollinator diversity, abundance and effectiveness in response to fire'. *Oikos* **92**: 71–80.

Pozzi, C., L. Rossini, Vecchietti, and F. Salamini (2004). 'Genes and genome change during domestication of cereals'. In *Cereal genomics*. P.K. Gupta and R.K. Varshney (eds) Kluwer Academic Publishers: in press.

Preiss, E., J.-L. Martin, and M. Debussche (1997). 'Rural depopulation and recent landscape changes in the Mediterranean region: consequences to the breeding avifauna'. *Landscape Ecology* **12**: 51–61.

Prentice, H. (1976). 'A study of endemism, *Silene diclinis*'. *Biological Conservation* **10**: 15–30.

Prentice, H. (1984). 'Enzyme polymorphism, morphometric variation and population structure in a restricted endemic, *Silene diclinis* (Caryophyllaceae)'. *Biological Journal of the Linnaean Society* **22**: 125–143.

Prentice, H. and S. Andersson (1997). 'Genetic variation and population size in the rare dioecious plant *Silene diclinis* (Caryophyllaceae)'. In *The role of genetics in conserving small populations. Proceedings of a British Ecological Society Symposium*, T.E. Tew, T.J. Crawford, J.W. Spencer, D.P. Stevens, M.B.Usher, and J. Warren (eds.). Peterborough, JNCC: 65–72.

Prentice, I.C., J. Guiot, and S.P. Harrison (1992). 'Mediterranean vegetation, lake levels and palaeoclimate at the last glacial Maximum'. *Nature* **360**: 658–660.

Preston, F.W. (1962*a*). 'The cannonical distribution of commones and rarity: Part I'. *Ecology* **43**: 185–215.

Preston, F.W. (1962*b*). 'The cannonical distribution of commones and rarity: Part II'. *Ecology* **43**: 410–432.

Prieur-Richard, A.-H., S. Lavorel, K. Grigulis, and A. Dos Santos. (2000). 'Plant community diversity and invasibility by exotics: invasion of Mediterranean old fields by *Conyza bonariensis* and *Conyza canadensis*'. *Ecology Letters* **3**: 412–422.

Puech, S. (1986). 'Production des diaspores et potentialités de germination chez quelques espèces à fruits charnus, ornithochores, dans le sud-est de la France'. *Ecologia Mediterranea* **XII(1–2)**: 148–154.

Puerto, A., M. Rico, M.D. Matias, and J.A. Garcia (1990). 'Variation in structure and diversity in Mediterranean grasslands related to trophic status and grazing intensity'. *Journal of Vegetation Science* **1**: 445–452.

Pugnaire, F.I., C. Armas, and F. Valladares (2004). 'Soil as a mediator in plant–plant interactions in a semi-arid community'. *Journal of Vegetation Science* **15**: 85–92.

Pugnaire, F.I. and M.T. Luque (2001). 'Changes in plant interactions along a gradient of environmental stress'. *Oikos* **93**: 42–49.

Quézel, P. (1978). 'Analysis of the flora of Mediterranean and Saharan Africa'. *Annals of the Missouri Botanical Garden* **65**: 479–534.

Quézel, P. (1985). 'Definition of the Mediterranean region and the origin of its flora'. In *Plant conservation in the Mediterranean area*, C. Gomez-Campo (ed.). Dordrecht, Dr W. Junk Publishers: 9–24.

Quézel, P. (1995). 'La flore du bassin méditerranéen: origine, mise en place, endémisme'. *Ecologia Mediterranea* **21**: (19–39.

Quézel, P. (1998). 'La végétation des marais transitoires à Isoetes en région méditerranéenne, intérêt patrimonial et conservation'. *Ecologia Mediterranea* **24**: 111–117.

Quézel, P. and M. Barbero (1982). 'Definition and characterization of Mediterranean-type ecosystems'. *Ecologia Méditerranéa* **8**: 15–29.

Quézel, P. and F. Médail (2003). *Ecologie et biogéographie des forêts du bassin méditerranéen*. Paris, Elsevier.

Quézel, P., J. Gamisans, and M. Gruber (1980). 'Biogéographie et mise en place des flores Méditerranéennes'. *Naturalia Monspeliensia* **Hors Série**: 41–51.

Quézel, P., M. Barbero, G. Bonnin, and R. Loisel (1990). 'Recent plant invasions in the Circum-Mediterranean region'. In *Biological invasions in Europe and the Mediterranean Basin*, F. di Castri, A.J. Hansen, and M. Debussche (eds.). Dordrecht, Kluwer Academic Press: 51–60.

Quilichini, A., M. Debussche, and J.D. Thompson (2001). 'Evidence for local outbreeding depression in the Mediterranean island endemic *Anchusa crispa* Viv. (Boraginaceae)'. *Heredity* **87**: 190–197.

Quilichini, A., M. Debussche, and J.D. Thompson (2004). 'Geographic differentiation in the Mediterranean island endemic *Anchusa crispa*: implications for the conservation of a protected species'. *Biological Conservation* **118**: 651–660.

Ramadan, A.A., A. El-keblawy, K.H. Shaltout, and J. Lovett-Doust (1994). 'Sexual polymorphism, growth, and reproductive effort in Egyptian *Thymelaea hirsuta* (Thymeleaeceae)'. *American Journal of Botany* **81**: 847–857.

Ramade, F.E.A. (1997). *Conservation des écosystèmes méditerranéennes*. Paris, Economica.

Rambal, S. (1984). 'Water balance and pattern of root water uptake by a *Quercus coccifera* L. evergreen shrub'. *Oecologia* **62**: 18–25.

Rambal, S. (2001). 'Hierarchy and productivity of Mediterranean-type ecosystems'. In *Global terrestrial productivity*, J. Roy, B. Saugier, and H.A. Mooney (eds.). London, Academic Press: 315–344.

Rathgeber, C., J. Guiot, P. Roche, and L. Tessier. (1999). 'Augmentation de productivité du chêne pubescent en région méditerranéenne française'. *Ann. For. Sci.* **56**: 211–219.

Raven, P.H. (1964). 'Catastrophic selection and edaphic endemism'. *American Naturalist* **98**: 336–338.

Raven, P.H. (1971). 'The relationships between "Mediterranean" floras'. In *Plant life of south-west Asia*, P.H. Davies, P. Harper, and I.C. Hedges (eds.). Edingburg, Botanical Society of Edingburg: 119–134.

Raven, P.H. (1972). 'Plant species disjunction, a summary'. *Annals of the Missouri Botanical Garden* **59**: 234–246.

Raven, P.H. (1973). 'The evolution of Mediterranean floras'. In *Mediterranean-type ecosystems. Origin and structure*, F. di Castri and H.A. Mooney. Heidelberg; Berlin, Springer-Verlag: 213–224.

Ravid, U. and E. Putievski (1983). 'Constituents of essential oils from *Majorana syriaca*, *Coridothymus capitatus* and *Satureja thymbra*'. *Planta Medica* **49**: 248–249.

Reille, M. (1992). 'New pollen analytical researches in Corsica: the problem of *Quercus ilex* L. and *Erica arborea* L.; the origin of *Pinus halepensis* miller forests'. *New Phytologist* **122**: 359–378.

Reille, M. and A. Pons (1992). 'The ecological significance of sclerophylous oak forests in the western part of the Mediterranean Basin: a note on pollen analytical data'. *Vegetatio* **99–100**: 13–17.

Reille, M., V. Andrieu, and J.-L. de Beaulieu (1996). 'Les grands traits de l'histoire de la végétation des montagnes méditerranéennes occidentales'. *Ecologie* **27**: 153–169.

Reille, M., J. Gamisans, J.L.d. Beaulieu, and V. Andrieu (1997). 'The late-glacial at Lac de Creno (Corsica, France): a key site in the western Mediterranean Basin'. *New Phytologist* **135**: 547–559.

Reille, M., J. Gamisans, J.L.d. Beaulieu, and V. Andrieu (1999). 'The Holocene at Lac de Creno (Corsica, France): a key site for the whole island'. *New Phytologist* **141**: 291–307.

Rejmánek, M. (1989). 'Invasivability of plant communities'. In *Ecology of biological invasions: a global perspective*, J.A. Drake, F. Di Castri, and R.H. Groves, F.J. Kruger, H.A. Mooney, M. Rajmanek, and M.H. Williamson (eds.). New York, Wiley & Sons: 369–388.

Rejmánek, M. and D.M. Richardson (1996). 'What attributes make some plant species more invasive?' *Ecology* **77**: 1655–1661.

Ren, Z. and R.J. Abbott (1991). 'Seed dormancy in Mediterranean *Senecio vulgaris*'. *New Phytologist* **117**: 673–678.

Rey, P.J. (1995). 'Spatio-temporal variation in fruit and frugivorous bird abundance in olive orchards'. *Ecology* **76**: 1625–1635.

Rey, P.J. and J. Alcantára (2000). 'Recruitment dynamics of fleshy-fruited plant (*Olea europaea*): connecting patterns of seed dispersal to seedling establishment'. *Journal of Ecology* **88**: 622–633.

Rey, P.J., J.E. Gutiérrez, J. Alcantára, and F. Valera (1997). 'Fruit size in wild olives: implications for avian seed dispersal'. *Functional Ecology* **11**: 611–618.

Rhazi, M., P. Grillas, A. Charpentier, and F. Médail (2004). 'Experimental management of Mediterranean temporary pools for conservation of the rare quillwort *Isoetes setacea*'. *Biological Conservation* **118**: 675–684.

Rice, E.L. (1979). 'Allelopathy—an update'. *Botanical Review* **45**: 15–109.

Richardson, D.M. and W.J. Bond (1991). 'Determinants of plant distribution; evidence from pine invasion'. *American Naturalist* **137**: 639–668.

Richardson, J.E., F.M. Weitz, M.F. Fay, Q.C.B. Cronk, H.P. Linder, G. TReeves, and M.W. Chase (2001). 'Rapid and recent origin of species richness in the Cape flora of South Africa'. *Nature* **412**: 181–183.

Rieseberg, L.H. (1997). 'Hybrid origins of plant species'. *Annual Review of Ecology and Systematics* **28**: 359–389.

Rieseberg, L.H. (2001). 'Chromosomal rearrangements and speciation'. *Trends in Ecology and Evolution* **16**: 351–358.

Rieseberg, L.H. and L. Brouillet (1994). 'Are many plant species paraphyletic?' *Taxon* **43**: 21–32.

Rieseberg, L.H. and J.F. Wendel (1993). 'Introgression and its consequences in plants'. In *Hybrid zones and the evolutionary process*, R.G. Harrison (ed.). New York, Oxford University Press: 70–109.

Rieseberg, L.H., O. Raymond, D.M. Rosenthal, Z. Lai, K. Livingstone, T. Nakazato, J.L. Durphy, A.E. Schwarbach, L.A. Donovan, and C. Lexer (2003). 'Major ecological transitions in wild sunflowers facilitated by hybridization'. *Science* **301**: 1211–1216.

Robertson, A.H.F. and M. Grasso (1995). 'Overview of the Late Tertiary—recent tectonic and palaeo-environmental development of the Mediterranean region'. *Terra Nova* **7**: 114–127.

Robles, C., G. Bonon, and S. Garzino (1999). 'Potentialités autotoxiques et allélopathiques de *Cistus albidus* L'. *Comptes Rendus de l'Académie des Sciences Paris, Sciences de la Vie* **322**: 677–685.

Robles, C., A. Bousquet-Mélou, S. Garzino, and G. Bonin (2003). 'Comparison of essential oil composition of two varieties of *Cistus ladanifer*'. *Biochemical Systematics and Ecology* **31**: 339–343.

Roiron, R. (1992). Flores, végétations et climats du Néogène méditerranéen: apports des macroflores du sud de la France et di nord-est de l'Espagne. Thése de Doctorat d'Etat. Université de Montpellier II: Montpellier.

Rosenbaum, G. and G.S. Lister (2004). 'Neogene and Quaternary Rosenbaum, G. and G.S. Lister (2004). 'Neogene and Quaternary rollback evolution of the Tyrrhenian Sea, the Apennines and the Sicilian Maghrebides'. *Tectonics* **23**: in press.

Rosenbaum, G., G.S. Lister, and C. Duboz (2002*a*). 'Reconstruction of the tectonic evolution of the western Mediterranean since the Oligocene'. *Journal of the Virtual Explorer* **8**: 107–130.

Rosenbaum, G., G.S. Lister, and C. Duboz (2002*b*). 'Relative motions of Africa, Iberia and Europe during Alpine orogeny'. *Tectonophysics* **359**: 117–129.

Rosenbaum, G., G.S. Lister, and C. Duboz (2004). 'The Mesozoic and Cenozoic motion of Adria (central Mediterranean): a review of constraints and limitations'. *Geodinamica Acta* in press.

Ross, J.D. and C. Sombrero (1991). 'Environmental control of essential oil production in Mediterranean plants'. In *Ecological chemistry and biochemistry of plant terpenoids*, J.B. Harborne and F.A. Tomas-Barberan (eds.). Oxford, Clarendon Press: 83–94.

Rousset, O. and J. Lepart (1999). 'Evaluer l'impact du pâturage sur le maintien des milieux ouverts. Le cas des pelouses sèches'. *Fourrages* **159**: 223–235.

Rousset, O. and J. Lepart (2000). 'Positive and negative interactions at different life stages of a colonizing species (*Quercus humilis*)'. *Journal of Ecology* **88**: 401–412.

Roy, J. (1981). 'Intraspecific variation in the physiological characteristics of perennial grasses of the Mediterranean region'. *Ecologia Mediterranea* **VIII**: 435–448.

Roy, J. and L. Sonié (1992). 'Germination and population dynamics of *Cistus* species in relation to fire'. *Journal of Applied Ecology* **29**: 647–655.

Ruiz-Rejon, M. and J.L. Oliver (1981). 'Genetic variability in *Muscari comosum* (Liliaceae): I. A comparative analysis of chromosome polymorphisms in Spanish and Aegean populations'. *Heredity* **47**: 403–407.

Rundel, P.W. (1988). 'Leaf structure and nutrition in Mediterranean climate sclerophyllys'. In *Mediterranean-type ecosystems*, R.L. Specht (ed.). Dordrecht, Kluwer: 157–167.

Rundel, P.W. (1995). 'Adaptive significance of some morphological and physiological characteristics in Mediterranean plants: facts and fallacies'. In *Time scales of biological responses to water constraints*, J. Roy, J. Aronson, and F. Di Castri (eds.). Amsterdam, SPB Academic Publishing: 119–139.

Runemark, H. (1969). 'Reproductive drift, a neglected principle in reproductive biology'. *Botanica Notiser* **122**: 90–129.

Runemark, H. (1971). 'The phytogeography of the central Aegean'. *Opera Botanica* **30**: 20–28.

Sack, L. and P.J. Grubb (2002). 'The combined impacts of deep shade and drought on the growth and biomass allocation of shade-tolerant woody seedlings'. *Oecologia* **131**: 175–185.

Sage, T.L., F. Strumas, W.W. Cole, and S.C.H. Barrett (1999). 'Differential ovule development following self- and cross-pollination: the basis of self-sterility in *Narcissus triandrus* (Amaryllidaceae)'. *American Journal of Botany* **86**: 855–870.

Salamini, F. (2003). 'Hormones and the green revolution'. *Science* **302**: 71–72.

Salamini, F., H. Özkan, A. Brandolini, R. Schäfer-Pregl, and W. Martin (2002). 'Genetics and geography of wild cereal domestication in the Near East'. *Nature Reviews Genetics* **3**: 429–441.

Salleo, S., A. Nardini, and M.A. Lo Gullo (1997). 'Is sclerophylly of Mediterranean evergreens an adaptation to drought'. *New Phytologist* **135**: 603–612.

Salvador, L., R. Ala, D. Agúndez, and L. Gil (2000). 'Genetic variation and migration pathways of maritime pine (*Pinus pinaster* Ait.) in the Iberian peninsula'. *Theoretical and Applied Genetics* **100**: 89–95.

San Miguel, E. (2003). 'Rue (Ruta L., Rutaceae) in traditional Spain: frequency and distribution of its medicinal and symbolic applications'. *Economic Botany* **57**: 231–244.

Sánchez-Lafuente, A.M., P.J. Rey, J.M. Alcantara, and F. Valera (1999). 'Breeding system and the role of floral visitors in seed production of a "few-flowered" perennial herb, *Paeonia broteroi* Boiss. & Reut. (Paeoniaceae)'. *Ecoscience* **6**(2): 163–172.

Sang, T., D.J. Crawford, and T.F. Stuessy (1997). 'Chloroplast DNA phylogeny, reticulate evolution, and biogeography of *Paeonia* (Paeoniaceae)'. *American Journal of Botany* **84**: 1120–1136.

Sans, F.X., J. Escarré, and J. Lepart (1998). 'Persistence of *Picris hieracioides* populations in old fields: an example of facilitation'. *Oikos* **83**: 283–292.

Schemske, D. and C.C. Horvitz (1984). 'Variation among floral visitors in pollination ability: a precondition for mutualism specialization'. *Science* **225**: 519–521.

Schemske, D.W., B.C. Husband, M.H. Ruckelshaus, C. Goodwillie, I.M. Parker, and J.G. Bishop (1994). 'Evaluating approaches to the conservation of rare and endangered plants'. *Ecology* **75**: 584–606.

Schiller, G. and C. Grunwald (1987). 'Resin monoterpenes in range-wide provenanced trials of *Pinus halepensis* Mill. in Israel'. *Silvae Genetica* **36**: 109–114.

Schluter, D. and R.E. Ricklefs (1993). 'Species diversity. An introduction to the problem'. In *Species diversity in ecological communities. Historical and geographical perspectives*, R.E. Ricklefs and D. Schluter (eds.). Chicago, IL, University of Chicago Press: 1–12.

Seddon, G. (1974). 'Xerophytes, xeromorphs, and scerophyllys: the history of some concepts in ecology'. *Biological Journal of the Linnaen Society* **6**: 65–87.

Seigler, D.S. and P.W. Price (1976). 'Secondary compounds in plants: primary functions'. *American Naturalist* **110**: 101–105.

Selvi, F. (1998). 'Floral biometrics of the *Anchusa undulata* L. group (Boraginaceae) from the central-eastern Mediterranean'. *Botanical Journal of the Linnaean Society* **128**: 251–270.

Sengör, A.M.C. (1979). 'Mid-Mesozoic closure of Permo-Triassic Tethys and its implications'. *Nature* **279**: 590–593.

Seufert, G., D. Kotzias, C. Spartá, and B. Versino (1995). 'Volatile organics in Mediterranean shrubs and their potential role in a changing environment'. In *Global change and Mediterranean type ecosystems*, J.M. Moreno and W.C. Oechel (eds.). New York, Springer: 343–370.

Sharkey, T.D. and E.L. Singsaas (1995). 'Why plants emit isoprene'. *Nature* **374**: 769.

Shmida, A. (1981). 'Mediterranean vegetation in California and Israel: similarities and differences'. *Israel Journal of Botany* **30**: 105–123.

Shmida, A. and A. Dafni (1989). 'Blooming strategies, flower size and advertising in the 'lily-group' geophytes in Israel'. *Herbertia* **45**: 111–123.

Shmida, A. and R. Dukas (1990). 'Progressive reduction in the mean body sizes of solitary bees active during the flowering season and its correlation with the sizes of the bee flowers of the mint family (Lamiaceae)'. *Israel Journal of Botany* **39**: 133–141.

Shmida, A. and M.V. Wilson (1985). 'Biological determinants of species diversity'. *Journal of Biogeography* **12**: 1–20.

Siljak-Yakovlev, S. (1996). 'La dysploïdie et l'évolution du caryotype'. *Bocconea* **5**: 211–220.

Siljak-Yakovlev, S., S. Peccenini, E. Muratovic, V. Zoldos, O. Robin, and J. Vallés (2003). 'Chromosomal differentiation and genome size in three European mountain *Lilium* species'. *Plant Systematics and Evolution* **236**: 165–173.

Simeon de Bouchberg, M., J. Allegrini, C. Bessiere, M. Attisso, J. Passet, and R. Granger (1976). *Propriétés microbiologiques des huiles essentielles de chimiotypes de Thymus vulgaris Linnaeus*. Estratto dalla Rivista Italiana Essenze, Profumi, Piante officinali, Aromi, Saponi, Cosmetici, Aerosol.

Simmons, M.T. and R.M. Cowling (1996). 'Why is the Cape Peninsula so rich in plant species? An analysis of the independent diversity components'. *Biodiversity and Conservation* **5**: 551–573.

Snaydon, R.W. (1970). 'Rapid population differentiation in a mosaic environment: I. The response of *Anthoxanthum odoratum* populations to soils'. *Evolution* **24**: 257–269.

Snogerup, S. (1967). 'Studies in the Aegean flora IX. *Erysimum* sect. *Cheiranthus*. B. Variation and evolution in the small population system'. *Opera Botanica* **14**: 5–86.

Snogerup, S. (1971). 'Evolutionary and plant geographical aspects of chasmophytic communities'. In *Plant life of south-west Asia*, P.A. Davis, P.C. Harper, and I.C. Hedgel (eds.). Edingburg, Botanical Society of Edingburg.

Soliva, M. and A. Widmer (2003). 'Gene flow across species boundaries in sympatric, sexually deceptive *Ophrys* (Orchidaceae) species'. *Evolution* **57**: 2252–2261.

Soltis, D.E., P.S. Soltis, M.W. Chase, M.E. Mort, D.C. Albach, M. Zanis, V. Savolainen, W.H. Hahn, S.B. Hoot, M.F. Fay, M. Axtell, S.M. Swensen, L.M. Prince, W.J. Kress, K.C. Nixon, and J.S. Farris (2000). 'Angiosperm phylogeny inferred from 18S rDNA, *rbcL*, and *atpB* sequences'. *Botanical Journal of the Linnaean Society* **133**: 381–461.

Specht, R.L. (1969*a*). 'A comparison of the sclerophyllous vegetation characteristics of Mediterranean type climates in France, California, and southern Australia: I. Structure, morphology, and succession'. *Australian Journal of Botany* **17**: 277–292.

Specht, R.L. (1969*b*). 'A comparison of the sclerophyllous vegetation characteristics of Mediterranean type climates in France, California, and southern Australia: II. Dry matter energy, and nutrient accumulation'. *Australian Journal of Botany* **17**: 293–308.

Specht, R.L. and E.J. Moll (1983). 'Mediterranean-type heathlands of the world: an overview'. In *Mediterranean-type ecosystems. The role of nutrients*, F.J. Kruger, D.T. Mitchell, and J.U.M. Jarvos (eds.). New York, Springer-Verlag: 41–73.

Specht, R.L. and P.W. Rundel (1990). 'Sclerophylly and foliar nutrient status of Mediterranean-climate plant communities in southern Australia'. *Australian Journal of Botany* **38**: 459–474.

Stahl-Biskup, E. (2002). 'Essential oil chemistry of the genus *Thymus*—a global view'. In *Thyme: the genus Thymus*, E. Stahl-Biskup and F. Sáez (eds.). London, Taylor & Francis: 75–124.

Staudt, M., N. Mandl, R. Joffre, and S. Rambal (2001). 'Intraspecific variability of monoterpene composition emitted by *Quercus ilex* leaves'. *Canadian Journal of Forest Research* **31**: 174–180.

Stebbins, G.L. (1942). 'The genetic approach to problems of rare and endemic species'. *Madrono* **74**: 241–272.

Stebbins, G.L. (1950). *Variation and evolution in plants*. New York: Columbia University Press.

Stebbins, G.L. (1952). 'Aridity as a stimulus to evolution'. *American Naturalist* **86**: 33–44.

Stebbins, G.L. (1970). 'Adaptive radiation in angiosperms: I. Pollination mechanisms'. *Annual Review of Ecology and Systematics* **1**: 307–326.

Stebbins, G.L. (1971). *Chromosomal evolution in higher plants*. London, E. Arnold Ltd.

Stebbins, G.L. (1974). *Flowering plants, evolution above the species level*. Cambridge, Belknap Press.

Stebbins, G.L. (1980). 'Rarity of plant species: a synthetic viewpoint'. *Rhodora* **82**: 77–86.

Stebbins, G.L. (1985). 'Polyploidy, hybridization, and the invasion of new habitats'. *Annals of the Missouri Botanical Garden* **72**: 824–832.

Stebbins, G.L. and J.C. Dawe (1987). 'Polyploidy and distribution in the European Flora: a reappraisal'. *Jahrbücher für Systematik, Pflanzengeschichte and Pflanzengeographie* **108**: 343–354.

Stebbins, G.L. and L. Ferlan (1956). 'Population variability, hybridization, and introgression in some species of *Ophrys*'. *Evolution* **10**: 32–46.

Stebbins, G.L. and J. Major (1965). 'Endemism and speciation in the California flora'. *Ecological Monographs* **35**: 1–35.

Stehilik, I. (2000). 'Nunataks and peripheral refugia for alpine plants during Quaternary glaciation in the middle part of the Alps'. *Botanica Helvetica* **110**: 25–30.

Sternberg, M., M. Gutman, A. Perevolotsky, E.D. Ungar, and J. Kigel (2000). 'Vegetation response to grazing management in a Mediterranean herbaceous community: a functional group approach'. *Journal of Applied Ecology* **37**: 224–237.

Strauss, S.Y. and W.S. Armbruster (1997). 'Linking herbivory and pollination: new perspectives on plant and animal ecology and evolution'. *Ecology* **78**: 1617–1618.

Strid, A. (1969). 'Evolutionary trends in the breeding system of *Nigella* (Ranunculaceae)'. *Botanika Notiser* **122**: 380–397.

Strid, A. (1970). 'Studies in the Aegean flora XVI. Biosystematics of the *Nigella arvensis* complex with special reference to the problem of non-adaptive radiation'. *Opera Botanica* **28**: 1–169.

Strid, A. (1972). 'Some evolutionary and phytogeographical problems in the Aegean'. In *Taxonomy, phytogeography, and evolution*, D.H. Valentine (ed.). London, Academic Press: 289–300.

Strid, A. and K. Papanicolaou (1985). 'The Greek mountains'. In *Plant conservation in the Mediterranean area*, C. Gomez-Campo (ed.). Dordrecht, Dr W. Junk Publishers: 89–111.

Suc, J.-P. (1984). 'Origin and evolution of the Mediterranean vegetation and climate in Europe'. *Nature* **307**: 429–432.

Suc, J.-P. (1989). 'Distribution latitudinale et étagement des associations végétales au Cenozoïque supérieure dns l'aire ouest-méditerraanéenne'. *Bullétin de la Société Géologique de la France* **8**: 541–550.

Suc, J.-P. and E. Bessais (1990). 'Perennité d'un climat thermo-xérique en Sicile avant, pendant et apès la crise de salinité messinienne'. *Comptes Rendus de l'Académie des Sciences Paris* **310**: 1701–1707.

Suc, J.-P. and J. Cravatte (1982). Etude palynologique du pliocène de Catalogne (nord-est de l'Espagne). *Paléobiologie Continentale* **13**: 1–31.

Suc, J.-P., G. Clauzon, M. Bessedik, S. Leroy, Z. Zheng, A. Drivaliari, P. Roiron, P. Ambert, J. Martinell, R. Domenech, I. Matias, R. Julia, and R. Anglada (1992). 'Neogene and Lower Pleistocene in southern France and northeastern Spain. Mediterranean environments and climate'. *Cahiers de micropaléontologie* **7**: 165–186.

Suc, J.-P., A. Bertini, N. Combourieu-Nebout, F. Diniz, S. Leroy, E. Russo-Ermoli, E. Bessais, and J. Ferrier (1995). 'Structure of west Mediterranean vegetation and climate since 5.3 ma'. *Acta Zoologika Cracovia* **38**: 3–16.

Suehs, C., F. Médail, and L. Affre (2001). 'Ecological and genetic features of the invasion by the alien *Carpobrotus* (Aizoaceae) plants in Mediterranean island habitats'. In *Plant invasions; species ecology and ecosystem management*, G. Brundu, J. Brock, I. Camarda, L. Child, and M. Wade (eds.). Leiden, Backhuys Publishers.

Suehs, C., F. Médail, and L. Affre (2004*a*). 'Invasion dynamics of two alien *Carpobrotus* (Aizoaceae) taxa on a Mediterranean island: I. Genetic diversity and introgression'. *Heredity* **92**: 31–40.

Suehs, C., F. Médail, and L. Affre (2004*b*). 'Invasion dynamics of two alien *Carpobrotus* (Aizoaceae) taxa on a Mediterranean island: II. Reproductive alternatives'. *Heredity* **92**: 550–556.

Suehs, C., L. Affre, and F. Médail (2004*c*). 'Unexpected insularity effects in invasive plant mating systems: the case of *Carpobrotus* (Aizoaceae) taxa in the Mediterranean Basin'. *Biological Journal of the Linnaean Society* in press.

Talavera, S., F. Bastida, P.L. Ortiz, and M. Arista (2001). 'Pollinator attendance and reproductive success in *Cistus libanoticus*'. *International Journal of Plant Science* **162**: 343–352.

Tan, K. (1980). 'Studies in the Thymeleaceae: a revision of the genus *Thymelaea*'. *Notes from the Royal Botanic garden Edingburgh* **38**: 189–246.

Tantaoui-Elaraki, A., N. Lattaoui, A. Errifi, and B. Benjilali (1993). 'Comparison and antimicrobial activity of the essential oils of *Thymus broussonettii*, *T. zygis* and *T. satureioides*'. *Journal of Essential Oil Research* **5**: 45–53.

Tapias, R., L. Gil, P. Fuentes-Utrilla, and J.A. Pardos (2001). 'Canopy seed banks in Mediterranean pines of southeastern Spain: a comparison between *Pinus halepensis* Mill., *Pinus pinaster* Ait., *P. nigra* Arn. and *P. pinea* L'. *Journal of Ecology* **89**: 629–638.

Tarayre, M. and J.D. Thompson (1997). 'The population genetic structure of the gynodioecious *Thymus vulgaris* (Labiateae) in southern France'. *Journal of Evolutionary Biology* **10**: 157–174.

Tarayre, M., J.D. Thompson, J. Escarré, and Y.B. Linhart (1995). 'Intra-specific variation in the inhibitory effects of *Thymus vulgaris* (Labiatae) monoterpenes on seed germination'. *Oecologia* **101**: 110–118.

Tarayre, M., P. Saumitou-Laprade, J. Cuguen, D. Couvet, and J.D. Thompson (1997). 'The spatial genetic structure of cytoplasmic (cpDNA) and nuclear (allozyme) markers within and among populations of the gynodioecious *Thymus vulgaris* (Labiatae) in southern France'. *American Journal of Botany* **84**: 1675–1684.

Tatoni, T., F. Magnin, G. Bonin, and J. Vaudour (1994). 'Secondary successions on abandoned cultivation terraces in calcareous Provence. 1. Vegetation and soils'. *Acta Oecologica* **15**: 431–447.

Tenhunen, J.D., A. Sala Serra, P.C. Harley, and R.L. Dougherty (1990). 'Factors influencing carbon fixation and water use by Mediterranean sclerophyll shrubs during summer drought'. *Oecologia* **82**: 381–393.

Terral, J.F. (2000). 'Exploitation and management of the olive tree during prehistoric times in Mediterranean France and Spain'. *Journal of Archaeological Sciences* **27**: 127–133.

Terral, J.F. and G. Arnold-Simard (1996). 'Beginnings of olive cultivation in eastern Spain in relation to Holocene bioclimatic changes'. *Quaternary Research* **46**: 176–185.

Terral, J.F., N. Alonso, C.R.M., N. Chatti, L. Fabre, G. Fiorention, P. Marinval, J.G.P., B. Prada, N. Rovira, and P. Alibert (2004). 'Historical biogeography of olive domestication (*Olea europaea* L.) as revealed by geometrical morphometry and applied to biological and archaeological material'. *Journal of Biogeography* **31**: 63–77.

Thanos, C.A. and K. Georghiou (1988). 'Ecophysiology of fire-stimulated seed germination in *Cistus creticus* and *C. salvifolius*'. *Plant Cell & Environment* **11**: 841–849.

Thanos, C.A. and P.W. Rundel (1995). 'Fire-followers in chaparral: nitrogenous compounds trigger seed germination'. *Journal of Ecology* **83**: 207–216.

Thanos, C.A., C.C. Kadis, and F. Skarou (1995). 'Ecophysiology of germination in the aromatic plants thyme, savory and oregano (Labiatae)'. *Seed Science Research* **5**: 161–170.

Thébaud, C. and M. Debussche (1991). 'Rapid invasion of *Fraxinus ornus* L. along the Hérault River system in southern France: the importance of seed dispersal by water'. *Journal of Biogeography* **18**: 7–12.

Thébaud, C. and M. Debussche (1992). 'A field test of the effects of infructescence size on fruit removal by birds in *Viburnum tinus*'. *Oikos* **65**: 391–394.

Thébaud, C., A.D. Finzi, L. Affre, M. Debussche, and J. Escarre (1996). 'Assessing why two introduced *Conyza* differ in their ability to invade Mediterranean old fields'. *Ecology* **77**: 791–804.

Thébaud, C. and D. Simberloff (2001). 'Are plants really larger in their introduced range?' *The American Naturalist* **157**: 231–236.

Thompson, J.D. (1991). 'The biology of an invasive plant: what makes *Spartina anglica* so successful?' *Bioscience* **41**: 393–401.

Thompson, J.D. (1999). 'Population differentiation in Mediterranean plants: insights into colonization history and implications for species diversification'. *Heredity* **82**: 229–236.

Thompson, J.D. (2001). 'How do visitation patterns vary among pollinators in relation to floral display and floral design in a generalist pollination system'. *Oecologia* **126**: 386–394.

Thompson, J.D. (2002). 'Population structure and the spatial dynamics of genetic polymorphism in thyme'. In *Thyme: the genus Thymus*, E. Stahl-Biskup and F. Sáez (eds.). London, Taylor & Francis: 44–74.

Thompson, J.D. and R. Lumaret (1992). 'The evolutionary dynamics of polyploid plants: origins, establishment and persistence'. *Trends in Ecology and Evolution* **7**: 302–307.

Thompson, J.D. and B. Dommée (1993). 'Sequential variation in the components of reproductive success in the distylous *Jasminum fruticans* (Oleaceae)'. *Oecologia* **94**: 480–487.

Thompson, J.D. and B. Dommée (2000). 'Morph specific stigma height variation in distylous *Jasminum fruticans*'. *New Phytologist* **148**: 303–314.

Thompson, J.D. and M. Tarayre (2000). 'Exploring the genetic basis and proximate causes of variation in female fertility advantage in gynodioecious *Thymus vulgaris*'. *Evolution* **54**: 1510–1520.

Thompson, J.D., T. McNeilly, and A.J. Gray (1993). 'The demography of clonal plants in relation to successional habitat change: the case of *Spartina anglica*'. In *Primary Succession on Land*, J. Miles and D.W.H. Walton (eds.). Oxford, Blackwell Scientific. **12**: 193–207.

Thompson, J.D., D. Manicacci, and M. Tarayre (1998). 'Thirty five years of thyme: a tale of two polymorphisms'. *Bioscience* **48**: 805–815.

Thompson, J.D., A.-G. Rolland, and F. Prugnolle (2002). 'Genetic variation for sexual dimorphism in flower size within and between populations of gynodioecious *Thymus vulgaris*'. *Journal of Evolutionary Biology* **15**: 362–372.

Thompson, J.D., S.C.H. Barrett, and A.M. Baker (2003*a*). 'Frequency-dependent variation in reproductive success in *Narcissus*: implications for the maintenance of stigma-height dimorphism'. *Proceedings of the Royal Society Ser. B* **270**: 949–953.

Thompson, J.D., J.-C. Chalchat, A. Michet, Y.B. Linhart, and B. Ehlers (2003*b*). 'Qualitative and quantitative variation in monoterpene co-occurrence and composition in the essential oil of *Thymus vulgaris* chemotypes'. *Journal of Chemical Ecology* **29**: 859–880.

Thompson, J.D., M. Tarayre, P. Gauthier, I. Litrico, and Y.B. Linhart (2004). 'Multiple genetic contributions to plant performance in *Thymus vulgaris*'. *Journal of Ecology* **92**: 45–56.

Thompson, J.N. (1994). *The coevolutionary process*. Chicago, IL, The University of Chicago Press.

Thuiller, W., J. Vayreda, J. Pino, S. Sabate, S. Lavorel, and C. Gracia (2003). 'Large-scale environmental correlates of forest tree distributions in Catalonia (NE Spain)'. *Global Ecology and Biogeography* **12**: 313–325.

Tilman, D. (1990). 'Constraints and tradeoffs: toward a predictive theory of competition and succession'. *Oikos* **58**: 3–15.

Tilman, D. (1994). 'Competition and biodiversity in spatially structured habitats'. *Ecology* **75**: 2–16.

Tilman, D. and S. Pacala (1993). 'The maintenance of species richness in plant communities'. In *Species diversity in ecological communities. Historical and geographical perspectives*, R.E. Ricklefs and D. Schluter (eds.). Chicago, University of Chicago Press: 13–25.

Todorovic, B. and B. Stevanovic (1994). 'Adaptive characteristics of the endemic species *Satureja horvatii* Silic (Lamiaceae) in mountain-Mediterranean and Mediterranean habitats'. *Botanical Journal of the Linnaean Society* **114**: 367–376.

Toumi, L. and R. Lumaret (1998). 'Allozyme variation in cork oak (*Quercus suber* L.): the role of phylogeography, genetic introgression by other Mediterranean oak species and human activities'. *Theoretical and Applied Genetics* **97**: 647–656.

Toumi, L. and R. Lumaret (2001). 'Allozyme characterisation of four Mediterranean evergreen oak species'. *Biochemical Systematics and Ecology* **29**: 799–817.

Trabaud, L. (1987). 'Dynamics after fire of sclerophyllous plant communities in the Mediterranean Basin'. *Ecologia Mediterranea* **13**: 25–37.

Trabaud, L. (1992). 'Community dynamics after fire disturbance: short-term change and long-term stability'. *Ekistics*: 287–292.

Trabaud, L. and J.-F. Galtié (1996). 'Effects of fire frequency on plant communities and landscape pattern in the Massif des Aspres (southern France)'. *Landscape Ecology* **11**: 215–224.

Trabaud, L. and J. Lepart (1980). 'Diversity and stability in garrigue ecosystems after fire'. *Vegetatio* **43**: 49–57.

Trabaud, L. and J. Oustric (1989). 'Heat requirements for seed germination of three *Cistus* species in the garrigue of southern France'. *Flora* **183**: 321–325.

Traveset, A. (1992). 'Production of galls in *Phillyrea angustifolia* induced by cecidmyiid flies'. In *Plant-animal interactions in Mediterranean-type ecosystems*, C.A. Thanos (ed.). Athens, University of Athens: 198–204.

Traveset, A. (1994*a*). 'Cumulative effects on the reproductive output of *Pistacia terebinthus* (Anacardiaceae)'. *Oikos* **114**: 361–367.

Traveset, A. (1994*b*). 'Reproductive biology of *Phillyrea angustifolia* L. (Oleaceae) and effect of galling insects on its reproductive output'. *Botanical Journal of the Linnaean Society* **114**: 153–166.

Traveset, A. (1995). 'Reproductive ecology of *Cneorum tricoccon* L. (Cneoraceae) in the Balearic Islands'. *Botanical Journal of the Linnaean Society* **117**.

Traveset, A. and E. Sáez (1997). 'Pollination of *Euphorbia dendroides* by lizards and insects: spatio-temporal variation in patterns of flower visitation'. *Oecologia* **111**: 241–248.

Traveset, A., J. Gulias, N. Riera, and M. Mus (2003). 'Transition probabilities from pollination to establishment in a rare dioecious shrub species (*Rhamnus ludovici-salvatoris*) in two habitats'. *Journal of Ecology* **91**: 427–437.

Triat-Laval, H. (1979). 'Histoire de la forêt provençale depuis 15 000 ans d'après l'analyse pollinique'. Forêt *Méditerranéenne* **1**: 19–24.

Turner, C. and G.E. Hannon (1988). 'Vegetational evidence for late quaternary climatic changes in south west Europe in relation to the influence of the North Atlantic Ocean'. *Philosophical Transactions of the Royal Society London Series B* **318**: 451–485.

Tzedakis, P.C. (1993). 'Long-term tree populations in northwest Greece through multiple Quaternary climatic cycles'. *Nature* **364**: 437–440.

Valdeyron, G. and D.G. Lloyd (1979). 'Sex differences and flowering phenology in the common fig, *Ficus carica*'. *Evolution* **33**: 673–685.

Valdés, B. (1991). 'Andalucia and the Rif. Floristic links and a common flora'. *Botanika Chronika* **10**: 117–124.

Valdeyron, G., B. Dommée, and P. Vernet (1977). 'Self-fertilisation in male-fertile plants of a gynodioecious species: *Thymus vulgaris* L'. *Heredity* **39**: 243–249.

Van Campo, M. (1984). 'Relations entre la végétation de l'Europe et les températures de surface océaniques après le dernier maximum glaciaire'. *Pollen & Spores* **26**: 497–518.

Van Dijk, P., M. Hartog, and W. Van Delden (1992). 'Single cytotype areas in autopolyploid *Plantago media* L'. *Biological Journal of the Linnaean Society* **46**: 315–331.

Vargas, P. (2003). 'Molecular evidence for multiple diversification patterns of alpine plants in Mediterranean Europe'. *Taxon* **52**: 463–476.

Vargas, P., C.M. Morton, and S.L. Jury (1999). 'Biogeographic patterns in Mediterranean and Macaronesian species of *Saxifraga* (Saxifragaceae) inferred from phylogenetic analyses of ITS sequences'. *American Journal of Botany* **86**: 724–734.

Varinard, O. (1983). Contribution à l'étude du polymorphisme chimique et sexuel chez les espèces végétales. Montpellier, Université de Montpellier 2, DEA thesis.

Vassiliadis, C., J. Lepart, P. Saumitou-Laprade, and P. Vernet (2000*a*). 'Self-incompatibility and male fertilisation success in *Phillyrea angustifolia* L. (Oleaceae)'. *International Journal of Plant Science* **161**: 393–402.

Vassiliadis, C., M. Valéro, P. Saumitou-Laprade, and B. Godelle (2000*b*). 'A model for the evolution of high frequencies of males in an androdioecious plant based on a cross-compatibility advantage of males'. *Heredity* **85**: 413–422.

Vassiliadis, C., P. Saumitou-Laprade, J. Lepart, and F. Viard (2002). 'High male reproductive success of hermaphrodites in the androdioecious *Phillyrea angustifolia*'. *Evolution* **56**: 1362–1373.

Vaughan, H.E., S. Taylor, and J.S. Parker (1997). 'The ten cytological races of the *Scilla autumnalis* species complex'. *Heredity* **79**: 371–379.

Verdaguer, D. and F. Ojeda (2002). 'Root starch storage and allocation patterns in seeder and resprouter seedlings of two Cape *Erica* (Ericaceae) species'. *American Journal of Botany* **89**: 1189–1196.

Verdú, J.R., M.B. Crespo, and E. Galante (2000). 'Conservation strategy of a nature reserve in Mediterranean ecosystems: the effects of protection from grazing on biodiversity'. *Biodiversity and Conservation* **9**: 1707–1721.

Verdú, M. (2000). 'Ecological and evolutionary differences between Mediterranean seeders and resprouters'. *Journal of Vegetation Science* **11**: 265–268.

Verdú, M. (2004). 'Physiological and reproductive differences between hermaphrodites and males in the androdioecious plant *Fraxinus ornus*'. *Oikos* **105**: 239–246.

Verdú, M. and P. García-Fayos (1996). 'Nucleation processes in a Mediterranean bird-dispersed plant'. *Functional Ecology* **10**: 275–280.

Verdú, M., P. Dávila, P. García-Fayos, N. Flores-Hernández, and A. Valiente-Baunet (2003). '"Convergent" traits of Mediterranean woody plants belonging to pre-Mediterranean lineages'. *Biological Journal of the Linnaean Society* **78**: 415–427.

Verdú, M., A.I. Montilla, and J.R. Pannell (2004). Paternal effects on functional gender account for cryptic dioecy in a perennial plant. *Proceedings of the Royal Society of London B* **271**: 2017–2023.

Verhoeven, J.T.A. (1979). 'The ecology of *Ruppia*-dominated communities in Western Europe. I. Distribution of *Ruppia* representatives in relation to their autoecology'. *Aquatic Botany* **6**: 197–268.

Verlaque, R. (1984). 'A biosystematic and phylogenetic study of the Dipsacaceae'. In *Plant Biosystematics*, W.F. Grant (ed.). Toronto, Academic Press: 307–320.

Verlaque, R. and J. Contandriopoulos (1990). 'Analyse des variations chromosomiques en région méditerranéenne: polyploïdie, différenciation et adaptation'. *Ecologia Mediterranea* **XVI**: 93–112.

Verlaque, R., A. Aboucaya, M.A. Cardona, and J. Contandriopoulos (1991). 'Quelques exemples de spéciation insulaire en Méditerranée occidentale'. *Botanika Chronika* **10**: 137–154.

Verlaque, R., J. Contandriopoulos, and A. Aboucaya (1995). 'Cytotaxonomie et conservation de la flore insulaire: les espèces endémiques ou rares de Corse'. *Ecologia Mediterranea* **21**: 257–268.

Verlaque, R., F. Médail, P. Quézel, and J.-F. Babinot (1997). 'Endémisme végétale et paléogéographie dans le bassin méditerranéen'. *Geobios* **M.S. no. 21**: 159–166.

Vernet, J.-L. (1997). L'homme et la forêt méditerranéenne de la Préhistoire à nos jours. Paris, Editions Errance.

Vernet, P., J.L. Guillerm, and P.H. Gouyon (1977*a*). 'Le polymorphisme chimique de *Thymus vulgaris* L. (Labiée): I. Repartition des formes chimiques en relation avec certains facteurs écologiques'. *Oecologia Plantarum* **12**: 159–179.

Vernet, P., J.L. Guillerm, and P.H. Gouyon (1977*b*). 'Le polymorphisme chimique de *Thymus vulgaris* L. (Labiée): II. Carte à l'echelle 1/25000 des formes chimiques dans la région de Saint-Martin-de-Londres (Herault-France)'. *Oecologia Plantarum* **12**: 181–194.

Vernet, P., P.H. Gouyon, and G. Valdeyron (1986). 'Genetic control of the oil content in *Thymus vulgaris* L.: a case of polymorphism in a biosynthetic chain'. *Genetica* **69**: 227–231.

Vilà, M. and C.M. D'Antonio (1998*a*). 'Fitness of invasive *Caprobotus* (Aizoaceae) hybrids in coastal California'. *Ecoscience* **5**: 191–199.

Vilà, M. and C.M. D'Antonio (1998*b*). 'Fruit choice and seed dispersal of invasive vs. noninvasive *Caprobotus* (Aizoaceae) in coastal California'. *Ecology* **79**: 1053–1060.

Vilà, M. and C.M. D'Antonio (1998*c*). 'Hybrid vigor for clonal growth in *Carpobrotus* (Aizoaceae) hybrids in coastal California'. *Ecological Applications* **8**: 1196–1205.

Vilà, M. and F. Lloret (2000). 'Seed dynamics of the mast seeding tussock grass *Ampledesmos mauritanica* in Mediterranean shrublands'. *Journal of Ecology* **88**: 479–491.

Vilà, M. and I. Muñoz (1999). 'Patterns and correlates of exotic and endemic plant taxa in the Balearic Islands'. *Ecologia Mediterranea* **25**: 153–161.

Vilà, M. and J. Weiner (2004). 'Are invasive species better competitors than native plant species?—evidence from pair-wise experiments'. *Oikos* **105**: 229–238.

Vilà, M., E. Weber, and C.M. D'Antonio (2000). 'Conservation implications of invasion by plant hybridization'. *Biological Invasions* **2**: 207–217.

Villani, F., M. Pigliucci, S. Benedettelli, and M. Cherubini (1991). 'Genetic differentiation among Turkish chestnut (*Castanea sativa* Mill.) populations'. *Heredity* **66**: 131–136.

Villani, F., M. Pigliucci, M. Lauteri, and M. Cherubini (1992). 'Congruence between genetic, morphometric, and physiological data on differentiation of Turkish chestnut (*Castanea sativa* Mill.)'. *Genome* **35**: 251–256.

Villani, F., M. Pigliucci, and M. Cherubini (1994). 'Evolution of *Castanea sativa* Mill. in Turkey and Europe'. *Genetic Research Cambridge* **63**: 109–116.

Villani, F., A. Sansotta, M. Cherubini, D. Cesaroni, and V. Sbordoni (1999). 'Genetic structure of natural populations of *Castanea sativa* Mill. in Turkey: evidence of a hybrid zone'. *Journal of Evolutionary Biology* **12**: 233–244.

Vitousek, P.M. (1990). 'Biological invasions and ecosystem processes: towards an integration of population biology and ecosystem studies'. *Oikos* **57**: 7–13.

Vokou, D. and N.S. Margaris (1982). 'Volatile oils as allelopathic agents'. In *Aromatic plants: basic and applied aspects*, N. Margaris, A. Koedam, and D. Vokou (eds.). The Hague, Boston, London, Martinus Nijhoff Publishers: 59–72.

Vokou, D. and N.S. Margaris (1984). 'Effects of volatile oils from aromatic shrubs on soil microorganisms'. *Soil Biology and Biochemistry* **16**: 509–513.

Vokou, D. and N.S. Margaris (1986*a*). 'Autoallelopathy of *Thymus capitatus*'. *Acta Oecologica—Oecologia Plantarum* **7**: 157–163.

Vokou, D. and N.S. Margaris (1986*b*). 'Variation of volatile oil concentration of Mediterranean aromatic shrubs *Thymus capitatus* Hoffmag and Link, *Satureja thymbra* L., *Teucrium polium*, L., and *Rosmarinus officinalis*'. *International Journal of Biometeorology* **30**: 147–155.

Vokou, D. and N.S. Margaris (1988). 'Decomposition of terpenes by soil microorganisms'. *Pedobiologia* **31**: 413–419.

Vokou, D., S. Kokkini, and J.M. Bessiere (1988). '*Origanum onites* (Lamiaceae) in Greece; distribution, volatile oil yeild, and composition'. *Economic Botany* **42**: 407–412.

Vokou, D., T. Petanidou, and D. Bellos (1990). 'Pollination ecology and reproductive potential of *Jankaea heldreichii* (Gesneriaceae): a Tertiary relict on Mt Olympus, Greece'. Biological Conservation **52**: 125–133.

Vokou, D., K. Katradi, and S. Kokkini (1993*a*). 'Ethnobotanical survey of Zagori (Epirus, Greece), a renowned centre of folk medicine in the past'. *Journal of Ethnopharmacology* **39**: 187–196.

Vokou, D., S. Kokkini, J.M. Bessière (1993*b*). 'Geographic variation of Greek Oregano (*Origanum vulgare* ssp. *hirtum*) essential oils'. *Biochemical Systematics and Ecology* **21**: 287–295.

Vokou, D., P. Douvli, and G.J. Blionis (2003). 'Effects of monoterpenoids, acting alone or in pairs, on seed germination and subsequent seedling growth'. *Journal of Chemical Ecology* **29**: 2281–2301.

Volaire, F. (1995). 'Growth, carbohydrate reserves and drought survival strategies of contrasting *Dactylis glomerata* populations in a Mediterranean environment'. *Journal of Applied Ecology* **32**: 56–66.

Volis, S., S. Mendlinger, Y. Turuspekov, and U. Esnazarov (2002*a*). 'Phenotypic and allozyme variation in Mediterranean and desert populations of wild barley, *Hordeum spontaneum* Koch'. *Evolution* **56**: 1403–1415.

Volis, S., S. Mendlinger, and D. Ward (2002*b*). 'Adaptive traits of wild barley plants of Mediterranean and desert origin'. *Oecologia* **133**: 131–138.

Wacquant, J.P. (1990). 'Biogeographical and physiological aspects of the invasion by *Dittrichia (ex-Inula) viscosa* W. Greuter, a ruderal species in the Mediteranean Basin'. In *Biological invasions in Europe and the Mediterranean Basin*, F. Di Castri, A.J. Hansen, and M. Bebussche (eds.). Dordrecht, Kluwer Academic Press: 353–364.

Waddington, C.H. (1965). 'Introduction to the symposium'. In *The genetics of colonizing species*, H.G. Baker and G.L. Stebbins (eds.). New York, Academic Press: 1–6.

Wallace, A.R. (1880). Island life. London, Macmillan & Co.

Waser, N.M., L. Chittka, M.V. Price, N.M. Williams, and J. Ollerton (1996). 'Generalization in pollination systems, and why it matters'. *Ecology* **77**: 1043–1060.

Wells, P.V. (1969). 'The relation between mode of reproduction and extent of speciation in woody genera of the California chaparral'. *Evolution* **23**: 264–267.

Werker, E., U. Ravid, and E. Putievsky (1985). 'Structure of glandular hairs and identification of the main components of their secreted material in some species of the Labiatae'. *Israel Journal of Botany* **34**: 31–45.

Westphal, M., J. Orsini, and P. Vellutini (1976). 'Le microcontinent corse-sarde, sa position initiale, données paléomagnétiques et raccords géologiques'. *Tectonophysics* **30**: 141–157.

Whittaker, R.H. and P.O. Feeney (1971). 'Allelochemicals: chemical interactions between species'. *Science* **171**: 757–770.

Williams, K.S., D.E. Lincoln, and P.R. Ehrlich (1983). 'The coevolution of *Euphydryas chalcedona* butterflies and their larval host plants: I. Larval feeding behavior and hostplant chemistry'. *Oecologia* **56**: 323–329.

Wolfe, L.M. (2001). 'Associations among floral polymorphisms in *Linum pubescens* (Linaceae), a heterostylous plant'. *International Journal of Plant Science* **162**: 335–342.

Wolfe, L.M. and A. Shmida (1997). 'The ecology of sex expression in a gynodioecious Israeli desert shrub (*Ochradenus baccatus*)'. *Ecology* **78**: 101–110.

Worley, A.C., A.M. Baker, J.D. Thompson, and S.C.H. Barrett (2000). 'Floral display in *Narcissus*: variation in flower size and number at the species, population and individual level'. *International Journal of Plant Science* **161**: 69–79.

Wright, S.I. and S.C.H. Barrett (1999). 'Size-dependent gender modification in a hermaphroditic perennial herb'. *Proceedings of the Royal Society London B* **266**: 225–232.

Zamir, D., N. Navot, and J. Rudich (1984). 'Enzyme polymorphism in *Citrullus lanatus* and *C. colocynthis* in Israel and Sinai'. *Plant Systematics and Evolution* **146**: 163–170.

Zamora, R. (1995). 'The trapping success of a carnivorous plant, *Pinguicula vallisnerifolia*: the cumulative effects of availability, attraction, retention and robbing of prey'. *Oikos* **73**: 309–322.

Zamora, R. and R. Gómez (1996). 'Carnivorous plant-slug interaction: a trip from herbivory to kleptoparasitism'. *Journal of Animal Ecology* **65**: 154–160.

Zamora, R., J.M. Gómez, and J. Hódar (1998). 'Fitness responses of a carnivorous plant in contrasting ecological scenarios'. *Ecology* **79**: 1630–1644.

Zavala, M.A., J.M. Espelta, and J. Retana (2000). 'Constraints and trade-offs in Mediterranean plant communities: the case of holm oak—Aleppo pine forests'. *Botanical Review* **66**: 119–149.

Zedler, P.H. (1990). 'Life histories of vernal pool vascular plants'. In *Vernal pool plants—their habitat and biology*, D.H. Ikeda and R.A. Schlising (eds.). Chico, California State University. Studies from the Herbarium No. **8**: 123–146.

Zedler, P.H., R. Gautier, and G.S. McMaster (1983). 'Vegetation change in response to extreme events: The effect of a short interval between fires in California chapparal and coastal scrub'. *Ecology* **64**: 809–818.

Zohary, D. (1965). 'Colonizer species in the wheat group'. In *The genetics of colonizing species*, H.G. Baker and G.L. Stebbins (eds.). New York, Academic Press: 403–419.

Zohary, D. and M. Feldman (1962). 'Hybridization between amphidiploids and the evolution of polyploids in the wheat (*Aegilops-Triticum*) group'. *Evolution* **16**: 44–61.

Zohary, D. and M. Hopf (2000). *Domestication of plants in the old world: the origin and spread of cultivated plants in West Asia, Europe, and the Nile Valley*. Oxford, Oxford University Press.

Zohary, D. and U. Nur (1959). 'Natural triploids in the orchard grass, *Dactylis glomerata* L., polyploid complex and their significance for gene flow from diploid to tetraploid levels'. *Evolution* **13**: 311–317.

Zohary, D. and U. Plitmann (1979). 'Chromosome polymorphism, hybridization and colonisation in the *Vicia sativa* group (Fabaceae)'. *Plant Systematics and Evolution* **131**: 143–156.

Zohary, D. and P. Spiegel-Roy (1975). 'Beginnings of fruit growing in the old world'. *Science* **187**: 319–327.

Zohary, M. (1973). *Geobotanical foundations of the middle east*. Stuttgart, Gustav Fischer Verlag.

Index

Adaptation 2, 75, 78, 86, 87, 91, 92, 94, 101–103, 109–113, 115, 117, 119, 121, 139–141, 143–145, 147, 151, 160, 161, 166, 167, 177, 182, 192, 197, 208, 225, 228, 229, 238, 240, 243
Adria (see Apulian microplate)
Aegean Sea 14, 34–35, 48, 49, 54, 87, 89
Aeolian Islands 44
African plate 11, 12, 13
Alboran Sea 14
Algeria 46, 189
Allelopathy 147, 148
Alpine orogeny 11, 13, 33, 34
Anatolia (see Turkey)
Androdioecy 180, 185–189
Andromonoecy 180, 188, 189
Apulian microplate 12, 13
Aromatic plants 111, 144, 146, 150, 166, 223, 238, 243, 245
Atlas Mountains 13

Balearic Islands 15, 39, 41, 44, 51–54, 58, 68, 70, 75–76, 80, 83, 84, 86, 92, 95, 96, 171, 177, 183, 189, 225
Balkans 10, 13, 18, 24, 31–35, 55, 58, 70, 71, 73, 74, 76, 105, 107, 115, 215, 217
Basin 2, 3, 6, 10–18, 22, 25, 32, 36, 37
Betic Cordillera 13–15, 33, 64

Calabria 14, 15, 18, 22, 52, 55, 74, 102, 221
Cape Floristic Region 63, 64, 71, 141
Carnivorous plant
Chemical defence 112, 147, 162
Chemotype 154–165
Chile 8
Chromosomal rearrangements 107
Cliffs 4, 6, 7, 15, 24, 34, 47, 48, 55, 58, 65, 76, 87–88, 241, 242
Climatic history 18, 54, 74,
Clonality 99, 185, 253
Coastal habitats 138, 231
Coevolution
Colonization 44, 47, 48, 58, 59, 62, 65, 71, 74, 75, 78, 83, 95, 96, 100, 101, 104, 110, 111, 116, 130, 131–136, 138, 142–144, 159, 182, 183, 185, 187, 193, 195, 203, 204, 207, 220, 240, 242, 243
Community ecology 44, 49, 164, 166, 221, 224–230, 232, 237,
Community structure (see community ecology)
Competition 51, 58, 116, 117, 120, 125, 135–137, 147, 148, 152, 164, 166, 182, 193, 229
Conservation 8, 9, 26, 41, 56, 57, 74, 107, 117, 222, 224, 239, 241–245
Convergent evolution 8, 113
Corsica 4, 5, 15, 17, 19, 25, 26–28, 32, 34, 44, 70, 74, 77, 79, 83–86, 95, 100, 158, 183, 215, 218, 221, 228, 232, 236, 237
Cretaceous 5, 6, 11, 13, 36
Crete 4, 7, 17, 18, 19, 28, 31, 34–36, 44, 47–50, 55, 69, 70, 80, 83, 88, 90, 95, 104, 131, 132, 228
Cultivation 11, 26
Cyclades 49, 83, 90
Cyprus 17, 32, 35, 36, 39, 44, 215
Cyrno-Sardinian microplate 15

Deceit pollination 97
Dehesa 121
Devonian 5
Dichogamy 190
Dioecy 180, 185–187, 189, 190, 194
Disjunct distribution 4, 68, 83
Dispersal 38, 43, 44, 46, 48, 56, 58–60, 63–65, 69, 71, 74, 76, 79, 92, 93,,95, 103, 110, 111, 133–143, 149, 165, 166, 169, 178, 185, 200, 204, 206–209, 211–215, 221, 223–225, 233, 237, 242, 245
Dispersal limitation 46, 48, 65, 133, 134, 138
Disturbance (see human activities)
Distyly 175, 196–200
Domestication 8, 11, 207–220, 223, 238, 239
Domestication syndrome 212–213
Dormancy 87, 110, 114, 120, 123, 142, 148, 149
Drift-ecological 46–50
Drift-genetic 50, 78–81, 87–91, 158
Dune 5

Ecological differentiation 63, 78, 86–88, 92, 107, 241
Egypt 46
Endemism 3, 4, 8, 13, 15, 18, 32, 33, 35–76, 78, 79, 88, 97, 100, 107, 240–245
Erosion 17
European plate 13, 19
Evolutionary potential 9, 238, 241, 243
Extinction 17, 21–23, 46–48, 50, 58, 59, 78, 207, 217, 241, 242

Facilitation 134–137, 164, 166, 242
Fertile crescent 209–211, 214–216, 238
Fire 26–28, 113, 138, 140–143, 148, 166
Fleshy fruits 111, 128–130
Floral fragrance 149, 150
Floral traits 57–59, 92, 120, 122, 167, 168, 175, 178, 179, 195, 204
Flower colour 79
Flower size 61, 79, 84, 125, 174, 181, 184, 191, 195

Flowering phenology 122, 123, 125, 194
Freezing temperature 10, 33, 159–162
Frequency-dependent selection 186, 200, 201
Furanocoumarins 150, 164
Fynbos 62–64

Gall 162, 163
Garrigue 7, 118, 121, 141, 148, 150, 152, 158, 164, 192, 193
Gender variation 183–185, 190, 191, 194
Generalist pollination 169, 204
Genetic variation 68–72, 74, 94, 99, 104, 208, 211, 214, 218, 221, 222, 229, 238
Geographic isolation 3
Geological history 1, 3, 9, 11, 12, 14, 17, 36, 37, 43, 53, 63, 101, 102, 107
Geophyte 75, 115, 120, 222
Germination 109, 117, 120, 123, 133, 134, 136, 142, 147–149, 161, 164, 211, 214, 225, 231, 237, 238
Glaciation 1, 18, 20, 23–25, 31, 32, 71–76, 96, 102
Grazing 58, 65, 130, 136, 138, 139, 221, 244
Greece 13, 17, 18, 24, 26, 32, 34, 35, 39, 55, 58, 72, 76, 88, 107, 132, 149, 150, 153, 156–158, 164, 170, 171, 178, 215, 219, 220, 223
Gynodioecy 180, 188–195
Gynomonoecy 180, 188

Heavy metal tolerance 143–145
Herbiovory 58, 112, 113, 130, 136–137, 147, 151–152, 161–162, 166, 167,-168, 177–180, 181, 204, 208, 224, 226, 227, 229, 231–234
Hermaphrodism 180, 185, 187, 189
Heterostyly 194, 196–199
Holocene 11, 25, 26, 28
Human activities 1–4, 11, 18, 25–29, 37, 39, 56, 58, 76, 101, 102, 109–111, 165, 188, 194, 207, 223, 228, 237, 239, 242, 243
Hybridization 38, 67, 78, 86, 91–98, 100, 102, 103, 107, 152, 208, 210, 215–218, 221–224, 229, 232–235, 238, 239, 240, 243
Hyères Islands 53, 104
Hyperaccumulation 144
Hysteranthy 122

Iberian microplate 12
Iberian peninsula 15, 18, 19, 24, 31–33, 35, 52, 54–55, 57, 58, 64, 69, 70, 73–75, 77, 88, 92–95, 102, 107, 113, 115, 119, 121, 130, 171, 183, 191, 196, 219, 221–223, 233
Inbreeding 68, 76–79, 83, 91, 101, 180, 182, 187, 188, 200, 206
Inbreeding depression 180, 182, 185, 186, 191, 192, 204, 206
Insularity 46
Invasive species 208, 223–239, 241, 243, 244
Ionian Sea 14
Irano-Turanian floristic province 35, 36, 94, 209
Irradiance 100, 111, 114, 117, 140
Isoprene 150, 152

Jurassic 5, 5 6, 11–14

Kabylies 15
Karpathos 47, 49, 77, 83, 95
Karyotype variation (see chromosome rearrangements)

Land-bridge connections 1, 17, 34, 40, 49, 54, 83, 87, 91, 187
Laurel forest 19
Leaf traits 91, 112, 115
Libya 46
Litter decomposition 118, 120–121, 148

Maghrebides 13
Male-female conflicts 200, 206
Malta 44, 54, 228
Mangrove 20
Marginal populations 69, 77, 86, 88, 107, 243
Maritime Alps 39, 43, 73, 76, 105, 130
Mediterranean climate region 2, 3
Mediterranean Sea 1–3, 10, 12–17, 29, 30
Messianian salinity crisis 11, 15–17, 20
Meta-population 59, 62, 182, 192, 204
Microplate 11–15, 31, 36, 39, 50, 52, 53, 65
Miocene 5, 11, 12, 14–20, 22, 26, 31, 33, 36, 50, 52, 53, 55, 63, 64
Modifier genes 175
Monoecy 180, 185–190
Monoterpenes 113, 144, 146–148, 150–154, 156–159
Morocco 15, 40, 33, 54–56, 70, 72, 74, 79, 81, 88, 92, 102, 187–189, 219, 242
Mosaic 4–8, 16, 22, 29, 36, 37, 58, 59, 63, 89, 108–110, 112, 115, 116, 118, 119, 121, 129, 132–134, 137, 139–141, 159, 165, 166, 168, 170, 172, 173, 175, 182, 185, 189, 192–194, 204, 205, 229, 240–242, 244
Mountains 1–3, 7, 10, 13, 17, 21, 24, 25, 30, 39, 41, 43, 45, 49–50, 44, 63–65, 74, 76, 102, 105, 126, 129, 143, 211, 220
Mt Etna 17
Mt Liban 13
Mutualism 169

Narrow endemism 3, 4, 37–66, 243
Natural selection 68, 86, 101, 107, 109, 119, 121, 143, 161, 177, 208, 217, 238
Nectar reward 171, 196
Neolithic 11, 26, 207, 209, 210, 213–215, 238
Nuclear-cytoplasmic interaction 191, 193–195
Nucleation 133–135, 137

Oligocene 11–13, 15, 16, 19, 20, 31, 36, 52, 64
Ovule discounting 199–201

Pathogens 162, 232
Peloponnese 17, 34, 35, 77, 83
Peripheral populations 81
Permian 12, 14
Phenotypic plasticity 100, 109, 123, 139, 225
Photosynthesis 58, 111, 114, 140, 151, 152
Phrygana 48, 125, 149, 164
Pleistocene 11, 13, 15, 21, 23, 31, 34, 36, 71, 73, 74, 88, 89, 93, 95, 100, 119, 211, 222
Pliocene 5, 11, 13–18, 21–23, 31, 34, 36, 48, 56, 63, 74, 83, 87, 88, 93, 111, 123, 128, 129, 143, 169
Pollen limitation 170, 177, 179, 180, 183, 204
Pollen transfer 168, 172, 174, 196, 197, 199, 200
Polyploidy 38, 78, 91, 92, 97, 98, 100, 101, 103, 107, 183, 215, 229

Population differentiation 69, 71, 73–76, 78, 79, 81, 87, 101, 103, 104, 106, 110, 116, 133, 161, 167, 217, 240
Population size 76, 79
Pyrenees 13, 15, 25, 76, 77, 100, 105, 235, 236

Quaternary 1, 11, 13, 14, 17, 23, 24, 26, 34, 43, 48, 54, 55, 71, 76, 77, 94, 95, 102

Rainfall, predictability and variation 10, 18, 20, 22–24, 30, 35–36, 110, 114, 119, 122, 130, 149
Refugia 23–25, 33, 71, 73–77, 102, 104, 105, 218
Regeneration 64, 111, 118, 120, 133, 134, 136, 137, 139, 141, 142, 148, 164–166, 184, 240, 242
Reproductive assurance 78, 179, 182, 183
Reproductive effort 116
Reproductive isolation 41, 67, 78, 91–94, 96, 98, 101, 105, 210, 217, 232, 235, 236
Resource allocation 151, 180, 183, 184, 186, 187
Resprouting 141–143
Rhodes 17, 34, 35, 39, 49, 83
Rif 13, 27, 32, 33, 54
Rural depopulation 129, 130

Sardinia 15, 17, 34, 39, 40, 44, 50, 52–54, 74, 77, 79, 83, 95, 104, 115, 153, 165, 183, 215, 218, 221, 228, 237
Sclerophylly 8
Secondary compounds 120, 146, 147, 150, 151, 166, 213, 214
Seed bank 117, 142
Seed dispersal 2, 6
Seed rain 133, 134, 136, 137
Seed shadow (see seed rain)
Self-compatibility 203, 213
Self-incompatibility 189, 197, 199
Self-interference 199, 201
Self-pollination 179, 180, 182, 183, 191, 198, 200, 201, 203, 206
Serotiny 142
Sex ratio 184, 190, 191–195, 204
Sexual dimorphism 191
Sicily 10, 14, 16, 17, 18, 20, 22, 31, 39, 44, 52–56, 218, 221, 237
Sierra de Cazorla 7, 171
Sierra Nevada (see Betic Cordillera)
Soils: nutrient and water stress: 4, 26–28, 37, 56, 62–65, 109, 110, 116, 118–121, 138, 143, 145, 148, 159, 161, 164
Southern France 3, 6, 7, 15, 20, 23–27, 43, 47, 54, 59, 60, 68, 70, 73, 76, 79, 80, 83, 86, 88, 110, 121, 129–148, 151–160, 170–179, 185, 189–199, 202, 203, 215, 220–223, 226, 227, 230, 233, 237, 243, 244
South-west Australia 8
Speciation 2, 67, 71, 78, 81, 83, 86, 87, 91, 93–95, 97, 100, 106, 107
Species diversity 3, 37–39, 44, 46, 50, 51, 62, 117, 138, 141, 224, 227
Species-area curve 44–46
Steppe vegetation 22–25, 35
Stigma-anther separation 61, 83–85, 176, 182
Stigma-height dimorphism 196–203
Storage organ 120, 146
Straits of Gibraltar 10, 14, 17, 39, 64, 102
Stress tolerance 56, 59
Succession 58, 65, 110, 133–141, 192, 193, 225, 242
Summer drought 1, 10, 18, 36, 110–112, 123, 124, 128, 141, 144, 191
Synanthy 122

Taurus Mountains 3, 13, 30
Temporary marsh 117, 185
Terra rossa 118, 119
Terrace 130, 138
Tertiary 13, 14, 19, 31, 33–34, 40, 49, 50, 55–56, 65, 113, 123, 128, 143
Tertiary endemic 56
Tethys 12, 19, 35, 65
Trias 5, 6
Trioecy 188, 190
Triploid bridge 98, 99
Tristyly 196, 198
Tunisia 10, 14, 17, 187, 189
Turkey 29, 49, 54, 55, 74, 83, 92, 95, 189, 211, 215, 216, 219, 220, 221
Tyrrhenian Sea 15

Unreduced gametes 91, 98

Volcanic activity 17, 35

Western California 8